Rare-Earth-Activated Phosph

Rare-Earth-Activated Phosphors

Chemistry and Applications

Edited by

Vikas Dubey

Department of Physics, BIT Raipur, Kendri, Chhattisgarh, India

Neha Dubey

Department of Physics, Govt. VYT PG Auto. College Durg, Durg, Chhattisgarh, India

Marta Michalska Domańska

Institute of Optoelectronics, Military University of Technology, Jarosław Dąbrowskiego, Poland

M. Jayasimhadri

Department of Applied Physics, Delhi Technology University, India

Sanjay J. Dhoble

Department of Physics, Rashtrasant Tukadoji Maharaj Nagpur University, Nagpur, Maharashtra, India

ELSEVIER

Elsevier
Radarweg 29, PO Box 211, 1000 AE Amsterdam, Netherlands
The Boulevard, Langford Lane, Kidlington, Oxford OX5 1GB, United Kingdom
50 Hampshire Street, 5th Floor, Cambridge, MA 02139, United States

ISBN: 978-0-323-89856-0

For information on all Elsevier publications visit our website at https://www.elsevier.com/books-and-journals

Publisher: Susan Dennis
Acquisitions Editor: Charles Bath
Editorial Project Manager: Hilary Carr
Production Project Manager: Kumar Anbazhagan
Cover Designer: Miles Hitchen

Typeset by TNQ Technologies

Contents

List of contributors

Sadhana Agrawal Physics, National Institute of Technology Raipur, Raipur, Chhattisgarh, India

Irfan Ayoub Department of Physics, National Institute of Technology Srinagar, Srinagar, Jammu and Kashmir, India

Dhritiman Banerjee Department of Physics, Indian Institute of Technology (Indian School of Mines), Dhanbad, Jharkand, India

Smriti Bhatia Functional Material Lab, School of Physics and Materials Science, Thapar Institute of Engineering & Technology, Patiala, Punjab, India

Rituraj Chandrakar Department of Mechanical Engineering, NMDC Polytechnic College, Dantewada, Chhattisgarh, India

Subrata Das Material Science and Technology Division, CSIR- National Institute for Interdisciplinary Science and Technology, Thiruvananthapuram, Kerala, India; Academy of Scientific and Innovative Research (AcSIR), Ghaziabad, Uttar Pradesh, India

Sanjay J. Dhoble Department of Physics, Rashtrasant Tukadoji Maharaj Nagpur University, Nagpur, Maharashtra, India

Neha Dubey Department of Physics, Govt. V.Y.T. PG. Autonomous College Durg, Durg, Chhattisgarh, India

Vikas Dubey Department of Physics, Bhilai Institute of Technology Raipur, Naya Raipur, Chhattisgarh, India

Karan Kumar Gupta Department of Chemical Engineering, National Taiwan University, Taipei, Taiwan

Nisar Hussain Department of Physics, National Institute of Technology Srinagar, Srinagar, Jammu and Kashmir, India

Abhilasha Jain Department of Metallurgical and Materials Engineering, VNIT, Nagpur, Maharashtra, India

Megha Jain Department of Physics, Chungbuk National University, Cheongju, Republic of Korea

Abhijeet R. Kadam Department of Physics, Rashtrasant Tukadoji Maharaj Nagpur University, Nagpur, Maharashtra, India

Jagjeet Kaur Department of Physics, Govt. V.Y.T. PG. Autonomous College Durg, Durg, Chhattisgarh, India

Sandeep Kaur Functional Material Lab, School of Physics and Materials Science, Thapar Institute of Engineering & Technology, Patiala, Punjab, India

Ram Krishna Department of Physics, School of Science, ISBM University, Gariyaband, Chhattisgarh, India

D.S. Kshatri Department of Applied Physics, Shri Shankaracharya Institute of Professional Management and Technology, Raipur, Chhattisgarh, India

Vijay Kumar Department of Physics, National Institute of Technology Srinagar, Srinagar, Jammu and Kashmir, India; Department of Physics, University of the Free State, Bloemfontein, South Africa

Puja Kumari Department of Physics, Darbhanga College of Engineering, Darbhanga, Bihar, India

Chung-Hsin Lu Department of Chemical Engineering, National Taiwan University, Taipei, Taiwan

Shubhra Mishra Department of Applied Physics, Shri Shankaracharya Institute of Professional Management and Technology, Raipur, Chhattisgarh, India

Jelena Mitrić Laboratory for Electronic Materials, Institute of Physics Belgrade, Belgrade, Serbia

Umer Mushtaq Department of Physics, National Institute of Technology Srinagar, Srinagar, Jammu and Kashmir, India

O.P. Pandey Functional Material Lab, School of Physics and Materials Science, Thapar Institute of Engineering & Technology, Patiala, Punjab, India

Yatish R. Parauha Department of Physics, Rashtrasant Tukadoji Maharaj Nagpur University, Nagpur, Maharashtra, India

Ruby Priya Department of Physics, Chandigarh University, Mohali, Punjab, India

Teerath Ram Department of Physics, Govt. V.Y.T. PG. Autonomous College Durg, Durg, Chhattisgarh, India

Seemin Rubab Department of Physics, National Institute of Technology Srinagar, Srinagar, Jammu and Kashmir, India

Janita Saji Department of Sciences and Humanities, Faculty of Engineering, Christ (Deemed to be University), Bangalore, Karnataka, India

Rakesh Sehgal Department of Mechanical Engineering, National Institute of Technology Srinagar, Srinagar, Jammu and Kashmir, India

Rishabh Sehgal Department of Electrical and Computer Engineering, University of Texas at Austin, Austin, TX, United States

Anju Singh Central Institute of Plastics Engineering and Technology, Korba, Chhattisgarh, India

Palwinder Singh Department of Physics, Chungbuk National University, Cheongju, Republic of Korea

Sudipta Som Department of Chemical Engineering, National Taiwan University, Taipei, Taiwan

N. Kumar Swamy Department of Physics, School of Science, ISBM University, Gariyaband, Chhattisgarh, India

Hendrik C. Swart Department of Physics, University of the Free State, Bloemfontein, South Africa

Raunak Kumar Tamrakar Department of Applied Physics, Bhilai Institute of Technology (Seth Balkrishan Memorial), Durg, Chhattisgarh, India

Vikrant Tapas Department of Mechanical Engineering, NMDC Polytechnic College, Dantewada, Chhattisgarh, India

Neeraj Verma Department of Physics, Govt. V.Y.T. PG. Autonomous College Durg, Durg, Chhattisgarh, India

Praveen Kumar Yadaw Department of Physics, School of Science, ISBM University, Gariyaband, Chhattisgarh, India

About the editors

Dr. Vikas Dubey obtained his master's degree in physics from the Govt. V.Y.T.PG. Auto. College Durg, India. He received gold medals for securing first rank in his master's degree. He completed his Ph.D. from the National Institute of Technology Raipur. He is currently working as Associate Professor and Head and Dean of Research and Development in the Department of Physics, Bhilai Institute of Technology Raipur, India. During the last 10 years of his research career, he has published more than 135 research papers in various reputed international journals, with more than 1800 citations. He has received 10 patent grants and 20 patents published at the national and international levels. He has published six authored books and 10 edited books and several book chapters. He is also the editor of various reputed journals and edited numerous special issues.

Dr. Neha Dubey is Principal Investigator at Govt.V.Y.T.PG. Autonomous College, where she began work following completion of her Ph.D. at RDVV Jabalpur in 2016. She has coauthored three books and three book chapters and published more than 35 articles in reputed journals. She is an editorial board member for the journal Advance Physics Letters and has received awards for her presentations at both national and international conferences. She has received 10 patent grants and 20 patents published at the national and international levels.

Dr. Marta Michalska Domańska graduated in physical chemistry at the University of Warsaw and completed her Ph.D. at the Military University of Technology (MUT, Poland). She took part in the COST project (Action MP 1302) as part of an internship at the University of Tübingen, Germany. She completed her postdoc in corrosion science (AlMagic grant) at TU Delft, Netherlands. She works in the materials science field and focuses especially on the synthesis and characterization of nanomaterials for photovoltaics and biomedical and spectroscopic applications. She received two awards for her Ph.D. thesis: MUT Rector's Award for distinguished Ph.D. thesis and the Prime Minister Award for the best Ph.D. thesis from all the defended theses in all fields in Poland in 2015. In 2016 she received the Scholarship for Young, Outstanding Researchers from the Polish Ministry of Science and Higher Education for her entire academic work. In 2020 she was nominated for "Scientist of Future 2020," by the Polish Award of Intelligent Development. She is a member of many international societies, like the International Electrochemical Society or the Polish Society of Materials Science. She takes part in many international scientific conferences, also as an Invited and Key Note Speaker. She is a coauthor of more than 30 scientific publications.

Mula Jayasimhadri is Assistant Professor in the Department of Applied Physics at Delhi Technology University, India, having previously held positions at Changwon National University and Pohang Univ. of Science and Technology (South Korea) and Sri Venkateswara University (India). He has published over 100 papers in peer-reviewed journals, presented over 60 papers at international conferences, organized five conferences, and won many awards for his work.

Sanjay J. Dhoble is Professor of Physics at Rashtrasant Tukadoji Maharaj Nagpur University, India. His research focuses on the synthesis and characterization of solid-state lighting nanomaterials as well as the development of radiation dosimetry phosphors using thermoluminescence, mechanoluminescence, and lyoluminescence techniques.

Section I

Fundamentals of rare-earth-activated phosphors

Introduction to phosphors and luminescence

1

Nisar Hussain [1], Irfan Ayoub [1], Umer Mushtaq [1], Rishabh Sehgal [2], Seemin Rubab [1], Rakesh Sehgal [3], Hendrik C. Swart [4] and Vijay Kumar [1,4]
[1]Department of Physics, National Institute of Technology Srinagar, Srinagar, Jammu and Kashmir, India; [2]Department of Electrical and Computer Engineering, University of Texas at Austin, Austin, TX, United States; [3]Department of Mechanical Engineering, National Institute of Technology Srinagar, Srinagar, Jammu and Kashmir, India; [4]Department of Physics, University of the Free State, Bloemfontein, South Africa

1. Introduction

In our day-to-day life, light plays a fundamental part, and it is a significant portion of research in the present century. The illumination-related innovations have changed different aspects of science and technology [1]. With the evolution of time, the requirement for better lighting systems led to the evolution of incandescent and fluorescent lights. For the first time, the incandescent lamp was fabricated by Thomas Edison in 1879, which was a huge accomplishment in human advancement. A large number of drawbacks were found in these light sources, for example, voltage contortion, issues identified with affectability, lower lifetime, low-efficiency values, greater expense, etc. [2]. To overcome such drawbacks, a huge amount of research has led us to the design of light-emitting diodes (LEDs) that were seen to have greater efficiency, high radiance, viability, high reliability, and eco-enjoyableness. These lighting systems have also been found to be highly reliable. The advancement in solid-state lighting (SSL) systems has effectively replaced incandescent bulbs [3]. The research in the pursuit of lighting systems has led us to the discovery of phosphors, which are solid substances that emit light after being excited by different excitation sources discussed later in the chapter. The luminescent materials are composed of two parts one is a host and the other is a dopant that is generally rare earth (RE) or a transition metal ion added in very small amounts [4]. These luminescent materials have proved to be of primary importance for the generation of SSL [5–9]. Recently, the research on RE or transition metal-doped SSL systems has increased at a very faster rate as they can act as the technology to significantly reduce energy consumption. In comparison to the metal ion-doped phosphors, the RE-doped phosphors are quite advantageous because of their high emission efficiency as well as sharp and controlled emission profile [10–13]. The term luminescence can be further defined as the spontaneous emission of light by a substance not resulting from heat because of this reason it is also called cold light, which makes it a form of cold body radiation. The process of luminescence is activated by different methods such as chemical responses, electrical energy, subatomic

Rare-Earth-Activated Phosphors. https://doi.org/10.1016/B978-0-323-89856-0.00008-0

movements, or pressure on crystals. This differentiates luminescence from incandescence, which is light radiated by a substance because of heating. Luminescence is fundamentally categorized into two types of emission called fluorescence and phosphorescence that are different from one another on the basis of delay in response to outer electromagnetic (EM) radiation. In the case of fluorescence, luminescence is generated very quickly after the incident radiation is absorbed. Subsequently, it stops nearly as fast as it starts. Generally, the wavelength of the emitted radiation is longer than the wavelength of the incident radiation [14]. As far as phosphorescence is concerned the emission of radiation is delayed with respect to the incidence of the radiation. This delay can range from several minutes to a couple of hours depending on the type of host and activator used [15−18]. For the majority of cases, the dopant concentration remains very low due to which the luminescence cycle is generally contracted. This process is known as concentration quenching [19]. The absorption of the photon that is used to energize the phosphor is brought upon by absorption of a photon by the host material itself or due to the doped impurities. Much of the time, the impurity ions also called activator ions are the ones responsible for creating the ideal luminescence. At the point when the activator particles show generally weak absorption, another sort of impurities can be codoped (included) and these impurities retain the energy and therefore transfer their energy to the activator particles and thus result in the generation of light [20].

The luminescence term was introduced by Eilhard Weidemann in the year 1888. This term has Latin origin and is derived from the word Lumin meaning light [21]. The emission of radiation by phosphors is a distinctive case of interaction between matter and energy. Such materials are also called luminophores [22]. The main property of such materials is that they convert energy from absorbed photons into visible light. The normal size of a phosphor ranges from 1 μm to 1 nm. The dopants used in a phosphor are generally 3d or 4f metal ions, and these dopants along with the host matrix are responsible for the release of photons [23]. The study of phosphors has a very extensive background but during the current decade, the research on phosphors has caught pace because of their incredible properties in the emission of visible light. In this pursuit, the first host matrix analyzed was barium sulfide (BaS), which was discovered in the 17th century by researcher Vincenzo Casciarlo. This luminescence was found in a barite crystal ($BaSO_4$), which produced a red emission on the absorption of sunlight [24,25]. The host matrices calcium sulfide (CaS) and strontium sulfide synthesized by F. Hoffmann and J.F. John, respectively, kick-started the research on phosphors [25]. Subsequently, comparative discoveries were reported from numerous spots in Europe, and these light-emitting crystals were named phosphors. This word signifies "light carrier" in Greek and shows up in Greek fantasies as the representation of the morning star Venus [26]. Phosphors can thus be defined as solid luminous materials, which after absorption could subsequently generate light that is visible to the human eye. Countless phosphors have been synthesized over the course of time, every one of which is having its characteristic property of generation of light and the time for which the material stays illuminated. Phosphors in which the emission of light is triggered by electron excitation, the luminescence is called as electroluminescence. Such phosphors are utilized in the production of TV screens and PC screens. Phosphors energized by bright, noticeable, and infrared radiation are utilized basically in fluorescent lights ordinarily utilized for general lighting purposes [27]. Lighting characterizes around 19% of power consumption around the world and thus production of improved, naturally pleasant and efficient lighting tools is desirable. It is generally acknowledged that (SSL) will displace the timeworn innovations (incandescent lamps) and others that are harmful from an ecological perspective (fluorescent lamps) [7].

The properties of the light are determined by the color possessed by the light. However, the color emitted by the phosphor depends on the activator/dopant ions used, as every dopant possesses its unique characteristic color. For the determination of the color possessed by the phosphor, the photoluminescence excitation and the photoluminescence emission spectrums are recorded. Similarly, the luminescence performance is determined by the characteristic properties like Color coordinates of the Commission Internationale de l'Eclairage (CIE), correlated color temperature (CCT), color rendering index (CRI), luminescence quality, quantum efficiency, and so on. For the better efficiency of the optoelectronic materials fabricated from the phosphor materials, careful monitoring of the above-mentioned characteristic parameters is necessary. This is accomplished by developing control over the peak position and full-width half maxima of the phosphor spectra. Keeping in view their importance they are briefly explained below:

CIE coordinates and CCT: For determining the applicability of the phosphor in the optoelectronic materials for various applications, the best criterion for evaluating the suitability is the chromaticity coordinated values. Different color models have been created by the International Commission on Illumination for the description of the color space. The color scheme of the CIE is based on a combination of tri stimulus values (RGB) represented in three dimensions [7]. The human eye already possesses three different color cones, and among them, each color is created by intermixing of three different cones. The color space came first in existence in 1931 for the approximation of all the colors developed by the human eye [28]. The spectral sensitive peaks for the different ranges such as short, medium, and longer exhibited by the different cones of the eye are 420–440 nm, 530–540 nm, 560–580 nm, respectively. So far, the CIE is considered to be the most precise and accurate than the rest of the available models as for determining the color parameters it uses the spectral power distribution (SPD) of the emitted light. It is the tri-stimulus function that tells us about the nature of the color emitted by the light. The tri-stimulus function that is the function of red, green, and blue color wavelengths helps us in determining the X, Y, and Z parameters. In the CIE model, the color is defined by the luminance parameter "Y" and by CIE coordinated (x,y) values that represent a specific location in the chromaticity chart [29]. The transformation equations used for obtaining the CIE coordinates from the tri-stimulus data are as follows [30]:

$$x = \frac{X}{X+Y+Z} \quad and \quad y = \frac{Y}{X+Y+Z}$$

A particular color is created by combining the three primary colors (RGB) and is illustrated on the CIE chart by a triangle that connects all three points. However, the color-coordinate corresponding to the nonmonochromatic source also lies inside the color gamut [31].

CCT stands for the color correlated temperature, which is a luminescence metric used to assess the color and temperature of the lighting sources. It is defined as an appropriate temperature in kelvin at which the black body radiator must be maintained so that it will achieve the chromaticity of the light source [32]. As the incandescent is considered a black body radiator because of its thermal emission, the color temperature is considered to be the temperature of the filament. As all the other light sources emit light that is not thermally emitted, thus the color temperature values for light sources other than incandescent are correlated in nature, called CCT [32]. The warmer light is generally yellowish with a lower CCT value while the cooler light is bluish with a higher CCT value [33]. There does not exist any universal source of light for the illumination rather it depends upon the requirements. Different light sources possess different CCT values depending upon their usage, the value of CCT ranges from 2200 to 6500 K for the different light sources as shown in Fig. 1.1 [34].

CRI and luminous efficiency: CRI or color rendering index is a pure number that is used to measure up to what extent an object is perceived when it is illuminated by a source of light in comparison with a reference source for given conditions [35]. This characteristic of light is based on the fact about objects look very different in color when viewed under different lamps. The CRI values range from minimum zero to a perfect 100. The CRI value of 100 would mean that the object to be illuminated looks the same as that it looks when illuminated with a reference light source.

Luminous efficacy is a method to measure the efficiency of a light source to convert electrical energy into visible photons. The unit of luminous efficacy is lumens per watt. The definition of luminous efficacy is the ratio of total luminous flux and power [35].

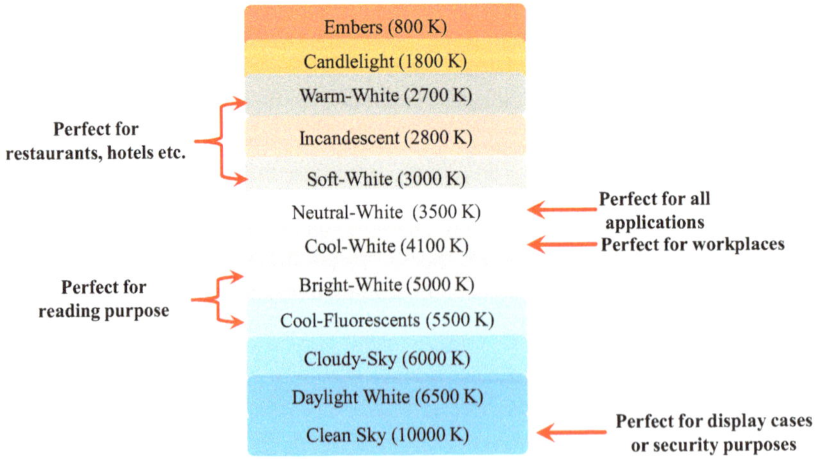

Figure 1.1 CCT values for different lightning applications.

Depending on how the power is measured, the measured power can be radiant flux or total electrical energy output. The measurement of the radiant flux gives luminous efficacy of radiation while the measurement of the total electrical energy output gives total luminous efficacy [32,34,36]. A light source changes electrical energy into visible light as well as heat. The luminous efficacy is calculated by the equation [37] given as follows:

$$\text{Efficacy} = (\text{luminous flux})/\text{power}$$

Charge transfer luminescence: Charge transfer luminescence refers to a transition where an electron shifts into the 4f energy level from surrounding atoms or ions. Such transitions that involve configurational change are completely allowed by the selection rules thus a broad emission spectrum is obtained named as charge transfer band [22]. It is well known that rare-earth ions are easily reducible and hence they are the best agents to exhibit such phenomenon [38]. This phenomenon has a wide range of applicative parts like in the case of solid lasers IR to visible light converters and IR quantum converters. One such example is Y_2O_3 doped with Eu^{3+} that displays a charge transfer band of 260 nm, which falls in the region of light in the EM spectrum. In this nanophosphor, the oxide host is excited by ultraviolet radiation that then transfers that absorbed energy to the activator ion and thus following the phenomenon of charge transfer.

2. Luminescence and its types

The term "luminescence" has been coined first by the German physicist Eilhard Weidman in 1888 for describing all the phenomena of light that do not arise because of the change in temperature [39]. This term is basically Latin in nature derived from the root word "lumen," which means light. It is defined as the release of photons caused by the transition of a phosphor from an excited state to lower energy levels. Being independent of the temperature, therefore for the distinction from other usual sources of light emission such as incandescent light emission, it is also pronounced as "Cold-light." That is why sometimes it also defined as the spontaneous emission of EM radiation by electronically energized species, that is, luminescence is the unrestricted outflow of light by material not coming about because of "hotness, or cold light" [21]. In general, the materials usually emit visible radiations, but on appropriate doping such as rare-earth ions, they can also emit UV or infrared radiations. The ability of a luminescence medium to convert excitation energy into visible light determines its effectiveness. There are two different ways for the classification of luminescence. Classification based on both modes is depicted in Fig. 1.2.

Among the first one is the mode of excitation, which is used to classify luminescence into different forms. PL involves the study of the absorption of a photon caused because of the stimulation of the electron. The well-known modes of this are fluorescence and phosphorescence. The other forms of luminescence that arise because of the

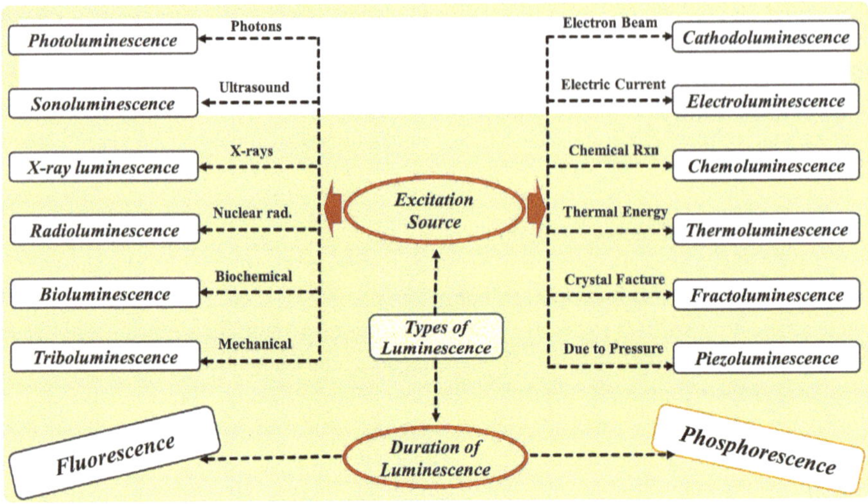

Figure 1.2 Various forms of luminescence.

different sources of the excitation include chemoluminescence, mechanoluminescence, bioluminescence, radioluminescence, electroluminescence, cathodoluminescence, and so on. The second one is based on the time taken by the system for the deactivation after the excitation radiation is either removed or switched off. In this case, the well-known modes of luminescence are fluorescence and phosphorescence (Fig. 1.3).

The emission that occurs within the time interval of 10^{-8} s soon after the excitation is known as the fluorescence and the emission that takes place in the latter interval of the time is known as the phosphorescence or afterglow. They differ from each other as phosphorescence is associated with the phenomenon of the trapping and radiative recombination of the carriers. The materials that exhibit the phenomenon of luminescence are generally known as a phosphor. The phosphor usually compromises a host material and an intentionally added impurity called an activator. The percentage of the activator ion remains in ppm for donor—acceptor systems. Energy must be delivered to

Figure 1.3 Classification of luminescence phenomenon based on lifetime.

phosphors whose properties are determined by the type of host matrix and dopant used. The phenomenon is observed in semiconductors whose bandgap has been found to be greater than zero. Metals with zero band differences do not show the phenomenon of luminescence. Therefore, the suitable materials for the phosphor should be either a wide band-gap semiconductor or an insulator. The different forms of luminescence are described in the next subsections [24,40–42].

2.1 Chemiluminescence

Chemiluminescence is the type of luminescence where light is emitted as a result of chemical reactions. The molecule in a subsequent higher vibrational state decays to a lower vibrational energy state and emits photons over the course of this reaction. Emission due to chemiluminescence takes place at a variety of wavelengths, ranging from UV to IR range. This reaction takes place by assigning at least one photon to every reacting molecule for the stimulation. The energy of the products created during the reaction is lower than those of the reactants. It is a tool used for a wide range of applications like it is used for tracking of contaminations in the air around us, for genetic analysis, glow sticks, enzyme-based biosensors, and so on.

2.2 Electroluminescence

It is a phenomenon in which electrical energy is converted into nonthermal visible photons, thus considered optoelectronic in nature. In this process, the used materials emit visible photons when an electric current of sufficient strength is passed through them. The recombination region of materials experiences a radiative mixture of charge carriers, thus resulting in this phenomenon. Electroluminescent compounds may also produce light by accelerating high-energy electrons in response to a strong electric stimulus. It possesses a wide range of applications especially in the field of lighting and flat-screen display technology.

2.3 Cathodoluminescence

When an electron beam hits a phosphor material, it triggers the generation of photons of various wavelengths to be emitted into the visible region of the EM spectrum. Cathodoluminescence is opposite to the photoelectric effect where visible photons are produced by the phosphor when electrons of suitable energy are incident on it. The main reason for this phenomenon to occur is the coupling of electrons from conduction bands with holes in the valence band. The main application of cathodoluminescence is in the field of manufacture of screens, such as LCDs and traditional cathode-ray tubes.

2.4 Thermoluminescence

When a phosphor material is subjected to a temperature, as it exceeds a certain threshold value the stored energy is reemitted in the forms of photons. This phenomenon of emission of photons by phosphor material is termed as thermoluminescence. However, in most cases the material used to emit light after getting cooled, there is another name for thermoluminescence termed as cryoluminescence. Ecological surveillance and archaeological dating are the main benefits offered by this phenomenon.

2.5 Photoluminescence

The phenomenon of photoluminescence is defined as the subsequent excitation and de-excitation (emission) of an electron residing in the ground state due to absorption of the photon. During the process of de-excitation (emission) an additional photon is released in the visible region of the EM spectrum. In other words, it is also defined as the optical radiation discharged by a physical framework coming about because of excitation to a nonequilibrium state by illumination with light. This process takes place in three cycles, which are recognized as follows: (1) production of electron–hole pair sets by retention of the energizing light, (2) radiative recombination of electron-hole pair sets, and (3) getaway of the recombination radiation [43]. The excitation and emission are the two types of luminescence spectra that are depicted by the pictorial representation in Fig. 1.4 [34].

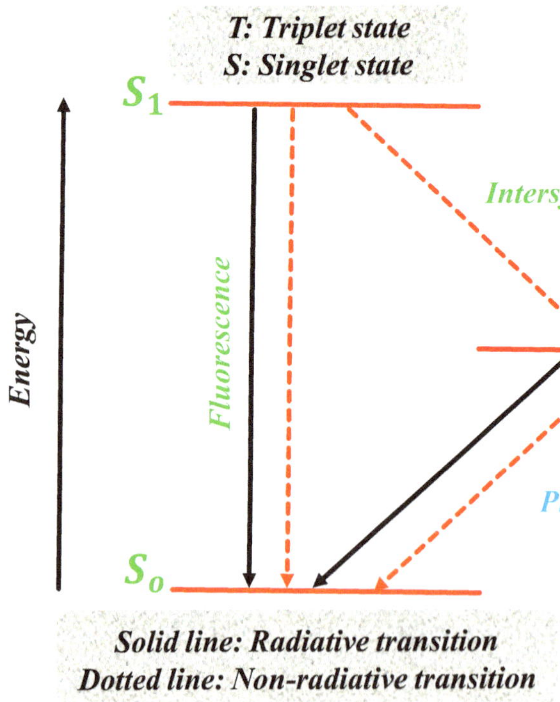

Figure 1.4 Fluorescence and phosphorescence mechanism. Reused with permission from Ref. [34].

In the case of the former, while varying the wavelength of the excitation sources, the amplitude of the emitted light at a given emission wavelength is determined. This spectrum reveals the information regarding the possible excited states of the phosphor and also provides the value of the band-gap possessed by the phosphor. The different light sources that are being used for the purpose of excitation in different regions include incandescent lamps for visible and infrared light, arc lamps for UV/visible/infrared light, and lasers for various wavelength regions. The latter "emission spectrum" is defined as the radiative transition of an electron from higher excited states to lower ones. It provides us with information about the spectral spectrum emitted by the sample. It is an incredible strategy for testing the discreteness of energy levels and to separate significant data about semiconductor test creation, quantum well thickness, for checking the mono dispersity of different materials, and so on [44]. PL has been further categorized into two types namely fluorescence and phosphorescence, which will be briefly explained later [45,46].

2.5.1 Fluorescence

It is a spin-allowed transition having a shorter lifetime in which a molecule absorbs ultraviolet or visible light and then results in the emission of light with a shorter frequency (Stokes shift). In the case of fluorescence, a molecule absorbs a certain wavelength of light, which causes an electron to be excited from its lower energy state (singlet) to a higher energy state (singlet). The fluorescence lifetime is the amount of time it takes for an electron in a higher state to return to the ground state. The electron's spin would not change during this process. Each electronic energy level corresponds to lower energy vibrational sublevels, which are made up of the lowest energy rotational sublevels. During the process of fluorescence, the electrons are generally excited from lower energy states to the higher excited ones. Due to the collision or emission, the electron at first gets down into the lower vibrational state of an excited state via nonradiative transition, and then from there, a radiative transition takes place through which the electron comes back into the ground state [34]. This radiative transition is called fluorescence. Fluorescence is a temperature-independent, instantaneous mechanism that produces light before the excitation source is switched-off. Fluorescence lamps are a well-known example of a device that emits UV light on passing a current through it. The other applications include luminescent paints, fluorescent dyes, biological detectors, gemology, and so on. Fluorescence occurs when a molecule de-excites from a singlet excited state to a singlet energy state via a radiative transition. This phenomenon is generally a two-step process represented as follows:

$$\text{Excitation}: \quad S_o + h\nu_{ex} = S_1$$

$$\text{Fluorescence}: \quad S_1 = S_o + h\nu_{em} + \text{heat}$$

where S_o represents the ground state, S_1 the excited state, h stands for planks constant, and ν represents the frequency of photon. The efficiency of any fluorescent material is measured in terms of the quantum yield, which reaches a maximal value of "1" only when every absorbed photon gets reemitted [34].

2.5.2 Phosphorescence

When a surface is irradiated with EM radiations, phosphorescence happens, which is a spin forbidden mechanism in which a photon is produced. Re-emission is sluggish in this process and also continues well after the cause of excitation gets removed. In the case of phosphorescence, the radiative transition takes place from meta-stable state to ground state. Thus, the interstate crossing takes place from singlet excited state to triplet ground state as depicted earlier in Fig. 1.3. The transformation from singlet to singlet is more noticeable than the transition from singlet to triplet as singlet to triplet state nonradiative transition is forbidden but is mediated by LS coupling. Light persists for a longer time in phosphorescence than it does in fluorescence, and the length of time can be prolonged by increasing the heating temperature. Furthermore, unlike fluorescence in the case of phosphorescence a longer wavelength photon is absorbed. Phosphorescence is the incremental release of light in the form of photons as a result of the absorption of radiation by a material. The theory of phosphorescence is used to charge materials that shine in the dark. Stickers, paints, wristwatches, and other items that glow in the dark are examples of phosphorescence. Spontaneous emission is considered to be the basic principle of luminescence. It is a process to emanate a quantized measure of energy in the form of a photon when an atom, a molecule, or a subatomic molecule travels from an excited energy state to a lower energy state, this process is known as the spontaneous emission. This phenomenon is pictorially represented in Fig. 1.5 [47]. Spontaneous emission is considered to be the main reason for the greater part of the light one observes, it is universal to the point that there are numerous names given to what in particular is a similar cycle [47]. On the off chance that atoms (or particles) are energized by certain methods other than heat, the unconstrained discharge is called luminescence. For instance, just like fireflies are luminescent many other phenomena such as electroluminescence, chemoluminescence, and so on, are various other types of luminescence relying upon how energized particles are created. Lasers start through spontaneous emission, at that point during persistent activity work by stimulated emission [47]. Spontaneous emission cannot be explained by

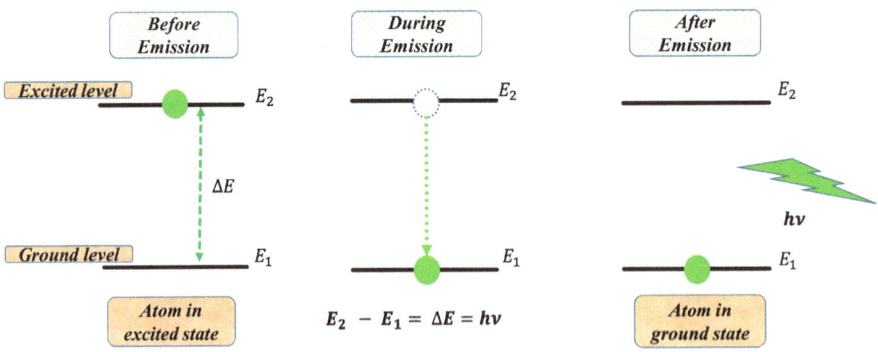

Figure 1.5 Pictorial representation of spontaneous emission.

the classical EM hypothesis and is of a very basic level a quantum cycle. Spontaneous emission is derived by Dirac from his quantum theory of radiation. It is the predecessor to the hypothesis, which he later called quantum electrodynamics [48].

3. Synthesis methods

Phosphors, which are incredibly pure inorganic compounds, have been prepared in a variety of ways by various researchers. For novel research on phosphors, one needs an incredibly strong research laboratory for the synthesis and characterization, as well as an extremely clean room for synthesizing very pure materials along with furnaces capable of reaching temperatures greater than 1600°C [49–51]. The method opted for preparing particular phosphors is chosen based on criteria the particular phosphor should contain. As most phosphor formulations contain more than four elements, obtaining uniform, single-phase powders is challenging in such type of phosphor materials. The luminescence efficiency of the materials is determined by their characteristics, which are determined by the atomic structure, composition, microstructure, defects, and interfaces, all of which are influenced by the synthesis process [34]. Moreover, different types of host lattices are used for the phosphors such as oxides, aluminates, silicates, sulfides, alumino-silicates, nitrides, and so on [42,50–54]. Besides different types of hosts, a small amount of activator ion is also introduced into it. It is the activator that determines the spectroscopic characteristics of the phosphors. Therefore, the incident radiation received by the host lattice is transferred directly to the activator ion. The most popular activator ions responsible for producing lattice phosphors of different colors are europium, terbium, samarium, cerium, and so on. The lattice that contains europium as an activator ion is generally red in color and thus emits orange-red light, similarly, the green and blue light-emitting lattice phosphors contain terbium and cerium, respectively, as activator ions. Chemistry plays a vital role in studying these complexes. The usual synthesis methods include sol-gel/Pechini method, solid-state reaction, solvothermal, coprecipitation, sonochemical synthesis, microwave-assisted techniques, and combustion synthesis methods. Among all the mentioned methods, the solid-state reaction is considered to be convenient of all [55,56]. A comparative study of some of these methods has been done by Mckittric et al. [57]. Table 1.1 gives a comparison of synthesis methods in terms of particle sizes, particle size distribution, morphological control, cost, purity, etc.

3.1 Solid-state method

The most commonly used method for preparing phosphor from a mixture of solid starting materials is a solid-state method. This simple process, which is ideal for mass manufacturing is used to make phosphor materials such as oxides, oxyfluorides, oxychlorides, oxynitrides, and so on. In general, the phosphors are made using a solid-state synthetic process based on high temperatures. It has the most impact on the

Table 1.1 Comparison of synthesis methods in terms of the particle sizes, morphology cost, purity, synthesis time, and limitations.

Synthesis method	Particle size	Particle size distribution	Morphological control	Cost	Purity	Synthesis time	Suitable phosphors	Limitations	References
Solid-state method (SS)	>5 μm	Narrow band	Poor good	Low	Poor good	Short long	All compounds	Extensive grinding and milling are required	[32,57–60]
Sol-gel method (SG)	10 nm–2 μm	Narrow	Medium	Good	Medium	Medium	All compounds except nitrides	Requires a soluble precursor	[32,57–59]
Combustion method (CS)	500 nm–2 μm	Medium	Poor	Low-medium	Medium good	Short	All compounds except nitrides	Requires a soluble precursor	[32,57,58,60,61]
Coprecipitation (CP)	10 nm–1 μm	Narrow	Very good	Medium	Medium	Medium	Oxides and fluorides	Requires a soluble precursor	[32,57,58,62]
Hydrothermal (HT)	10 nm–1 μm	Narrow	Good	Medium good	Medium-high	Very long	All compounds	Requires a soluble precursor	[32,57]

physicochemical properties of prepared materials. This approach is mostly used to synthesize solid light-emitting materials that cannot be synthesized at room temperature but need a higher temperature for their synthesis. The main reason due to which the rate of a reaction and probability of its occurrence change is free energy, surface area of the used reactants, reaction conditions, reactivity of chemicals used, and structure of material synthesized [63]. By employing this process most silicate materials such as $La_2Si_2O_7$ were fabricated besides many other materials viz. strontium silicates added with europium impurity [64,65]. The steps involved in the synthesis are that the reagents, such as metal oxides, silicon, and impurities are first measured, compressed, and combined in a mortar. After this, all the chemical reagents are combined with the double-distilled water and stirred at a lower temperature [66]. Different fluxes such as Li_2CO_3, NH_4F, NH_4Cl, BaF_2, YF_3, H_3BO_3, and so on are applied as precursors for lowering the temperature, because of which the rate of reaction is increased, and a better crystalline structure is obtained [67]. Besides these, the fluxes also enhance the material's morphological and luminescence properties [68,69]. Acetone is often also used to ensure the uniform mixing of the reagents. As the reaction proceeds slowly at lower temperatures, solid-state synthesis is progressed at increased temperature. At a higher temperature the reactants diffuse at a higher rate, activation energy is lowered and the speed of the reaction is also increased. As a result, high temperatures (1200−1600°C) are used to carry out the solid-state reaction. Following all the mentioned steps, finally results in the production of required materials. The precursors are readily available, and the reaction also does not produce any radioactive waste. The biggest benefit of this approach is that it is an environmentally safe process that yields thermodynamically stable materials. However, the reaction takes a long time to complete and requires a high temperature, which is the method's biggest disadvantage. In addition, because of impurities and inhomogeneity, the process is less effective [70]. This method also fails to meet the grain size and phase homogeneity of the required standard. Another disadvantage of this process is the prohibitively costly machinery used to fire the phosphors at extremely high temperatures. Because of frequent chemical cleaning, crystal structures and luminescence tend to deteriorate. The pictorial representation of the whole process is shown in Fig. 1.6.

3.2 Sol−gel method

For the processing of bright nanophosphors, this approach is very flexible, accurate, and cost-effective [71,72]. This technique is mainly used for the production of light-emitting silica-based phosphors. Besides that, this technology has also been used to make other light-emitting materials possessing lattices like that of sulfides, oxysulfides, borates, tungstates, halides, phosphates, and other compounds [73]. This process is commonly used to make nano/microsized phosphors with a consistent particle distribution. This approach entails combining precursor solutions with a chelating reagent and a surfactant in pure water. The most commonly preferred chelating agent and surfactant for synthesizing the phosphors are citric acid and polyethylene glycol, respectively. The resulting solution is then stirred at about 80°C for a while, and after some time, condensation occurs, yielding a polymeric network that leads to the

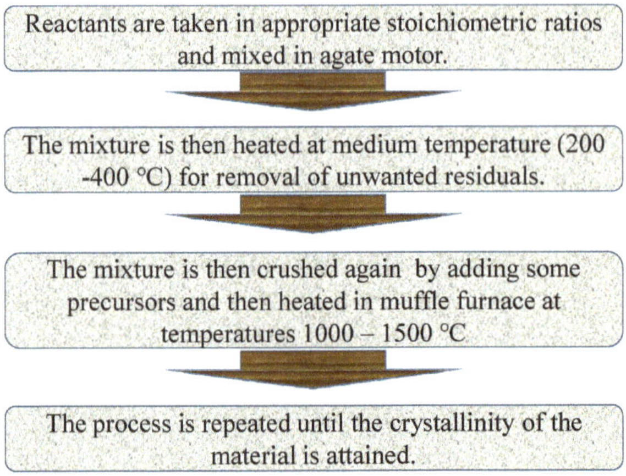

Figure 1.6 Schematic flow chart of the solid-state reaction.

formation of sol that eventually gets transformed into a gel. Gelation happens when a solution loses its fluidity suddenly and gets transformed into an unstable solid [74]. Heat treatments are used to extract the organic material of the phosphor from the final product. However, before the extraction of the material, it becomes sometimes important to anneal the phosphors at temperatures higher than 500°C. The size and morphology of the final product can be varied by analyzing the ratio of chelating agent, surfactant, and metal ions. On the other hand, in cases where the lanthanides remain present in the form of oxides, they are usually converted into nitrites first. This process is done by dissolving the stoichiometric volume of lanthanide oxide in the concentrated solution of the nitric acid (HNO_3). The mixture is then heated with continuous stirring, and a small amount of distilled water is added to the solution for the dilution. Despite the fact that the sol–gel is recognized for its high purity and high-quality product phosphors for the solid-state lighting technology, it lags behind due to its lengthy procedure of fabrication. The sol–gel method is also known as the Pechini method. Bharat et al. [75] have fabricated barium orthovanadate phosphor by employing this method. It has been experienced that the observed phosphor possesses a CRI value of 72 and a CCT value of 4875 K, which is found to be in good agreement with the criteria met for the lighting technology. It has been further observed that the CRI value can be further improved up to 80 by combining it with the red-emitting phosphor such as $Y_2Mo_4O_{15}$. The schematic diagram of the synthesis method is shown in Fig. 1.7. Few other examples of the phosphor that were prepared by utilizing this method are $BaGd_2O_4{:}Eu^{3+}/Tb^{3+}/Dy^{3+}$ [76], $LaAlO_3{:}Tb^{3+}/Tm^{3+}$ [77], $LaOCl{:}Tb^{3+}/Dy^{3+}$, etc. [78].

3.3 Spray pyrolysis method

This method of synthesis is an efficient, fast, easy, and low-cost method for producing powdered particles with a specified morphology. This method has proved to be very

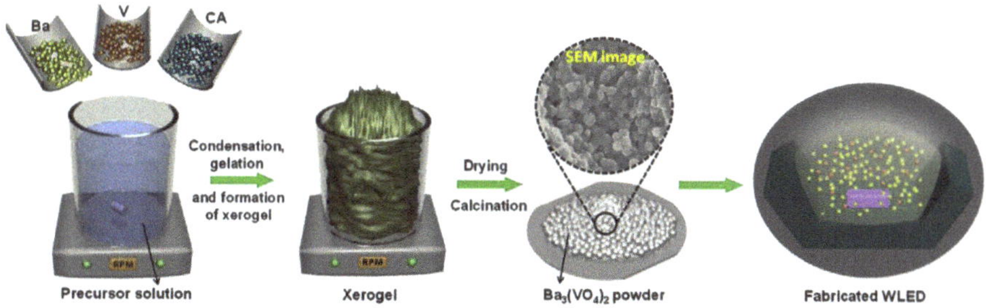

Figure 1.7 Sol–gel method for the synthesis of $Ba_3(VO_4)_2$ phosphor. Reused with permission from Ref. [75].

important for the fabrication of submicron and spherically shaped phosphors [79]. Besides different types of phosphors, nano-powders, chloro-silicates, and multicomponent oxides are all part of this process [80–83]. There exist two approaches for the preparation of the phosphor by this method, that is, aerosol spray pyrolysis (ASP) or flame spray pyrolysis (FSP). Among the two, the FSP, a modified version of ultrasonic spray pyrolysis is the most common method for the fabrication of silicate-based nanophosphors. The emission core in silicate materials is accompanied by a silica coating, which aids in surface stability and luminescence enhancement [208]. The Mn-doped Zn_2SiO_4 phosphor has been fabricated by using the FSP along with the aid of a filter expansion aerosol generator [84]. Utilizing ammonium chloride flux, $Ca_8Mg\ (SiO_4)_4C_{12}$ phosphor doped with Eu^{3+} was also successfully prepared by this method. Eu^{3+}-doped barium strontium silicate luminescent materials have been prepared from colloidal solution in the presence as well as the absence of the flex [85]. For the proper synthesis of the targeted nanophosphors, the spray pyrolysis process entails four stages. The first step involves the synthesis of droplets from the precursors, after which the drop is made to shrink in size by the process of evaporation. After that the precursor is converted to oxide, resulting in the formation of a solid substance and hence the execution of the fourth stage [79]. The schematic setup for the spray pyrolysis method is shown in Fig. 1.8. The experimental setup includes a filter precipitator along with the burner ultrasonic generator for the production of droplets. To form precursor aerosol, metal nitrates, activator ions, and silica are dissolved in distilled water and then placed on an ultrasonic vibrator [86]. The oxidation of the aerosol is done by passing oxygen and fuel gases through it, which are then isolated by the vacuum pump and latterly gets deposited at silicon fibers. The prepared aerosols are first preheated at a lower temperature for the purpose of solvent evaporation before being heated in the furnace at a high temperature for decomposition. In this way, the chemical composition and the morphology of the obtained materials are extremely pure and homogeneous. The only disadvantage of this method is that it yields a substance with high porosity. It has been observed that the hollowness and porosity have a major impact on morphology and thermal stability. To solve this challenge, a modified spray pyrolysis technique is now used for material synthesis using an artificial colloidal solution [87].

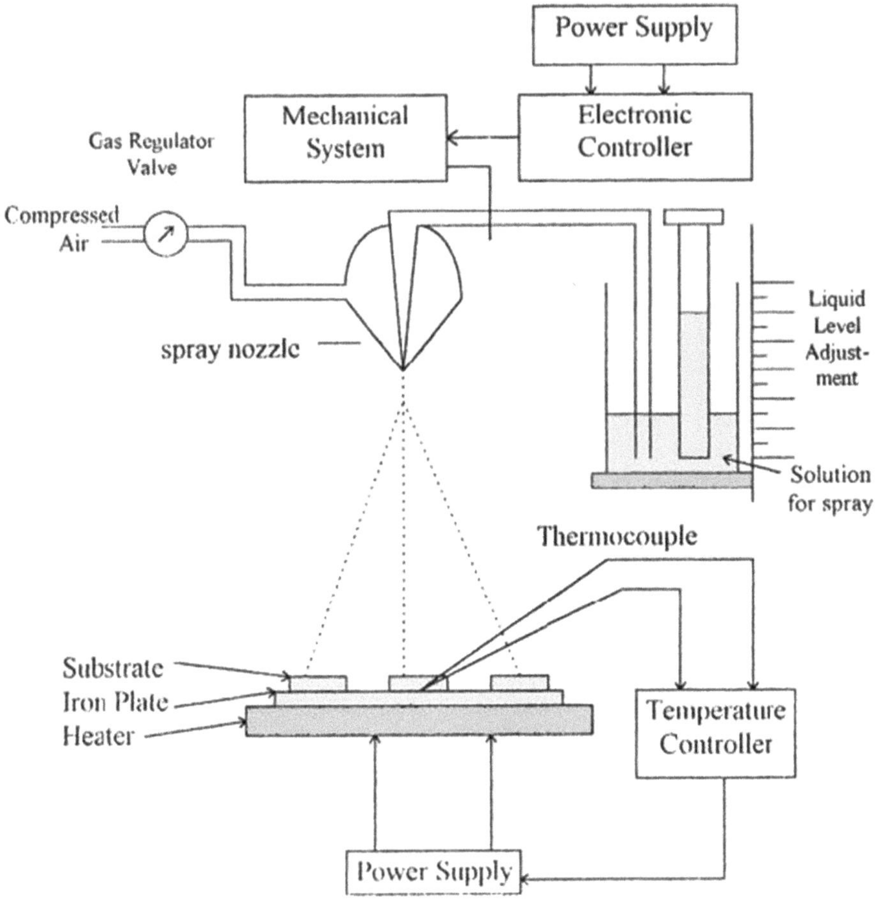

Figure 1.8 Schematic arrangement of the spray pyrolysis method.
Reused with permission from Ref. [79].

3.4 Combustion method

This method is among the wet chemical methods of the synthesis that does not require any further calcination. It is being considered as one of the convenient methods for synthesis as it is more cost-effective and energy-efficient. It is an exothermic reaction that produces heat and light and acts as a common way for the fabrication of the multi-constituent oxides in a short duration of time. In comparison to solid-state synthesized powders, the materials are found to possess the properties like chemical homogeneity, have fewer impurities, and also possess higher surface areas. To do the combustion synthesis, fuel and oxidizer are the primary requirements. Generally, nitric acid and urea are taken as fuel and reducing agents respectively. The basic procedure involves igniting an aqueous solution of precursors that includes metal nitrate and fuel also at a low temperature of approximately up to 500°C. This results in the occurrence of an extremely exothermic oxidative reaction, that too normally with flame [72,73]. The heat produced raises the reaction temperature up to 1200°C and increases the reaction rate to such an extent that the process gets completed in less than 10 min. Due to the lesser time required for the completion of the process, nanocrystalline powder is also produced by this method. However, it results in the production of weak clusters of submicron-sized powder that is latterly ground to a fine powder in an easy way. By utilizing the urea and europium as fuel and oxidizer, respectively. $SrMg_2Al_{16}O_{27}$: Eu^{2+} was prepared by burning 99.99% pure precursors, that is, strontium nitrate, magnesium nitrate, and aluminum nitrate at 550°C [88]. Besides this, it has also been used for the fabrication of long persistent phosphors like $CaAl_4O_7$:Eu^{2+}/Dy^{3+} [89]. Few other samples of the phosphors synthesized by using the combustion route are $Y_3Al_5O_{12}$:Ce^{3+} [90] and $NaBaPO_4$:Eu^{2+} [91]. Besides these methods, there exist a large number of other synthesizing methods also as discussed earlier. A short summary of different types of phosphors prepared by different methods is given in Table 1.2.

Table 1.2 Various methods of preparation of phosphors and their applications.

Phosphors	Synthesis method	Results cum applications	References
$NaSrO_4$:Dy^{3+}	Combustion method	Mentioned phosphor codoped with La^{3+} was found to exhibit luminescence properties. Also, it has been observed that near-UV-excited $NaSr_{0.9425}VO_4$:$0.05Dy^{3+}$, $0.0075La^{3+}$ phosphors proved to be an efficient material for w-LEDs.	[92]
Sr_3MoO_6:Eu^{3+}	Solid-state reaction method	The Eu^{3+}-activated Sr_3MoO_6 phosphor has two excitation peaks at 353 and 467 nm, on the behalf of which PL emission spectra are observed. This phosphor sample emits a bright red color. The purity of the colors was estimated to be about 94.53% and 94.78%, respectively. Thus, it acts as an efficient red-emitting phosphor for w-LEDs and versatile display film.	[93]

Continued

Table 1.2 Various methods of preparation of phosphors and their applications.—cont'd

Phosphors	Synthesis method	Results cum applications	References
$SrSO_4$:Eu	Chemical precipitation method	The TL properties of phosphor synthesized were investigated. These phosphors have linearity ranging from 0.1 to 20 Gy. The properties of PL were also investigated. In contrast to commercial uses, the sample phosphor demonstrated high sensitivity. For short-term radiation detection, the phosphor was found to be stable.	[94]
$CaMgP_2O_7$:Gd^{3+}	Solid-state, sol–gel, wet-chemical and solution combustion methods	A Gd^{3+}-activated $CaMgP_2O_7$ phosphor was synthesized using various synthesis methods in this study. Within a 274 nm excitation frequency, all the samples of this phosphor obtained by different approaches have been observed to shows PL at 314 nm. These findings indicated that this phosphor may be a viable option for phototherapy lamps.	[95]
$CaYAl_3O_7$: Ce^{3+}, Tb^{3+}, Eu^{3+} and Sm^{3+}	Polymeric precursor method	The authors investigated various lanthanide (Ln)-doped $CaYAl_3O_7$ phosphors in this study. These phosphors are found to emit the dopant's signature radiation.	[96]
$LaPO_4$:Gd^{3+}	Recrystallization method	The phosphor displayed that 275 nm under the emission wavelength of 314 nm in the PLE spectrum. It has been found that the 314 nm emission is because of 6 PJ-8S7/2 transitions of the dopant. The studies revealed that it could be used as a phototherapy lamp phosphor to treat a variety of skin diseases.	[97]

Table 1.2 Various methods of preparation of phosphors and their applications.—cont'd

Phosphors	Synthesis method	Results cum applications	References
$LaAlO_3$:Ce,Dy	Hydrothermal method	The TL property of the phosphor for UV dosimetry has been studied. The TL glow curves of the phosphor enabled with 5.0 mol% Ce^{3+} and 1.0 mol% Dy^{3+} reveal two dominant peaks at 151 and 213°C. Besides high TL performance, the substance has a linear response varying from 0.04 to 1.20 mJ/cm^2. As a result of these findings, prepared phosphor appears to be a promising candidate for use as a UV dosimeter.	[98]

4. Host lattice for phosphors

To synthesize an efficient nanophosphor, the choice of the perfect host matrix is very important. The chosen host matrix must possess lower lattice phonon energy along with high radiative transition rates. Besides these, the host should also possess high emission ability as well as a regulated emission spectrum with excellent chemical stability. The environment has a major impact on the luminescent properties of the activator ions. Also, the absorption and emission characteristics of phosphors are influenced by the interaction of host matrix and activator ions, such as covalence, crystal field, and inhomogeneous broadening. Furthermore, the interaction among electrons decreases as covalency increases, thus as covalency increases the electronic transitions shift to lower energy [34]. A crystal field is a type of electric field generated by the air surrounding the luminescent nucleus. The optical excitations of the ions usually take place in the quasi-states of the atom. These states are formed by removal of the degeneracy due to the perturbation done by asymmetric crystal field. The symmetry of these ionic interactions is determined by the crystallographic conditions surrounding the host cation. The crystal field strength is determined by measuring the nearest neighboring distance of cation and by the nature of the bonding. The degree of intermixing determines the speed of these forced electric dipole transitions. However, a significant increase is found in the luminescence intensity of the phosphor due to the elimination of the nonradiative recombination effects, quenching sites, surface defects, and so on [34,99]. Up to now a large number of new and promising phosphors have been synthesized, developed, and updated. Some examples of different types of hosts that have been identified so far are given in Table 1.3.

Table 1.3 The emission wavelength of rare-earth-doped phosphor in various host materials.

Host lattice	Phosphors	Emission wavelengths	References	Host lattice	Phosphors	Emission wavelengths	References
Oxides	$Y_2BaZnO_5:Yb^{3+}, Tm^{3+}$	479, 654, and 796 nm	[100]	Aluminates	$SrMg_2Al_{16}O_{27}:Eu^{2+}$	453 nm	[61]
	$Y_2O_3:Eu^{3+}, Tb^{3+}$	485, 545, 585, and 625 nm	[101]		$BaAl_2O_4:Eu^{2+}$, RE^{3+} (Y, Pr)	495 and 512 nm	[102]
	$ZnO:Eu^{3+}$	590 and 600 nm	[103]		$Ba_4Al_{14}O_{25}:Eu^{2+}$, RE^{3+} (Nd, Dy)	512 nm	[104]
	$SnO_2:Eu^{3+}$	–	[105]		$SrAl_2O_4:Eu^{2+}$, RE^{3+} (Dy, Pr, Y)	506 nm	[106]
Silicate	$CaMgSi_2O_6:Eu^{3+}, Tb^{3+}$	543 and 612 nm	[107]	Phosphate	$NaCaPO_4:Eu^{2+}$	500 and 506 nm	[108]
	$La_2SiO_5:Eu^{3+}, Tb^{3+}$	614 nm	[109]		$KCaPO_4:Eu^{2+}$	473 nm	[110]
	$SrMgSi_2O_6:Eu^{2+}, Nd^{3+}$	470 nm	[111]		$Na_2Zn_5(PO_4)_4:Eu^{3+}$	487, 546, and 611 nm	[112]
Sulfide	$CaS:Eu^{2+}$	635 nm	[113]	Zirconate	$CaZrO_3:Er^{3+}, Yb^{3+}$	523, 544, and 662 nm	[114]
	$SrS:Eu^{2+}$	600 nm	[113]		$ZrO_2:Er^{3+}, Yb^{3+}$	410, 525, and 544 nm	[115]
	$ZnS:RE^{3+}$ (RE = Tb, Sm, ho)	–	[116,117]		$ZrO_2:Eu^{3+}$	605 nm	[118]

5. Some important applications of phosphors

Rare-earth-activated nano phosphor ions are innovative potential materials for the development of different lighting applications. Some of the most significant uses of them are discussed later:

5.1 Field emission displays

In the current scenario, the field emission displays (FEDs) act as the emerging technological output of the phosphor materials. It is because of their excellent system output and low power consumption; FEDs have proved to be very useful for advanced flat panel displays (FEPs). Their working principle relies on electroluminescent vacuum fluorescent displays (VFDs) that operate at lower voltages and make use of a field emission cathode as an electron source. The displays of high-definition material are becoming more popular in the present culture. The conical electron emitters located opposite to the phosphor anode, packaging of the device, and the types of phosphor material used are all important factors in the production of FEDs [34]. Phosphors are regarded as a necessary part of FEDs. Thus, for manufacturing highly effective lower voltage-controlled FEDs, the identification of nanophosphors in terms of their specifications such as morphology, particle size distribution, particle structure, and deposition process, physical as well as chemical is vital [119,120]. Some of the examples of the most widely used RE-doped phosphor for different emission systems are $NaBiF_4:Eu^{3+}$ [121], $Y_3Al_5O_{12}:Tb^{3+}$ [122], and so-on.

5.2 3D display technology

Rare-earth-doped nanophosphors have been extensively studied for the development of innovative three-dimensional (3D) display structures for a decade. Different scientists approached for its development but the first successful attempt was made in 1966. In the mentioned year, researchers were able to create a 3D visualization by employing rare-earth-doped fluorine glass that possesses a size of one cubic centimeter as a projection medium by the application of a variety of rare-earth ions like Pr^{3+}, Er^{3+}, and Tm^{3+} [123]. Besides this, Er^{3+} and Tm^{3+} have also been used as light emission centers along with Yb^{3+} as a sensitizer ion for the creation of rare-earth-based multidoped full-color displays [124]. Transparent media with RE-activated up-converting phosphors have also been developed for 3D projection technology. Laine and others [125] mentioned the method efficient for the fabrication of blue and red light-emitting (LE) polycrystalline $Y_2O_3:Yb^{3+}$ ceramics, which have shown a lot of potentiality for 3D display technology. Similarly, $NaYF_4:Yb^{3+}/Er^{3+}$ nano phosphor embedded in poly (methyl methacrylate) has also been demonstrated as a strong candidate for the said technology [126].

5.3 Solar cells for solar spectral converters

The most challenging issue in today's global energy scenario is the creation of renewable and affordable energy supplies. The photovoltaic conversion serves as a possible candidate for achieving this target because it can directly convert solar energy into electric energy. The industry has been dominated by silicon wafer-based photovoltaic. The mechanisms of down-conversion and down-shift applied to the rare-earth-doped phosphor that comprises the ions pair (RE_s-RE_a) are expected to be the most effective spectral converters to change the incident solar spectrum [127]. RE_s ions act as a sensitizer, thereby allowing high-energy photons to be used more easily, while RE_a ions act as dopants, converting higher-energy photons into lower-energy photons that are then absorbed by silicon-based solar cells. As a result, the spectral response of solar cells is matched to the incident solar spectrum, resulting in a substantial increase in Si solar cell performance. Among the silicon-based solar cells, the most commonly used rare-earth-doped solar spectral converters are SrF_2:Yb^{3+} and $CaLaGa_3S_6O$:Yb^{3+}, which are near-infrared down-converting and down-shifting, respectively [128,129].

5.4 Generation of tricolor lamps

The development of the tricolor lamps is mainly based on the concept of a combination of three different narrow-band emissive wavelengths about 610 nm (red), 540 nm (green), and 450 nm (blue). This combination should be done in such a way so that a tricolor system formed should possess a high color rendering index [130]. In general, tricolor phosphors are typically present in the form of thin films within the fluorescent bulb. The excitation of the phosphors is done by the absorption of the ultraviolet radiation released by mercury and electron interaction that gets eventually transformed into visible light [131]. A typical fluorescent lamp is usually made up of red (55%), green (35%), and blue (15%) phosphor materials. For these types of lamps, emissions from Eu(III) (red), Tb(III) (green), and Eu(II) (blue) are probably treated as the most preferable emissions. Ronda et al. [132] use Y_2O_3:Eu^{3+}, $BaMgAl_{10}O_{17}$:Eu^{2+}, and $CeMgAl_{11}O_{19}$:Tb^{3+} for red, blue, and green emissions, respectively. Som et al. [133] reported defect correlated fluorescent quenching and electron–phonon coupling in Eu^{3+} doped $CaTiO_3$ phosphors for red emission in a display application. The well-known examples for tricolor phosphors are phosphate, aluminate, borate, and silicate systems [134,135]. But as aluminate possesses some important properties such as high thermal stability, high luminous potency, and so on, it is most commonly used [136].

5.5 White light-emitting devices

White light-emitting devices (WLEDs) that possess increased conversion performance of electrical energy into optical energy, power effectiveness, long operating life, and do not pose a threat to the environment can be used in a variety of settings and can potentially replace conventional lighting sources. The most readily used WLEDs are on the blue light-emitting InGaN-based LED or YAG:Ce^{3+} phosphor that emits yellow light [137]. However, in commercial phosphor, weak CRI <75, high CCT

>6200 K, and high thermal quenching have severely affected the quality of white light produced by complementary blue and yellow emission [34,138−141]. This can be attributed to the lack of red light-emitting portion in the YAG:Ce^{3+} phosphors, which makes WLEDs unsuitable for general illumination [34]. Exciting new phosphors with a high red-light output are particularly desirable The effective host and other materials that are found to be suitable for the red light are categorized into oxides [142], silicates [143], borates [144], phosphates [145], sulfides [146], and so on. The most prominent rare-earths that are used as activated luminescent centers for the production of red emission are Eu^{3+}, Pr^{3+}, and Sm^{3+}. Besides these most of the times, codopants can also be inserted into desirable host lattices in addition to these activators for the attainment of better red light. Few examples of the rare-earth activated red phosphors used to boost white light efficiency are Y_2SiO_5:Eu^{3+} [147], Sr_2CaMoO_6:Sm^{3+} [148], $LiBaBO_3$:Eu^{3+} [149], Y_2WO_6:Pr^{3+} [150], and so on. Some other examples of phosphors for pc-LEDs along with their characteristics are given in Table 1.4.

5.6 Luminescent paints

Paint is a luminescent substance that transforms into a rigid thin film after being applied to an object's surface in the shape of a continuous thin sheet. The phosphor substances that possess the property of persistent luminescence act as an important pigment in these paints. These paints are formed by doping phosphor with varnish to form the required paints. These paints shine in the dark and have the efficient quality of standard colors when exposed to sunlight. Luminescent paints can glow for a few minutes to several hours after being exposed to UV or visible photons. These paints have a wide range of applications due to their nontoxicity, dispersive transparency, and high color luminous properties, including road marking, decoration, rescue roads, posters, signboards, currency notes, and safety markers. The well-known example of the popular long afterglow persistent luminescent phosphor employed in luminescent paint is $SrAl_2O_4$:Eu^{2+}/Dy^{3+} [160].

5.7 Agricultural purposes

It is due to the characteristic feature of phosphors such as high performance and durability that they are being now widely applicable as a light conversion agent in practical plastic films used to cover greenhouses [137,161]. For light conversion agent (LCA), the nanophosphors should be excited by wavelengths ranging either from 290 to 350 nm or 510−580 nm, while simultaneously emitting 400−480 nm (blue light) and 600−680 nm (red light). The emission range of matter follows the excitation distribution of chlorophyll [161], and thus results in a significant increment in plant photosynthesis. According to reports, RE-doped phosphor LCA has a greater impact on improving crop quality, dominance, and prematureness in crops that rely on photosynthetic efficiency [150,162].

Table 1.4 Characteristics of some phosphors applicable for pc-WLEDs.

Phosphor	Excitation (nm)	Emission (nm)	CIE (*x*, *y*)	LE (lm/W)	CCT (K)	CRI	References
$GdAl_3(BO_3)_4$:Dy^{3+},Ce^{3+}	400	480, 575, 620	(0.31, 0.33)	5	6480	78	[151]
$(Gd,Lu)_3Al_5O_{12}$:Ce^{3+}	561	475–800	(0.449, 0.527)	–	3567	–	[152]
$(Tb,Gd)_3Al_5O_{12}$:Ce^{3+}	550	460	(0.3681, 0.3691)	97.2	3681	73	[153]
$Sr_3MgSi_2O_8$:Eu^{2+},Mn^{2+}	375	470, 570, 680	(0.35, 0.33)	–	4494	92	[154]
Sr_2SiO_4:Eu^{2+}	400	550	(0.39, 0.41)	–	3.8	68	[155]
$SrLa_2Al_2O_7$:Dy^{3+}	575	426, 452, 468	(0.32, 0.29)	–	8480	–	[156]
$Tb_3Al_5O_{12}$:Ce^{3+}	460	545, 555	(0.3681, 0.3691)	9–23	3500–4500	80–86	[157]
Ca_2SiS_4:Eu^{2+}	450	560, 660	–	–	3000	67	[158]
$Ca_2Gd_8Si_6O_{26}$:Sm^{3+}	405	600, 650	(0.599, 0.399)	–	–	–	[159]

5.8 Anticounterfeiting

Counterfeiting has become a challenging concern for companies and governments as it poses a serious threat to the quality of brands and the protection of brands [163,164]. Anticounterfeiting is a technique with the help of which government and companies can protect the authenticity of brands and prevent people from snooping and copying. In this technique, the charge carriers in deep traps are released for photo-stimulated phosphors when near-infrared (NIR) light is incident on it. The luminescent centers under NIR excitation produce the same or different frequency to hide real information, which is the main principle for anticounterfeiting. In the year 2010, Zhang and others [165] reported a nano-phosphor Mg_2SnO_4 that was used for this purpose. In this nano phosphor, the charge carriers are not released via conduction band but directly through the emitting center with the help of the tunneling effect to generate photo-stimulated luminescence (PSL). Lin and coworkers synthesized an undoped $CaZrO_3$ [166] in which the PSL spectrum showed a considerable difference in the persistent luminescence spectrum. They claimed that this property of undoped $CaZrO_3$ is because the charge carriers are not able to reach the P_1 levels due to the tunneling effect under NIR excitation, causing the luminescence emission occurrence as a result of the transition of P_2 and P_3 energy levels. In the year 2014, Xu and his coworkers [167] found that because of the introduction of Dy^{3+} and Tm^{3+} in α-Sr_2SiO_4 doped with Eu^{2+} the defects in the structure were corrected along with the change in the color generated during the PSL. In the year 2013, Zhang and others [168] used Tm^{3+} as a codopant in $Ba_4(Si_3O_8)_2$ doped with Eu^{2+} to generate deep traps that were able to hold trapped charge carriers at room temperature. In this phosphor, some charge carriers were seen to return to the emitting centers via conduction band while other charge carriers produced luminescence via tunneling effect when excited by NIR photons this led to the difference between persistent luminescence emission and PSL emission that could be effectively used for anticounterfeiting.

5.9 In vivo bioimaging

In prevalent imaging techniques, damage to the normal tissue is an inevitable problem [169–171]. The use of persistent luminescence and photo-stimulated phosphors have made it possible to generate images with a high signal-to-noise ratio higher contract, longer lifetime, less toxicity, and higher efficiency [172–177]. Due to the presence of defects in the host matrix and defects produced because of the addition of the dopants and dopants these nano-phosphors are able to trap charge carriers that are then released at later times. Such nano-phosphors that generate luminescence after the excitation ends are called as persistent luminescence (PL) phosphors. For in vivo bioimaging, these PL phosphors are first charged with photons of shorter wavelength and then are injected into the body where after activation they produce luminescence to generate an image. Such imaging techniques where precharged luminescence materials are used for imaging is called in vivo bioimaging. Comparative in vivo imaging results (photographs, luminescence images, and long-lasting phosphorescence (LLP) decay curves) after NIR-induced LLP are displayed in Fig. 1.9A–E for Zn-s sample

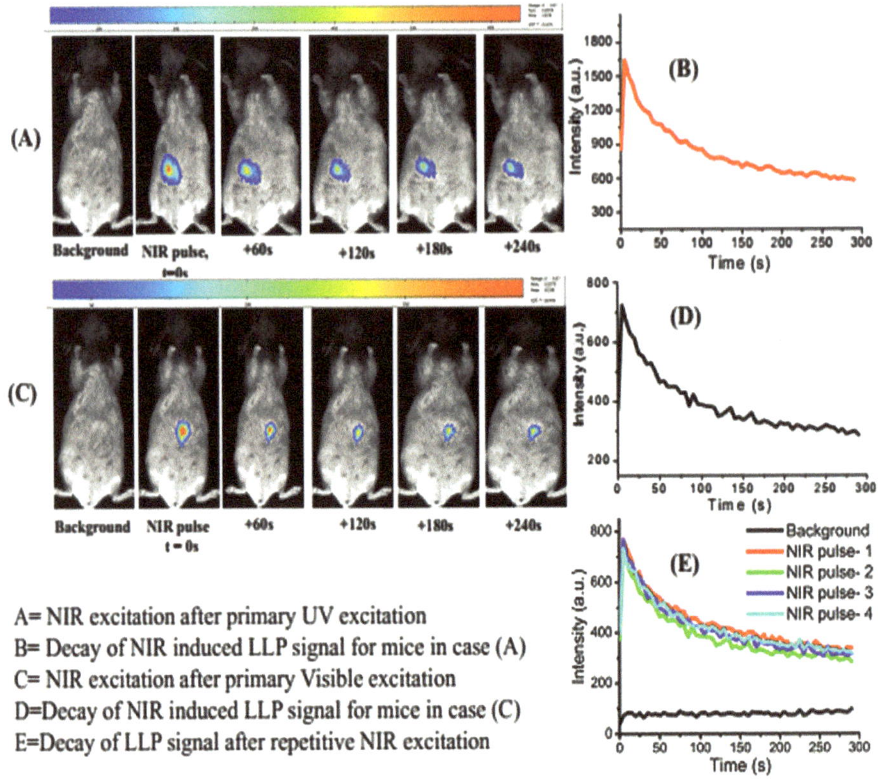

Figure 1.9 In vivo NIR-induced persistent luminescence (LLP) results for Zn-s sample. (A) Images recorded for primary UV (254 nm) excitation; (B) in vivo decay of the NIR-induced LLP after primary UV excitation; (C) images recorded after primary visible (580 nm) excitation and NIR photostimulation; (D) in vivo decay of the NIR-induced LLP after primary visible excitation; (E) effect of multiple NIR pulses (30 s duration each) after primary visible excitation of suspension of nanoparticles ex vivo.
Reused with permission from Ref. [178].

[178]. In the year 2013, Basavaraju and others [179] synthesized a novel red persistent phosphor $MgGa_2O_4$ and doped it with Cr^{3+} and used for imaging in vivo bioimaging. They found that this nano-phosphor proved to be a better imaging probe for in vivo bioimaging but its persistent luminescence time was inadequate, which limits its use for long-term imaging. In the year 2017, Sharma and others [178] found that $ZnGa_2O_4$ doped with Cr^{3+} nanoparticles could be activated in situ through biological applications. They found that charge carriers are generated in deep traps and then transferred to shallow traps where they produce luminescence. The emission wavelength of these materials is mostly located in the NIR window 1 (650–950 nm). Such materials have a low signal-to-noise ratio and weak penetration. Thus, the materials that have emission wavelength in the second optical window (950–1500 nm) and third optical window (1500–1800 nm) have been found to be more suitable for deep tissue [180].

6. Conclusions

This chapter focused on the different types of luminescence, mechanism of fluorescence, and phosphorescence from phosphors. The use of phosphors for different applications has helped mankind in many ways. In current times, the use of phosphors has led us to generate efficient lighting systems and better imaging probes. Their use has also led us to multicolor (or white) lighting systems. Our life has become a lot easier with the help of nano-phosphors but still, more research needs to be done to find more efficient and highly reliable phosphors. As phosphors are termed as the material of the future, the growth of improved and effective phosphors is very important.

References

[1] V. Kumar, S. Som, S. Dutta, S. Das, H.C. Swart, Red-light-emitting inorganic La_2CaZnO_5 frameworks with high photoluminescence quantum efficiency: theoretical approach, Mater. Des. 93 (2016) 203–215, https://doi.org/10.1016/j.matdes.2015.12.153.

[2] P. Gluck, J. King, Physics of incandescent lamp Burnout, Phys. Teach. 46 (2008) 29–35, https://doi.org/10.1119/1.2823997.

[3] D. Poelman, J.E. Van Haecke, P.F. Smet, Advances in sulfide phosphors for displays and lighting, in: J. Mater. Sci. Mater. Electron., Springer, 2009, pp. 134–138, https://doi.org/10.1007/s10854-007-9466-3.

[4] H. Hu, W. Zhang, Synthesis and properties of transition metals and rare-earth metals doped ZnS nanoparticles, Opt. Mater. 28 (5) (2006) 536–550.

[5] A. Birkel, K.A. Denault, N.C. George, C.E. Doll, B. Héry, A.A. Mikhailovsky, C.S. Birkel, B.C. Hong, R. Seshadri, Rapid microwave preparation of highly efficient Ce^{3+}-substituted garnet phosphors for solid state white lighting, Chem. Mater. 24 (2012) 1198–1204, https://doi.org/10.1021/cm3000238.

[6] A.K. Vishwakarma, K. Jha, M. Jayasimhadri, A.S. Rao, K. Jang, B. Sivaiah, D. Haranath, Red light emitting $BaNb_2O_6$:Eu^{3+} phosphor for solid state lighting applications, J. Alloys Compd. 622 (2015) 97–101.

[7] Y.-C. Lin, M. Karlsson, M. Bettinelli, Inorganic Phosphor Materials for Lighting, Springer, Cham, 2017, pp. 309–355, https://doi.org/10.1007/978-3-319-59304-3_10.

[8] N.C. George, K.A. Denault, R. Seshadri, Phosphors for solid-state white lighting, Annu. Rev. Mater. Res. 43 (2013) 481–501, https://doi.org/10.1146/annurev-matsci-073012-125702.

[9] N. Hirosaki, T. Takeda, S. Funahashi, R.J. Xie, Discovery of new nitridosilicate phosphors for solid state lighting by the single-particle-diagnosis approach, Chem. Mater. 26 (2014) 4280–4288, https://doi.org/10.1021/cm501866x.

[10] V.B. Pawade, N.S. Dhoble, S.J. Dhoble, Promising blue emitting $Ca_{3.5}Mg_{0.5}Si_3O_8Cl_4$: Eu^{2+} nano-phosphor for near UV-excited white-LEDs, Optoelectron. Advanc. Mater.-Rapid Commun. 5 (2011) (2011) 208–210.

[11] D.J. Naczynski, M.C. Tan, M. Zevon, B. Wall, J. Kohl, A. Kulesa, S. Chen, C.M. Roth, R.E. Riman, P.V. Moghe, Rare-earth-doped biological composites as in vivo shortwave infrared reporters, Nat. Commun. 4 (2013) 2199.

[12] V.B. Pawade, S.J. Dhoble, Optical properties of blue emitting Ce^{3+} activated $XMg_2Al_{16}O_{27}$ (X = Ba, Sr) phosphors, Opt Commun. 284 (18) (2011) 4185–4189.
[13] D.P. Dutta, V. Sudarsan, P. Srinivasu, A. Vinu, A.K. Tyagi, Indium oxide and europium/dysprosium doped indium oxide nanoparticles: sonochemical synthesis, characterization, and photoluminescence studies, J. Phys. Chem. C 112 (2008) 6781–6785, https://doi.org/10.1021/jp800576y.
[14] J. Knight, N. Schlager, Science of Everyday Things, vol. 1, 2002 (United States of America).
[15] K.G. Tshabalala, S.H. Cho, J.K. Park, S.S. Pitale, I.M. Nagpure, R.E. Kroon, H.C. Swart, O.M. Ntwaeaborwa, Luminescent properties and X-ray photoelectron spectroscopy study of $ZnAl_2O_4$:Ce^{3+}, Tb^{3+} phosphor, J. Alloys Compd. 509 (2011) 10115–10120, https://doi.org/10.1016/j.jallcom.2011.08.054.
[16] I. Mindru, G. Marinescu, D. Gingasu, L. Patron, L. Diamandescu, C. Ghica, B. Mironov, Doped aluminium based spinels synthesized by a soft chemistry method, Mater. Sci. Eng. B Solid-State Mater. Adv. Technol. 170 (2010) 99–106, https://doi.org/10.1016/j.mseb.2010.02.036.
[17] K.G. Tshabalala, Synthesis and Characterization of Down–Conversion Nanophosphors, University of the Free State, 2014. http://scholar.ufs.ac.za/xmlui/handle/11660/984. (Accessed 29 October 2020).
[18] B.K. Gupta, A. Kumar, P. Kumar, J. Dwivedi, G.N. Pandey, G. Kedawat, Probing on green long persistent Eu^{2+}/Dy^{3+} doped $Sr_3SiAl_4O_{11}$ emerging phosphor for security applications, J. Appl. Phys. 117 (2015), https://doi.org/10.1063/1.4922983.
[19] D.L. Dexter, J.H. Schulman, Theory of concentration quenching in inorganic phosphors, J. Chem. Phys. 22 (1954) 1063–1070, https://doi.org/10.1063/1.1740265.
[20] D.L. Dexter, A theory of sensitized luminescence in solids, J. Chem. Phys. 21 (1953) 836–850, https://doi.org/10.1063/1.1699044.
[21] B. Valeur, M.N. Berberan-Santos, A brief history of fluorescence and phosphorescence before the emergence of quantum theory, J. Chem. Educ. 88 (2011) 731–738, https://doi.org/10.1021/ed100182h.
[22] C.R. Ronda (Ed.), Luminescence: From Theory to Applications, John Wiley & Sons, 2007.
[23] K. Shinde, S. Dhoble, H. Swart, K. Park, Phosphate Phosphors for Solid State Lighting, Springer series, 2012, https://doi.org/10.1007/978-3-642-34312-4.
[24] E.N. Harvey, A History of Luminescence from the Earliest Times until 1900, American Philosophical Society, Philadelphia, 2011, https://doi.org/10.5962/bhl.title.14249.
[25] H. Singh, V. Sri, G. Granth, History of Luminescence from Ancient to Modern Times, Trans Tech Publ. (n.d.). https://doi.org/10.4028/www.scientific.net/DDF.361.1.
[26] S. Shionoya, W.M. Yen, H. Yamamoto (Eds.), Phosphor Handbook, CRC press, 2018.
[27] E. Radkov, A. Setlur, Z. Brown, J. Reginelli, High CRI phosphor blends for near-UV LED lamps, in: I.T. Ferguson, N. Narendran, S.P. DenBaars, J.C. Carrano (Eds.), Fourth Int. Conf. Solid State Light., SPIE, 2004, p. 260, https://doi.org/10.1117/12.565832.
[28] K. Masaoka, Y. Nishida, Metric of color-space coverage for wide-gamut displays, Opt Express 23 (2015) 7802, https://doi.org/10.1364/oe.23.007802.
[29] Commission Internationale de l'Éclairage, in: Commission Internationale de l'Éclairage Proceedings. Commission Internationale de l'Éclairage Proceedings, 1924.
[30] Y. Ohno, CIE fundamentals for color measurements, in: NIP & Digital Fabrication Conference (Vol. 2000, No. 2, Pp. 540-545). Society for Imaging Science and Technology, January 2000.

[31] C.H. Huang, T.M. Chen, $Ca_9La(PO_4)_7$:Eu^{2+}, Mn^{2+}: an emission-tunable phosphor through efficient energy transfer for white light-emitting diodes, Opt Express 18 (5) (2010) 5089–5099.
[32] S. Ye, F. Xiao, Y.X. Pan, Y.Y. Ma, Q.Y. Zhang, Phosphors in phosphor-converted white light-emitting diodes: recent advances in materials, techniques and properties, Mater. Sci. Eng. R Rep. 71 (2010) 1–34, https://doi.org/10.1016/j.mser.2010.07.001.
[33] P. Narwal, M.S. Dahiya, A. Agarwal, A. Hooda, S. Khasa, Compositional dependence of white light emission in Dy^{3+} doped NaCl–BaO bismuth borate glasses, J. Lumin. 209 (2019) 121–128, https://doi.org/10.1016/j.jlumin.2019.01.042.
[34] I. Gupta, S. Singh, S. Bhagwan, D. Singh, Rare earth (RE) doped phosphors and their emerging applications: a review, Ceram. Int. 47 (2021) 19282–19303, https://doi.org/10.1016/j.ceramint.2021.03.308.
[35] J.E. Kaufman (Ed.), Handbook, I. L., Illum. Eng. Soc., New York, 1972.
[36] H. Zhu, C.C. Lin, W. Luo, S. Shu, Z. Liu, Y. Liu, J. Kong, E. Ma, Y. Cao, R.S. Liu, X. Chen, Highly efficient non-rare-earth red emitting phosphor for warm white light-emitting diodes, Nat. Commun. 5 (2014) 1–10, https://doi.org/10.1038/ncomms5312.
[37] J.M.P.J. Verstegen, D. Radielović, L.E. Vrenken, A new generation of "deluxe" fluorescent lamps, combining an efficacy of 80 lumens/W or more with a color rendering index of approximately 85, J. Electrochem. Soc. 121 (1974) 1627, https://doi.org/10.1149/1.2401757.
[38] M. Balestrieri, S. Colis, M. Gallart, G. Schmerber, M. Ziegler, P. Gilliot, A. Dinia, Photoluminescence properties of rare earth (Nd, Yb, Sm, Pr)-doped CeO_2 pellets prepared by solid-state reaction, J. Mater. Chem. C. 3 (2015) 7014–7021, https://doi.org/10.1039/c5tc00075k.
[39] E.N. Harvey, (1957) A History of Luminescence from the Earliest Times until, 1900.
[40] S. Zhang, H. Liang, C. Liu, Increased 1D2 red emission of Pr^{3+} in $NaGdTiO_4$:Pr^{3+} due to temperature-assisted host sensitization and its color variation, J. Phys. Chem. C 117 (2013) 2216–2221, https://doi.org/10.1021/jp3120258.
[41] N.A. Atari, Piezoluminescence phenomenon, Phys. Lett. 90 (1982) 93–96, https://doi.org/10.1016/0375-9601(82)90060-3.
[42] P.F. Smet, I. Moreels, Z. Hens, D. Poelman, Luminescence in sulfides: a rich history and a bright future, Materials 3 (2010) 2834–2883, https://doi.org/10.3390/ma3042834.
[43] H.B. Bebb, E.W. Williams, Photoluminescence I: theory, semicond, Semimetals 8 (1972) 181–320, https://doi.org/10.1016/S0080-8784(08)62345-5.
[44] T.H. Gfroerer, Photoluminescence in analysis of surfaces and interfaces, in: Encycl. Anal. Chem., John Wiley & Sons, Ltd, Chichester, UK, 2006, https://doi.org/10.1002/9780470027318.a2510.
[45] J.S. Sparks, R.C. Schelly, W.L. Smith, M.P. Davis, D. Tchernov, V.A. Pieribone, D.F. Gruber, The covert world of fish biofluorescence: a phylogenetically widespread and phenotypically variable phenomenon, PLoS One 9 (2014) 83259, https://doi.org/10.1371/journal.pone.0083259.
[46] A. Köhler, J.S. Wilson, R.H. Friend, Fluorescence and phosphorescence in organic materials, Adv. Mater. 14 (10) (2002) 701–707.
[47] P.W. Milonni, Why spontaneous emission? Am. J. Phys. 52 (1984) 340–343, https://doi.org/10.1119/1.13886.
[48] C.H. Henry, Theory of spontaneous emission noise in open resonators and its application to lasers and optical amplifiers, J. Lightwave Technol. 4 (1986) 288–297, https://doi.org/10.1109/JLT.1986.1074715.

[49] F.A. Kröger, An introduction to luminescence of solids by H. W. Leverenz, Acta Crystallogr. 4 (1951), https://doi.org/10.1107/s0365110x51000313, 80–80.

[50] H. Terraschke, C. Wickleder, UV, blue, green, yellow, red, and small: newest developments on Eu^{2+}-doped nanophosphors, Chem. Rev. 115 (2015) 11352–11378, https://doi.org/10.1021/acs.chemrev.5b00223.

[51] G.B. Nair, H.C. Swart, S.J. Dhoble, A review on the advancements in phosphor-converted light emitting diodes (pc-LEDs): phosphor synthesis, device fabrication and characterization, Prog. Mater. Sci. 109 (2020) 100622, https://doi.org/10.1016/j.pmatsci.2019.100622.

[52] R.J. Xie, N. Hirosaki, Y. Li, T. Takeda, Rare-earth activated nitride phosphors: synthesis, luminescence and applications, Materials 3 (2010) 3777–3793, https://doi.org/10.3390/ma3063777.

[53] J. McKittrick, M.E. Hannah, A. Piquette, J.K. Han, J.I. Choi, M. Anc, M. Galvez, H. Lugauer, J.B. Talbot, K.C. Mishra, Phosphor selection considerations for near-UV LED solid state lighting, ECS J. Solid State Sci. Technol. 2 (2013) R3119–R3131, https://doi.org/10.1149/2.017302jss.

[54] K.N. Shinde, S.J. Dhoble, Europium-activated orthophosphate phosphors for energy-efficient solid-state lighting: a review, Crit. Rev. Solid State Mater. Sci. 39 (2014) 459–479, https://doi.org/10.1080/10408436.2013.803456.

[55] X. Ma, W. Zhuang, H. Guo, R. Liu, Y. Liu, Y. Hu, X. Wen, Effect of Si-N substituting for Al-O bonds on luminescence $Sr_3AlO_4F:Ce^{3+}$ phosphor, J. Rare Earth. 32 (2014) 399–403, https://doi.org/10.1016/S1002-0721(14)60084-3.

[56] G. Zhu, Z. Ci, Y. Shi, Y. Wang, Synthesis and photoluminescence properties of a red emitting phosphor $NaSr_2(NbO_3)_5:RE^{3+}$ (RE = Sm, Pr) for white LEDs, J. Rare Earth. 31 (2013) 1049–1052, https://doi.org/10.1016/S1002-0721(12)60401-3.

[57] J.K. Han, J.I. Choi, A. Piquette, M. Hannah, M. Anc, M. Galvez, J.B. Talbot, J. McKittrick, Phosphor development and integration for near-UV LED solid state lighting, ECS J. Solid State Sci. Technol. 2 (2013) R3138–R3147, https://doi.org/10.1149/2.014302jss.

[58] Y. Pan, M. Wu, Q. Su, Comparative investigation on synthesis and photoluminescence of YAG:Ce phosphor, Mater. Sci. Eng. B Solid-State Mater. Adv. Technol. 106 (2004) 251–256, https://doi.org/10.1016/j.mseb.2003.09.031.

[59] E.-H. Kang, S.-W. Choi, S.E. Chung, J. Jang, S. Kwon, S.-H. Hong, Photoluminescence characteristics of $Sr_3SiO_5:Eu^{2+}$ yellow phosphors synthesized by solid-state method and Pechini process, J. Electrochem. Soc. 158 (2011) J330, https://doi.org/10.1149/2.016111jes.

[60] S. Ekambaram, K.C. Patil, Synthesis and properties of Eu^{2+} activated blue phosphors, J. Alloys Compd. 248 (1997) 7–12, https://doi.org/10.1016/S0925-8388(96)02622-9.

[61] S. Singh, V. Tanwar, A.P. Simantilleke, H. Kumar, D. Singh, Synthesis and photoluminescence behavior of of $SrMg_2Al_{16}O_{27}:Eu^{2+}$ nanocrystalline phosphor, Optik 225 (2021) 165873, https://doi.org/10.1016/j.ijleo.2020.165873.

[62] L. Qi, B.I. Lee, J.M. Kim, J.E. Jang, J.Y. Choe, Synthesis and characterization of ZnS:Cu, Al phosphor prepared by a chemical solution method, J. Lumin. 104 (2003) 261–266, https://doi.org/10.1016/S0022-2313(03)00079-6.

[63] A.R. West, Solid state chemistry and its applications. John Wiley and sons, 1985 paper back, price £ 14.95, US $ 19.95, Cryst. Res. Technol. 21 (1986), https://doi.org/10.1002/crat.2170210140, 166–166.

[64] S.K. Misra, S.I. Andronenko, A variable-temperature X-band EPR study of the Gd^{3+} ion in a $La_2Si_2O_7$ crystal characterized by monoclinic site symmetry, Appl. Magn. Reson. 32 (2007) 377−384, https://doi.org/10.1007/s00723-007-0020-5.
[65] J.H. Lee, Y.J. Kim, Photoluminescent properties of Sr_2SiO_4:Eu^{2+} phosphors prepared by solid-state reaction method, Mater. Sci. Eng. B Solid-State Mater. Adv. Technol. 146 (2008) 99−102, https://doi.org/10.1016/j.mseb.2007.07.052.
[66] Y. Tan, Y. Liu, Z. Zhang, M. Hofmann, L. Grover, Comparing three methods for the synthesis of pure ß-dicalcium silicate, in: 2010 4th Int. Conf. Bioinforma. Biomed. Eng. ICBBE 2010, 2010, https://doi.org/10.1109/ICBBE.2010.5515290.
[67] B.K. Park, S.S. Lee, J.K. Kang, S.H. Byeon, Single-step solid-state synthesis of $CeMgAl_{11}O_{19}$:Tb phosphor, Bull. Kor. Chem. Soc. 28 (2007) 1467−1471, https://doi.org/10.5012/bkcs.2007.28.9.1467.
[68] H. Guo, H. Zhang, J. Li, F. Li, Blue-white-green tunable luminescence from $Ba_2Gd_2Si_4O_{13}$:Ce^{3+}, Tb^{3+} phosphors excited by ultraviolet light, Opt Express 18 (26) (2010) 27257−27262.
[69] N. Taghavinia, G. Lerondel, H. Makino, T. Yao, Europium-doped yttrium silicate nanoparticles embedded in a porous SiO_2 matrix, Nanotechnology 15 (2004) 1549−1553, https://doi.org/10.1088/0957-4484/15/11/031.
[70] L.Y. Bao, W. Gao, Y.F. Su, Z. Wang, N. Li, S. Chen, F. Wu, Progression of the silicate cathode materials used in lithium ion batteries, Chin. Sci. Bull. 58 (2013) 575−584, https://doi.org/10.1007/s11434-012-5583-3.
[71] Y. Li, B. You, W. Zhang, M. Yin, Luminescent properties of β-$Lu_2Si_2O_7$:RE^{3+}(RE=Ce, Tb) nanoparticles by sol-gel method, J. Rare Earth. 26 (2008) 455−458, https://doi.org/10.1016/S1002-0721(08)60117-9.
[72] S. Célérier, C. Laberty, F. Ansart, P. Lenormand, P. Stevens, New chemical route based on sol-gel process for the synthesis of oxyapatite $La_{9.33}Si_6O_{26}$, Ceram. Int. 32 (2006) 271−276, https://doi.org/10.1016/j.ceramint.2005.03.001.
[73] S. Sheoran, S. Singh, A. Mann, A. Samantilleke, B. Mani, D. Singh, Novel synthesis and Optical investigation of trivalent Europium doped $MGd_2Si_3O_{10}$ (M = Mg^{2+}, Ca^{2+}, Sr^{2+} and Ba^{2+}) nanophosphors for full-color displays, J. Mater. NanoSci. 6 (2) (2019) 73−81.
[74] J. Zha, H. Roggendorf, in: C.J. Brinker, G.W. Scherer (Eds.), Sol-gel science, the physics and chemistry of sol-gel processing, xiv, Academic Press, Boston, 1990, ISBN 0-12-134970-5, p. 908, https://doi.org/10.1002/adma.19910031025. Adv. Mater. 3 (1991) 522−522.
[75] L. Krishna Bharat, G. Nagaraju, S. Khaja Hussain, J.S. Yu, Sol-gel derived barium orthovanadate phosphors for white light-emitting diodes, Dye. Pigment. 150 (2018) 44−48, https://doi.org/10.1016/j.dyepig.2017.11.006.
[76] G. Seeta Rama Raju, E. Pavitra, J.S. Yu, Pechini synthesis of lanthanide (Eu^{3+}/Tb^{3+}or Dy^{3+}) ions activated $BaGd_2O_4$ nanostructured phosphors: an approach for tunable emissions, Phys. Chem. Chem. Phys. 16 (2014) 18124−18140, https://doi.org/10.1039/c4cp02266a.
[77] X. Liu, L. Yan, J. Lin, Synthesis and luminescent properties of $LaAlO_3$:RE^{3+} (RE) Tm, Tb) nanocrystalline phosphors via a sol-gel process, J. Phys. Chem. C 113 (2009) 8478−8483, https://doi.org/10.1021/jp9013724.
[78] G. Li, C. Li, C. Zhang, Z. Cheng, Z. Quan, C. Peng, J. Lin, Tm^{3+} and/or Dy^{3+} doped LaOCl nanocrystalline phosphors for field emission displays, J. Mater. Chem. 19 (2009) 8936−8943, https://doi.org/10.1039/b912115c.

[79] P.S. Patil, Versatility of chemical spray pyrolysis technique, Mater. Chem. Phys. 59 (1999) 185–198, https://doi.org/10.1016/S0254-0584(99)00049-8.
[80] K.Y. Jung, K.H. Han, Y.C. Kang, H.K. Jung, Preparation of $CaMgSi_2O_6$:Eu blue phosphor particles by spray pyrolysis and its VUV characteristics, Mater. Chem. Phys. 98 (2006) 330–336, https://doi.org/10.1016/j.matchemphys.2005.09.032.
[81] H.S. Kang, S.K. Hong, Y.C. Kang, K.Y. Jung, Y.G. Shul, S. Bin Park, The enhancement of photoluminescence characteristics of Eu-doped barium strontium silicate phosphor particles by co-doping materials, J. Alloys Compd. 402 (2005) 246–250, https://doi.org/10.1016/j.jallcom.2005.04.143.
[82] H.Y. Koo, S.K. Hong, J.M. Han, Y.C. Kang, Eu-doped $Ca_8Mg(SiO_4)_4Cl_2$ phosphor particles prepared by spray pyrolysis from the colloidal spray solution containing ammonium chloride, J. Alloy. Compd. 457 (2008) 429–434, https://doi.org/10.1016/j.jallcom.2007.02.138.
[83] B. Thiébaut, Flame spray pyrolysis: a unique facility for the production of nanopowders, Platin. Met. Rev. 55 (2011) 149–151, https://doi.org/10.1595/147106711X567680.
[84] Y.C. Kang, S.B. Park, Zn_2SiO_4:Mn phosphor particles prepared by spray pyrolysis using a filter expansion aerosol generator, Mater. Res. Bull. 35 (2000) 1143–1151, https://doi.org/10.1016/S0025-5408(00)00306-8.
[85] H.S. Kang, Y.C. Kang, K.Y. Jung, S. Bin Park, Eu-doped barium strontium silicate phosphor particles prepared from spray solution containing NH_4Cl flux by spray pyrolysis, Mater. Sci. Eng. B Solid-State Mater. Adv. Technol. 121 (2005) 81–85, https://doi.org/10.1016/j.mseb.2005.03.013.
[86] H. Hasegawa, T. Ueda, T. Yokomori, $Y_2Si_2O_7$:Eu/SiO_2 core shell phosphor particles prepared by flame spray pyrolysis, Proc. Combust. Inst. 34 (2013) 2155–2162, https://doi.org/10.1016/j.proci.2012.07.067.
[87] H.S. Kang, Y.C. Kang, H.D. Park, Y.G. Shul, Green-emitting yttrium silicate phosphor particles prepared by large scale ultrasonic spray pyrolysis, Kor. J. Chem. Eng. 20 (2003) 930–933, https://doi.org/10.1007/BF02697301.
[88] V.B. Pawade, S.J. Dhoble, Novel blue-emitting $SrMg_2Al_{16}O_{27}$:Eu^{2+} phosphor for solid-state lighting, Luminescence 26 (2011) 722–727, https://doi.org/10.1002/bio.1304.
[89] A.N. Yerpude, S.J. Dhoble, Combustion synthesis of blue-emitting submicron $CaAl_4O_7$:Eu^{2+}, Dy^{3+} persistence phosphor, Luminescence 27 (2012) 450–454, https://doi.org/10.1002/bio.1373.
[90] M. Upasani, B. Butey, S.V. Moharil, Synthesis, characterization and optical properties of $Y_3Al_5O_{12}$:Ce phosphor by mixed fuel combustion synthesis, J. Alloys Compd. 650 (2015) 858–862, https://doi.org/10.1016/j.jallcom.2015.08.076.
[91] J. Sun, X. Zhang, H. Du, Combustion synthesis and luminescence properties of blue $NaBaPO_4$:Eu^{2+} phosphor, J. Rare Earth. 30 (2012) 118–122, https://doi.org/10.1016/S1002-0721(12)60006-4.
[92] M. Chen, K. Qiu, P. Zhang, W. Zhang, Q. Yin, Enhancement of $NaSrVO_4$:Dy^{3+}-white-phosphor photoluminescence via La^{3+} co-doping, Ceram. Int. 45 (2019) 22547–22552, https://doi.org/10.1016/j.ceramint.2019.07.282.
[93] Y. Hua, S.K. Hussain, J.S. Yu, Eu3+-activated double perovskite Sr3MoO6 phosphors with excellent color purity for high CRI WLEDs and flexible display film, Ceram. Int. 45 (2019) 18604–18613, https://doi.org/10.1016/j.ceramint.2019.06.084.

[94] S. Jayasudha, K. Madhukumar, C.M.K. Nair, R.G. Nair, S. Rajesh, T.S. Elias, V.M. Anandakumar, N. Gopakumar, TL dosimetric characterization of gamma irradiated $SrSO_4$:Eu phosphors, J. Lumin. 183 (2017) 259−265, https://doi.org/10.1016/j.jlumin.2016.11.044.
[95] R.G. Kunghatkar, V.L. Barai, S.J. Dhoble, Synthesis route dependent characterizations of $CaMgP_2O_7$:Gd^{3+} phosphor, Result. Phys. 13 (2019) 102295, https://doi.org/10.1016/j.rinp.2019.102295.
[96] G.F. da C. Bispo, S.M.V. Novais, V.C. Teixeira, Z.S. Macedo, M.E.G. Valerio, Luminescence characterization and the energy level scheme for lanthanide doped $CaYAl_3O_7$, J. Lumin. 203 (2018) 486−491, https://doi.org/10.1016/j.jlumin.2018.06.073.
[97] A.O. Chauhan, A.B. Gawande, S.K. Omanwar, Narrow band UVB emitting phosphor $LaPO_4$:Gd^{3+} for phototherapy lamp, Optik 127 (2016) 6647−6652, https://doi.org/10.1016/j.ijleo.2016.04.131.
[98] V.H. Oliveira, N.M. Khaidukov, E.C. Silva, L.O. Faria, Study of TL properties of $LaAlO_3$:Ce, Dy crystals for UV dosimetry, Radiat. Meas. 46 (2011) 1173−1175, https://doi.org/10.1016/j.radmeas.2011.08.022.
[99] A.N. Belsky, J.C. Krupa, Luminescence excitation mechanisms of rare earth doped phosphors in the VUV range, Displays 19 (1999) 185−196, https://doi.org/10.1016/S0141-9382(98)00049-3.
[100] I. Etchart, I. Hernndez, A. Huignard, M. Bérard, M. Laroche, W.P. Gillin, R.J. Curry, A.K. Cheetham, Oxide phosphors for light upconversion; Yb3 and Tm3 co-doped Y_2BaZnO_5, J. Appl. Phys. 109 (2011) 063104, https://doi.org/10.1063/1.3549634.
[101] S. Som, S.K. Sharma, Eu^{3+}/Tb^{3+}-codoped Y_2O_3 nanophosphors: rietveld refinement, bandgap and photoluminescence optimization, J. Phys. D Appl. Phys. 45 (2012) 415102, https://doi.org/10.1088/0022-3727/45/41/415102.
[102] D. Singh, V. Tanwar, A.P. Simantilleke, B. Mari, P.S. Kadyan, I. Singh, Rapid synthesis and enhancement in down conversion emission properties of $BaAl_2O_4$:Eu^{2+}, RE^{3+} (RE^{3+} = Y, Pr) nanophosphors, J. Mater. Sci. Mater. Electron. 27 (2016) 2260−2266, https://doi.org/10.1007/s10854-015-4020-1.
[103] N. Wu, Q. Luo, X. Qiao, C.Q.I. Ma, The preparation of a Eu^{3+}-doped ZnO bi-functional layer and its application in organic photovoltaics, Mater. Res. Express 2 (2015) 125901, https://doi.org/10.1088/2053-1591/2/12/125901.
[104] S. Singh, V. Tanwar, A. Simantilleke, D. Singh, N.-S.& Nano, U, Structural and photoluminescent investigations of $SrAl_2O_4$:Eu^{2+}, RE^{3+} improved nanophosphors for solar cells, Elsevier, 2020, https://doi.org/10.1016/j.nanoso.2020.100427 (n.d.).
[105] V. Kiisk, T. Kangur, M. Paalo, T. Tätte, S. Lange, S. Pikker, I. Sildos, Structural and luminescence characteristics of SnO_2:Eu and SnO_2:Eu, Sb nanophosphors upon annealing at high temperatures, Mater. Chem. Phys. 130 (2011) 293−298, https://doi.org/10.1016/j.matchemphys.2011.06.042.
[106] D. Singh, V. Tanwar, A.P. Simantilleke, S. Bhagwan, B. Mari, P.S. Kadyan, K.C. Singh, I. Singh, Synthesis and enhanced luminescent characterization of $SrAl_4O_7$:Eu^{2+}, RE^{3+} (RE = Nd, Dy) nanophosphors for light emitting applications, J. Mater. Sci. Mater. Electron. 27 (2016) 5303−5308, https://doi.org/10.1007/s10854-016-4428-2.
[107] S. Singh, V. Tanwar, A. Simantilleke, H. Kumar, D. Singh, Optik, U, Structural and spectroscopic properties of $CaMgSi_2O_6$:RE^{3+} (Eu^{3+} and Tb^{3+}) nanophosphors under UV-illumination, Elsevier, 2020, https://doi.org/10.1016/j.ijleo.2020.165364.

[108] C. Qin, Y. Huang, L. Shi, G. Chen, X. Qiao, H.J. Seo, Thermal stability of luminescence of $NaCaPO_4$:Eu^{2+} phosphor for white-light-emitting diodes, J. Phys. D Appl. Phys. 42 (2009) 185105, https://doi.org/10.1088/0022-3727/42/18/185105.

[109] S. Singh, A.P. Simantilleke, D. Singh, Synthesis, structural and photoluminescence behaviour of novel La_2SiO_5:Eu^{3+}/Tb^{3+} nanomaterials for UV-LEDs, Optik 221 (2020) 165324, https://doi.org/10.1016/j.ijleo.2020.165324.

[110] D.K. Yim, I.S. Cho, C.W. Lee, J.H. Noh, H.S. Roh, K.S. Hong, Preparation and photoluminescence properties of γ-KCaPO4:Eu^{2+} phosphors for near UV-based white LEDs, Opt. Mater. 33 (2011) 1036–1040, https://doi.org/10.1016/j.optmat.2011.02.031.

[111] D. Jia, W. Jia, Y. Jia, Long persistent alkali-earth silicate phosphors doped with Eu^{2+}, Nd^{3+}, J. Appl. Phys. 101 (2007) 023520, https://doi.org/10.1063/1.2409767.

[112] K.N. Shinde, S.J. Dhoble, A. Kumar, Ce and Eu activated $Na_2Zn_5(PO_4)_4$, a new promising novel phosphor, Bull. Mater. Sci. 34 (2011) 937–942, https://doi.org/10.1007/s12034-011-0218-x.

[113] K. Suresh, K.V.R. Murthy, C. Atchyutha Rao, N.V. Poornachandra Rao, Rare earth doped alkali earth sulfide phosphors for white-light LEDs, ISRN Condens. Matter Phys. 2011 (2011) 1–3, https://doi.org/10.5402/2011/392917.

[114] A. Maurya, A. Bahadur, A. Dwivedi, A.K. Choudhary, T.P. Yadav, P.K. Vishwakarma, S.B. Rai, Optical properties of Er^{3+}, Yb^{3+} co-doped calcium zirconate phosphor and temperature sensing efficiency: effect of alkali ions (Li+, Na+ and K+), J. Phys. Chem. Solid. 119 (2018) 228–237, https://doi.org/10.1016/j.jpcs.2018.04.004.

[115] P. Salas, C. Angeles-Chávez, J.A. Montoya, E. De La Rosa, L.A. Diaz-Torres, H. Desirena, A. Martiínez, M.A. Romero-Romo, J. Morales, Synthesis, characterization and luminescence properties of ZrO_2:Yb^{3+}-Er^{3+} nanophosphor, in: Opt. Mater. (Amst)., Elsevier, 2005, pp. 1295–1300, https://doi.org/10.1016/j.optmat.2004.10.019.

[116] J. van den Bossche, K.A. Neyts, P. de Visschere, D. Corlatan, H. Pauwels, R. Vercaemst, L. Fiermans, D. Poelman, R.L. van Meirhaeghe, W.H. Laflére, F. Cardon, XPS study of TbF3 and TbOF centres in ZnS, Phys. Status Solidi 146 (1994) K67–K70, https://doi.org/10.1002/pssa.2211460248.

[117] T. Tohda, Y. Fujita, T. Matsuoka, A. Abe, New efficient phosphor material ZnS:Sm, P for red electroluminescent devices, Appl. Phys. Lett. 48 (1986) 95–96, https://doi.org/10.1063/1.96946.

[118] H. Hobbs, S. Briddon, E. Lester, The synthesis and fluorescent properties of nanoparticulate ZrO_2 doped with Eu using continuous hydrothermal synthesis, Green Chem. 11 (2009), https://doi.org/10.1039/b812149d, 484–49.

[119] Q.H. Zhang, J. Wang, C.W. Yeh, W.C. Ke, R.S. Liu, J.K. Tang, M.B. Xie, H.B. Liang, Q. Su, Structure, composition, morphology, photoluminescence and cathodoluminescence properties of $ZnGeN_2$ and $ZnGeN_2$:Mn^{2+} for field emission displays, Acta Mater. 58 (2010) 6728–6735, https://doi.org/10.1016/j.actamat.2010.08.038.

[120] M. Itoh, L. Ozawa, Cathodoluminescent phosphors, Annu. Reports Prog. Chem. - Sect. C. 102 (2006) 12–42, https://doi.org/10.1039/b417154n.

[121] P. Du, X. Huang, J.S. Yu, Facile synthesis of bifunctional Eu3+-activated NaBiF4 red-emitting nanoparticles for simultaneous white light-emitting diodes and field emission displays, Chem. Eng. J. 337 (2018) 91–100, https://doi.org/10.1016/j.cej.2017.12.063.

[122] P. Psuja, D. Hreniak, W. Strek, Low-voltage cathodoluminescence properties of $Y_3Al_5O_{12}$:Tb^{3+} nanopowders, J. Alloys Compd. 451 (2008) 571–574, https://doi.org/10.1016/j.jallcom.2007.04.136.

[123] E. Downing, L. Hesselink, J. Ralston, R. Macfarlane, A three-color, solid-state, three-dimensional display, Science 273 (1996) 1185–1189, https://doi.org/10.1126/science.273.5279.1185.

[124] Z. Yang, Z. Feng, Z. Jiang, Upconversion emission in multi-doped glasses for full colour display, J. Phys. D Appl. Phys. 38 (2005) 1629–1632, https://doi.org/10.1088/0022-3727/38/10/019.

[125] T.R. Hinklin, S.C. Rand, R.M. Laine, Transparent, polycrystalline upconverting nanoceramics: towards 3-D displays, Adv. Mater. 20 (2008) 1270–1273, https://doi.org/10.1002/adma.200701235.

[126] C. Li, J. Lin, Rare earth fluoride nano-/microcrystals: synthesis, surface modification and application, J. Mater. Chem. 20 (2010) 6831–6847, https://doi.org/10.1039/c0jm00031k.

[127] T. Trupke, M.A. Green, P. Würfel, Improving solar cell efficiencies by down-conversion of high-energy photons, J. Appl. Phys. 92 (2002) 1668–1674, https://doi.org/10.1063/1.1492021.

[128] B.M. Van Der Ende, L. Aarts, A. Meijerink, Near-infrared quantum cutting for photovoltaics, Adv. Mater. 21 (2009) 3073–3077, https://doi.org/10.1002/adma.200802220.

[129] G. Zhang, C. Liu, J. Wang, X. Kuang, Q. Su, An intense charge transfer broadband sensitized near-infrared emitting $CaLaGa_3S_6O:Yb^{3+}$ phosphor suitable for solar spectral convertor, Opt Express 19 (24) (2011) 24314–24319.

[130] Y. Zhao, X. Wang, Y. Zhang, Y. Li, X. Yao, Optical temperature sensing of up-conversion luminescent materials: fundamentals and progress, J. Alloys Compd. 817 (2020) 152691, https://doi.org/10.1016/j.jallcom.2019.152691.

[131] C. Tunsu, T. Retegan, C. Ekberg, Sustainable Processes Development for Recycling of Fluorescent Phosphorous Powders-Rare Earths and Mercury Separation A Literature Report, Gothenburg, Sweden, 2011.

[132] C.R. Ronda, T. Jüstel, H. Nikol, Rare earth phosphors: fundamentals and applications, J. Alloys Compd. 275–277 (1998) 669–676, https://doi.org/10.1016/S0925-8388(98)00416-2.

[133] S. Som, A.K. Kunti, V. Kumar, V. Kumar, S. Dutta, M. Chowdhury, S.K. Sharma, J.J. Terblans, H.C. Swart, J. Appl. Phys. 115 (2014) 193101.

[134] C.R. Ronda, Phosphors for lamps and displays: an applicational view, J. Alloys Compd. 225 (1995) 534–538, https://doi.org/10.1016/0925-8388(94)07065-2.

[135] D. Singh, Europium doped silicate phosphors: synthetic and characterization techniques, Adv. Mater. Lett. 8 (2017) 656–672, https://doi.org/10.5185/amlett.2017.7011.

[136] W. Bin Hung, T.M. Chen, Efficiency enhancement of silicon solar cells through a downshifting and antireflective oxysulfide phosphor layer, Sol. Energy Mater. Sol. Cells 133 (2015) 39–47, https://doi.org/10.1016/j.solmat.2014.11.011.

[137] E.F. Schubert, J.K. Kim, Solid-state light sources getting smart, Science 308 (2005) 1274–1278, https://doi.org/10.1126/science.1108712.

[138] N. Kimura, K. Sakuma, S. Hirafune, K. Asano, N. Hirosaki, R.J. Xie, Extrahigh color rendering white light-emitting diode lamps using oxynitride and nitride phosphors excited by blue light-emitting diode, Appl. Phys. Lett. 90 (2007) 051109, https://doi.org/10.1063/1.2437090.

[139] H.S. Jang, H. Yang, S.W. Kim, J.Y. Han, S.G. Lee, D.Y. Jeon, White light-emitting diodes with excellent color rendering based on organically capped CdSe quantum dots and $Sr_3SiO_5:Ce^{3+}$, Li+ phosphors, Adv. Mater. 20 (2008) 2696–2702, https://doi.org/10.1002/adma.200702846.

[140] K.A. Denault, J. Brgoch, M.W. Gaultois, A. Mikhailovsky, R. Petry, H. Winkler, S.P. Denbaars, R. Seshadri, Consequences of optimal bond valence on structural rigidity and improved luminescence properties in $SrxBa_2$-$xSiO_4$:Eu^{2+} orthosilicate phosphors, Chem. Mater. 26 (2014) 2275−2282, https://doi.org/10.1021/cm500116u.

[141] Q. Wang, D. Deng, Y. Hua, L. Huang, H. Wang, S. Zhao, G. Jia, C. Li, S. Xu, Potential tunable white-emitting phosphor $LiSr_4(BO_3)_3$:Ce^{3+}, Eu^{2+} for ultraviolet light-emitting diodes, J. Lumin. 132 (2012) 434−438, https://doi.org/10.1016/j.jlumin.2011.09.003.

[142] J.G. Li, X. Li, X. Sun, T. Ishigaki, Monodispersed colloidal spheres for uniform y_2O_3:Eu^{3+} red-phosphor particles and greatly enhanced luminescence by simultaneous Gd^{3+} doping, J. Phys. Chem. C 112 (2008) 11707−11716, https://doi.org/10.1021/jp802383a.

[143] S. Singh, D. Singh, Synthesis and spectroscopic investigations of trivalent europium-doped M_2SiO_5 (M = Y and Gd) nanophosphor for display applications, J. Mater. Sci. Mater. Electron. 31 (2020) 5165−5175, https://doi.org/10.1007/s10854-020-03076-5.

[144] R. Shrivastava, J. Kaur, V. Dubey, White light emission by Dy^{3+} doped phosphor matrices: a short review, J. Fluoresc. 26 (2016) 105−111, https://doi.org/10.1007/s10895-015-1689-8.

[145] Y. Zhu, Z. Sun, Z. Yin, H. Song, W. Xu, Y. Wang, L. Zhang, H. Zhang, Self-assembly, highly modified spontaneous emission and energy transfer properties of $LaPO_4$:Ce^{3+}, Tb^{3+} inverse opals, Dalton Trans. 42 (2013) 8049−8057, https://doi.org/10.1039/c3dt50390a.

[146] G.A. Kumar, D.X. Liu, Y. Tian, M.G. Brik, D.K. Sardar, 5d-4f emission of Eu^{2+} and electron-vibrational interaction in several alkaline earth sulfides doped with Eu^{2+} and Er^{3+}, Opt. Mater. 50 (2015) 199−203, https://doi.org/10.1016/j.optmat.2015.10.024.

[147] B. Lauritzen, N. Timoney, N. Gisin, M. Afzelius, H. De Riedmatten, Y. Sun, R.M. MacFarlane, R.L. Cone, Spectroscopic investigations of Eu^{3+}:Y_2SiO_5 for quantum memory applications, Phys. Rev. B Condens. Matter 85 (2012) 115111, https://doi.org/10.1103/PhysRevB.85.115111.

[148] L. Wang, H.M. Noh, B.K. Moon, S.H. Park, K.H. Kim, J. Shi, J.H. Jeong, Dual-mode luminescence with broad near UV and blue excitation band from Sr_2CaMoO_6:Sm^{3+} phosphor for white LEDs, J. Phys. Chem. C 119 (2015) 15517−15525, https://doi.org/10.1021/acs.jpcc.5b02828.

[149] S. Xu, P. Li, Z. Wang, T. Li, Q. Bai, J. Sun, Z. Yang, Luminescence and energy transfer of Eu^{2+}/Tb^{3+}/Eu^{3+} in $LiBaBO_3$ phosphors with tunable-color emission, J. Mater. Chem. C. 3 (2015) 9112−9121, https://doi.org/10.1039/c5tc01577d.

[150] M. Suchea, S. Christoulakis, M. Androulidaki, E. Koudoumas, CaS:Eu, Sm and CaS:Ce, Sm films grown by embedding active powder into an inert matrix, Mater. Sci. Eng. B Solid-State Mater. Adv. Technol. 150 (2008) 130−134, https://doi.org/10.1016/j.mseb.2008.03.017.

[151] Q.Y. Zhang, C.H. Yang, Y.X. Pan, Enhanced white light emission from $GdAl_3(BO_3)_4$:Dy^{3+}, Ce^{3+} nanorods, Nanotechnology 18 (2007) 145602, https://doi.org/10.1088/0957-4484/18/14/145602.

[152] Q. Meng, J.G. Li, Q. Zhu, X. Li, X. Sun, The effects of Mg^{2+}/Si^{4+} substitution on crystal structure, local coordination and photoluminescence of $(Gd,Lu)_3Al_5O_{12}$: Ce garnet phosphor, J. Alloys Compd. 797 (2019) 477−485, https://doi.org/10.1016/j.jallcom.2019.05.086.

[153] J. Chen, Y. Tang, Y. Tian, D. Hao, Y. Lin, S. Zhou, Fabrication of $(Tb,Gd)_3Al_5O_{12}$:Ce^{3+} phosphor ceramics for warm white light-emitting diodes application, Opt. Mater. Express 9 (2019), https://doi.org/10.1364/OME.9.003333.

[154] J.S. Kim, P.E. Jeon, Y.H. Park, J.C. Choi, H.L. Park, G.C. Kim, T.W. Kim, White-light generation through ultraviolet-emitting diode and white-emitting phosphor, Appl. Phys. Lett. 85 (2004) 3696–3698, https://doi.org/10.1063/1.1808501.
[155] J.K. Park, M.A. Lim, C.H. Kim, H.D. Park, J.T. Park, S.Y. Choi, White light-emitting diodes of GaN-based Sr_2SiO_4:Eu and the luminescent properties, Appl. Phys. Lett. 82 (2003) 683–685, https://doi.org/10.1063/1.1544055.
[156] P. Sehrawat, A. Khatkar, S. Devi, A. Hooda, S. Singh, R.K. Malik, S.P. Khatkar, V.B. Taxak, An effective emission of characteristic cool white light from Dy^{3+} doped perovskite type $SrLa_2Al_2O_7$ nanophosphors in single-phase pc WLEDs, Chem. Phys. Lett. 737 (2019), https://doi.org/10.1016/j.cplett.2019.136842.
[157] Y. Chen, M. Gong, G. Wang, Q.S. A.P, High efficient and low color-temperature white light-emitting diodes with phosphor, Aip.Scitation.Org., 2007. https://aip.scitation.org/doi/abs/10.1063/1.2771538. (Accessed 27 May 2021).
[158] P.F. Smet, K. Korthout, J.E. Van Haecke, D. Poelman, Using rare earth doped thiosilicate phosphors in white light emitting LEDs: towards low colour temperature and high colour rendering, Mater. Sci. Eng. B Solid-State Mater. Adv. Technol. 146 (2008) 264–268, https://doi.org/10.1016/j.mseb.2007.07.042.
[159] G. Seeta Rama Raju, J.S. Yu, J.Y. Park, H.C. Jung, B.K. Moon, Photoluminescence and cathodoluminescence properties of nanocrystalline $Ca_2Gd_8Si_6O_26:Sm^{3+}$ phosphors, J. Am. Ceram. Soc. 95 (2012) 238–242, https://doi.org/10.1111/j.1551-2916.2011.04762.x.
[160] A. Kumar, G. Kedawat, P. Kumar, J. Dwivedi, B.K. Gupta, Sunlight-activated Eu^{2+}/Dy^{3+} doped $SrAl_2O_4$ water resistant phosphorescent layer for optical displays and defence applications, New J. Chem. 39 (2015) 3380–3387, https://doi.org/10.1039/c4nj02333a.
[161] X. Liu, B. Lei, Y. Liu, The application of phosphor in agricultural field, in: Phosphors, up Convers. Nano Part. Quantum Dots Their Appl., Springer, Singapore, 2016, pp. 119–137, https://doi.org/10.1007/978-981-10-1590-8_4.
[162] J. Wu, D. Newman, I.V.F. Viney, Study on relationship of luminescence in CaS:Eu, Sm and dopants concentration, J. Lumin. 99 (2002) 237–245, https://doi.org/10.1016/S0022-2313(02)00342-3.
[163] Z. Sun, J. Yang, L. Huai, W. Wang, Z. Ma, J. Sang, J. Zhang, H. Li, Z. Ci, Y. Wang, Spy must Be spotted: a multistimuli-responsive luminescent material for dynamic multimodal anticounterfeiting and encryption, ACS Appl. Mater. Interface. 10 (2018) 21451–21457, https://doi.org/10.1021/acsami.8b08977.
[164] Z. Liu, L. Zhao, W. Chen, X. Fan, X. Yang, S. Tian, X. Yu, J. Qiu, X. Xu, Multiple anti-counterfeiting realized in $NaBaScSi_2O_7$ with a single activator of Eu^{2+}, J. Mater. Chem. C. 6 (2018) 11137–11143, https://doi.org/10.1039/c8tc04018d.
[165] J. Zhang, M. Yu, Q. Qin, H. Zhou, M. Zhou, X. Xu, Y. Wang, The persistent luminescence and up conversion photostimulated luminescence properties of nondoped Mg_2SnO_4 material, J. Appl. Phys. 108 (2010) 123518, https://doi.org/10.1063/1.3524280.
[166] X. Liu, J. Zhang, X. Ma, H. Sheng, P. Feng, L. Shi, R. Hu, Y. Wang, Violet-blue up conversion photostimulated luminescence properties and first principles calculations of a novel un-doped $CaZrO_3$ phosphor for application in optical storage, J. Alloys Compd. 550 (2013) 451–458, https://doi.org/10.1016/j.jallcom.2012.10.145.

[167] X. Xu, X. Zhang, T. Wang, J. Qiu, X. Yu, Color variation of photo-stimulated luminescence in strontium ortho-silicate with the assistance of trap centers, Mater. Lett. 127 (2014) 40–43, https://doi.org/10.1016/j.matlet.2014.04.092.

[168] X. Zhang, X. Xu, Q. He, J. Qiu, X. Yu, Significant improvement of photo-stimulated luminescence of $Ba_4(Si_3O_8)_2:Eu^{2+}$ by Co-doping with Tm^{3+}, ECS J. Solid State Sci. Technol. 2 (2013) R225–R229, https://doi.org/10.1149/2.010311jss.

[169] K. Takahashi, J. Miyahara, Y. Shibahara, Photostimulated luminescence (PSL) and color centers in $BaFX:Eu^{2+}$ (X = Cl, Br, I) phosphors, J. Electrochem. Soc. 132 (1985) 1492–1494, https://doi.org/10.1149/1.2114149.

[170] H. Von Seggern, T. Voigt, W. Knüpfer, G. Lange, Physical model of photostimulated luminescence of x-ray irradiated $BaFBr:Eu^{2+}$, J. Appl. Phys. 64 (1988) 1405–1412, https://doi.org/10.1063/1.341838.

[171] M.K. Crawford, L.H. Brixner, K. Somaiah, X-ray excited luminescence spectroscopy of barium fluorohalides, J. Appl. Phys. 66 (1989) 3758–3762, https://doi.org/10.1063/1.344037.

[172] J. Wang, H. Zhang, B. Lei, H. Dong, H. Zhang, Y. Liu, N. Lai, Y. Fang, Z. Chen, Red persistent and photo-stimulated luminescence properties of $SrCaSi_5N_8:Eu^{2+}$, Tm^{3+} solid solution, in: Opt. Mater. (Amst)., Elsevier B.V., 2014, pp. 1855–1858, https://doi.org/10.1016/j.optmat.2014.04.048.

[173] H. Liu, B. Feng, L. Luo, C. Han, P.A. Tanner, Near infrared photostimulated persistent luminescence and information storage of $SrAl_2O_4:Eu^{2+}$, Dy^{3+} phosphor, Opt. Mater. Express 6 (2016) 3375, https://doi.org/10.1364/ome.6.003375.

[174] Q. He, G. Qiu, X. Xu, J. Qiu, X. Yu, Photostimulated luminescence properties of Eu^{2+}-doped barium aluminate phosphor, Luminescence 30 (2015) 235–239, https://doi.org/10.1002/bio.2720.

[175] Z. Zou, X. Tang, C. Wu, D. Wang, J. Zhang, Z. Ci, S. Du, Y. Wang, How to tune trap properties of persistent phosphor: photostimulated persistent luminescence of $NaLuGeO_4:Bi^{3+}$, Cr^{3+} tailored by trap engineering, Mater. Res. Bull. 97 (2018) 251–259, https://doi.org/10.1016/j.materresbull.2017.09.011.

[176] F. Lin, S. Leilei, Research on the competition game between net about car and traditional taxi under internet pus background, in: Proc. 2017 6th Int. Conf. Comput. Sci. Netw. Technol. ICCSNT 2017, Institute of Electrical and Electronics Engineers Inc., 2018, pp. 47–50, https://doi.org/10.1109/ICCSNT.2017.8343475.

[177] T. Matsuzawa, Y. Aoki, N. Takeuchi, Y. Murayama, A new long phosphorescent phosphor with high Brightness, $SrAl_2O_4:Eu^{2+}$, Dy^{3+}, J. Electrochem. Soc. 143 (1996) 2670–2673, https://doi.org/10.1149/1.1837067.

[178] S.K. Sharma, D. Gourier, E. Teston, D. Scherman, C. Richard, B. Viana, Persistent luminescence induced by near infra-red photostimulation in chromium-doped zinc gallate for in vivo optical imaging, Opt. Mater. 63 (2017) 51–58, https://doi.org/10.1016/j.optmat.2016.06.053.

[179] N. Basavaraju, S. Sharma, A. Bessiere, Red persistent luminescence in $MgGa_2O_4:Cr^{3+}$; a new phosphor for in vivo imaging, J. of P.D., undefined, Iopscience.Iop.Org., 2013. (Accessed 21 May 2021).

[180] J. Xu, D. Murata, J. Ueda, B. Viana, S. Tanabe, Toward rechargeable persistent luminescence for the first and third biological windows via persistent energy transfer and electron trap redistribution, Inorg. Chem. 57 (2018) 5194–5203, https://doi.org/10.1021/acs.inorgchem.8b00218.

Further reading

[1] G.B. Nair, S.J. Dhoble, Orange light-emitting $Ca_3Mg_3(PO_4)_4:Sm^{3+}$ phosphors, Luminescence 32 (2017) 125–128, https://doi.org/10.1002/bio.3194.

[2] S.A. Pardhi, G.B. Nair, R. Sharma, S.J. Dhoble, Investigation of thermoluminescence and electron-vibrational interaction parameters in $SrAl_2O_4:Eu^{2+}$, Dy^{3+} phosphors, J. Lumin. 187 (2017) 492–498, https://doi.org/10.1016/j.jlumin.2017.03.028.

Instrumentation techniques for radiation dosimetry by using luminescence spectroscopy techniques

2

Ram Krishna[1], *Praveen Kumar Yadaw*[1], *Vikas Dubey*[2] *and N. Kumar Swamy*[1]
[1]Department of Physics, School of Science, ISBM University, Gariyaband, Chhattisgarh, India; [2]Department of Physics, Bhilai Institute of Technology Raipur, Naya Raipur, Chhattisgarh, India

1. Introduction

Most biomedical applications rely on radiation therapies. Presently, radiation is used for carrying out radiotherapeutic or imaging procedures like CT scans in the field of medicine [1]. Radiation has an intensity to penetrate through human skin, and as a result of radiation tissue interaction, skin tissues could be damaged [2].

It is essential to control and tune according to the requirement of intentional and unintentional radiation exposures.

Radiation can be classified into two categories, that is, charged particle radiation that consists of fast electrons and heavy charged particles, and uncharged radiation that comprises electromagnetic radiation and neutrons [3]. In radiation spectroscopy, both these radiation types interact with matter in different ways, hence, the need for radiation measurement and monitoring to control its effects in the matter accordingly.

However, during its applications, there has to be precaution since excessive radiation attenuation by human tissue may result in high absorbed dose values [4]. This could be a root cause of secondary malignancies [5] such as skin redness, skin swelling, dry, peeling skin, moist desquamation, and skin ulcers.

When a dosimeter is irradiated with natural or artificial ionizing radiation, part of this irradiation is stored in the material, specifically in point defects [6,7]. The radiation stored in the material can be released, stimulating it in different ways. Two ways that are well known are (1) by heating (thermal stimulation) and (2) by exposure to appropriate wavelength light (optical stimulation). In both cases, the luminescence phenomenon from the material is observed. The luminescence which produced by thermal stimulation is called as thermoluminescence (TL), whereas the luminescence which produced by chemical reactions is called as chemiluminescence and finally if luminescence is achieved by optical stimulation, such luminescence is known as optically stimulated luminescence (OSL) [8−11]. The luminescent intensity of the material is related to exposure to radiation and is therefore used to evaluate radiation doses (dosimetry).

Rare-Earth-Activated Phosphors. https://doi.org/10.1016/B978-0-323-89856-0.00014-6

Table 2.1 Types of dosimeters present in the market.

S No	Type	Abbreviation
1	TLDs	Thermoluminescence dosimetry
2	OSLDs	Optically stimulated luminescence dosimetry
3	CLDs	Chemicoluminescent dosimetry
4	MLDs	Mechanically heated stimulated luminescence dosimetry
5	NeD	Neutron dosimetry
6	PD	Photon dosimetry

1.1 Dosimetry

Dosimetry is a more sophisticated extension of the preceding application, which is generally a safety directed system. For many years, Si detectors were felt to be the best choice owing to the low atomic number (Z) of Si, which is the closest to the low Z of human tissue in comparison to the high Z of CdTe. On the other hand, this high Z allows high X-ray and γ ray absorption and thus higher sensitivity to lower doses. The absolute dose measurement required correction of the absorption in relation to the energy and then conversion to actual tissue dose. This can be done easily now using microprocessors, analog circuits, or more simply by filter correction. Such dosimeters are now commonly present in the market (Table 2.1). A complete dosimeter was constructed with particle-photon differentiation capability to monitor the irradiation of space flight crew members, who are exposed to natural radiation X-rays, γ rays, and protons.

1.1.1 Process diagram of dosimetry

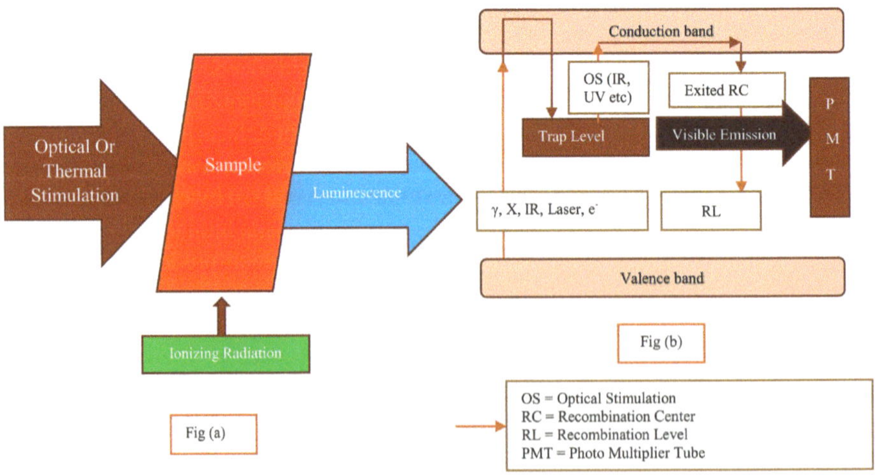

Figure 2.1 (A) Luminescence detector dosimeter (operational principle), (B) Band diagram of light rays emission.

1.2 Need of dosimeter

Medical radiation dosimetry involves measurement, calculation, and assessment of the quantity and quality of ionizing radiation exposed to and attenuated by the human body. Various gas, liquid, and solid-state dosimeters are used to quantify radiation; these are predominantly grouped under the ionization chamber, semiconductor, and diamond detector types [6,7]. These detectors are used to measure radiation delivered internally by ingesting or inhaling radioactive substances and externally using external beam radiation therapy (Fig. 2.1A and B).

1.2.1 Dosimetry analysis setup

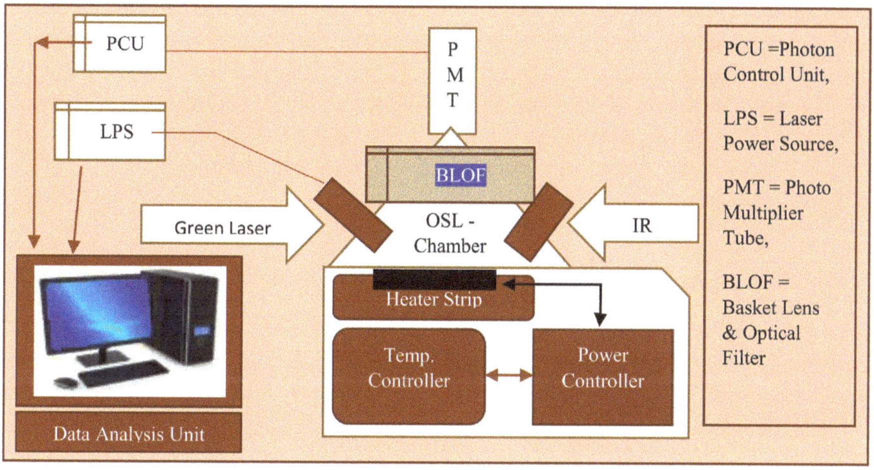

Figure 2.2 TL/OSL reader for dosimetry.

1.3 Types of dosimetry

There is a wide range of dosimeters available for commercial application. It can be further classified based on size, working principle, and type of source. The list of the type is presented as follows (Fig. 2.2).

1.3.1 Electron Paramagnetic Resonance Dosimetry (EPRDs),
1.3.2 Chemical Dosimetry or radiation Dosimetry (CDs or RDs),
1.3.3 Thermoluminescent dosimetry (TLDs) and
1.3.4 Optically stimulated luminescence dosimetry (OSLDs)
1.3.5 Alanine/electron paramagnetic resonance dosimetry system (AEPRDs)
1.3.6 Gas ionization detectors (GIDs)
1.3.7 Fluorescent nuclear track detector (FNTD)
1.3.8 Plastic nuclear track detectors (PNTD) in neutron dosimetry
1.3.9 Three-dimensional radiation (TDR) dosimetry (or gel polymer dosimetry)

1.3.1 Electron paramagnetic resonance (EPR) dosimetry

Dosimetry based on the detection by electron paramagnetic resonance (EPR) spectroscopy of ionizing radiation-induced radicals is an established method for the retrospective dosimetry of past exposures and the dosimetry of potentially exposed persons in radiological emergencies. The dose is generally estimated by measuring the physical damage induced in materials contained in objects placed on or next to the potentially exposed person.

EPR dosimetry is based on the detection and the measurement of the concentration of stable radiation-induced radicals obtained from reactions triggered in the matter by ionizing radiation using EPR spectrometry. In solids, the radicals become trapped (i.e., they are not "free") and in some cases, they can be stable over long periods. EPR is then the choice of method for direct detection of radiation-induced radicals in solids. Ionizing radiations induce radicals in most materials, but the yield of production varies significantly. Therefore, the suitability of materials as dosimeters is mainly related to their radiation sensitivity and signal stability in time. Materials that have already been investigated are, for example, sugar, plastics, glass, wool, and cotton. Stable radicals are also generated in solid biological tissues (nails, hair, and calcified tissues). The calibration of EPR signal intensity as a function of absorbed dose can be done for each sample by delivering additional radiation doses with a laboratory source and then EPR measurement [12].

1.3.2 Chemical dosimetry or radiation dosimetry

The study of physical or chemical changes due to irradiation of the substance is called radiation dosimetry, and it depends on the nature of radiations, dose of radiation, and time of exposure. Also, we classify the chemical dosimetry, namely **Fricke Dosimetry** and the second subtype is **Cerike Dosimetry**.

1.3.2.1 Fricke dosimetry

Frike → Ferrric → Fe (Iron)

In this dosimetry, the oxidation of ferrous to ferric ions is studied, that is, $Fe^{2+} \rightarrow Fe^{3+} + e^{-}$; for this, Fricke solution is prepared by mixing the following reagents with a wavelength of 304 nm

0.4 g $FeSO_4.6H_2O$ + 0.06 g NaCl in 22 mL of 95.98% H_2SO_4 and making volume up to 1 L using D.W.	→ → → 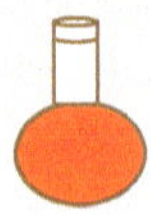

1.3.2.2 Cerike dosimetry or ceric sulfate dosimetry

In ceric sulfate dosimetry, the conversion of Ce host material ions are studied, it is found that $Ce^{4+} \rightarrow Ce^{3+}$ conversion took place; for this study purpose, 1−10 ml ceric sulfate solution (acidified) is exposed to the radiation and Ce^{4+} is converted to Ce^{3+}, at wavelength is 320 nm.

1.3.3 Thermoluminescent dosimeters

A thermoluminescent dosimeter, or TLD, is a type of radiation dosimeter, consisting of a piece of a thermoluminescent crystalline material inside a radiolucent package. When a thermoluminescent crystal is exposed to ionizing radiation, it absorbs and traps some of the energy of the radiation in its crystal lattice and get provide emission an outlet.

1.3.4 Optically stimulated luminescence dosimeters

The first commercial OSL dosimeter, promoted as Luxel, based on the pulse optically stimulated luminescence dosimeters (OSL) technique and Al_2O_3:C was developed for occupational personnel dosimetry and introduced by Landauer in 1998. The sensitive element of the OSL dosimeter is Al_2O_3:C powder mixed with an organic binder and printed on a transparent plastic substrate. The latest modification depicted in the OSL sensor is sandwiched between two parts of a clamshell filter pack having five detector areas with different filtration: open window, plastic, copper-aluminium, aluminum, and a central imaging element made from a perforated copper plate. The imaging element allows a service provider to distinguish between static and dynamic exposures using a patented image processing algorithm.

1.3.5 Alanine/electron paramagnetic resonance dosimetry system

Alanine, one of the amino acids, pressed in the form of rods or pellets with an inert binding material, is typically used for high-dose dosimetry. The dosimeter can be used at a level of about 10 Gy or more with sufficient precision for radiotherapy dosimetry. The radiation interaction results in the formation of alanine radicals, the concentration of which can be measured using an electron paramagnetic resonance (known also as electron spin resonance) spectrometer. The intensity is measured as the peak to peak height of the central line in the spectrum. The readout is nondestructive [13].

(i) Alanine is tissue equivalent and requires no energy correction within the quality range of typical therapeutic beams. It exhibits very little fading for many months after irradiation. The response depends on environmental conditions during irradiation (temperature) and storage (humidity).

(ii) At present, alanine's potential application for radiotherapy is in dosimetry comparisons among hospitals.

1.3.6 Gas ionization detectors dosimetry

Dosimetry ultimately aims at reflecting the radiation dose absorbed by the organism rather than by the dosimeter. To understand the damage induced by radiation, fundamental understanding of radiobiological processes is a prerequisite, which allows the description of radiation action at a microscopic level. Accordingly, much effort is put into bioequivalent dosimetry.

1.3.7 FNTD

As always, there is no one perfect dosimeter for all applications. One of the most challenging dosimetry problems is neutron dosimetry. Thermal (<0.5 eV) and intermediate (0.5 eV−10 keV) neutrons are relatively easy to measure using TL and OSL detectors containing 6Li and 10B isotopes [14,15], but dosimetry of fast neutrons (>10 keV) still remains one of the most difficult tasks. Novel fluorescent nuclear track detector (FNTD) technology is a new luminescent dosimetry technique suitable for replacing plastic nuclear track detectors (PNTD) in neutron dosimetry and LET spectroscopy of heavy charged particles [16−24].

1.3.8 PNTD

The new technology combines a new luminescent aluminum oxide crystal, doped with carbon and magnesium (Al_2O_3: C, Mg) [25,26], with a laser scanning confocal fluorescence microscopy technique [27]. Images of tracks are formed as bright objects on dark background in fluorescent contrast and digitally processed.

1.3.9 Three-dimensional radiation dosimetry

Here, we define a 3D dosimeter as a dose measurement device, which can record the 3D dose distribution in a continuous medium. Consequently, the spatial resolution of this dosimeter is mostly determined by the readout technique used with this dosimeter. In contrast, a dosimeter that is made of many point-like detectors at discrete points and a nonnegligible distance among those can be called as pseudo-3D dosimeter. In this review, we will minimize the discussion on the pseudo-3D dosimeter. The following 3D dosimeters are currently available commercially or in research laboratories: The polymer gel dosimetry (PGD); the Fricke gel dosimeter (FGD); the solid plastic dosimetry (SPD). Furthermore, we cannot ignore two other 3D dosimeters, scintillators, and Cherenkov-radiation detectors. The former is a rather old technology, but it has not gained much attention as a 3D dosimetry tool mostly because the material has mass density and an effective atomic number very different from those of the water [28]. The Cherenkov detector has been studied by one group for the last few years and has the potential to be an excellent 3D dosimeter [29], but it needs further extensive studies by many different investigators before clinical applications. Therefore, we focus on PGD, FGD, and SPD in the rest of this article. A comparative analysis of dosimetry has been listed in Table 2.2.

Table 2.2 Comparative table on various types of dosimetry.

S No.	Parameter	Full name	Definition	Function	Types	Application	Dose rate	References
1	PDs	Pocket dosimeters	A small ionization detection instrument that indicates ionizing radiation exposure directly	Pocket dosimeters are used to provide the wearer with an immediate reading of his or her exposure to X-rays and γ rays.	**1.** Passive dosimeters **2.** Film badge	Immediate reading	1,000–3,000 grains	[30]
2	CDs	Chemiluminescence dosimeters and RDs (or radiation dosimetry)	A chemical dosimeter is a material that undergoes a measurable chemical change that can be calibrated to a known radiation source	A dosimeter is designed to monitor exposure and serve as a tool to alert an individual if radiation dose readings exceed unhealthy levels of radiation exposure	**1.** Fricke dosimetry **2.** Cerike dosimetry	**1.** Monitor the effectiveness of biocides, **2.** Effect of drugs, and **3** .Detection of microorganisms in foods	50,000 grains	[31]

Continued

Table 2.2 Comparative table on various types of dosimetry.—cont'd

S No.	Parameter	Full name	Definition	Function	Types	Application	Dose rate	References
3	TLDs	Thermoluminescent dosimetry	A thermoluminescent dosimeter, abbreviated as TLD, is a passive radiation dosimeter, which measures ionizing radiation exposure by measuring the intensity of visible light emitted from a sensitive crystal in the detector when the crystal is heated	A thermoluminescent dosimeter, abbreviated as TLD, is a passive radiation dosimeter, which measures ionizing radiation exposure by measuring the intensity of visible light emitted from a sensitive crystal in the detector when the crystal is heated.	**1.** Calcium fluoride TLDs and 2. Lithium fluoride TLDs	**1.** Potential applications in radiotherapy **2.** Medicine	Personal dose equivalent, Hp(0.07). The Hp(0.07)	[7]
4	OSLDs	Optically stimulated luminescence dosimetry	Optically stimulated luminescence (OSL) is a process in which a preirradiated (exposed to ionizing radiation) material when subjected to an appropriate optical stimulation, emits a light signal proportional to the absorbed dose. The wavelength of the emitted light is the characteristic of the OSL material	Optically stimulated luminescent dosimeters, OSLDs, are plastic disks infused with aluminum oxide doped with carbon Al_2O_3:C. These disks are encased in a light-tight plastic holder. The intensity of the luminescence depends on the dose absorbed by the OSLD and the intensity of the stimulation light	**1.** OL excitation dosimetry. **2.** Time resolved OSL Dosimetry	**1.** Environmental controls in industry **2.** Laboratory or **3.** Hospital **4.** Medical procedures, that is, radiotherapy and scanners	**1.** X-ray energy Range 40 kVp) = dose (0.58 mGy)	[32,33]

5	EPRDs	Electron paramagnetic resonance dosimetry	EPR dosimetry is based on radiation-induced radicals in hydroxyapatite, which is the mineral phase of teeth and bones	In EPR dosimetry, solid dosimeter materials such as alanine (AL) or, more recently, lithium formate monohydrate (LFM) are typically used	**1.** Radiation-induced and **2** . Radiation insensitive signals	**1.** Radiotherapy based on their favorable dosimetric properties	**1.** Mineral phase of teeth and bones	[12]
6	GIDDs	Gas ionization dosimeter	Gaseous ionization detectors are radiation detection instruments used in radiation protection applications to measure ionizing radiation	Ionization, in chemistry and physics, any process by which electrically neutral atoms or molecules are converted to electrically charged atoms or molecules (ions)	It depends on ionizing gas	**1.** Biomedical applications, **2.** Medical application	Electron Beams with energies below 10 MeV	[34,35]
7	RPLD system	Radiophotoluminescent dosimetry	Radiophotoluminescence (RPL) is a phenomenon seen in luminescent materials in which the appearance of new photoluminescence (PL) emission is induced by an incident ionizing radiation such as X-rays; and the signal is stable even after the irradiation and during the PL measurement	The principle of radiophotoluminescence (RPL) is applied to the glass dosimeter, which is one of the most excellent solid-state dosimeters. The silver-activated phosphate glass irradiated with ionizing radiations emits luminescence when exposed to UV light. This phenomenon is called RPL	—	**1.** CCD-based spectrometer **2.** Individual and **3.** Environmental monitoring as well as in clinical dosimetry	Measurable dynamic range became from 0.1 mGy to 10 Gy [37].	[36]

1.4 Luminescence and types of luminescence

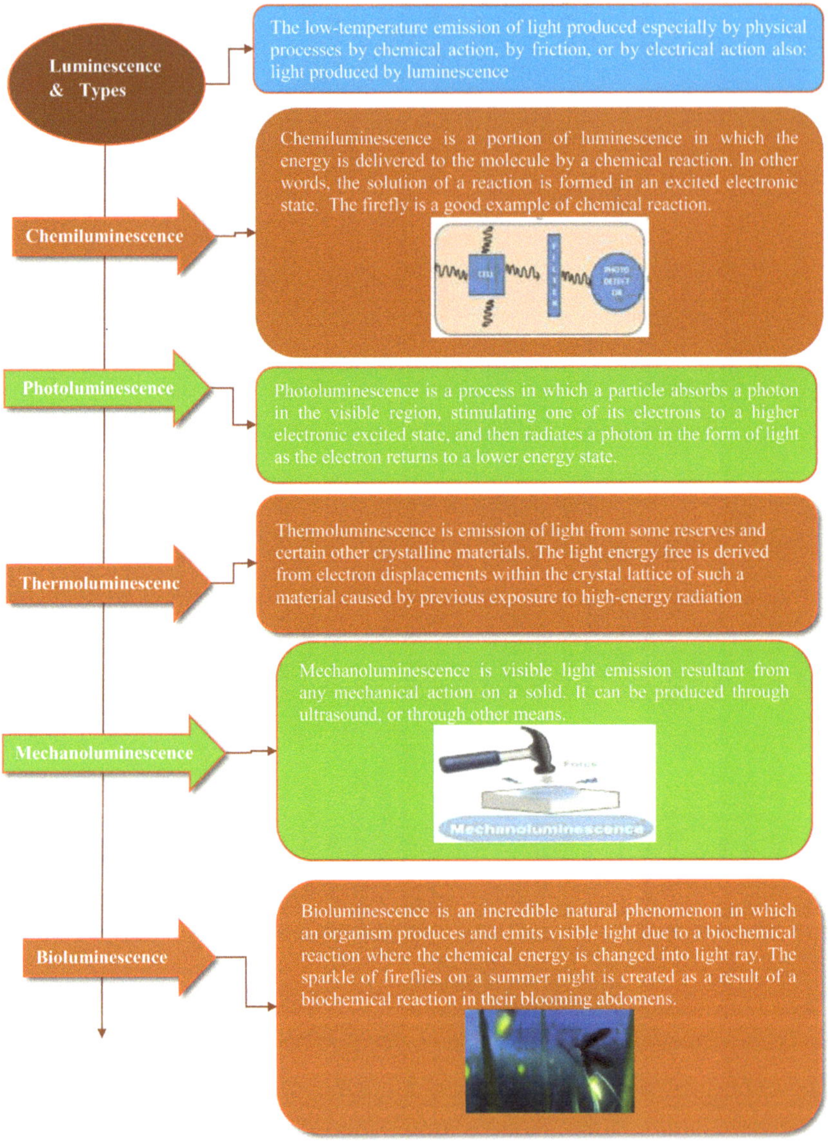

1.5 Luminescence relation

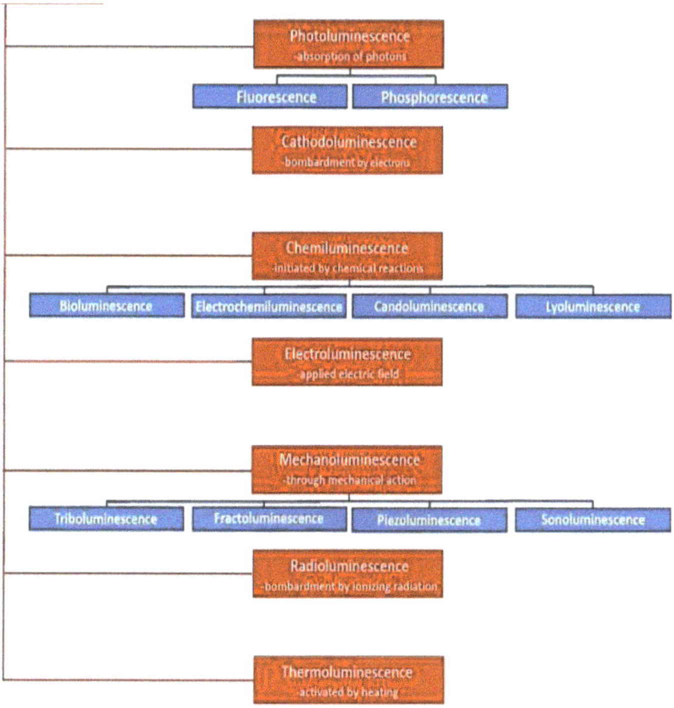

1.5.1 Chemiluminescence

Chemiluminescence is a subdivision of luminescence in which the energy is supplied to the molecule by a chemical reaction. In other words, the product of a reaction is formed in an excited electronic state. The "firefly" reaction is an excellent example. Luciferin reacts with adenosine triphosphate (ATP) in the presence of the enzyme luciferase (E) to form a complex. This complex then reacts with oxygen to form products with the emission of radiation. The amount of light emitted is proportional to the concentration of ATP if the concentrations of the other reactants are held constant for a series of runs.

A schematic diagram of an instrument useful for this measurement is shown in Fig. 2.3, where the essential components consist of an injection system for the reactants, a cell cavity, and a detector. By injecting a constant concentration of luciferase, magnesium sulfate, and luciferin into standard ATP samples, a calibration curve of luminescence intensity versus concentration can be made. This technique is especially

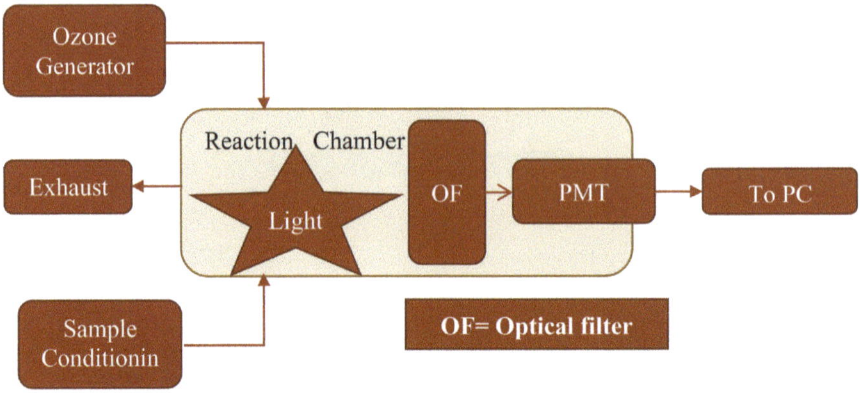

Figure 2.3 Diagram of an instrument used for observing chemiluminescence.

valuable because all living microorganisms contain ATP. Thus, the method may be used to monitor the effectiveness of biocides, the effect of drugs, and the detection of microorganisms in foods.

An application of chemiluminescence in the monitoring of parts per billion (PPB) level of NO in air. The chemiluminescence measurement is based on the gas-phase reaction of NO with OZONE.

$$NO + O_3 \rightarrow NO^*{}_2 + O_2$$

$$NO^*{}_2 \rightarrow NO_2 + \text{light}$$

1.5.2 Photoluminescence

Photoluminescence (PL) is the spontaneous emission of light from a material following optical excitation. It is a powerful technique to probe discrete energy levels and to extract valuable information about semiconductor sample composition, quantum well thickness, or quantum dot sample monodispersity.

Photoluminescence spectroscopy is a contactless, nondestructive method of probing the electronic structure of materials. Light is directed onto a sample, where it is absorbed and imparts excess energy into the material in a process called photo-excitation.

1.5.3 Thermoluminescence

Thermoluminescence is the process in which a mineral emits light while it is being heated: it is a stimulated emission process occurring when the thermally excited emission of light follows the previous absorption of energy from radiation.

1.5.4 Mechanoluminescence

Mechanoluminescence is a process of light emission from materials in response to external mechanical actions. As mechanical energy is ubiquitously available in nature, mechanoluminescence can provide sustainable solutions to challenging problems in the fields of biology and optoelectronics as well as energy and environmental sciences.

1.5.5 Bioluminescence

The light is produced by a compound called luciferin, which releases light when it reacts with oxygen. All bioluminescent animals contain luciferin, but some (like tiny dinoflagellate plankton) produce their own, whereas others (like squid and some fish) absorb bacteria that contain luciferin.

1.6 Luminescence techniques for dosimetry

Three methods of thermally (TSL), optically (OSL), and radio (RPL) stimulated luminescence are discussed and compared. A wide range of applications of TL detectors in individual and environmental dosimetry with an attention put on ultra-high dose measurements is presented and described [38].

1.6.1 Products for luminescence

To measure photoluminescence, a photoluminescence spectrometer is required and Edinburgh instruments offer a range of single-photon counting photoluminescence spectrometers to measure photoluminescence spectra, lifetimes, and anisotropy and quantum yields of samples.

2. Theoretical background

To understand how TL can be used to study luminescent materials, a theoretical background of the phenomenon is indispensable. The following discussion borrows heavily from Refs. [8,9,39–41].

The basic effect leading to the production of TL is the trapping of charge carriers, that is, electrons and holes, produced during exposure to an external source at defect sites in the material. Defect sites can be divided into two categories: (1) those inherently present in the material; and (2) those produced by external means, such as deliberately doping the sample with impurities. A well-known example of the first category is a negative ion vacancy. Such a vacancy can trap an electron and is then called an F center. An example of the second category is a lattice vacancy caused by a higher valence impurity ion at the position of a lattice ion. A trivalent cation impurity in a divalent lattice, for example, would induce the formation of a cation vacancy in the lattice to maintain charge neutrality. Cationic vacancies are potential sites for trapping holes. There are many other types of defects that can act as electron or hole trapping centers. Revealing the nature of those defects is part of the research of luminescent materials.

Heating of the material causes the release of the trapped charge carriers and the recombination of electrons and holes at a luminescent center. The release of some of the stored energy excites the center and relaxation may lead to the emission of light. The nature of the luminescent center can be revealed by measuring the emission spectrum that is in many cases characteristic for a specific element.

We note that not every recombination event leads to luminescence. If, for example, there is not enough energy available to excite the center, the recombination will be nonradiative. The process leading to recombination includes, in many cases, the transition of charge carriers through the conduction or valence band, but localized transitions may also take place. The theory of TSL consists of solving the relevant sets of coupled differential equations, describing the charge transfer [3]. In the following, we discuss the simplest model. This model shows all the characteristics of the TSL phenomenon.

2.1 Base papers suggesting TL material for dosimetry applications

Table 2.3 shows materials exhibiting thermoluminescence in response to ionizing radiation including calcium fluoride, lithium fluoride, calcium sulfate, lithium borate, calcium borate, potassium bromide, and feldspar (Table 2.4).

Table 2.3 Comparative table for various sensors.

S No.	Abstract	Material	Application	Conclusion	DOI
1	TL dosimetrists, research workers, and students, it includes information on glow curves, emission spectra, the mechanisms of TL production, trapping parameters associated with TL signals as well as a lot of information about the dosimetric properties of many materials when exposed to low LET radiation. TL dosimetry is often used by workers who do not have much knowledge of the underlying physical mechanisms causing emission. For those who only wish to use TL dosimetry as a means of measuring radiation dose and are content to remain uninformed about crystal lattice structures, vacancy and impurity centers, and the fundamental characteristics and properties of dosimeters, there is no need to purchase this book. It includes numerous original conventional glow curves as well as isometric and contour plots of the emission spectra of each of the specimens studied	**1.** Feldspar sample **2.** 4 mg mono-layer coarse grain quartz sample	**1.** Routine dating applications, **2.** Lamp element	The development of a single Grain OSL system has provided a measurement capacity of nearly two orders of magnitude higher than that of a conventional luminescence reader without compromising on sensitivity or reproducibility.	Thermoluminescent dosimetry materials: Properties and uses, J. Radiol. Protect. 16 (4) (July 2009). https://doi.org/10.1088/0952–4746/16/4/018

Continued

Table 2.3 Comparative table for various sensors.—cont'd

S No.	Abstract	Material	Application	Conclusion	DOI
2	TLD phosphors with low cost and simple method of preparation on large scale for TLD materials. One of the phosphors is LiF:Mg,Cu,P (LiF: MCP), and another one is $LiCaAlF_6$: Eu, which has the potential to replace conventionally used $CaSO_4$:Dy TL dosimeter. LiF: MCP and $LiCaAlF_6$: Eu phosphors are potential candidates for TL dosimetry and could be good replacement for commercially available phosphors.	We have discussed in detail all three types of TLD materials. First, our study includes complete detail of material properties, methods, and dosimetric characterizations of LiF: MCP phosphor; second, it includes a new TL dosimeter, $LiCaAlF_6$: Eu and its dosimetric characterizations, and lastly on some TL properties of Li_5AlO_4: Mn and MgB_4O_7: Dy,Na. In this review, we discuss some recent developments in radiation dosimetry with regard to measurement techniques and material preparations.	**1.** Thermoluminescence dosimetry is widely used for dosimetric measurements, especially for personnel dose monitoring. **2.** Thermoluminescence dosimetry materials offer long-term information storage and versatile applications in different types of radiations including X-rays, gamma (γ) and beta (β) rays, and neutrons. **3.** Laser application	In the present review, we discussed only for radiation dosimetry and have come to the conclusion that LiF: MCP and $LiCaAlF_6$: Eu may replace $CaSO_4$: Dy in TL dosimetry.	Recent advances and opportunities in TLD materials: A review, Defect & Diff. Forum 347 (December 2013) 75–110. https://doi.org/10.4028/www.scientific.net/DDF.347.75

3	Two new types of "tissue equivalent" thermoluminescence dosimetry (TLDs) have aroused attention: LiF: Mg,Cu,Na,Si and $Li_2B_4O_7$:Cu,Ag,P. In this work, the characteristics of both detectors were compared with the characteristics of the well-known type LiF: Mg, Ti detector, TLD-100. The following properties were investigated: The glow curve structures, relative sensitivity, batch homogeneity and uniformity, detection threshold, reproducibility of the response, linearity in the wide dose range and fading. Also, the energy dependence for medium and low energy X-rays was determined in the range of mean energies between 33 and 116 keV. The most important is the five times higher sensitivity than that of TLD-100. This detector is also very promising, especially in medical dosimetry	Comparative study of LiF:Mg, Cu, Na, Si, and $Li_2B_4O_7$:Cu, Ag, P TL detectors	**1.** Inherent characteristic of the TLD. **2.** Used for LiF dosimeters	LiF:Mg,Cu,Na, Si detector has very high sensitivity (75 times higher than that of TLD-100) and is convenient for use in a very low range of doses. $Li_2B_4O_7$: Cu,Ag,P detector shows some improvements in comparison with the previously prepared types of lithium borate.	Comparative study of LiF:Mg, Cu, Na, Si and Li_2B4O_7:Cu, Ag, P TL detectors, Rad. Protect. Dosimet. 119 (1−4) (February 2006) 191−6 follow journal. https://doi.org/10.1093/rpd/nci557

Continued

Table 2.3 Comparative table for various sensors.—cont'd

S No.	Abstract	Material	Application	Conclusion	DOI
4	The optically stimulated luminescence, thermoluminescence, and photoluminescence in various Cu and Ag-doped zinc-based fluoropervoskites are studied. In all the samples, photoluminescence of Cu+ and Ag+ is observed, which could be attributed to 3d9 4 s1 ←→3d10 and 4d9 5 s1 ←→5d10 transitions, respectively. The observed emission is double humped and the main emission band shifts to the lower wavelength side with increasing ionic size of the alkali ion. All the studied fluoropervoskites show reasonable OSL sensitivity. Highest sensitivity is observed for Cu doped $ZnNaF_3$ and is 25% compared to commercial Al_2O_3:C. Weak thermoluminescence is observed in all samples. In all Ag doped samples the TL peak is observed at 200°C whereas for Cu-doped samples peak is observed at 150°C, and correlated with the OSL. The samples show good linear dose response in the 10 mGy−1.2 Gy range and show good reusability characteristics.	CW-OSL response of various doped zinc-based fluropervoskites. (a) BSL response of Cu-doped $ZnLi_2F_4$ Sample; (b) BSL response of Ag-doped $ZnLi_2F_4$ sample; (c) BSL response of Cu-doped $ZnNaF_3$ sample; and (d) BSL response of Ag doped $ZnNaF_3$ sample.	Radiation dosimetry using OSL	This study will lead to the development of zinc-based fluoropervoskite phosphors for the radiation dosimetry using OSL	Photoluminescence, thermoluminescence, and optically stimulated luminescence studies in zinc-based fluoropervoskites, Rad. Protect. Dosimet. 179 (1) (September 2017) 1−6 follow journal. https://doi.org/10.1093/rpd/ncx205

5.	On recent advances in the development of **luminescence measurement systems** and techniques at Riso. These include the following: (1) optical stimulation units based on new-generation powerful blue light (470 nm) emitting diodes providing up to 28 mW/cm^2 for OSL measurements; (2) an infrared (830 nm) laser diode unit providing up to 400 mW/cm^2 for stimulation of feldspars; and finally (3) an optical stimulation attachment based on a focused solid-state green (532 nm) laser for rapid OSL measurements of individual sand-sized single grains of a sample.	Sr/Y, feldspars;	**1.** Useful luminescence signals from feldspars in most applications. **2.** Flexible measurement procedures for quartz and feldspar samples	The new powerful IR and blue light OSL light sources have facilitated the use of fast and flexible measurement procedures for quartz and feldspar samples and are especially useful with the single-aliquot regenerative-dose protocol.	L. Bùtter- Jensen, et al., Advances in luminescence instrument systems, Rad. Measur. 32 (2000) 523–528. https://doi.org/10.1016/S1350-4487(00)00039-1

Table 2.4 Medical photonic device dosimetry ranges.

S No	Dosimetry	Dose type	Photovoltaic sensors/Solar cells	Photodiodes/LEDs	Phototransistors/ MOSFETs	Radiation type
1.	Diagnostic radiology	Air Kerma	–	0.003–0.450 [42]	–	X-rays
2	Radiotherapy	Air Kerma	–	50–200 [43]	–	γ rays
3	Diagnostic radiology	Air Kerma	–	0.006–0.400 [44]	–	X & γ rays
4	Not specified	Air Kerma	–	47.2–330 [45]	–	γ rays
5	General medical dosimetry	Solid water Phantom	–	0.1–500 [46]	–	X-rays
6	Computed Tomography	Air Kerma	–	0.340–8.30 [47]		X-rays
7	Breast cancer Radiotherapy	Alderson Rando absorbed dose	–	200 [48]	–	X-rays
8	Not specified	Air Kerma	–	10,000–50,000 [49]	–	γ rays
9	Radiotherapy	Anthropomorphic phantom absorbed dose	–	–	4000–18,500 [50]	X-rays
10	Diagnostic radiology	Air Kerma	–	0.001–0.043 [51]	–	X-rays
11	Imaging	Human wrist and index finger phantom	–	(One pixel)	–	X- rays
12	Radiotherapy	Human body phantom	–	–	–	X-rays

3. Instrumentation for ML dosimeter

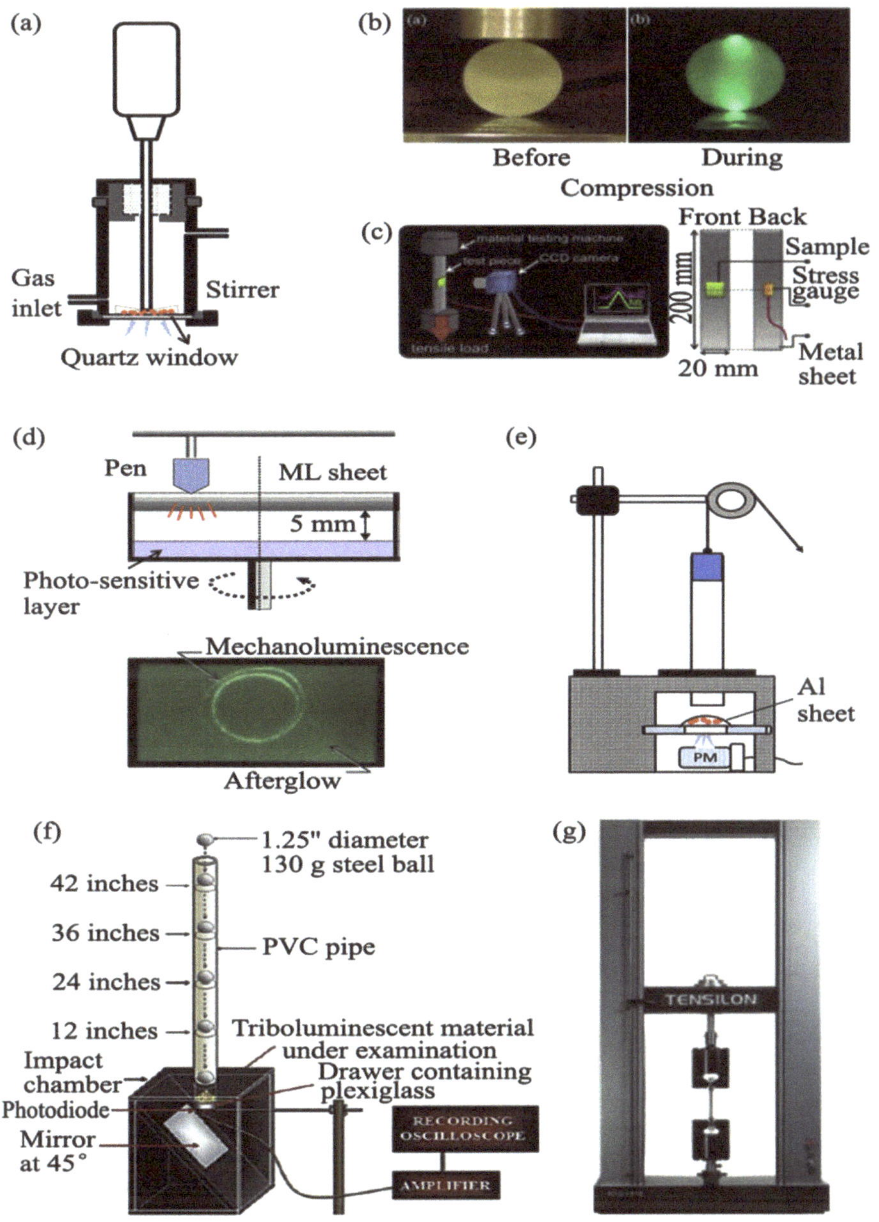

Figure 2.4 Deep trap-induced fracto-mechanoluminescence manual setup.

4. Conclusions

Techniques and methods applied in luminescence dating and dosimetry at many laboratories around the world have been reviewed, and an attempt has been made to describe the state of the art in instrument and method development as shown in Fig. 2.4.

There is one problem that remains to be addressed in the development of combined TL/OSL instrumentation using different stimulation light spectra. This is concerned with the design of a flexible optical detection filter changing system to allow for the rapid (automatic) selection of the optimal detection window whether using infrared or visible light stimulation. Changing of excitation or detection filters may, if not properly protected either by hardware or software, cause serious damage to the photomultiplier (PM) tube because of insufficient suppression of stray light from the stimulation light source.

The growing industrial interest in ultrabright LEDs as light indicators (e.g., from automobile manufacturers) may soon make visible LEDs commercially available with substantially higher emission power than is available today. These LEDs should provide sufficient power to be considered a real alternative to laser and incandescent lamp stimulation light sources in OSL. The immediate advantages of using LEDs over filtered broad band lamps are as follows: (a) reduced heat dissipation, with less effect on the stimulation optics and (b) no need for mechanical shutters to control stimulation exposure.

In the future, a major effort will no doubt be put into the development of sensitive systems capable of measuring luminescence from small aliquots, even down to single grains. The immediate advantages of this are that the accrued dose can be determined from only one aliquot and that variations in dose from grain to grain can be studied in detail. The latter feature will be especially valuable in studies of young, incompletely bleached materials and in the identification of sediment disturbance in natural deposits. Such improvements will continue to require increases in detection sensitivity.

Further developments and investigations of luminescence imaging systems for obtaining spatially resolved TL and OSL signals from multimineral samples are also foreseen. These systems give rapid and valuable information about the mineralogy of the sample and enable an individual analysis of luminescence signals from single grains of a sample.

Acknowledgment

The author has drawn review comparisons on publications kindly supplied by a number of authors who have contributed to the development of a wide variety of luminescence techniques and methods. However, having been in the field of developing luminescence instruments and methods for many years, the author of the present paper is inevitably biased toward the work carried out at the research center.

The author is grateful to Guide sir for going through the manuscript and for helpful comments and discussions.

References

[1] Elsevier Science Ltd, Rad. Measur. 27 (5/6) (1997) 749–768. # 1998.

[2] G.F. Knoll, Radiation Detection and Measurement, John Wiley and Sons, Hoboken, NJ, USA, 2010, ISBN 0-471-07338-5.

[3] M. Hofstetter, J. Howgate, I.D. Sharp, M. Stutzmann, S. Thalhammer, Development and evaluation of gallium nitride-based thin films for x-ray dosimetry, Phys. Med. Biol. 56 (2011) 3215 (CrossRef).

[4] T. Rivera-Montalvo, Radiation therapy dosimetry system, Appl. Radiat. Isot. 83 (2014) 204–209 (CrossRef).

[5] T. Kron, J. Lehmann, P.B. Greer, Dosimetry of ionising radiation in modern radiation oncology, Phys. Med. Biol. 61 (2016) R167 (CrossRef, PubMed).

[6] Ptw, Ionising Radiation Detectors Including Codes of Practice Freiburg, PTW, Germany, 2018, 2018/2019, (PTW Detector Mannual). Available online, www.ptw.de/fileadmin/data/download/catalogviewer/DETECTORS_Caten_16522900_11/blaetterkatalog/blaetter katalog/pdf/complete.pdf. (Accessed 30 November 2018).

[7] S. Vatnitsky, H. Jarvinen, Application of a natural diamond detector for the measurement of relative dose distributions in radiotherapy, Phys. Med. Biol. 38 (1993) 173 (CrossRef, PubMed).

[8] S.W.S. McKeever, Thermoluminescence of Solids, Cambridge University Press, Cambridge, England, 1988.

[9] R. Chen, S.W.S. McKeever, Theory of Thermoluminescence and Related Phenomena, World Scientific Publishing Co., Singapore, 1997.

[10] L. Bøtter-Jensen, S.W.S. McKeever, A.G. Wintle, Optically Stimulated Luminescence Dosimetry, Elsevier, Amsterdam, 2003.

[11] E.G. Yukihara, S.W.S. McKeever, Optically Stimulated Luminescence: Fundamentals and Applications, John Wiley & Sons Ltd., Wiley, 2011.

[12] C. Karunakaran, et al., In Spin Resonance Spectroscopy, 2018.

[13] J. Izewska, Division of Human Health, International Atomic Energy Agency, Vienna & G. RAJAN Medical Physics and Safety Section, Bhabha Atomic Research Centre, Mumbai, Maharashtra, India.

[14] J.C.R. Mittani, A.A.R. d Silva, F. Vanhavere, M.S. Akselrod, E.G. Yukihara, Investigation of neutron converters for production of optically stimulated luminescence (OSL) neutron dosimeters using Al_2O_3:C, Nucl. Instrum. Methods B 260 (2007) 663–671.

[15] E.G. Yukihara, J.C. Mittani, F. Vanhavere, M.S. Akselrod, Development of new optically stimulated neutron dosimeters, Radiat. Meas. 43 (2008) 309–314.

[16] G.M. Akselrod, M.S. Akselrod, E.R. Benton, N. Yasuda, A Novel Al_2O_3 fluorescent nuclear track detector for heavy charged particles and neutrons, Nucl. Instrum. Methods B 247 (2006) 296–306.

[17] M.S. Akselrod, A.E. Akselrod, New Al_2O_3:C,Mg crystals for radiophotoluminescent dosimetry and optical imaging of tracks produced by heavy charge particles, Radiat. Protect. Dosim. 119 (1–4) (2006) 218–221.

[18] M.S. Akselrod, R.C. Yoder, G.M. Akselrod, Confocal fluorescent imaging of tracks from heavy charged particles utilizing new Al_2O_3:C,Mg crystals, Radiat. Protect. Dosim. 119 (1–4) (2006) 357–362.

[19] G.J. Sykora, M.S. Akselrod, M. Salasky, S.A. Marino, Novel Al_2O_3:C,Mg fluorescence nuclear track detectors for passive neutron dosimetry, Radiat. Protect. Dosim. 126 (2007) 278–283.

[20] G.J. Sykora, M.S. Akselrod, Novel fluorescent nuclear track detector technology for mixed neutron-gamma fields, Radiat. Meas. 45 (3–6) (2010) 594–598.

[21] G.J. Sykora, M.S. Akselrod, Photoluminescence study of photochromically and radiochromically transformed Al_2O_3:C,Mg crystals used for fluorescent nuclear track detectors, Radiat. Meas. 45 (3–6) (2010) 631–634.

[22] G.J. Sykora, M.S. Akselrod, F. Vanhavere, Performance of fluorescence nuclear track detectors in monoenergetic and broad spectrum neutron fields, Radiat. Meas. 44 (2009) 988–991.

[23] G.J. Sykora, M. Salasky, M.S. Akselrod, Properties of novel fluorescent nuclear track detectors for use in passive neutron dosimetry, Radiat. Meas. 43 (2008) 1017–1023.

[24] G.J. Sykora, M.S. Akselrod, E.R. Benton, N. Yasuda, Spectroscopic properties of novel fluorescent nuclear track detectors for high and low LET charged particles, Radiat. Meas. 43 (2008) 422–426.

[25] M.S. Akselrod, A.E. Akselrod, S.S. Orlov, S. Sanyal, T.H. Underwood, Fluorescent aluminum oxide crystals for volumetric optical data storage and imaging applications, J. Fluoresc. 13 (2003) 503–511.

[26] S. Sanyal, M.S. Akselrod, Anisotropy of optical absorption and fluorescence in Al_2O_3: C,Mg crystals, J. Appl. Phys. 98 (2005) 033518.

[27] A. Diaspro (Ed.), Confocal and Two-Photon Microscopy: Foundations, Applications, and Advances, Wiley, New-York, 2002.

[28] A.S. Beddar, Water equivalent plastic scintillation detectors in radiation therapy, Radiat. Protect. Dosim. 120 (2006) 1–6, https://doi.org/10.1093/rpd/nci694 (PMID: 16882685).

[29] A.K. Glaser, R. Zhang, D.J. Gladstone, B.W. Pogue, Optical dosimetry of radiotherapy beams using Cherenkov radiation: the relationship between light emission and dose, Phys. Med. Biol. 59 (2014) 3789–3811, https://doi.org/10.1088/0031-9155/59/14/3789 (PMID: 24938928).

[30] Nuclear Technology Publishing, Rad. Protect. Dosimet. 84 (1–4) (1999) 325–330.

[31] B.L. Gupta, R.M. Bhat, G.R. Narayan, S.R. Nilekani, Chemical dosimetry techniques for various applications under different geometries 59 (1) (2000) 81–90.

[32] P.A. Jursinica, Characterization of optically stimulated luminescent dosimeters, OSLDs, for clinical dosimetric measurements, Med. Phys. 34 (No. 12) (December 2007) 4594–4604.

[33] M. Fahmi Mohd Yusof, et al., Dose measurement using Al_2O_3 dosimeter in comparison to LiF:Mg,Ti dosimeter and ionization chamber at low and high energy X-ray, Adv. Nucl. Sci. & Eng. Sustain. Nucl. Energy Knowl. AIP Conf. Proc. (1799) 040007-1–040007-7.

[34] A. Fridman. Plasma Chemistry, Cambridge University Press, 2008.

[35] J. Izewska, Division of Human Health, International Atomic Energy Agency, Vienna 71–99.

[36] G. Okada, et al., Sm-doped CsBr crystal as a new radio-photoluminescence (RPL) material, J. Rare Earths 34 (No. 8) (August 2016) 769–773.

[37] T. Yamamoto, RPL dosimetry: principles and applications, concepts and trends in medical radiation dosimetry, AIP Conf. Proc. 1345 (2011) 217–230. American Institute of Physics 978-0-7354-0901-9.

[38] M. Nowina Konopka, et al., Nonlinear Optics and Quantum Optics, vol. 48, 133–146. https://www.researchgate.net/publication/318686649.

[39] A.J.J. Bos, Theory of thermoluminescence, Radiat. Meas. 41 (2006) 45–56 (CrossRef).

[40] R. Chen, V. Pagonis, Thermally and Optically Stimulated Luminescence—A Simulation Approach, JohnWiley, Chichester, UK, 2011, ISBN 978-0-470-74927-2.

[41] C.M. Sunta, Unraveling Thermoluminescence; Springer Series in Materials Science, vol. 202, Springer, New Delhi, India, 2015, ISBN 978-81-322-1939-2.
[42] C. Romei, A. Di Fulvio, C.A. Traino, R. Ciolini, F. d'Errico, Characterization of a low-cost PIN photodiode for dosimetry in diagnostic radiology, Phys. Med. 31 (2015) 112–116 (CrossRef).
[43] H. Diab, A. Ibrahim, R. El-Mallawany, Silicon solar cells as a gamma ray dosimeter, Measurement 46 (2013) 3635–3639 (CrossRef).
[44] C.N. Oliveira, H.J. Khoury, E.J. Santos, PiN photodiode performance comparison for dosimetry in radiology applications, Phys. Med. 32 (2016) 1495–1501 (CrossRef, PubMed).
[45] M.S. Anelković, G.S. Ristić, Feasibility study of a current mode gamma radiation dosimeter based on a commercial PIN photodiode and a custom made auto-ranging electrometer, Nucl. Technol. Radiat. Protect. 28 (2013) 73–83 (CrossRef).
[46] P. Zygmanski, C. Abkai, Z. Han, Y. Shulevich, D. Menichelli, J. Hesser, Low-cost flexible thin-film detector for medical dosimetry applications, J. Appl. Clin. Med. Phys. 15 (2014) 311–326 (CrossRef, PubMed).
[47] C.M.M. Paschoal, D.N. Souza, L.A.P. Santo, Characterization of three photo detector types for computed tomography dosimetry, World Acad. Sci. Eng. Technol. 56 (2011) 92–95 (CrossRef).
[48] L.A. Santos, G.G. Araujo, F.L. Oliveira, E.F. Silva Jr., M.A. Santos, An alternative method for using bipolar junction transistors as a radiation dosimetry detector in breast cancer treatment, Radiat. Meas. 71 (2014) 407–411 (CrossRef).
[49] M.M. Pejovic, Application of p-channel power VDMOSFET as a high radiation doses sensor, IEEE Trans. Nucl. Sci. 62 (2015) 1905–1910 (CrossRef).
[50] M. Dybek, B. Kozłowska, Evaluation of the applicability of MOSFET detectors in radiotherapy, Radiat. Meas. 71 (2014) 412–415 (CrossRef).
[51] M. Nazififard, K.Y. Suh, A. Mahmoudieh, Experimental analysis of a novel and low-cost pin photodiode dosimetry system for diagnostic radiology, Rev. Sci. Instrum. 87 (2016) 073502 (CrossRef).
[52] A.S. Shionoya, W.M. Yen (Eds.), Phosphor Handbook, CRC Press, Boca Raton, Fl., USA, 1999.
[53] S.W.S. Mckeever, Thermoluminescence of Solids, Cambridge University, Cambridge, 1985.
[54] D.R. Vij (Ed.), Luminescence of Solids, Plenum Press, New York, 1998.
[55] G. Blasse, B.C. Grabmaeir, Luminescent Materials, Springer Verlag, Berlin, 1994.

Properties and characterization of rare-earth-activated phosphors

3

Jelena Mitrić
Laboratory for Electronic Materials, Institute of Physics Belgrade, Belgrade, Serbia

1. Properties of rare-earth-activated phosphors

Phosphors are defined mostly as inorganic, solid, crystalline materials consisting of a host lattice doped with some impurities. These materials are characteristic for showing luminescence while absorbing incident light in the near-ultraviolet or visible region and emitting light in the visible or near-infrared region.

Phosphors host lattice is suitable for the incorporation of divalent or trivalent rare-earth series of elements. Those are often referred as impurities or activator ions. Activator ions are exposed to UV or visible light (or some other excitation source, depending on a type of phosphor), absorb it, excite to a higher state, after which relaxation occurs. Then, a photon is emitted and the activator goes back to a lower state. During the relaxation, some energy is lost due to nonradiative processes, which can affect phosphor efficiency. The efficiency of phosphor depends on the amount of absorbed and emitted light (in ideal phosphors these amounts are equal). Other than phosphor efficiency, properties to look for in an ideal phosphor are its high thermal stability, color quality, color rendering index (CRI), and other optical properties such as emission wavelength, quenching concentrations, and luminescent lifetime.

This chapter will concern different properties and characterization of rare-earth-activated phosphors.

1.1 Luminescence

The word *luminescence* originates from the Latin word for light, which is *lumen*. The term luminescence is to describe the conversion of energy into (near)-visible light by some nonincandescent sources. Light, in general, can be classified as incandescence or luminescence (even though this is not completely true, as it is well known that some radiation can occur as reflection, scattering (Raman and Rayleigh), laser radiation, etc.). Incandescence light is one obtained from solid material due to heating. On the other hand, luminescence occurs when the material is irradiated by electromagnetic radiation, electrical or chemical excitation, or others. In that way, atoms of which luminescent material consists of, are energized and releasing their excess energy in the form of UV, visible, or infrared electromagnetic radiation. Depending on an excitation source, phosphors can produce different kinds of luminescence (see Table 3.1).

Rare-Earth-Activated Phosphors. https://doi.org/10.1016/B978-0-323-89856-0.00006-7

Table 3.1 Type of luminescence depending on an excitation source.

Type of luminescence	Excitation source
Photoluminescence	Electromagnetic radiation
Electroluminescence	Electric current
Cathode luminescence	Beam or electrons or ions
Chemiluminescence	Chemical reaction energy
X-ray luminescence	X-rays
Bioluminescence	Chemical reactions in living creatures
Triboluminescence	Mechanical energy

1.2 Mechanism of luminescence

To obtain the emission of visible light, the excitation source requires the minimum of energy given by Einstein's law:

$$E = h\nu = \frac{hc}{\lambda} \tag{3.1}$$

where E is energy, h is Planck's constant, ν is the frequency of light, c is the velocity of light in vacuum, and λ is its wavelength.

This energy is transferred to electrons responsible for luminescence, which then have enough energy to move from the ground state to higher energy levels. Those energy levels have strictly defined energies, specified by quantum mechanical laws. Electrons that play a part in the process of luminescence are the ones from the outermost shells. Upon electron's return to the ground state, an energy difference is emitted through the emission of ultraviolet, visible, or near-infrared light.

When luminescence is started by a photoexcitation, that is, photon absorption, it is referred as photoluminescence. Different photoluminescence processes are illustrated in Fig. 3.1, using Jablonski[1] diagram.

Jablonski diagram incorporates all possible photophysical electron transitions in a molecule. That includes internal conversion (IC), intersystem crossing (ICS), vibrational relaxation (VR) fluorescence (FL), and phosphorescence (PHO). Intersystem conversion, intersystem crossing, and vibrational relaxation belong to nonradiative transitions. Nonradiative transitions do not include the emission of radiation during the transition of a molecule from an excited to the ground or any other lower excited state. Fluorescence and phosphorescence are radiative transitions that include radiation emission during molecule transition from an excited state to the ground or any other lower excited state.

Vibrational relaxation is a nonradiative transition that occurs between vibrational levels in one electronic state. Internal conversion is a transition between vibrational

[1] Jablonski diagram illustrates the electronic states and transitions of a molecule. The states are represented vertically by energy and horizontally by spin multiplicity.

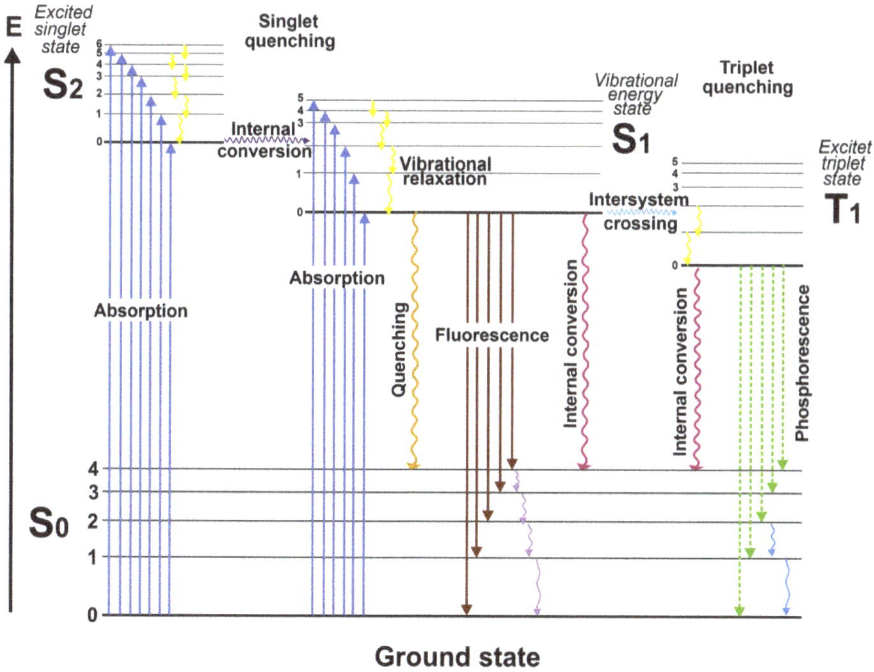

Figure 3.1 Jablonski diagram.

levels of two electronic states with the same multiplicity. Intersystem crossing is a nonradiative transition that occurs between two electronic states with different multiplicities.

Fluorescence is a radiative transition between electronic states with the same multiplicity. Fluorescent emission occurs when electron transfers form a ground vibrational level of a first ground ***singlet*** state to any other of vibrational levels of a ground electron state. The fluorescence rate constant is $\sim 10^6 - 10^9\ s^{-1}$. Phosphorescence is a radiative, (spin) forbidden transition between electronic states with different multiplicities. Phosphorescence occurs when electron transfers from a ground vibrational level of an excited ***triplet*** state of a molecule to any other of vibrational level of a ground electron state. Phosphorescence rate constant is $\sim 10^{-3} - 10^2\ s^{-1}$.

Luminescence of solid-state materials can be internal and external. Internal luminescence occurs due to electronic transitions of a matrix and external due to the presence of different impurities in a matrix, which are usually activators or dopant ions. Also, external luminescence can be divided into localized and nonlocalized ones. Localized luminescence happens when impurities are included intentionally, which represent luminescent centers, and nonlocalized luminescence includes electron–hole matrices. When localized centers of luminescence are made intentionally, by adding impurities, they are often called *activators*.

Mechanism of luminescence can be explained by Fig. 3.1 where S_0 represents ground (singlet) state and S_1, S_2, and T_1 excited singlet and triplet states, respectively.

The blue arrow represents the excitation of an atom to a higher energy level when absorbing the incident light. IC, ICS, and VR represent nonradiative transitions, while FL and PHO are radiative processes. Therefore, when an atom returns to a ground state after excitation it emits radiation that is longer than the excitation wavelength.

For a better understanding of the luminescence in phosphors, one must include the configurational coordinate diagram (CCD). CCD for radiative and nonradiative transitions in phosphors is shown in Fig. 3.2A and B, respectively. Diagram consists of two curves—bottom one represents ground state energy, while the top one is an excited energy state. The activator excites from a ground state to a higher energy level, when some of the absorbed energy is given to a host lattice after which an activator goes back to a lower energy excited state. After that, the activator goes back to a ground state by releasing energy as emitting light. This represents radiative transition with the energy of excitation greater than the energy of emission. Horizontal line denoted with *r* represents the average interatomic distance from the activator to its neighbors. Vertical axis is energy. When the activator goes to an excited state, the curve can shift to a different *r*, because of the interaction of the activator and neighbor atoms (due to bonding length change and electrostatic charges).

When luminescent material gets irradiated and activators absorb energy, the electron jumps to an excited state. When it gets to an excited state, the electron can relax until it reaches the position of the vibrational modes. At 0 K, for luminescence to occur, the electron must get to the minimum of the excited state. After that, it can go back to a ground state. Then, it relaxes somewhere on a ground state to get excited again and repeat the same process. At $T > 0$ K, luminescence can occur from any position and can relax to any of the ground state vibration modes. These vibration modes determine the width of the emission curves, as the width is determined by the number of modes—the less the modes, the smaller the width of emission peaks.

When it comes to nonradiative transitions (Fig. 3.2B), energy curves of the ground and excited state must be close enough to each other, so the electrons can be able to exchange the states nonradiatively. This nonradiative process includes an electron going back from a high vibrational state to the ground state curve, relaxing to the lowest energy level of the ground state.

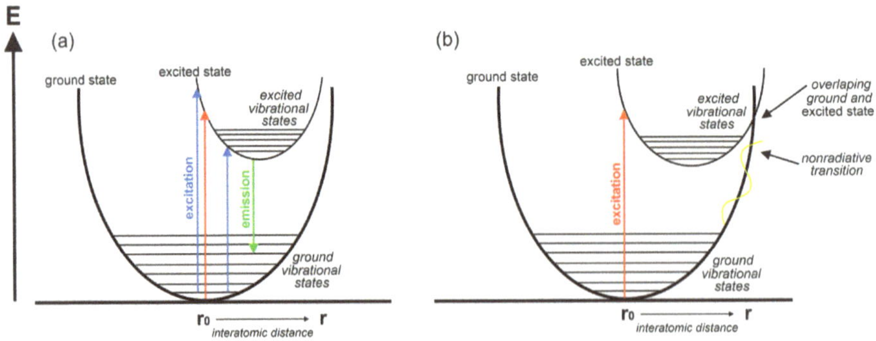

Figure 3.2 (A) Radiative transitions and (B) nonradiative transitions explained by the configurational coordinate diagram.

One more important property of phosphors can be explained by the configurational coordinate model. That is phosphor efficiency. From a configurational coordinate diagram, it can be concluded that phosphor efficiency is lower with amorphous phosphors rather than in crystalline phosphors. The reason for this is crystal surrounding in crystalline phosphors. While this surrounding in crystalline phosphors is strong, as atoms are positioned in their appropriate sites, the distances between ground and excited states are small, in amorphous phosphors this surrounding is weak and therefore the distance between the ground and excited states becomes larger, causing them to overlap. This contributes to nonradiative emission.

Luminescence and efficiency of phosphors are at their maximum when impurity centers that cause luminescence are at their optimal concentration. At the beginning, when impurity center concentration is increased, the luminescence also increases. At one point, luminescence usually saturates and further decreases. The reason for this is that impurity centers with increasing concentration become closer to other centers, and the energy from one center, instead of being emitted as light, can be transferred to another center. This can still produce light, but most probably nonradiatively. The other thing is the valence and size of the exchanging ions, which can, if not matched, lead to difficult incorporation to a host lattice. Therefore, the optimal concentration of dopants depends on the nature of phosphors and their luminescent centers.

Photoluminescence of rare-earth ionic centers and transition metals occurs due to matrix light absorption or light absorption of impurities. If impurities that absorb light emit it by radiative or nonradiative transitions, they are called activators, but if they use just to transfer the energy to an activator, those impurities are called sensitizers. In some cases, the host lattice acts as a sensitizer by transferring the energy to an activator ion. Sensitizers are often rare-earth ions or transition metals. The luminescence centers consist of crystalline (or microcrystalline) solid materials, which is a homogeneous host lattice with very high purity. In Fig. 3.2, a schematic diagram of different luminescent materials is shown. Like it was said earlier, when an impurity is intentionally added to a host, the lattice is called an activator. It is a small amount of a material that is added to a phosphor host lattice. Activator absorbs incident radiation during the process of luminescence and as a result, photons emit (Fig. 3.2A). On the other hand, when activator ions have a weak absorption coefficient, sensitizer ions are often incorporated into the phosphor lattice. These ions absorb incident radiation and transfer it efficiently to activator ions that then continue to absorb the energy (Fig. 3.2B). Sometimes, there are host phosphor lattices that are in no need for activators or sensitizers. These are called self-activated phosphors and are self-sufficient to absorb the incident radiation and then emit light (Fig. 3.2C).

2. Rare-earth-activated phosphor characterization

In this part of the chapter, some common techniques used for phosphor materials characterization will be presented. When it comes to phosphor characterization, techniques will be roughly divided into structural and optical characterization techniques. These types of characterization can discover a lot about rare – activated phosphor properties and their potential use in many fields of industry and technology [1–4]. As structural characterization techniques, scanning electron microscopy (SEM), X-ray diffraction analysis, and atomic force microscopy will be presented. And for optical characterization, photoluminescence spectroscopy (PL), Raman spectroscopy, and infrared spectroscopy will be shown. These are the most common techniques that can be found for rare-earth-activated phosphor characterization. Structural characterization techniques will be used mostly for determining the structure of a material, particle size and distribution, as well as morphology and topology of the sample. The optical characterization will be mainly used for vibrational spectroscopy and luminescent properties of a material.

2.1 Scanning electron microscopy

Scanning electron microscopy is a characterization technique used for investigation of:

- Morphology—the study of the size, arrangement, and shape of particles that make up the material surface.
- Topology—the study of the texture of the sample, the surface features that can be only up to several nanometers.
- Composition—the scanning electron microscope is often equipped with an energy dispersive X-ray spectrometer (EDX) that can give information about the quality and quantity of a sample—of which elements and compounds is the sample composed of.
- Crystallographic parameters—information about the arrangement of atoms in a sample and their degree of order.

A scanning electron microscope is an instrument that uses an electron beam to irradiate a sample that is investigated. Electrons interact with the sample, giving information about morphology, topology, composition, and crystal structure (Fig. 3.3).

When an electron beam falls onto an investigated sample, several types of electrons can be produced:

- Secondary electrons—electrons that are emitted from the surface occupying atoms. These atoms give an interpretable image of the surface and contrast is determined by the morphology of the sample. The resolution of the image will depend on a diameter of an electron beam. Today's instrumentation offers very narrow electron beams, offering to detect particles up to 1 nm in size. Magnification can be from 10 to 500,000 times. In Fig. 3.4, an SEM image of some phosphor is presented.
- Back-scattered electron (BSE)—electrons that reflect from atoms of the surface by elastic scattering. The contrast of the photograph is determined by the atomic number of the elements that made up the sample. These electrons occur due to elastic collisions between electrons from a primary beam and the nucleus of an atom in the sample. More BSE is emitted if higher the atomic number of an atom is. As a result, one obtains an image which is a distribution of chemical phases in a material investigated. In Fig. 3.5, a back-scattered SEM image of a phosphor is presented.
- Auger electron—electrons emitted when primary electron beam collides with a sample and removes inner shell electrons from the specimen, typically from the first 2 nm of a surface. In

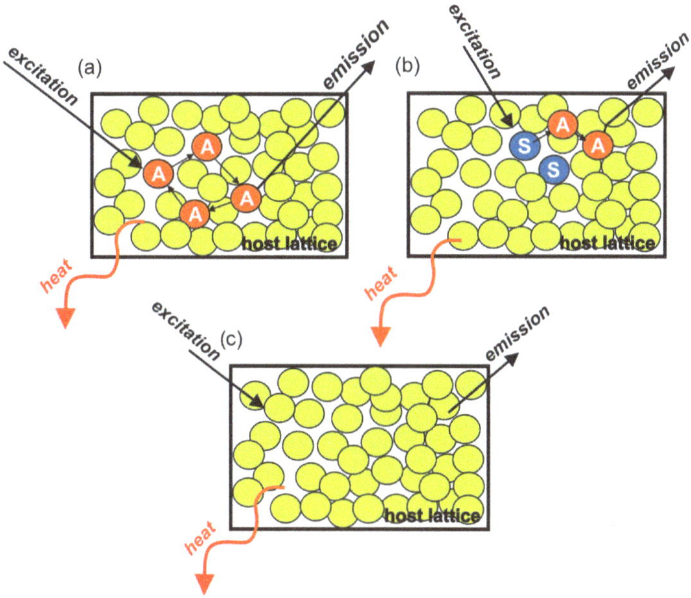

Figure 3.3 Schematic diagram of (A) luminescent material with activator ions, (B) luminescent material with activator and sensitizer ions, and (C) self-activated luminescent material.

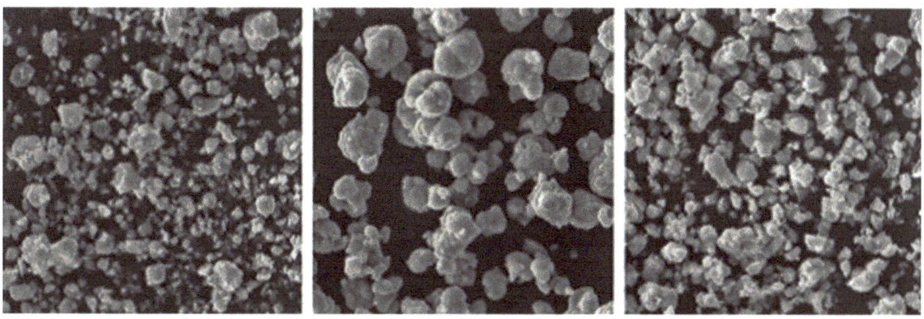

Figure 3.4 Scanning electron microscope images of the phosphors from left to right: cerium-doped calcium tungstate, cerium-doped strontium tungstate, and cerium-doped barium tungstate.

this case, a higher energy electron fills the shell and as a result, energy is released. Auger electrons analysis is very convenient for surface science.

- Characteristic X-rays—similar process to Auger electrons, but instead of emitting a weak bound outer shell electron, X-rays are emitted. These rays are used to identify the composition of a sample. Detection of X-rays is done by EDX spectrometry. In Fig. 3.6, an SEM image with an EDX analysis is presented.

Figure 3.5 SEM images in backscattered electrons of $YNbO_4:Eu^{3+}$.

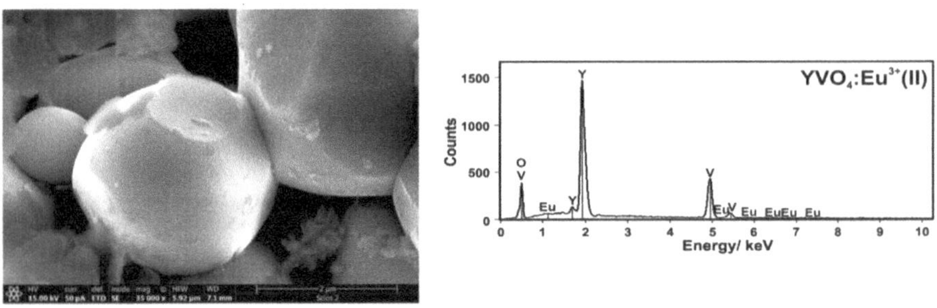

Figure 3.6 SEM image of a $YVO_4:Eu^{3+}$ (left) with corresponding EDX analysis (right).

- Cathodoluminescence—process similar to Auger and X-rays, but what is emitted is a photon in the UV–Vis or IR part of the electromagnetic spectrum. What is usually obtained is CL photographs over one region of a sample.

2.2 X-ray diffraction analysis

X-ray diffraction analysis is a type of analysis for structural characterization of solid materials. X-rays are part of electromagnetic waves that are in the range of 0.1–10 nm. These waves are convenient for exploring crystal structure as they are comparable to the size of atoms. X-rays with more energy are used to penetrate deeper into a material and offer information about bulk crystal structure (Fig. 3.7).

In general, X-rays can be produced in two ways: synchrotron radiation or in X-ray tubes, which are often used in laboratory conditions. In an X-ray tube, a focused electron beam is accelerated on a solid target under high voltage. When highly accelerated

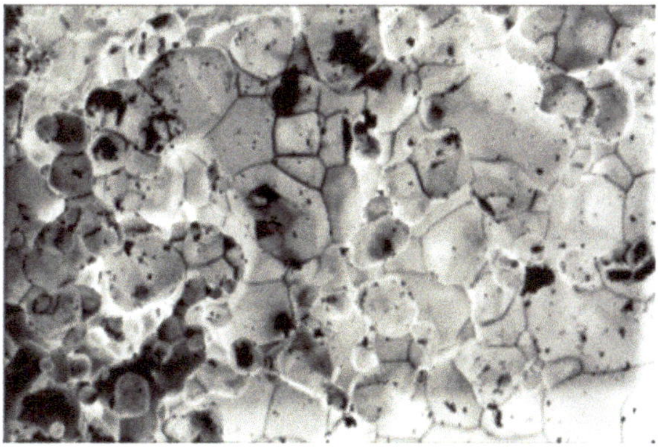

Figure 3.7 Cathodoluminescence image of ZnO:Zn at 495 nm.

electrons collide with atoms of a solid target, they tend to slow down and as a result, a continuous spectrum of X-rays is obtained (so-called *Bremsstrahlung radiation*). Also, high-energy electrons can throw out the inner shell electrons and take their place. After this process, characteristic X-ray radiation of material is emitted.

When a crystal is irradiated with X-ray radiation, its regular three-dimensional arrangement of atoms (i.e., crystal lattice) interacts with it in a way that most of the X-rays will interfere destructively with each other and totally cancel each other out. However, in some specific directions, constructive interference will happen. These rays are responsible for the X-ray diffraction pattern that is obtained and is used for crystal structure determination.

Namely, X-rays diffract in a way that they reflect from a family of planes within a crystal. That was shown by W.L. Bragg in the 19th century. Reflections from planes only occur under certain conditions that are strictly determined by Bragg's law:

$$n\lambda = 2d\sin(\theta) \tag{3.2}$$

where n is the diffraction order (n is an integer, $n = 1, 2, 3, \ldots$), λ is a wavelength of radiation, d is a distance between atomic planes, and θ is an incident radiation angle. This is represented in Fig. 3.8.

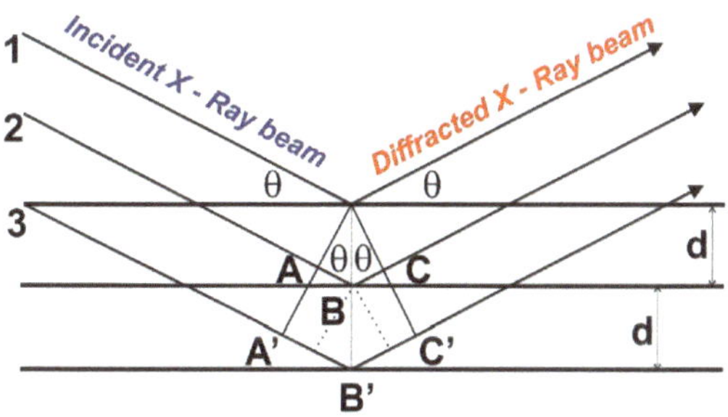

Figure 3.8 Schematic representation of Bragg's law.

Crystal structure (or crystal lattice) represents a unique arrangement of atoms, molecules, and ions in a crystal. Every crystal structure has a motif, which is a set of atoms, molecules, and ions arranged in a particular way. Motifs are duplicated through the lattices from a point of lattice, periodically in three dimensions. These points can be regarded as identical small units, called unit cells, which fill up the crystal space. The length and angles of unit cells are called lattice parameters that determine the symmetry of a crystal structure. An example of a crystal lattice of YVO_4 phosphor is presented in Fig. 3.9.

2.2.1 X-ray powder diffraction

X-ray powder diffraction (XRPD) is a widely used X-ray technique, especially for phosphor materials, as they are often in a form of powder. The term powder indicates that the sample consists of fine grains of crystalline material. Also, it considers that crystallites in a sample are randomly oriented. Therefore, a diffraction pattern obtained really represents concentric rings of the scattering peaks that correspond to d-spacing in a crystal. A diffraction pattern is a pattern that shows positions and the intensity of peaks. Every diffraction pattern is characteristic for each material and it is used for its identification. In Fig. 3.10, a diffraction pattern of YVO_4 doped with europium ions was shown, prepared by two different preparation methods. YVC was prepared by the solid-state reaction method and YVS by solution combustion synthesis.

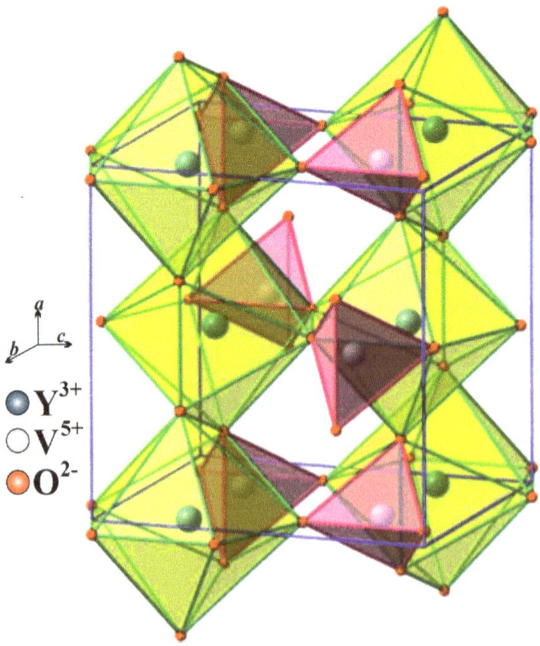

Figure 3.9 Unit cell YVO_4 crystal structure.

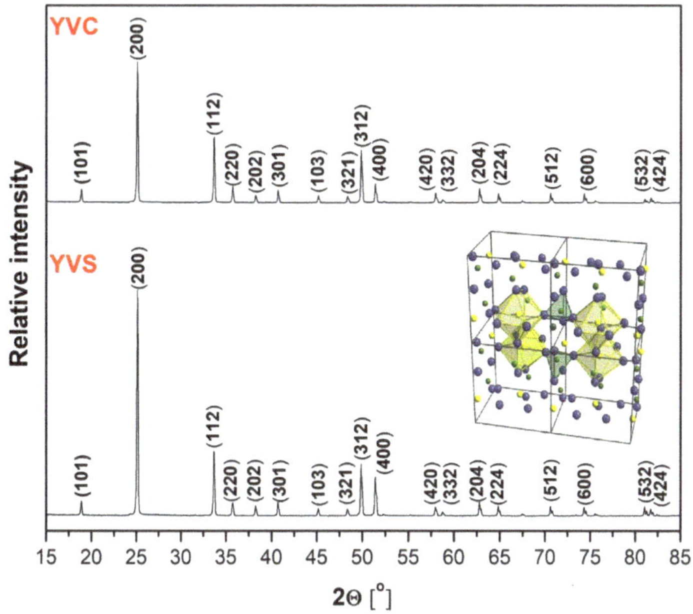

Figure 3.10 Diffraction patterns for $YVO_4:Eu^{3+}$ phosphor, made by two preparation methods: YVC by solid-state reaction method and YVS by solution combustion synthesis.

2.3 Photoluminescence spectroscopy

PL is a nondestructive and contactless optical method of probing the electronic structure of materials. When light is irradiated onto a sample, a process called photoexcitation is started, where light is absorbed and imparts the excess energy into a material. This excess energy can be dissipated by the sample in two ways: by the emission of light or through luminescence. The light can be analyzed spectrally, spatially, and temporally.

When photo excitation causes electrons to move from equilibrium states to excited states and to come back, the excess energy can be radiative or nonradiative, as it is explained detail in Section 1.2 and in Fig. 3.1, so further details on photoluminescence processes will not be further discussed.

PL spectrum is often mistaken for absorption spectrum. However, absorption spectrum measures transitions from the ground to the excited electron state and PL concerns with vice versa transitions—from the excited to the ground state. An excitation spectrum is very much alike to absorption spectrum, as it is represented as intensity versus excitation wavelength. The wavelength at which the molecule is absorbing energy can be equal to the excitation wavelength that can offer more intense emission at a red-shifted wavelength.

An example of PL emission spectrum for $GdVO_4$:Sm phosphor is presented in Fig. 3.11. From this spectrum, it can be seen that this nanopowder has luminescence in green, orange, and red regions. Also, all transitions from excited states can be obtained. It can be concluded from which groups of a molecule the strongest emission comes from.

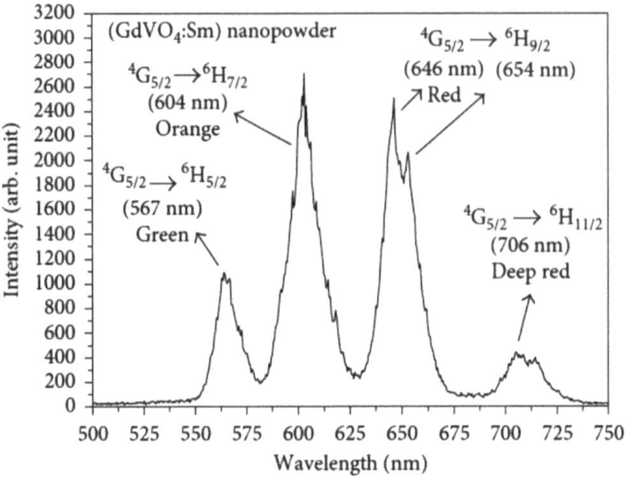

Figure 3.11 PL emission spectrum of $GdVO_4$:Sm nanopowder.

Also, what can be performed is the streak image of the photoluminescence spectrum. This is presented in Fig. 3.12. The horizontal scale corresponds to the wavelength, and the vertical scale corresponds to the time development of the spectrum. Images are in pseudocolor, which means that different color means different optical intensity.

2.4 Fourier transform infrared spectroscopy

Fourier transform (FT) is a mathematical technique to resolve complex waves into their frequency components. Nowadays, FT infrared spectroscopy is more frequently used than conventional infrared spectrometers, as their sources are weak and their detectors are of small sensitivity. In conventional spectroscopy, radiant power versus radiation frequency is recorded and in the time domain spectroscopy, the change in radiant power is plotted as a function of time. In FT spectroscopy, a time-domain plot is converted into a frequency domain spectrum, and it is done by computers with very high speed. FT infrared spectroscopy is a type of analysis that is used for obtaining information about chemical bonding in a material. It is mainly used as a qualitative method, to identify materials in a sample.

Fourier transform infrared spectroscopy (FTIR) is used to measure the ability of the sample to absorb or transmit light at each different wavelength. In conventional IR spectrometers, this is done by shining a monochromatic light beam at a sample and measuring how much light is absorbed. This process has to be repeated for every wavelength that wants to be analyzed. With the FTIR technique, instruments measure all wavelengths at once.

FTIR analysis uses the fact that all bonds and groups in a molecule vibrate at characteristic frequencies and a molecule that is irradiated with infrared radiation absorbs energy at frequencies that are characteristic of vibrations of a molecule. During FTIR

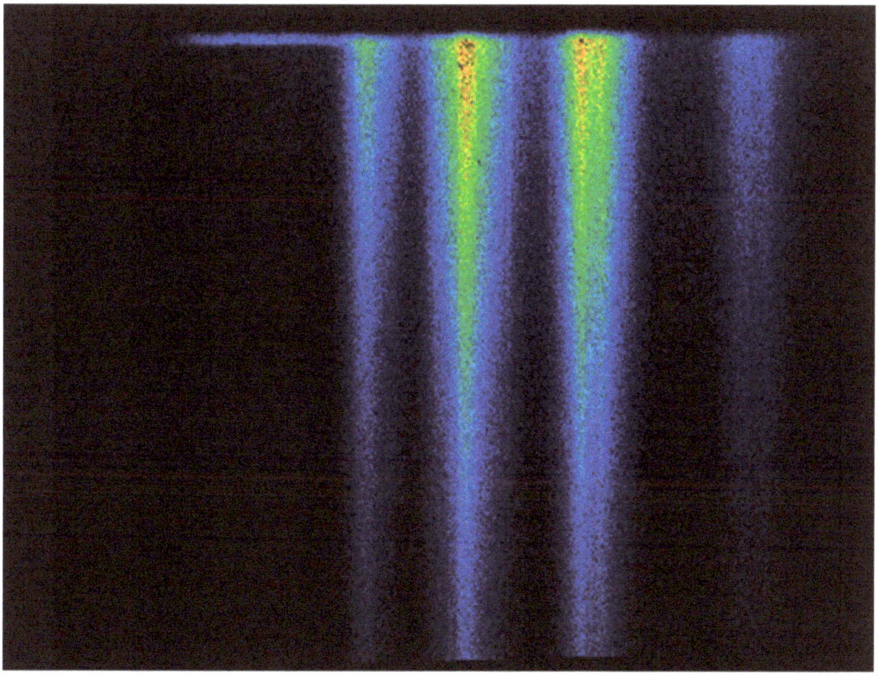

Figure 3.12 The streak image of PL spectrum of $GdVO_4$:Sm nanopowder.

analysis, transmittance and reflectance of a sample under infrared radiation at different frequencies are translated into an IR absorption plot consisting of reversed peaks. Obtained spectra were then used to assign all of the obtained peaks with characteristic vibrations in a molecule. The obtained spectra are compared with library data.

Analyzing the vibrations of a complex molecule is not easily done. The spectrum must be analyzed using library data, as well as the extrapolation of studies of similar molecules. As some bands in a molecule arise from a complex interaction within vibrations in a molecule, absorption bands generally represent single vibration modes. The structure of a molecule can be solved by analyzing the exact position of absorption bands, so therefore the analysis must be precise and careful. Especially, the change in an environment of a molecule can affect the shift of absorption peaks.

An example of using Fourier transform spectroscopy in phosphor analysis is given in Fig. 3.13. Namely, in Fig. 3.13, an FTIR spectrum of $Gd_2Zr_2O_7$:Eu^{3+} was presented. This phosphor is well known to have two characteristic isomeric crystal structures, pyrochlore and fluorite one. In this work, using XRPD analysis, the structure of this material was shown to be a fluorite one. But, the infrared spectrum is showing characteristic peaks for fluorite as well as for pyrochlore crystal structure. The thing is, X-ray diffraction spectrometer was used to analyze the bulk crystal structure, and indeed, the bulk structure of this phosphor is fluorite symmetry. But, infrared spectroscopy was used to analyze the surface of a material, and it was shown that this material

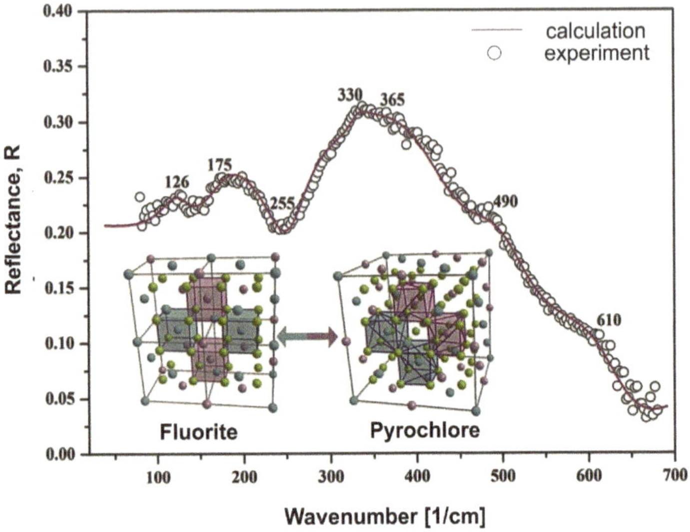

Figure 3.13 Infrared spectrum of $Gd_2Zr_2O_7$:Eu^{3+} showing modes of fluorite and pyrochlore structure.

has two phases, as some literature data also confirms—the fluorite one in bulk and pyrochlore structure on the surface. This is an excellent example of how infrared spectroscopy is used for structural characterization of phosphors, and how can be an excellent complementary technique to X-ray diffraction spectroscopy.

2.5 Raman spectroscopy

Indian physicist C.V. Raman discovered the inelastic scattering phenomenon that later got his name, at the beginning of the 19th century. For this discovery, Raman got the Nobel Prize. Namely, when monochromatic radiation interacts with the sample on which is irradiated, it may be reflected, absorbed, or scattered. Photons that interact with molecules in a sample mostly scatter elastically. This is well known as *Rayleigh scattering*. Photons that have scattered in a Rayleigh way have the same energy as the incident light. However, it happens for one in a million photons that scatter ***inelastically***. This scattering is called Raman scattering. These photons have a different energy from the radiation source that causes them. This difference originates from the different vibrational and rotational properties of a molecule.

When talking about Raman scattering, the incident radiation interacts with the sample, and the resulting scattered photon can have a higher or lower wavelength than the incident radiation. This means that it can have a blue or red shift, respectively. Red-shifted photons are described as Stokes shift. These are the more common ones. The blue-shifted photons are described as anti-Stokes shift. This is more representatively shown in Fig. 3.14.

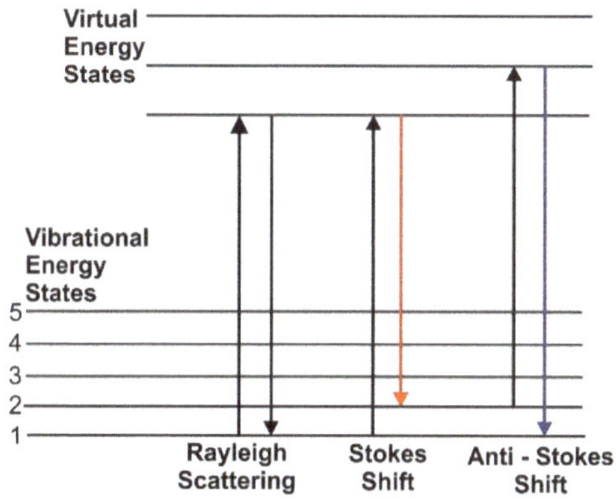

Figure 3.14 Rayleigh, stokes, and anti-stokes scattering diagram.

When a photon interacts with an electron cloud of a material, it excites its electrons into a virtual state. The virtual state is not a real state, and it lasts for a short period of time. It is a distortion of the electron cloud. An electron from a virtual state relaxes into an excited vibrational state. This conditions an electron to lose some energy, and it is described as Stokes shift or Stokes scattering. If an electron from some excited state

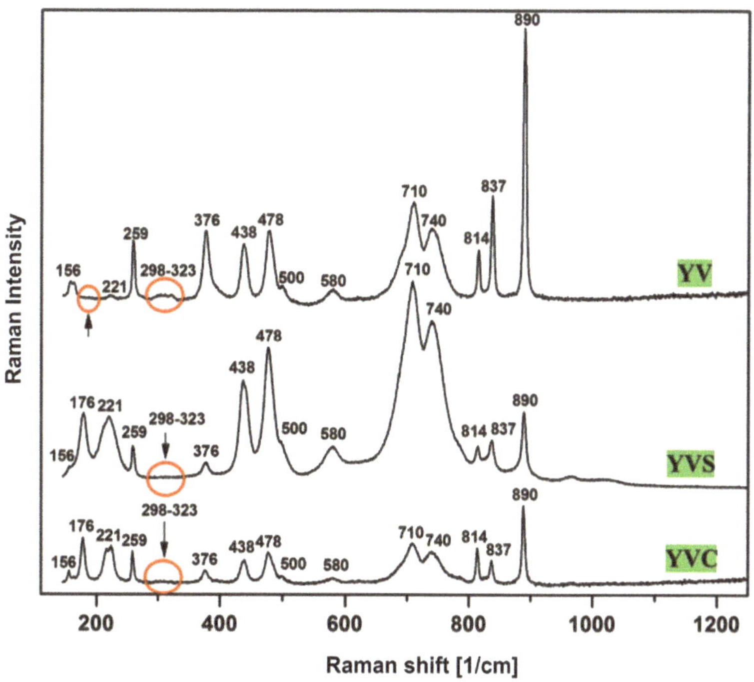

Figure 3.15 Raman spectra of undoped and europium-doped YVO_4.

was excited to a virtual state and then relaxed to a ground state, it has higher energy than the incident energy of radiation and it is regarded as blue shift or anti-Stokes shift (anti-Stokes scattering).

Raman spectroscopy is a technique used to determine the modes of vibration of a molecule. It is used for qualitative and quantitative analysis. Raman signal happens due to a change in polarizability of a molecule or in other words a measure of deformability of a molecule (its bonds) in an electric field. Unlike infrared spectroscopy, Raman spectroscopy offers a lot of advantages: it does not require sample preparation, it can be easily used for solids, liquids, or gasses, samples can be in glass chambers (unlike at infrared spectroscopy where glass causes interference), a nondestructive method, etc.

An excellent example of Raman spectroscopy of phosphors is on $YVO_4:Eu^{3+}$ phosphors and analysis of the incorporation of europium ions into a YVO_4 lattice. In Fig. 3.15, a Raman spectrum of doped and undoped Raman spectra of $YVO_4:Eu^{3+}$ was presented (prepared by two methods: solid-state reaction method (YVC) and solution combustion synthesis (YVS)).

References

[1] J. Mitrić, U. Ralević, M. Mitrić, J. Ćirković, G. Križan, M. Romčević, M. Gilić, N. Romčević, J. Raman Spectrosc. 50 (2019) 802–808.
[2] J. Mitrić, N. Paunović, M. Mitrić, J. Ćirković, M. Gilić, M. Romčević, N. Romčević, Physica E Low Dimens. Syst. Nanostruct. 134 (2021) 114923.
[3] J. Mitrić, N. Paunović, M. Mitrić, B. Vasić, U. Ralević, J. Trajić, M. Romčević, W.D. Dobrowolski, I.S. Yahia, N. Romčević, Physica E Low Dimens. Syst. And Nanostruct. 104 (2018) 64–70.
[4] J. Mitrić, J. Križan, J. Trajić, G. Križan, M. Romčević, N. Paunović, B. Vasić, N. Romčević, Opt. Mater. 75 (2018) 662–665.

Defects in rare-earth-doped inorganic materials

4

4.1 The role of defects in rare-earth-activated inorganic photoluminescent materials

Puja Kumari[1], *Dhritiman Banerjee*[2] *and Subrata Das*[3,4]
[1]Department of Physics, Darbhanga College of Engineering, Darbhanga, Bihar, India; [2]Department of Physics, Indian Institute of Technology (Indian School of Mines), Dhanbad, Jharkand, India; [3]Material Science and Technology Division, CSIR-National Institute for Interdisciplinary Science and Technology, Thiruvananthapuram, Kerala, India; [4]Academy of Scientific and Innovative Research (AcSIR), Ghaziabad, Uttar Pradesh, India

1. Introduction

Defects play significant roles in many luminescent processes, where the remarkable influence in the phosphor performance in various ways is observed. The inorganic photoluminescent materials known as phosphors are usually composed with a crystalline host lattice activated or doped with a minute quantity of activators, which switch light from a shorter wavelength to a longer wavelength (down conversion) via some allowed electronic transitions. Such conversion mechanisms of light are very popularly used in pc-LED lighting and display. A UV or a blue LED supplies the small wavelength excitation, and the resultant long-wavelength light generated from the system is used for many purposes of lighting and display, etc.

Defects are often electrically and optically active and introduce levels in the material's bandgap, which involve a transition between different charge states of the activator ions or defects. Rare-earth ions are most popularly used as activators due to their unique electronic configuration of lanthanide ions and shielding its $4f^n$ orbitals by filled $5s^2$ and $5p^6$ orbitals. These are comparatively less interacting with the ligand ions and thus have high excitation light absorption efficiency. The characteristic photoluminescence of these ions (such as Eu^{3+}, Tb^{3+}, Dy^{3+}, and Tm^{3+}) appears as the sharp emission lines upon suitable excitation. Incorporating various suitable Ln^{3+} ions in metal oxides and perovskites in the form of activators, charge compensators, or substituted ions leads to certain kinds of defects, which contribute to enhancement in photoluminescence properties.

Rare-Earth-Activated Phosphors. https://doi.org/10.1016/B978-0-323-89856-0.00004-3

Doping of activators or other ions is preferred to create vacancies in the semiconductor materials. It is a method through which intentionally doped ions substitute one element. The dopant can be anionic or cationic, and anionic or cationic vacancy can be created depending upon the type of substitution. In phosphor material, which is mostly powder, vacancy is usually created during the synthesis process. In the case of some well-known metallic oxides (TiO_2, ZnO, SnO_2, etc.), defects have been reported in their pure and doped forms. The doping procedures are carried out in these materials during the synthesis process or postsynthesis treatment. The anionic defects (oxygen vacancies and other anion dopants), which can be generated via the structural alteration in the crystal, are very effective in examining the population and the distribution of dopant ions.

Because of the doping of trivalent lanthanide ions (such as Eu^{3+}, Dy^{3+}, Nd^{3+}, and Pr^{3+}), the charge imbalance between the dopant and the host ion can create oxygen vacancies. It has been reported to have a similar phenomenon in CeO_2 nanoparticles when doped with Pr^{3+} ions. The formation of oxygen vacancies has been confirmed through Raman Spectra. It is known that the type of oxygen vacancies can be identified by using Raman spectroscopy as the variation of fabrication techniques can control them. The complicated illustration of optical modes is $\Gamma = A_{1g} + A_{2g} + B_{1g} + B_{2g} + E_g + A_{2u} + 2B_{1u} + 3E_u$. Herein, A_{2g} and $2B_{1u}$ are not categorized as Raman or infrared (IR) active. The A_{2u} and E_u modes are symmetrical and are IR-active.

Meanwhile, A_{1g}, B_{1g}, B_{2g}, and E_g modes are Raman active. For example, the identification of the presence of planar oxygen vacancy in SnO_2 nanowire has been demonstrated by Liu et al. [1]. They showed that IR active mode at $\sim 234\ cm^{-1}$ because of the relaxation of $k = 0$ selection rule as a result of subbridging of oxygen vacancy. Similarly, 573 and 618 cm^{-1} Raman mode has been seen in-plane and bridging oxygen vacancies, respectively. Similar to SnO_2 nanoparticles, similar techniques have also been applied to study the oxygen vacancies of ZnO nanoparticles. $\Gamma = A_1 + E_1 + 2E_2 + 2B_1$ can represent the optical mode of ZnO nanoparticles. In this, B_1 is known to be the silent mode. The A_1 and E_1 modes are polar and are Raman and infrared active, and E_2 is known to be Raman active modes and not infrared inactive. Eventually, E_2 (low) arises from Zn vibrations, while E_2 (high) arises from oxygen atoms. E_2 (high) mode on annealing was found to shift toward lower frequency on increasing oxygen vacancy, but on reoxidizing, the frequency of the E_2 (high) mode returns to its initial value. Therefore, the study of this model can help us understand the oxygen vacancy elaboration in ZnO. The oxygen vacancies could reduce the average atomic mass and might be the reason for the fluctuations in the oxygen positions. This could be the main reason for the redshift in the emission peak. Zeng et al. [2] have also reported that defect creation in TiO_2 is due to the Eu^{3+} doping and subsequent energy transfer between the host and defects to activator ions for the photoluminescence process. It is possible when the energy levels due to defects created in the material are comparable to those of the dopant ions. The exchange of Eu^{3+} for Ti^{4+} formed oxygen vacancies and triggered lattice distortions in TiO_2 crystal that resulted in deviation of site symmetry of Eu^{3+} ion from D_{2d} in anatase to D_{2h} in rutile phase.

Reportedly, the oxygen vacancies in TiO_2 semiconductors are generated by doping with suitable elements having metallic or nonmetallic nature. In most reports, the metallic ions doping can be achieved by facile synthesis route using inorganic precursors of the corresponding element. Substitutional defects are usually created when semiconductors are doped with an aliovalent dopant or dopant of a comparatively ionic radius than the substituted host ion. The defect creation also causes the change in crystal symmetry that further leads to splitting the energy levels and affects the luminescence properties of Ln^{3+} ions. The luminescence properties governed by the defects are validated via the electronic levels that are highly localized in the forbidden gap owing to the involvement of donors/acceptors in electronic transitions. The synthesis method, dopant type, and sintering are the main factors that create defects in oxide materials. Generally, the oxygen defects are situated in the region between the bottom of the conduction band to the central bandgap position of the oxides. They have functioned as donors on electronic excitation from the valence band or energy transfer mechanism. They can be triggered by excitations at/above absorption band edge (>5.2 eV) and followed by relaxation by rapid energy dissipation in the lattice.

The exchange of aliovalent ions is not easy due to the size mismatch, which usually generates elastic and electronic stress. However, these kinds of doping have significant modulating optoelectronic and catalytic properties. Among the several inorganic compounds, tungstate scheelite ($I4_1/a$ space group) has several properties that label them as luminescent hosts. Special optical characteristics have been observed when lanthanide ions are doped in AXO_4 (A = Ca/Sr, X = W/Mo). In general, energy transfer between host to dopant occurs in doping ions of similar valency and ionic radius. Whereas the optimum optical performance of AXO_4 becomes challenging when cation vacancies are formed or strained due to size and charge mismatch is created in the crystal due to Ln^{3+} doping [3] in Ca/Sr sites, as shown in Fig. 4.1.1.

A charge compensator is then helpful in further decreasing the nonradiative transitions due to these defects. Charge compensators sometimes create anionic vacancies such as oxygen vacancies that contribute to the enhancement of the optical performance of the material. Therefore, the earlier discussions indicate the requirement of clear perception on defects owing to doping, which would be beneficial for designing and developing efficient phosphors. This chapter uncovered the abnormal activities and relevant defects in the luminescence process of various rare-earth-doped metal

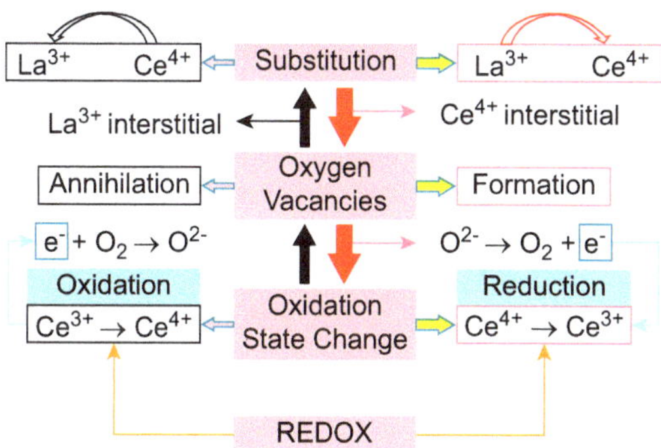

Figure 4.1.1 Oxygen vacancy formation and annihilation in lanthanum cerium oxide [3].

oxide phosphors. It revealed a new direction to realize the actual effects of various defect centers on their photoluminescence.

2. Role of defects in the phosphor doped inorganic materials

Inorganic luminescent materials are the most widely used source of lighting and display application. These materials can be broadly classified into two different kinds [4]: Activator-based and the other is nonactivator-based inorganic luminescent material. In the first type, luminescence is achieved through the transition between energy levels of activator ions like doping by phosphors, ions, etc. At the same time, the second type is based on semiconductors and defect-related materials. Both these types of materials have their unique advantages and disadvantages. In the case of activator-based luminescent material, the prohibitive factor is the wide use of expensive and less environmentally friendly materials. Whereas in the case of defect-related luminescent materials, they have intrinsic toxicity [5,6] and potential environmental hazards that can work against their employment.

In defect-related emission materials, the emission arises due to band-to-band transition and from the transitions between the defect levels. The defect-based emissive materials has some inherent strong points on their side like less toxicity, tunable emission by defect state tuning and cost-effective or low-cost materials [7]. The defects that are found in the inorganic materials are vacancies, impurities, antisite defects, donor-acceptor defects, etc. [8,9].

Due to the generation of defect states, charge imbalance is eliminated by the localization of electron-hole pairs. This gives rise to defect states inside the host material, leading to the generation of luminescence due to carrier trapping by the defect states. The defect states inside an inorganic material can be influenced by the morphology, surface functionalization, etc., technique during its synthesis. Therefore, the energy levels and the associated emission properties of Ln^{3+} may also be affected by the previously mentioned reasons. The local geometry of these phosphor activator ions determines the luminescence properties of the lanthanum doped or phosphor doped materials. The energy gap between the 5*d* and 4*f* levels, which also determines the wavelength of emission colors, can be achieved via two important processes [10] known as the centroid shift and the ligand field splitting. The centroid shift is facilitated by the polarizability of anions and a decrease in electronegativity of the cations present in the host. The second effect, that is, ligand field splitting, depends on bond lengths, molecular overlap, or degree of covalency between the activator ion and its ligands. The above effects in different matrices further tune the 5d energy level by several electron volts (eV) [11].

In oxide-based host matrices, Ln^{3+} activator ions usually occupy the well-defined positions in the matrix, which are identified based on the simple understanding of the charge and size of dopant and matrix ions. However, such mechanisms may not be universally valid. The host lattice arrangements can play either actively or passively in the luminescence process. Active matrices can efficiently module the shape of the energy levels involved in the emission process Via modulating the luminescence kinetics. Among the various defect states [12], oxygen vacancies (V_O) act as hole trapping sites. Excitation spectra study revealed that electrons are trapped at defect states that are neutral or ionized oxygen vacancies [13]. In these doped materials, electrons sometimes find a way to recombine to give rise to visible emission radiatively [14].

The emission behaviors of a host doped with rare-earth ions can be tuned by adjusting the valence ratio of Eu^{2+}: Eu^{3+}. Crystal field disparity around Eu^{2+} ions root a superposition of linear luminescence and encourages the emission tuning in $(1-x)$ CPO- x CYPO: y Eu^{2+}, $z Mn^{2+}$ ($x = 0-1.0$, $y = 0-0.20$, $z = 0-0.40$) where the energy transmission between $Eu^{2+} \rightarrow Mn^{2+}$ is active (as shown in Fig. 4.1.2) [19]. Using this strategy, the tunability in color has been achieved from the sequential emission color alteration from blue and cyan area and then to red-orange. So, tunability can be achieved by changing the environment and inducing energy transfer between ions inside a host matrix by doping them with lanthanide. Different models [15–18] have been proposed by research groups across the globe to explain the origin of luminescence in Ln^{3+}/Eu^{3+} doped metal oxide matrices, including resonant energy transfer, defect-related Auger transitions, bound exciton model, and shallow exciton model.

2.1 Resonance energy transfer/Auger transition model

The resonance energy transfer model was investigated using a-Si-H(Er) as model material [15]. It has been found to have pronounced luminescence around 1.54 μm. It was found that on applying excitation to the system, the amorphous matrix absorbs the energy first, leading to creating a free carrier as a primary step in the entire luminescence

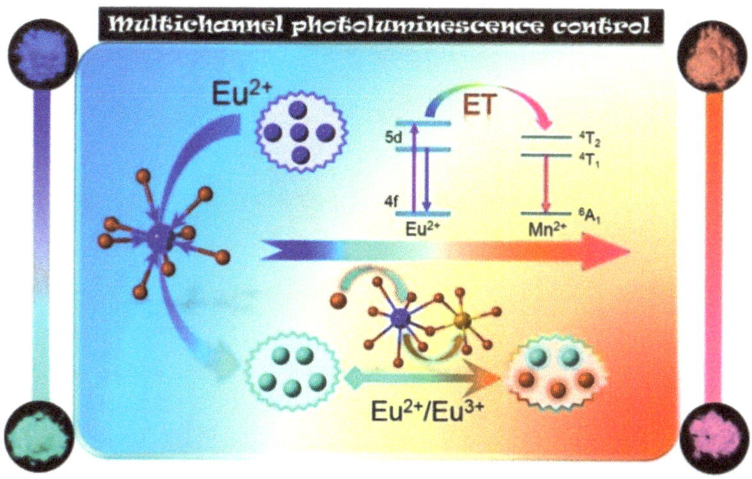

Figure 4.1.2 Schematic diagram to illustrate how different mechanisms [19] can be used to tune photoluminescence emission by multichannel control by $Eu^{2+} \rightarrow Mn^{2+}$ energy transfer and valence mixing of Eu^{2+}/Eu^{3+}.

mechanism. Then the energy transfer takes place through Auger excitation related to the seizure of band-tail electrons by unbiased defects (D^0). An electron-hole pair could also give energy (produced via recombination) to an electron located in the conduction band. Usually, resonant excitation of the *f* electrons happens via capturing electrons in D^0 states (Auger process). The Auger process is mediated by the electron capture situated at the tail of the CB via an unbiased defect state D^0 that converts it into D^-. Such defect centers are ascribed to the dangling bonds in the differently charged states (D^0, D^+, and D^-). The entire process has been represented schematically in Fig. 4.1.3.

2.2 Bound exciton model

The two-stage excitation theory [17] has later been investigated in further detail, and it was found that the Er^{3+} intermediate state may be crucial for producing bound exciton. Dissociation of the intermediate state has occurred within the experimentally available wavelength range, that is, between 7.5 and 17 μm. So, photodissociation of the bound exciton in the intermediate stage is vital for the two-stage excitation process. The detailed process is shown in Fig. 4.1.4.

Based on the earlier discussion, the mechanism of energy movement for rare-earth centers in the semiconductor can be depicted. The recombination step will be limited by the four-step by step processes, as shown in Fig. 4.1.5. In regimes I, III, and IV, multiple processes will compete for the flow of energy [17]. The fastest mechanism that is possible will dominate the mentioned regimes. In regime IV, both emission

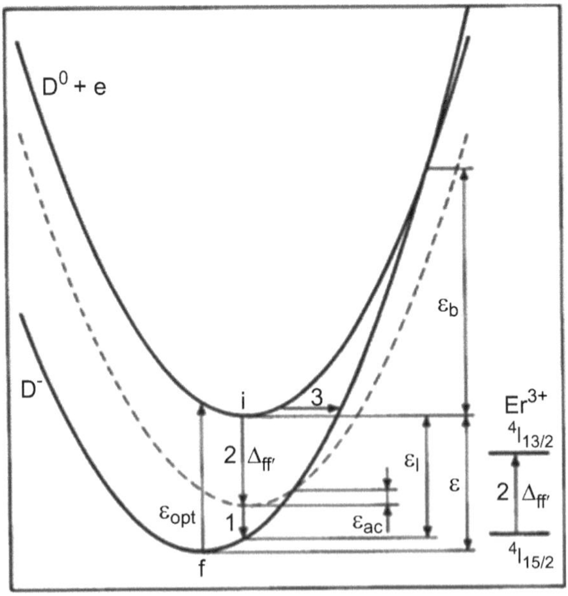

Figure 4.1.3 Electron capture configuration coordinate diagram of D^0 states: (1) transition occurring radiatively (0.8–0.9 eV), (2) Er^{3+} ions Auger excitation mediated by defect states, and (3) nonradiative electron captures through the multiphonon process by the D^0 defect. The *dashed curve* represents the Defect-related Auger excitation (DRAE) transition17 in the virtual state of the adiabatic potential.

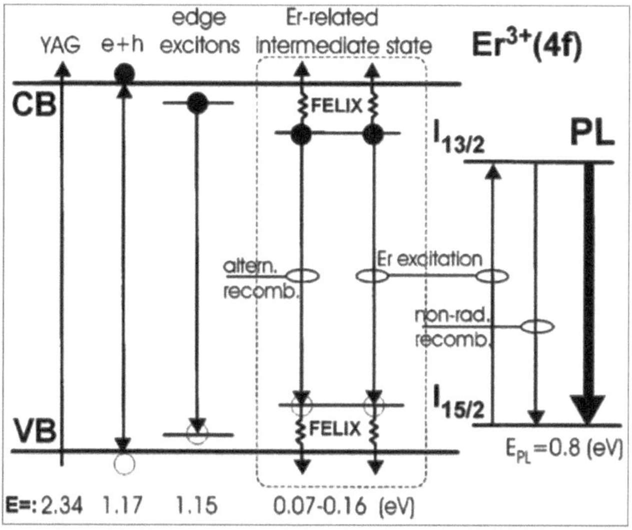

Figure 4.1.4 Schematic diagram showing excitation pathway of Er ion in the silicon matrix considering [17] the role of bound exciton in the intermediate state of Er.

intensity and decay time properties are controlled fastest among the two processes. The nonradiative energy flow rate is dominant for the fastback transfer procedure, and radiative emission is reduced. In contrast, in the case of slow nonradiative back transfer processes, a lifetime of the natural emission from the excited energy state of the rare earth is detected. All the processes explained earlier have been visualized in Fig. 4.1.5.

Defect states play a critical role along with the rare earth doped in these inorganic materials like $BaZrO_3$ in this context [20]. The electrons are first excited by applying excitation wavelength to the conduction band, which subsequently finds its way to the defect states through nonradiative emission. Few of these electrons recombine with the holes to emit radiation of a certain wavelength. In contrast, a fraction of excited electrons jumps to the excited states of the rare-earth atoms (La^{3+}, Eu^{3+}, etc.). This way, they populate the radiative energy levels like in the case of the $^5D_0$ level of Eu^{3+}. There is also a high probability that electrons recombine in a nonradiative way with the holes above the valence band. This nonradiative recombination energy is usually transferred to the nearest rare-earth ions and excites it, populating the radiative energy level like $^5D_0$ of Eu^{3+} ion. The entire mechanism has been shown schematically in Fig. 4.1.6.

Host to dopant energy transfer is another way to induce luminescence in rare-earth-doped metal oxides like ZnO quantum dots or sulfides, etc. This type of energy transfer [21] is facilitated due to the overlapping of the emission spectra of metal oxides with the absorption spectra of doped phosphor ions. In these systems, the excitonic and defect state emissions are used to excite the Eu^{3+} atoms. The electrons are excited and drop down to populate the energy level responsible for the visible wavelength

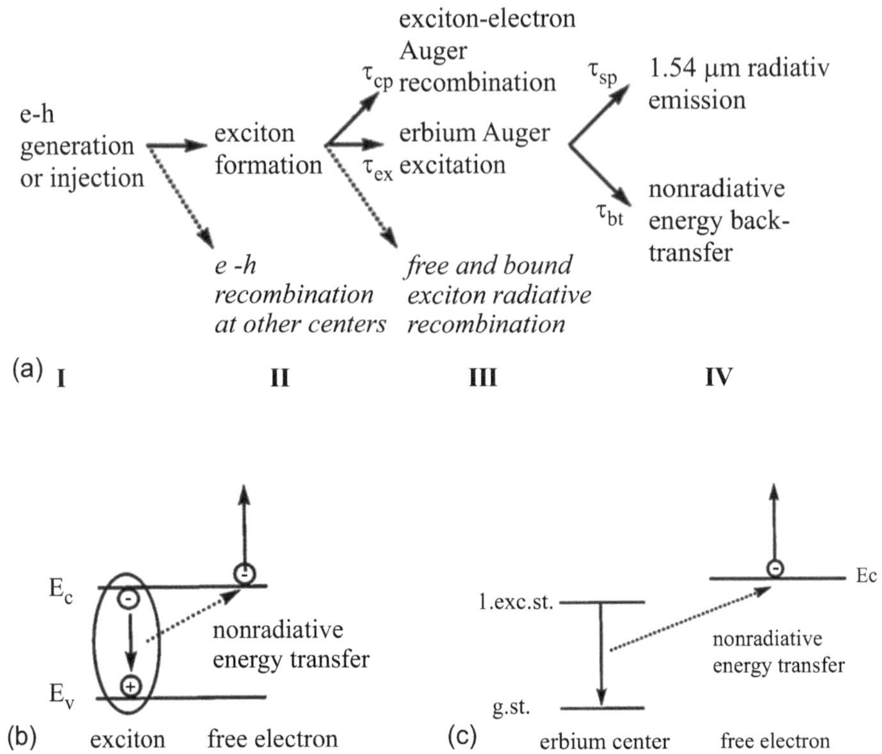

Figure 4.1.5 Deexcitation and excitation processes [17] taking place in the Er-doped Si system. (A) Er luminescence processes at low temperature have been represented in a model that shows all the processes described earlier, (B) Auger recombination process of Exciton-electron. (C) Back transfer of energy from the excited shell of Er, that is, 4f shell by an impurity Auger process involving free carriers.

region by relaxing to the stable state multiphonon processes. After these processes, the emission is observed by the transition from populated levels to the ground state. A suitable example of these kinds of systems as shown in Fig. 4.1.7 is CdS quantum dots doped by Europium [22]. Here the host lattice [23] acted as a sensitizer to Eu^{3+} ions.

Similar cases are applicable for metal oxides or metal oxides doped by the phosphor. Similarly, a significant energy transfer pathway by embedding the metal oxide quantum dots in the matrix of Eu^{3+} doped silica [24] is also reported, as shown in Fig. 4.1.8.

Because of embedding quantum dots (QDs) in the silica matrix, it showed strong ultraviolet emission. However, the ultraviolet luminescence of ZnO QDs decreases monotonically with an increase of Eu^{3+} doping content with the subsequent enhancement in the Eu^{3+} emission owing to the energy transmission from ZnO to the doped Eu^{3+}. The change in photoluminescence emission with the incorporation of Eu in the ZnO crystal has been shown in Fig. 4.1.9.

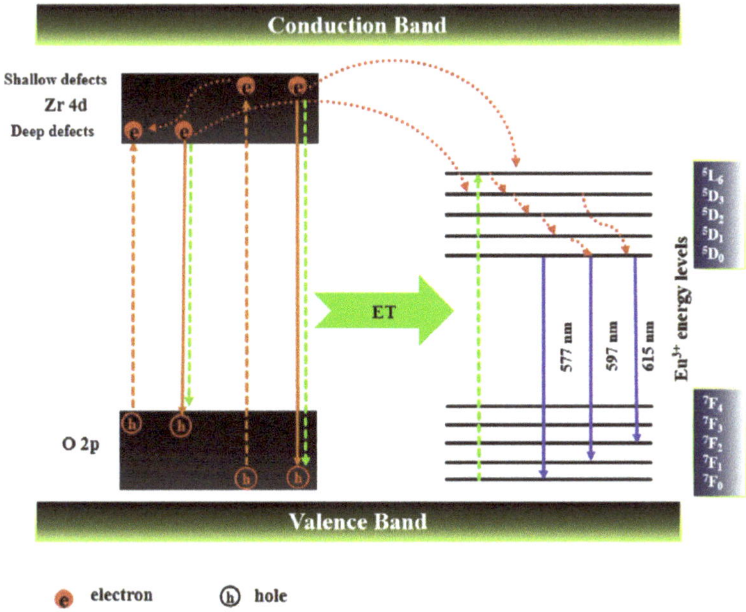

Figure 4.1.6 Schematic diagram to represent the photoluminescence mechanisms inside the $BaZrO_3:Eu^{3+}$ thin film [20].

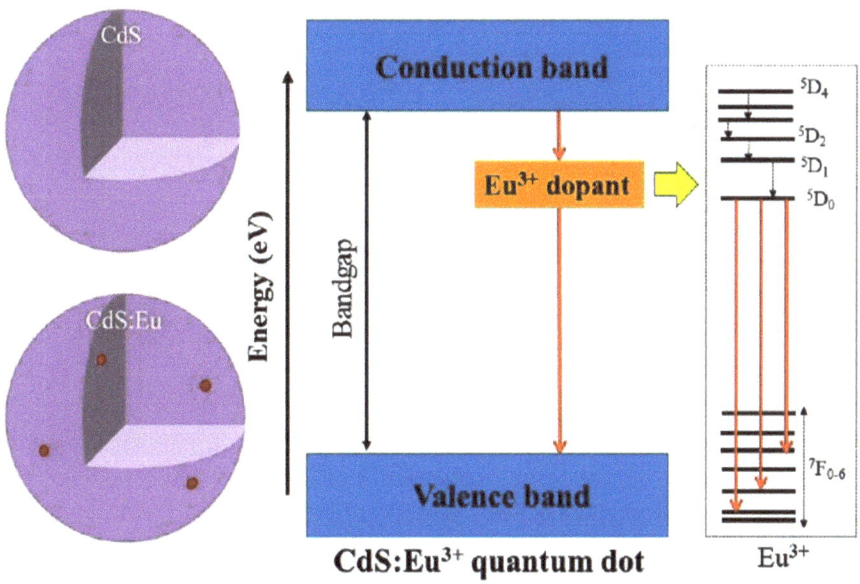

Figure 4.1.7 Schematic drawing and band alignments [22] of Eu-doped CdS QDs, where Eu^{3+} impurities have been shown by the *red dots*, and the energy transfer process between CdS to Eu^{3+} has been shown schematically.

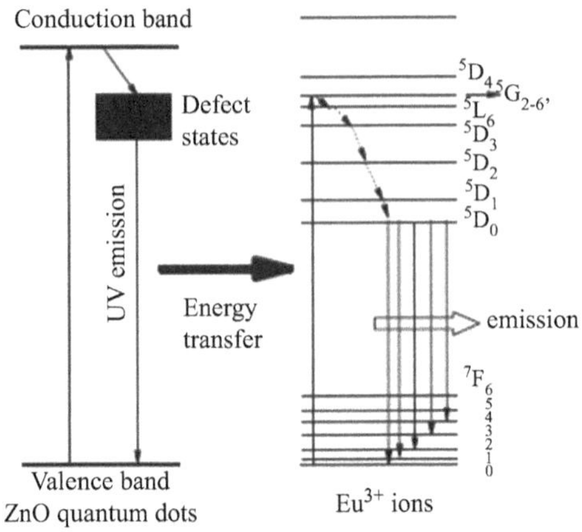

Figure 4.1.8 Schematic diagram demonstrating [24] the energy transfer mechanism responsible for the emissions from Eu^{3+} ions-doped ZnO QDs.

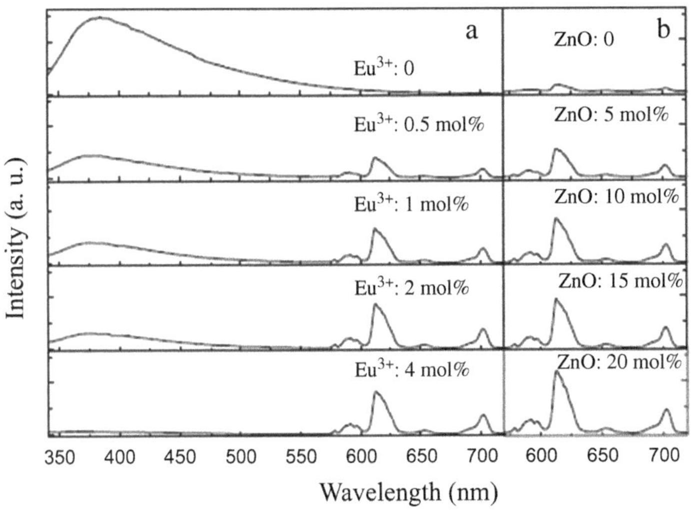

Figure 4.1.9 Photoluminescence emission spectra of (A) 10 mol % Eu^{3+}-doped ZnO nanoparticles and (B) samples with different content of ZnO in them [24].

It can be concluded from these studies that defect states in ZnO QDs might be passivated by silica capping. At the same time, variation in phosphor concentration, in this case, Eu^{3+} can be a strategy in tuning the efficiency of energy transmission between the matrix and the doped ions.

3. Types of defects and their formulations

The crystalline materials have an intrinsic property above 0 K; they contain imperfections in the crystal structure known as defects. These imperfections are of different types:

(i) Single-atom defects (point defects [25])
(ii) Defect involving lattice site or multiple atoms (extended defects).

The point defects [26,27] mainly consist of missing atoms on the lattice sites (vacancies), atoms that are additional and are not occupying the same lattice site (interstitials), or some other atoms occupying the actual position of the native atom (dopant or substitutional defects). White the extended defects consist of linear or planar mismatching of the lattice known as dislocations or stacking faults, grain boundaries, and clusters of point defects. The extended defects can exist in a single material or crystal but generally occur at the interface of two materials in abundance. These types of defects are prominent even when the other material is grown by the epitaxial growth technique. The reason behind this can be ascribed to the existence of lattice constant mismatch, which can give rise to screw dislocation.

The occurrence of these defect states can alter the optoelectronic and physical properties of the materials. Generally, points defects can alter the electronic properties and behavior of the materials carrier concentration, carrier mobility, carrier trapping, etc., while the effect of extended defects is mainly on the physical properties of the nanoparticles, that is, toughness or strength of the material, stress withstanding ability, etc. Though the extended defect states do not significantly affect the material's optoelectronic properties, certain indirect probabilities still exist by functioning as a source or sink for the point defects. Therefore, it can play an important role in generating nonradiative recombination centers for the excitons created during the photo-excitation. Due to the presence of such nonradiative centers, the performance of the optoelectronic devices is usually degraded. Since most of them come unwillingly, therefore, controlling them is a challenge.

Surface defect states in an inorganic semiconductor like ZnO or TiO_2 or ZnS or CdS mainly include the vacancy defect states such as oxygen vacancies whose formation on the surface is very difficult to control due to their low formation energy and abundance. Meanwhile, the crystalline surfaces can also support the formation of such defect states similar to that of the bulk crystal. Many analogical similarities, therefore, come into play while discussing these two aspects. Understanding these phenomena is very important in the context of nano-devices, which employ nanoscale phenomena to display superior performance. The surface phenomenon or defect states dominate the optoelectronic properties because with the decrease in particle size, surface to volume ratio of the material increases manifold. Besides optoelectronic properties at atomic scale resolutions, these materials perform well in catalysis at the molecular level. However, a precise understanding of surface defect states remained prominent to date. Recent studies have pointed out the role of surfaces or interfaces behind the appearance of point defect behavior in bulk. Further investigation revealed that bulk properties are strongly correlated with the surface defects present in the inorganic crystals. The reason behind their formation and its effect on the luminescence of the materials has been discussed as follows.

3.1 Vacancy defects

Vacancies are the result of the absence of a particular atom in the lattice. Vacancies [27] are one of the most abundant and stable defect states in metal oxides. These defects in a metal oxide may involve the metal atom or the oxygen vacancy. Among the two types of vacancy defects, oxygen vacancy (V_O) plays a vital character in influencing the properties of the metal oxides, which are often invisible due to the concentration of their presence in the sample. Some other metal oxides render superconducting properties [28], ionic conductivities [29], enhancing catalytic properties [30], etc. This type of defect can compensate for the electronic charges through ionic defects. It has important applications in ionic electronics as well. Oxygen vacancies have a higher diffusion coefficient in several oxides such as [31] $SrTiO_3$, and they can diffuse up to several nm distances inside the crystal. Due to these diffusion properties, it can be tailored dynamically via the application of magnetic field, electric field, pressure, temperature, and light, to name a few [32,33]. The presence of oxygen vacancy in the metal oxide materials is often linked with the materials' n-type conductivity and the prime source of green emission from such nanostructures.

The presence of these vacancies on the surface of the nanoparticles induces band bending in the metal oxide nanoparticles, giving rise to Fermi level pinning [13]. Due to this band bending, the green emission is supposed to get set inside the metal oxide. Doping of phosphors such as Eu^{3+} and Pd^{3+} in the metal oxides such as ZnO, TiO_2, or SnO_2 nanoparticles leads to the tuning of the emissions. Still, the vacancies could significantly reduce the emission by promoting nonradiative reaction pathways. Effect of oxygen vacancy in the case of one of the most promising solid-state light-emitting diodes using cerium-doped yttrium aluminum garnet (YAG: Ce^{3+}) has been reported [34]. The process has been schematically shown in Fig. 4.1.10.

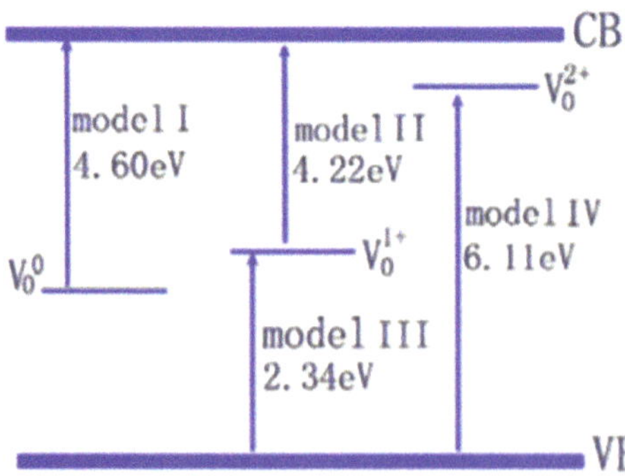

Figure 4.1.10 Proposed optical transition models induced by neutral and charged oxygen vacancies in YAG [34].

Model I: An electron transfer from the defect state, that is, V_O^0 to the conduction band, the excitation energy is around 4.60 eV.

$$V_O^0 + \text{photon} \rightarrow V_O^{1+} + e^-$$

Model II: Alteration of an electron from the defect state of the V_O^{1+} to the conduction band, where the excitation energy is around 4.22 eV.

$$V_O^{1+} + \text{photon} \rightarrow V_O^{2+} + e^-$$

Model III: Alteration of an electron from the top of the valence band to the V_O^{1+} defect state, having excitation energy of around 2.34 eV.

$$V_O^{1+} + e^- + \text{photon} \rightarrow V_O^0$$

Model IV: Alteration of an electron from the top of the valence band to the shallow unoccupied level of V_O^{2+}, having excitation energy 6.11 eV.

$$V_O^{2+} + e^- + \text{photon} \rightarrow V_O^{1+}$$

In many metal oxides, alongside YAG: Ce^{3+}, an oxygen vacancy is one of the most important defect states.

In wide bandgap semiconductors, oxygen vacancies usually feature deep charge transition levels (CTLs). They also display considerable structural relaxation among the various charged states. Oxygen vacancies can also act as electron traps during the ionization process and appear in neutral or charged states [15]. The most common charge states of the oxygen vacancies are +2, +1, or 0. Oxygen vacancies, which are positively charged, are the most credible donor in these materials. Charged oxygen vacancies have been emphasized as more important donor defects compared to interstitials or antisite defect states. Oxygen vacancies are therefore very crucial for the maintenance of charge neutrality and therefore are electrically active. Eventually, oxygen vacancies (V_O) can readily contribute to the charge transfer process with the doped activator ions such as Eu^{2+} and Ce^{3+} owing to the specific electronic configurations.

As the charge transfer is possible between the vacancies and the doped phosphor atom, there is a probability of energy transfer between the phosphor atom sites. This resonant energy transfer can occur until the phosphor atom comes near either a singly or doubly charged oxygen vacancy. This process is very likely because, in most cases, the concentration of the doped phosphor atom will surpass the concentration of the oxygen vacancies (V_O), so a huge portion of the oxygen vacancies will have phosphor atoms in their juxtaposition. Eventually, multiphonon processes may also occur between the ground and excited state of the doped ions between the ground state (Ce^{3+}: $4f^1 5d^0$) and excited states (Ce^{3+}: $4f^0 5d^1$) as in the case of Ce^{3+} doped YAG.

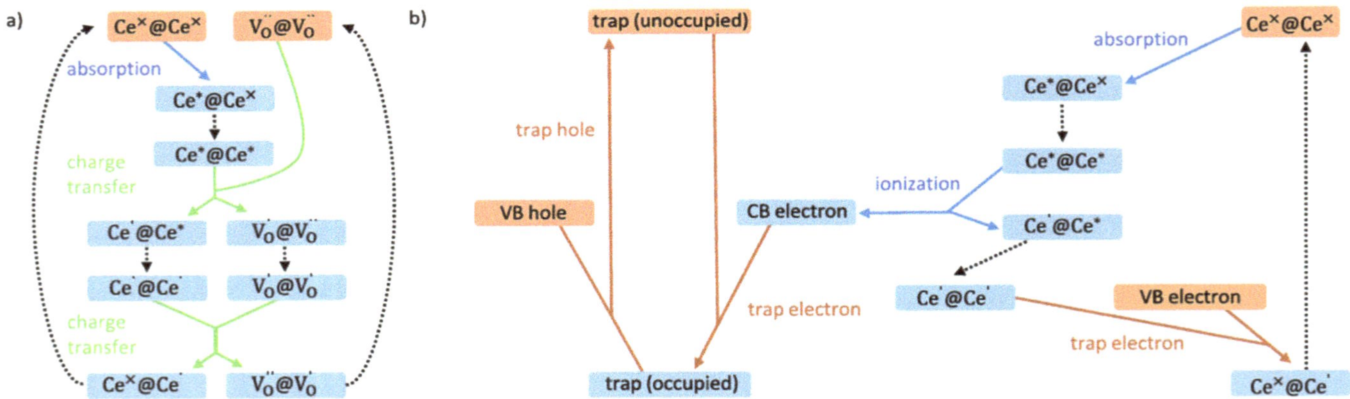

Figure 4.1.11 Reaction paths schematic diagram. (A) Oxygen vacancy induced thermal quenching via charge transfer. (B) Ce 5d state to the conduction band electron transfer based thermal quenching at a killer center [35] by subsequent trapping.

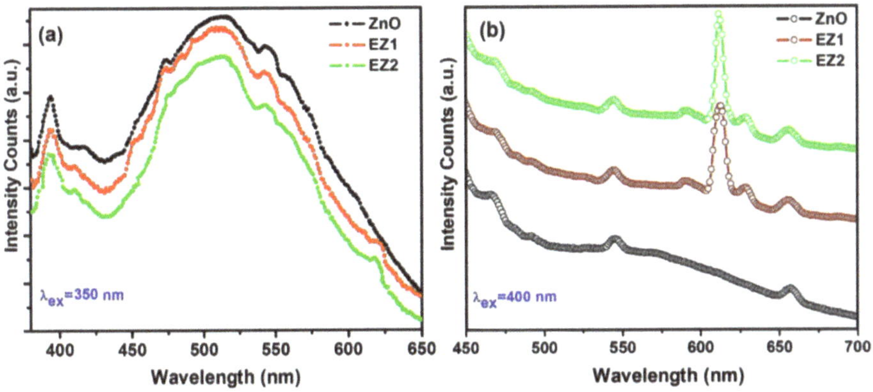

Figure 4.1.12 Change in photoluminescence emission [36] spectra with change in Eu^{3+} doping concentration in ZnO under different excitation wavelengths: (A) 350 nm and (B) 400 nm.

In this system, Debye temperature is found to increase with the concentration of Ce atoms. This leads to a large high-frequency phonon population at low temperatures, which contributes to multiphonon processes. The detailed process has been shown schematically in Fig. 4.1.11.

There is also a high chance of luminescence quenching through concentration quenching, where the energy is resonantly transferred between the activator ions until it is completely dissipated with no specified quenching site in the matrix. According to thermal ionization models, electrons are emitted from the excited state of the doped phosphor atom to the conduction band edge of the defect level, mainly oxygen vacancy (V_O) present in the metal oxides. However, the above-stated condition may dominate the quenching process that usually depends upon the temperature and concentration of the doped activator ions. Reportedly, doping concentration plays a key role in determining the luminescence from Europium doped ZnO, as shown in Fig. 4.1.12 where EZ_1 and EZ_2 correspond to 0.001 and 0.002 M doped Eu in ZnO [36].

The last two stated mechanisms require the presence of trap or defect states related to surface vacancies. So, these vacancies need to be controlled to get high luminescence efficiency from phosphor-doped metal oxides. V_O-vacancies have low formation enthalpy in ZnO, so they are abundant, while zinc vacancies (V_{Zn}) are less abundant than oxygen. Oxygen vacancy is placed 0.5−0.8 eV above the valence band. Zinc vacancies are amphoteric defects that mean it consists of two acceptor and donor levels, which mean it can accommodate both electrons and holes. Research groups present its position 1.39 eV above.

3.2 Interstitial defects

Interstitial defects are usually generated in the metal oxides either by the presence of the metal ion or due to the presence of nonmetal ions that form the core component atoms in the host lattice. The host lattice is also known to have a significant impact on the 4f energy level of the dopant ion. This is facilitated by the host lattice's

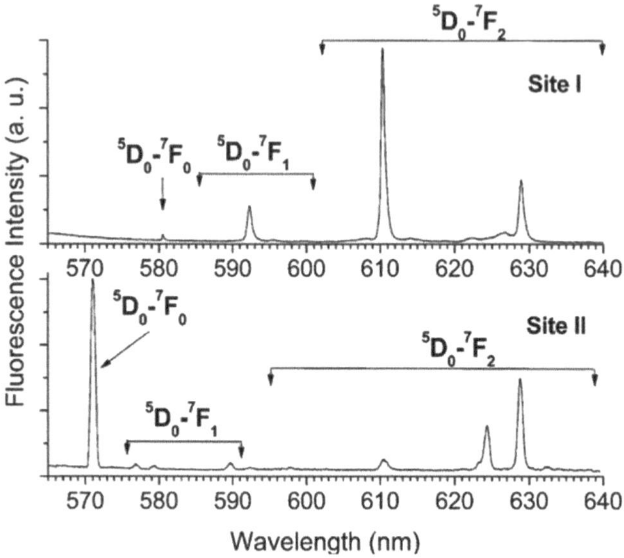

Figure 4.1.13 Comparison of fluorescence spectra [37] of Eu^{3+}:BaFCl microcrystals at Sites I and II in the range of 570–640 nm.

produced crystal field splitting effect, which can tune the probability of *f-f* transitions. When the Eu^{3+} ion is doped in BaFCl crystal characteristics, normal and irregular crystal field spectra of Eu^{3+} have been observed due to host influence over the ion. It was observed that the Eu^{3+} in the crystal site has both normal as well as anomalous behavior. The normal behavior can be explained by the standard crystal field splitting model. At the same time, the anomalous spectra or the fluorescence dynamics are due to charge inequity and oxygen compensators that encourage effective crystal field distortion, as shown in Fig. 4.1.13.

Sometimes the host has a protective effect on the luminescence property of the dopant ion in solution. Interstitial defects have an important link with the phosphor doping of inorganic lattice. The defect results from the mismatching of the dopant ions and the lattice atoms, which can open up the nonradiative energy loss pathway by dissipating the excitation energy of the dopant phosphor ions to the lattice. This reduces the optical emission efficiency of the system as a whole. It is, therefore, suggested to have a similar atomic radius for the dopant ions and the lattice atom. This is the logic used behind the development of rare-earth phosphor doping-based luminescent materials. The Ln^{3+} ions have an almost similar atomic radius, so it is suggested to be doped in the lattice containing alkaline earth ions such as Ca^{2+}, Sr^{2+}, or Ba^{2+} or some transition metal ions such as Ti^{4+} or Zr^{4+}. Phosphors having Ln^{3+} doping always give rise to the interstitial anionic or cationic vacancies. The interstitial defect states generate defect-related trapping centers due to the charge compensation effect [38].

The introduction of interstitial oxide ions in the $CdSiO_3$ crystal [41], as shown in Fig. 4.1.14 has been used as a convenient scheme that modifies R^{3+} type rare-earth

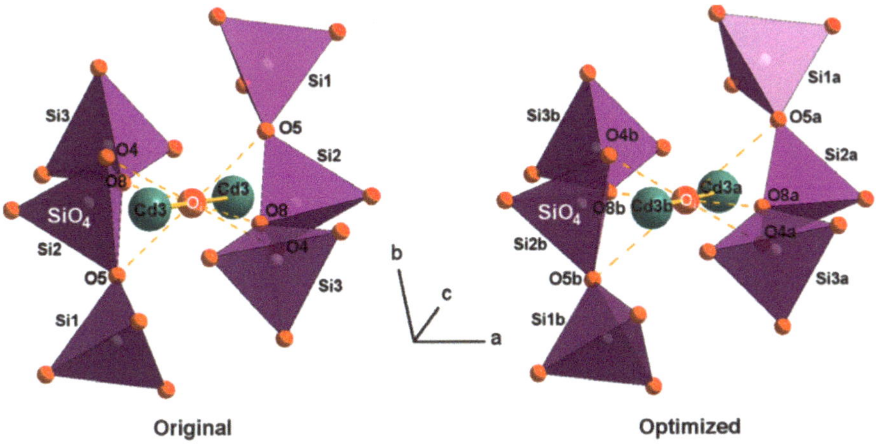

Figure 4.1.14 Surroundings of interstitial oxide ion [38] in the original ($^1/_2$, $^1/_2$, $^1/_2$) site (left) and the $CdSiO_3$ is the optimized structure (right).

atoms. The introduction of this interstitial oxygen (O_i) leads to strong electrostatic repulsion with the nearest oxygen atom, leading to the likely creation of electron traps in the defect sites. These schemes often induce persistent photoluminescence in other phosphor materials as the traps slowly release the electrons.

The formation of interstitial defects in the host lattice of metal oxides specifically is accompanied by the occupancy of the metal or the oxygen atom at the interstitial position. The interstitial defects are induced in the metal oxide host by the introduction of nonequilibrium processes. Moreover, these defects are not stable enough and have high diffusion compared to vacancy defect states that are more stable than the interstitial defect states. Strong nonequilibrium processes like high-temperature heating of the samples in the environment with rich vapors of the particular metal or oxygen induce metal or oxygen interstitial in the crystal. Therefore, interstitial vacancy plays an extremely important role in the host lattice and the rare-earth (RE)-doped host matrix. In these defects, the particular atom assumes a normally unoccupied site in the crystal structure.

3.3 Substitutional defects

In the case of substitutional doping of a host, the dopant replaces the original atom in the crystals. It is different from the interstitial defect because the dopant atom should only occupy the lattice site. Depending upon the size of the substituted phosphor atom, the stress or strain in the neighboring atom of the host can be understood. Also, in the last discussion about interstitial defect states, we have given insights into how the substitutional defect states enter the crystal as interstitial defects. On reaching vacancy, it gives rise to substitutional defect states. The understanding of the substitutional defect role on the metal oxides has already been carried upon SnO_2. It is equally likely to be of a similar type in other metal oxides.

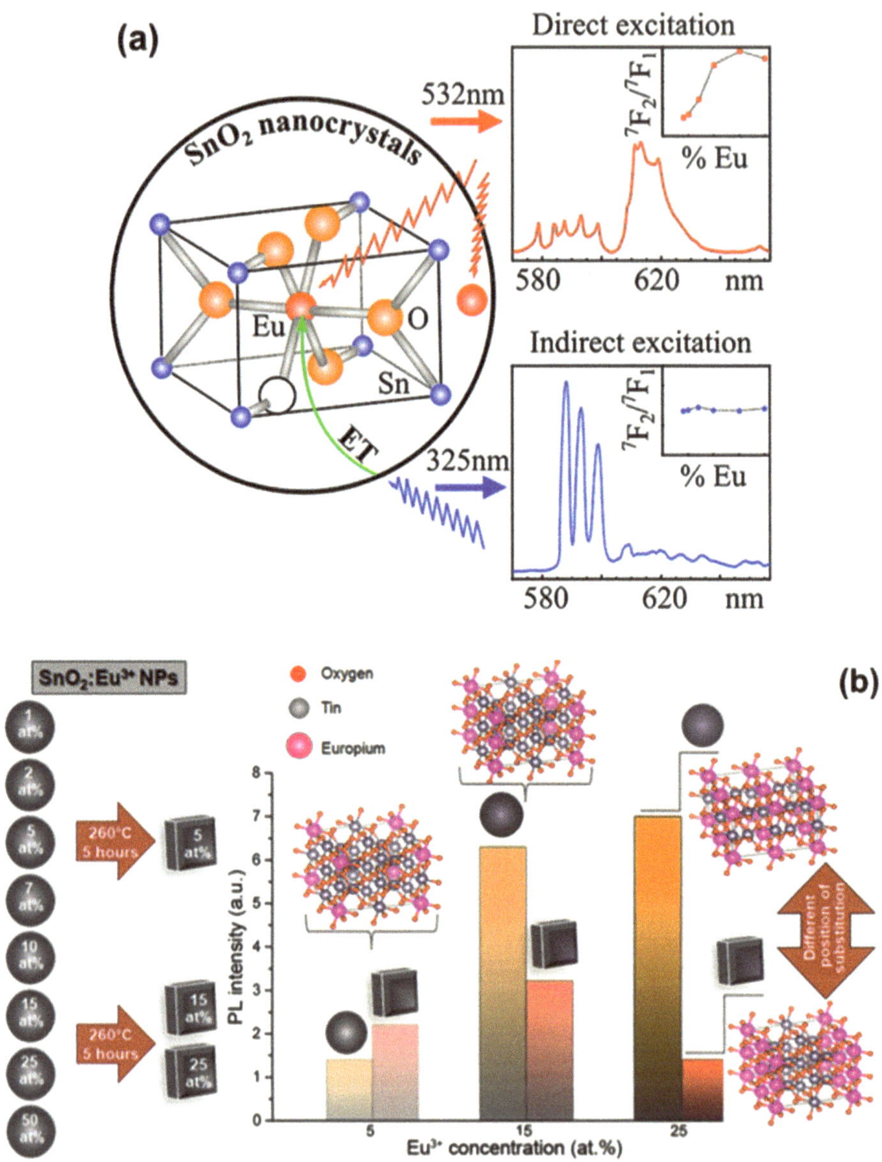

Figure 4.1.15 (A) Strong energy transfer [40] between SnO_2 and Eu^{3+} ions-based variation on PL emission spectra excited at 532 and 325 nm, (B) PL intensity variation with a doping concentration [41] of Eu in the SnO_2 nanoparticles.

It was reported previously that Eu^{3+} ions could easily substitute the Sn^{4+} ions [39] or similar metal ions from the metal oxide lattice and generate stress and strain in the crystal based on the size difference of the original atom and the dopant ions. Oxygen vacancies coupled with interstitial defect states are observed to promote the generation

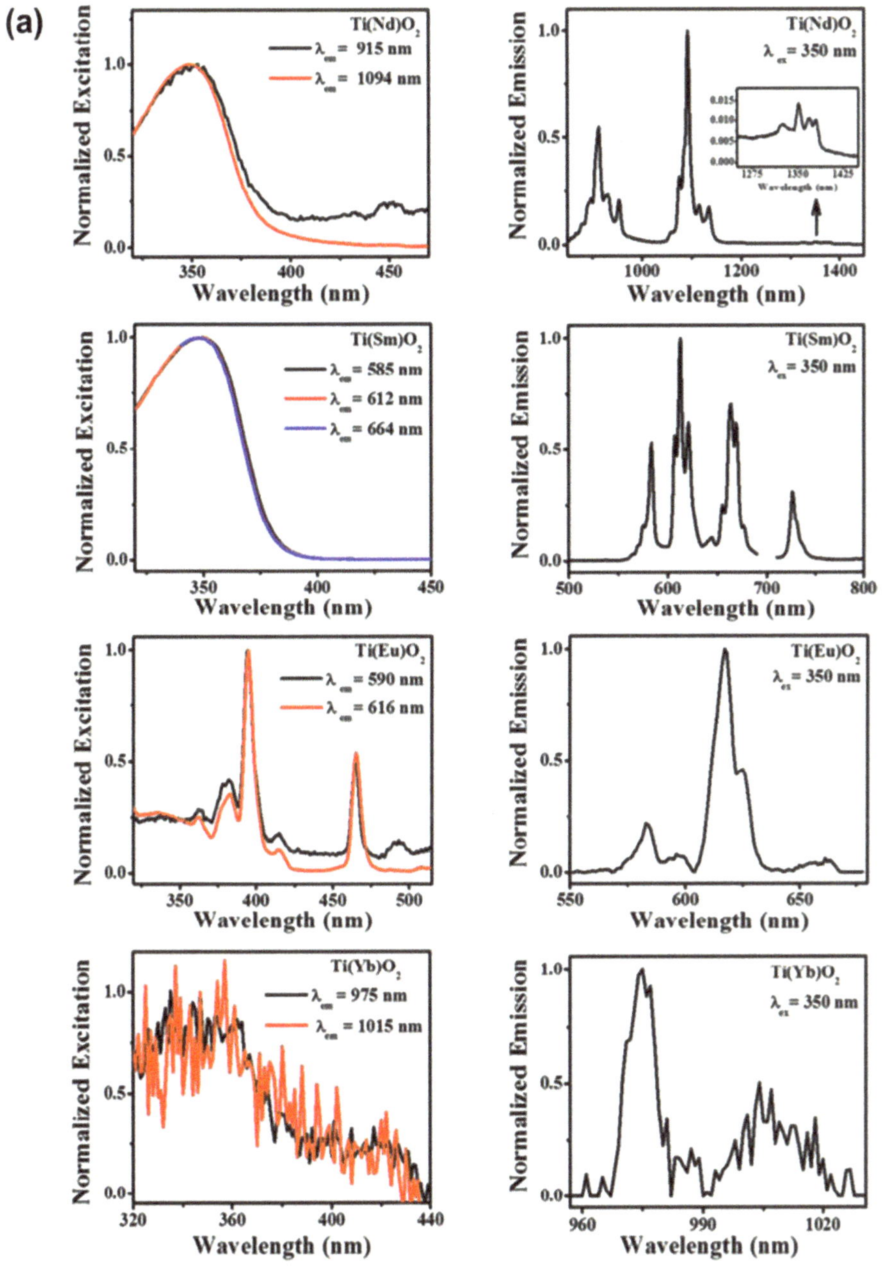

Figure 4.1.16 (A) Luminescence excitation (left) and emission spectra [42] (right) are shown for Ti (Ln) O_2 nanoparticles [Lanthanide = Yb, Sm, Nd, Eu] dispersed in neutral water. In the Ti (Sm) O_2 emission spectrum, the region near 700 nm is broken. (B) Energy level schemes for the Ln^{3+} [Ln = Yb, Eu, Nd, Sm] ground and excited states are shown w.r.t. the conduction and valence band of the host TiO_2 nanoparticles.

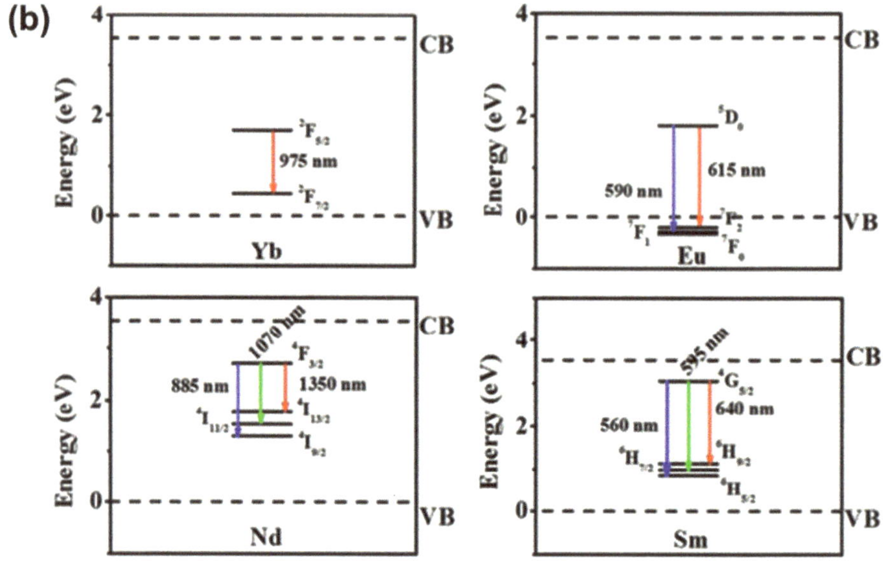

Figure 4.1.16 cont'd.

of substitutional defect states. The study also revealed that in the phosphor atom, Eu^{3+} can exist as associates or substitutional isolated centers, respectively. It has further been observed that Eu^{3+} emission is ruled by comparatively strong three magnetic dipoles (MD) emission lines at 590 nm ($^5D_0-^7F_1$) as shown in Fig. 4.1.15A. In these types of systems, the metal oxide host performs as an antenna for the absorption of the UV excitation energy and ultimately transfers the energy to the doped phosphor atoms, thus compensates for the low absorption ability of the parity forbidden *f-f* transition. Recent studies have also demonstrated that the afterglow behavior of the Eu-SnO_2 host is due to the coexistence of uniform and exponential trap states. The specific location of the defects in Eu^{3+}-doped SnO_2 can significantly[42] alter the functional properties of the system as a whole such as photoluminescence as shown in Fig. 4.1.15B.

While the literature on TiO_2 doped with La atom shows that the substitutional effect of La^{3+} ions in the TiO_2 crystal can donate to the alteration in the emission and absorption property of the host, as shown in Ref. [42] Fig. 4.1.16A. The energy level diagram after doping has been shown in Fig. 4.1.16B.

The substitutional doping mechanism for trivalent lanthanide ions (such as Nd^{3+}, Sm^{3+}, Eu^{3+}, and Er^{3+}) has advanced sufficiently while identifying the different doping sites the Sm^{3+} and Eu^{3+}; several lattice sites have already been identified. The doping site has been identified with the help of lattice site-selective luminescence spectroscopy study of host lattice doped with Sm^{3+} and Eu^{3+}. "Surface effect" is accountable for a wide luminescence visualized with a wide range of dopants, regardless of host and luminescent activator. The luminescence emission study details relationships between defects related to luminescence, changes in morphology, surface

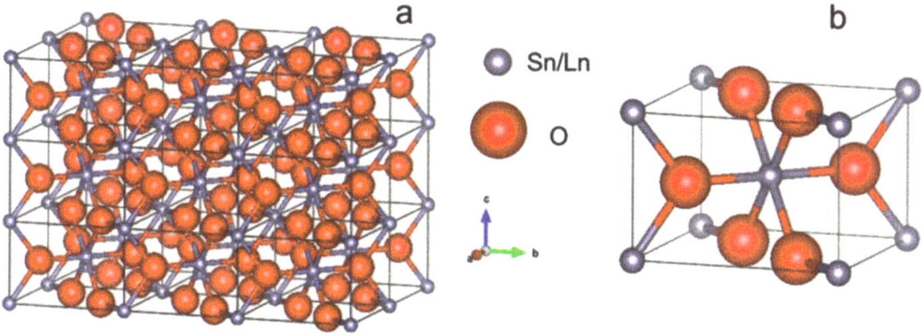

Figure 4.1.17 Crystal structure [39] of SnO_2.

effects doping, and particle size. The system studied to understand the impact of the "*surface effect*" is SnO_2 doped with lanthanide ions. The crystal structure of SnO_2 nanoparticles is shown in Fig. 4.1.17.

The Sn centers, marked as I to V, have been identified by analyzing the photoluminescence by Bogdan Cojocaru et al. [42], which confirms the role of surface effects. The emission profile is shown in Fig. 4.1.18. Emission from the center I is entirely excited via SnO_2. Oppositely, broader emission, excited exclusively by Eu's f−f absorptions, has been observed and denoted to surface type Eu center, labeled as center V as shown in Fig. 4.1.18A and B. Besides the substitutional isolated C_{2h} (I) and surface centers (V), at least three more Eu centers, labeled as II−IV (or non-C_{2h} centers) as shown in Fig. 4.1.18A and B. Upon studying the sharpness and nature of the emission, it was suggested that the center II−IV is the result of the intrinsic properties of SnO_2. The emission is the result of interaction between bulk defects and substituted Eu ions in the lattice. The interacting bulk defect is proposed to be oxygen vacancy that interacts with the substituted Eu ion in the lattice.

4. Outcome of defects on bandgap and photoluminescent properties

It is well known that the defects created in phosphor material due to lanthanide-doped ions incorporate energy levels situated in the bandgap of the host lattice. Siddique and Tripathi have reported defects formation in the NiO matrix by doping with lanthanum ions [43]. There is a blue shift in the photoluminescence emission spectra of NiO with an increase in the concentration of La^{3+} ions. The optical band gap (E_g) of NiO is reported to be increased with the increase in the La^{3+} concentrations as shown in Fig. 4.1.19. This widening in the bandgap is influenced by the creation of defects in the material since La^{3+} ions preferably occupied the Ni^{2+} sites in NiO, which resulted in the charge imbalance and released more electrons to confine the creation of defects. For degenerate semiconductors, the growth in bandgap energy is due to the Fermi level

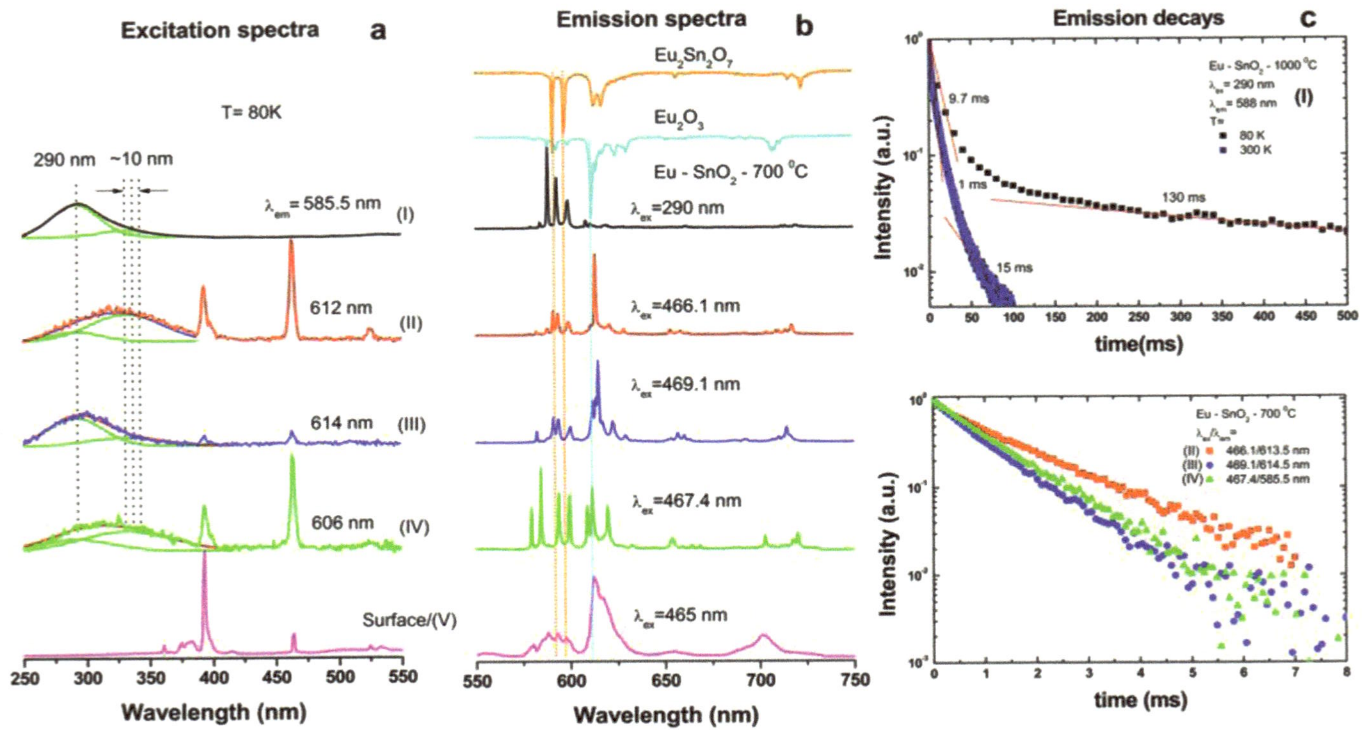

Figure 4.1.18 Summary of the emission properties [39] of Eu-SnO_2 calcined at 700°C/1000°C and measured at 80 and 300 K. Excitation spectra of Eu centers labeled (A), emission spectra Eu centers labeled (B) and emission decays (C) of as I−V centers (see text).

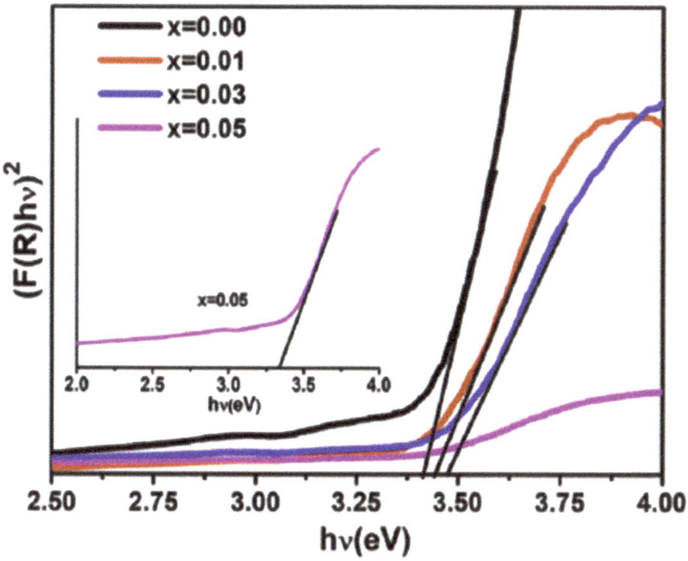

Figure 4.1.19 Tauc plot for bandgap calculation of NiO: La^{3+}[43].

movement into the conduction band [44]. Nevertheless, the increase in the concentration of La^{3+} ions in the lattice could increase the abundance of the oxygen vacancies in the lattice. Hence, the significant reduction in the bandgap value might be caused by incorporating the oxygen vacancies.

In other reported results, the presence of oxygen vacancies between the valence and conduction bands for $SrZnO_2$ doped with Eu^{3+} ions have been confirmed by XPS analysis, as shown in Ref. [45] Fig. 4.1.20. Oxygen vacancies and oxygen interstitials can interchange with each other. Some of the desorbed oxygen atoms from the interstitial sites can be merged and produce O_2 gas as per the following ($O_o \rightarrow (1/2)\, O_2\uparrow + V\ddot{o} + 2e$). While the others relax in interstitial sites via forming $Vo^{\times} - O_i$ pairs ($O_o \rightarrow Vo^{\times} + O_i$). Because of the enhancement in oxygen fractional pressure and $O_i \leftrightarrow (1/2)\, O_2\uparrow$, the absorption of oxygen atoms responds to the formation of $Vo^{\times} - O_i$ pairs and drives O_i to return to $Vo^{\times}$ resulting in a reduction in O_i atoms ($Vo^{\times} + O_i \rightarrow O_o$) and an escalation in oxygen vacancies ($O_o \rightarrow (1/2)\, O_2\uparrow + V_{\ddot{o}} + 2e$, $V\ddot{o} + e \rightarrow V_{\dot{o}}$, $V\dot{o} + e \rightarrow Vo^{\times}$). Thus, the amount of oxygen defects in samples fluctuates. In the oxygen atmosphere, oxygen defects are mostly oxygen interstitials, whereas, in inert (N_2 and Ar) gases, oxygen vacancies are leading among all oxygen defects.

UV−VIS absorption spectra of the $SrZnO_2$: 0.02 Eu^{3+} sample recorded in the presence of different gases, namely air, nitrogen, oxygen, and argon, are shown in Fig. 4.1.21. Calculation of bandgap showed a change in the value with change in different gases. A decrease in the bandgap value of $SrZnO_2$: 0.02 Eu^{3+} with respect to that measured in the presence of oxygen has been attributed to a change in oxygen vacancy near the valence band. On increasing oxygen vacancy, the defect states

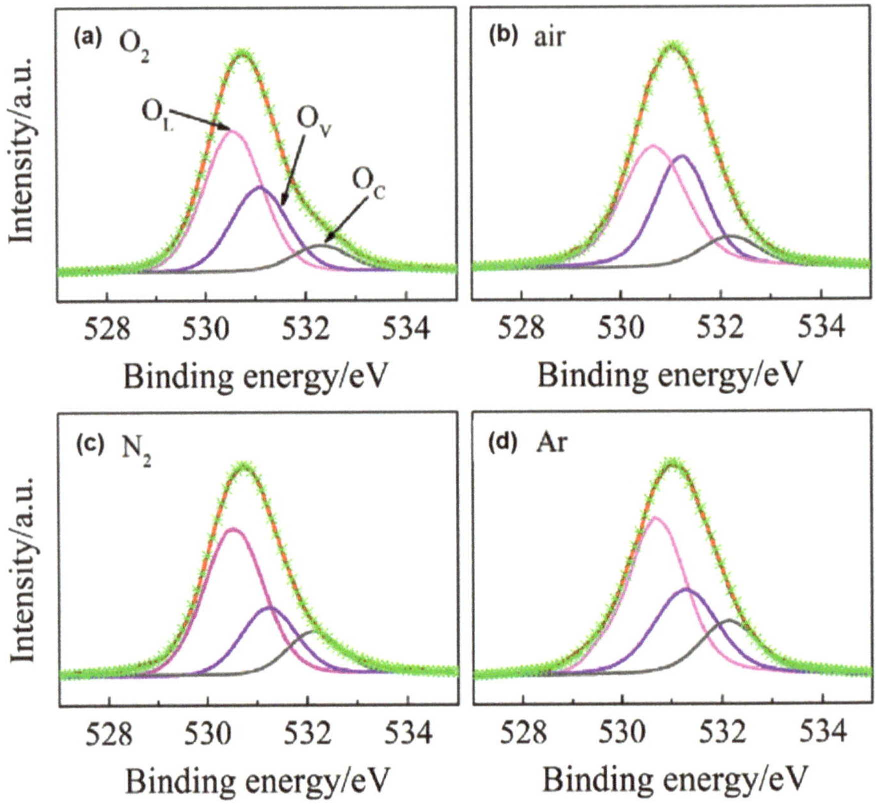

Figure 4.1.20 Gaussian deconvoluted XPS spectra of $SrZnO_2$: 0.02 Eu^{3+} phosphor prepared at different atmospheres [44].

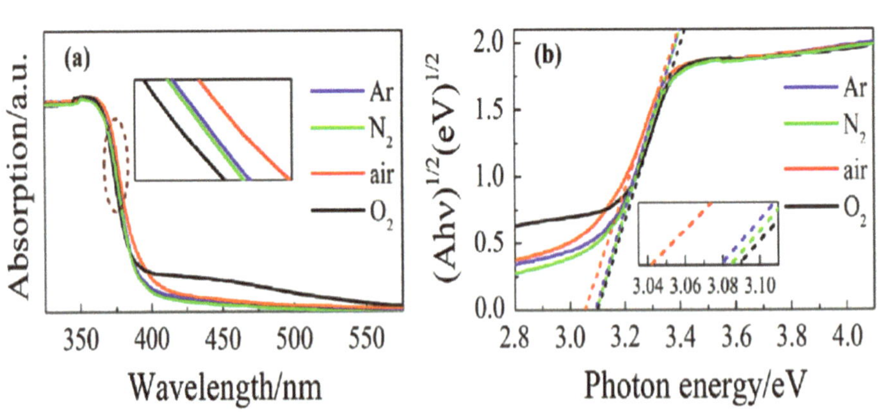

Figure 4.1.21 (A) UV−Vis spectra of $SrZnO_2$: 0.02Eu^{3+} and (B) band gap plot [44].

become delocalized and overlap with the valence band's edge. This leads to the raising of the position of the valence band [46]. Removal or missing the oxygen ions from the host lattice leads to the creation of oxygen vacancies that may be neutral oxygen vacancies or vacancies with electrons trapped in them. In addition to vacancies, oxygen interstitial (O_i) may also get introduced if the removed oxygen atom occupies the interstitial site in the lattice. So, on exposure to different gases, the oxygen vacancies are removed to a certain extent, reducing the oxygen vacancy defect states near the valence band. The decrease in oxygen vacancy defect states takes place maximum in air, enhancing the bandgap. Similar, oxygen vacancy-induced bandgap narrowing has been previously observed in ZnO or TiO_2 [47,48].

As it is clear from the earlier discussions, the defects in materials can be created by different synthesis techniques, and there are various defects. Among all the studied defects, oxygen vacancy defects created due to aliovalent cationic substitution and defects created due to doping of relatively bigger/smaller ionic radii atoms have a subsequent effect on the optical properties of the materials. In photoluminescence spectra of rare-earth ion-doped materials, excitation of the material is done by UV or IR radiation depending upon the presence of excitation levels [49]. Corresponding emissions are achieved in the visible region with the proper selection of activator ions. The energy provided for the excitation of the atoms is related to the bandgap of the material, which is the energy essential to excite the electrons from the top of the VB to the CB. The defects created due to doping bring changes in crystal structure sometimes, or it creates energy levels between the conduction band and valence

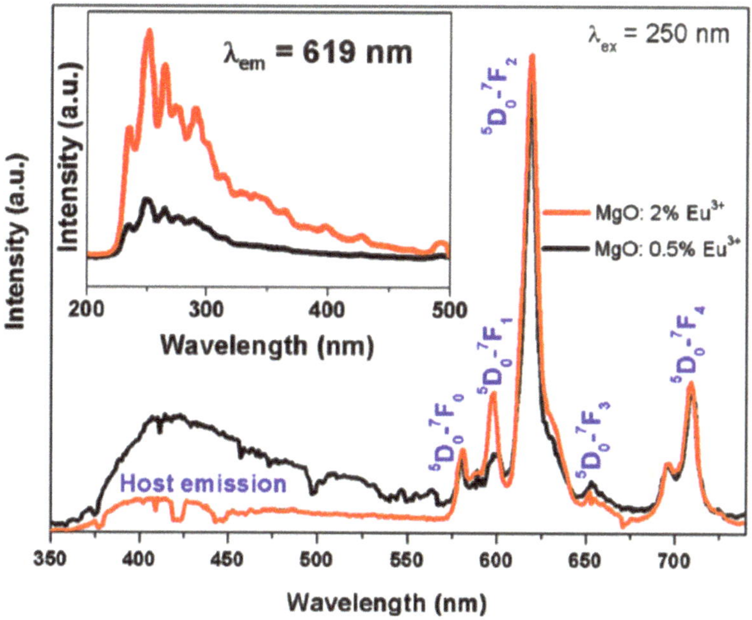

Figure 4.1.22 Emission spectra of MgO: Eu^{3+} phosphor [45].

band. Sometimes, it stimulates the emission process by subsequent energy transfer from the defect site to the activator center. Oxygen vacancies (V_O) are crucial in determining the physical properties of metal oxides and in demonstrating significant building blocks in many scientific and technological fields because of their unique properties.

Kadri et al. [50] have reported photoluminescence of MgO doped with Eu^{3+} ions and formation of defects due to aliovalent doping. Since there is a mismatch in ionic radius and valency of Eu^{3+} and Mg^{2+} ions, cationic defects are created during the doping process due to the replacement of Eu^{3+} ions into the Mg^{2+} sites. The following equation gives the charge balance mechanism:

$$2Eu^{\bullet\bullet} + 3Mg_{Mg}^{\bullet\bullet} \leftrightarrow 2Eu_{Mg}^{\bullet}Mg + V_{Mg}''$$

Cationic defects can affect the quantum yield of lanthanide contained phosphors via providing nonradiative transitions [51]. Fig. 4.1.22 depicts the emission spectra of MgO doped with various concentrations of Eu^{3+} ions. To remove these types of defects and to increase the quantum yield, a charge compensator is required to be doped. The codoping in MgO: Eu^{3+} phosphor with Li^{+} ions into the Mg^{2+} sites created $2Li_{Mg}{}'$ and gave rise to enormous concentrations of oxygen vacancies $V_O^{\bullet}$. The possible equation is given as follows.

$$Li^{\bullet} + 2Mg_{Mg}^{\bullet\bullet} \leftrightarrow 2Li_{Mg}' + O_o^x + V_O^{\bullet\bullet}$$

Here, oxygen vacancies act as sensitizers in transferring the energy from MgO to Eu^{3+} ion, increasing luminescence efficacy, as shown in Fig. 4.1.23.

Aliovalent doping creates defects due to charge imbalance in the crystal. Li^{+} ion has proven to be an efficient charge compensator in several oxide materials to compensate for such defects. ZnO is a host that when doped with rare-earth ions such as Eu^{3+} or Tb^{3+} creates a charge imbalance due to ionic valency mismatch between Zn^{2+} ions and RE^{3+} ions. On codoping with Li^{+} ions, a sufficient increase in luminescence characteristics has been reported. Fig. 4.1.24 shows the variation of PL emission intensity with the incorporation of Li^{+} ions in the ZnO host.

The part of Li^{+} ion in the photoluminescence can be clarified as follows [52]: The ionic size of a Li^{+} ion (0.6 Å) is fairly lesser than that of a Zn^{2+} ion (0.74 Å), which permits Li^{+} ions to be incorporated in ZnO without affecting the Zn^{2+} sites to form lattice defects. This improves the transition probability of rare-earth ions and their emission intensity. Nonetheless, only little Li^{+} ions contribute to the charge compensation. As Li^{+} ions' concentration enhances, it starts to substitute Zn^{2+} ions and subsequently generates negative charges via increasing oxygen vacancies for charge neutralization in ZnO. In addition, the formed oxygen vacancies extinguish the crystallinity of the samples, which is responsible for the emission quenching.

CeO_2 is another oxide that when doped with lanthanide ions, shows enhancement in its optical properties due to defect creation [53]. Ceria nanoparticles have high oxygen storage abilities because of the diffusion and reactivity of oxygen vacancies. The

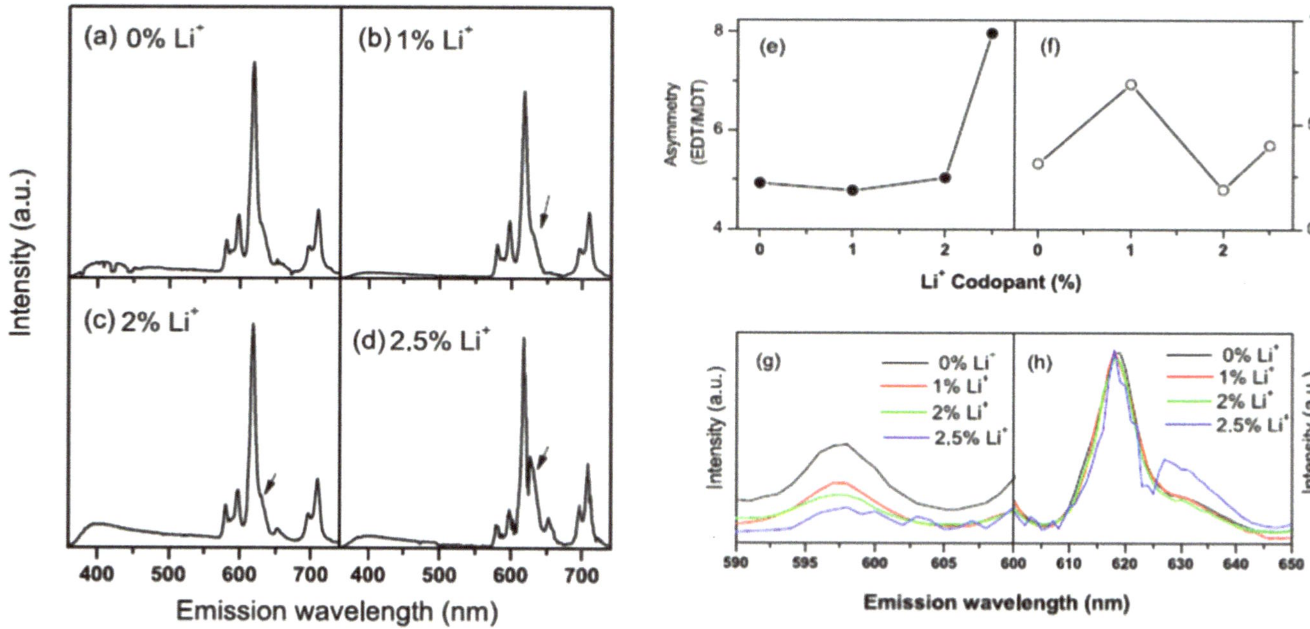

Figure 4.1.23 (A–D) 2% Eu^{3+}-doped MgO nanocrystals emission spectra codoped with different concentrations of Li^{+} ions. The asymmetry ratio variation of and (F) Eu^{3+} to the host emission ratio with Li^{+} codoping. Emission spectra by selective zoning the spectral range for (G) MDT ($^5D_0 \rightarrow {}^7F_1$) and (H) EDT ($^5D_0 \rightarrow {}^7F_2$) [45].

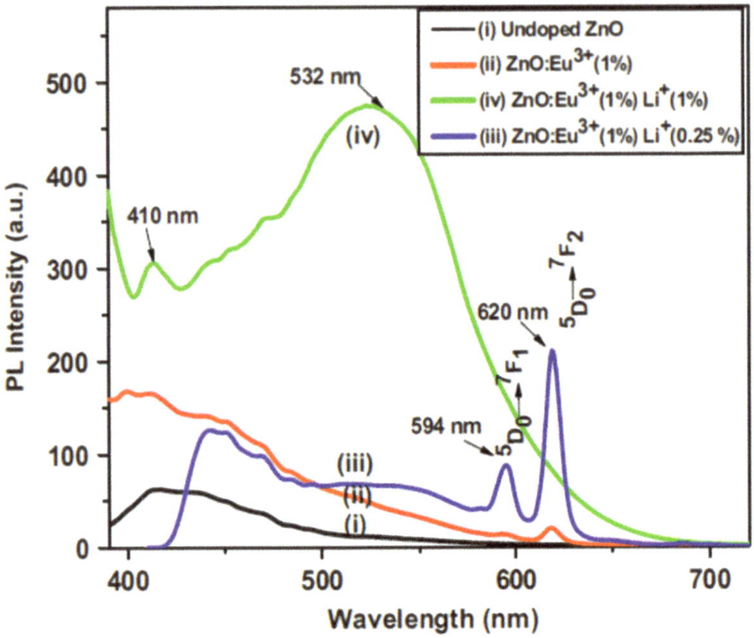

Figure 4.1.24 Room temperature PL emission spectra at excitation wavelength 370 nm of (i) undoped (ii) ZnO: Eu^{3+} (1%) (iii) ZnO: Eu^{3+} (1%), Li^{+} (1%) (iv) ZnO: Eu^{3+} (1%), Li^{+} (0.25%) [46].

higher the concentration of oxygen vacancies in ceria leads to greater conversion of Ce^{4+} states to Ce^{3+}. This conversion of Ce^{4+} states to Ce^{3+} is nonradiative. However, when the electrons from Ce^{3+} excited states relax to the ground state, the transition is radiative, leading to fluorescence signals centered around 520 nm. This emission is related to the oxygen vacancies induced conversion of Ce^{3+} to Ce^{4+}. When ceria is doped with trivalent lanthanide ions such as Sm^{3+} and Nd^{3+}, which have comparatively fewer activation energies between the dopant ions and O-vacancies, it enhances the concentration of oxygen vacancies. The doping of ceria nanoparticles with Sm and Nd enhances the sensitivity of the fluorescence signal from the Ln^{3+} doped ceria nanoparticles (LDC-NPs). The doping of Sm^{3+}, Nd^{3+} generates more oxygen vacancies in CeO_2 nanomaterial. Both the doping is more sensitive to the presence of vacancies than the undoped one. Fig. 4.1.25 shows the explanation with the schematic diagram.

La^{3+} ions do not contain any 4f electrons as well as 4f excited energy levels. In NiO: La^{3+}, a high concentration of La^{3+} ions are responsible for quenching the near band edge violet and blue emission, with enhanced green emission [46]. NiO: La^{3+} containing oxygen vacancies as confirmed via the XPS analysis. The green-emitting band at 543 nm (2.28 eV) can be ascribed to a trapped hole in the doubly ionized oxygen vacancy (V_o^{2+}) recombine with the trapped electron at a shallow level positioned below the CB. Remarkably, the green emission related to oxygen vacancies is slightly lower than the observed value at 2.48 eV related to Ni vacancies. An

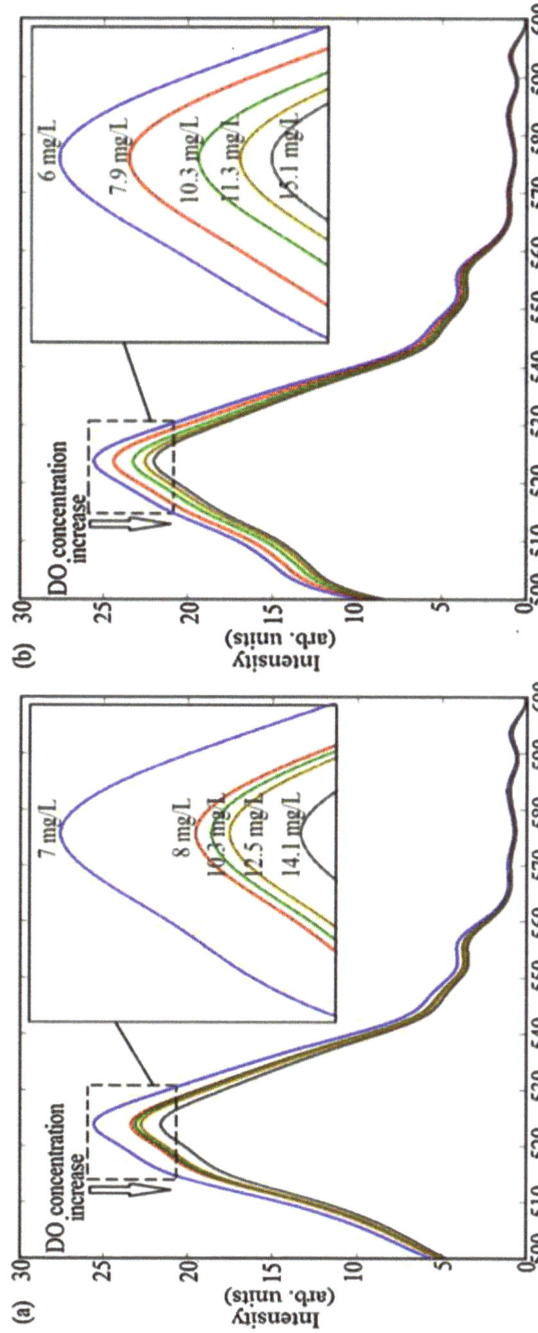

Figure 4.1.25 Fluorescence spectra with the variation of oxygen concentration in (A) Sm-doped and (B) Nd-doped CeO_2 samples [47].

improved green emission with some blue shift has been observed with the increment in La^{3+} content due to the band gap's widening. The increase in emission property is because of the elevation in the amount of doubly ionized oxygen vacancies (V_o^{2+}), making La^{3+}-doped NiO promising for fabricating light-emitting diodes.

It has been observed that in phosphor materials doped with lanthanide ions and codoped with alkali, alkaline ions, oxygen vacancy plays an important role. Several reports support that anionic vacancy contributes constructively to enhancing the luminescence characteristics of bulk and nanomaterials. One way to create it is by creating charge imbalance in the crystal with the incorporation of ions having greater valency than that of the substituted ion in the host. Another way is to increase the oxygen vacancy or antisites defect by adding a charge compensator such as an alkali ion. Das et al. [54] have reported similar work in $CaWO_4$ doped with Nd^{3+} ions and further compensating the charge imbalance in the crystal by codoping with Li^+ ions. Calcium tungstate doped with Nd^{3+} ions showed emission bands both in visible and NIR regions of the electromagnetic spectrum. On codoping with Li^+ ion, several-fold enhancements in emission level have been observed, which were generated from $^4F_{3/2}$ to $^4I_{11/2}$ (NIR region) level. Elevation of the NIR emission upon Li^+ doping might improve crystallinity and boosted asymmetry. Lithium is partially acted as the sensitizer via absorbing the host blue and green emissions and transferring them to Nd^{3+} ions. Eventually, the codoping of Li^+ ions in the Ca^{2+} sites lead to antisites defect $2Li'_{Ca}$ along with a considerable amount of oxygen vacancies $V_o^{\cdot}$ via as follows:

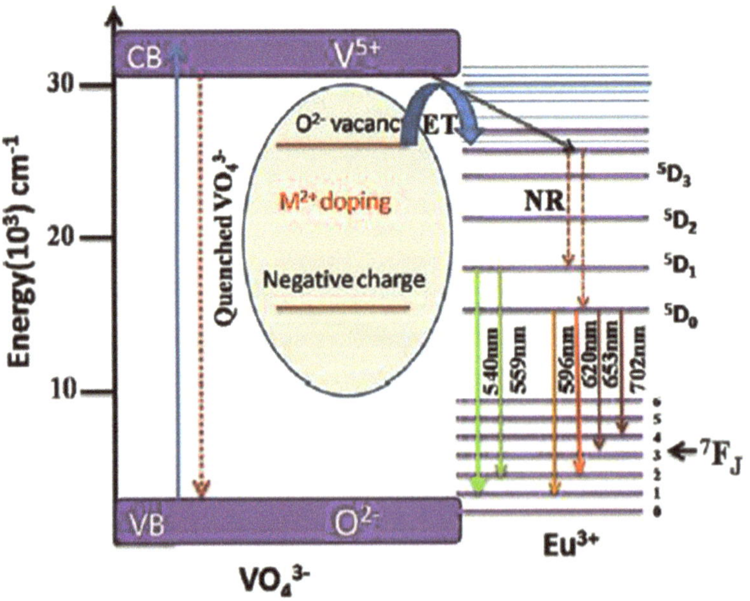

Figure 4.1.26 Schematic diagram showing energy transfer from oxygen vacancy to Eu^{3+} ion in YVO_4: Eu^{3+}, M^{2+} (M=Ca, Ba, Sr) [49].

$$Li^{\bullet} + 2Ca_{Ca}^{\bullet\bullet} \leftrightarrow 2Li'_{Ca} + O_o^x + V_o^{\bullet\bullet}$$

These oxygen vacancies might also have acted as the sensitizers via generating efficient radiative energy transfer to Nd^{3+} ions due to the strong mixing of CTB ensuing in improved luminescence. In our previous work, we have reported a similar report for yttrium vanadate doped with trivalent rare-earth ions and codoped with divalent alkaline earth ions [55]. YVO_4 is proven to be an excellent host when doped with any trivalent rare-earth ions such as Eu^{3+}, Dy^{3+}, and Tb^{3+} due to the similar ionic radius of yttrium and chemical property to the dopant. However, the concentration quenching mechanism can have only a low dilution capacity of dopant ions. To improve the photoluminescence property of YVO_4: Eu^{3+}, further codoping of M^{2+} (M = Ca, Ba, Sr) has been done, which creates defects (oxygen vacancies) in the phosphor and leads to a sufficient increase in the luminous yield. Fig. 4.1.26 shows the schematic diagram representing the role of these vacancies in energy transfer to the activator ions, increasing the intensity of YVO_4 doped with Eu^{3+} ions.

Due to the incorporation of M^{2+} ions in YVO_4:Eu^{3+}, an improvisation in energy transfer from the excited state of O^{2-} ion to Eu^{3+} ion appeared in the excitation mechanism. Meanwhile, the ratio of charge to the radius of M^{2+} is also crucial in luminescence elevation. This ratio of Ca^{2+} (1.1A), Sr^{2+} (1.26A), and Ba^{2+} (1.35A) ions are lower than that of Y^{3+} ions. When such alkaline ions are introduced in the Y^{3+} sites, the close-fitting binding between O^{2-} and Y^{3+} is predictable to be released, improving charge transfer excitation band and emission intensity [56]. Another conceivable reason for the improvement in PL intensity is the formation of oxygen vacancies because of the charge inequality in the crystal. The replacement of Y^{3+} ions through Ca^{2+}/Ba^{2+}/Sr^{2+} results in a negative charge Ca'_Y/Ba'_Y/Sr'_Y in the lattice in accordance of the Kröger-Vink defect notation and some deviations in lattice parameters that result in an improvement in PL intensity. The oxygen vacancies, $V_o^{\bullet\bullet}$, can compensate for these negative charges. It is recommended that in YVO_4:Eu^{3+}/M^{2+}, one oxygen vacancy can be created for every two Y^{3+} replaced by 2 M^{2+} for the charge compensation. Therefore, when M^{2+} ions are codoped to YVO_4: Eu^{3+}, the VO_4^{3-} to Eu^{3+} energy transfer and the energy transfer from oxygen vacancies to Eu^{3+} start to occur and the emission is enhanced subsequently.

5. Defect-related phosphorescence generation due to rare-earth doping

Phosphorescence is a process in which emission is received for a longer duration or after removing an excitation source. In this type of luminescence, the transition from triplet to singlet level takes place. Phosphorescence is achieved by deep and shallow trap creation by suitable doping of lanthanide ions. Phosphate, aluminate, silicate, and gallate are a few host materials that show persistent luminescence upon doping rare-earth ions. $SrAl_2O_4$: Eu^{2+}, Dy^{3+} is the most studied long-lasting phosphor. However, other compositions of this perovskite material have also emerged as

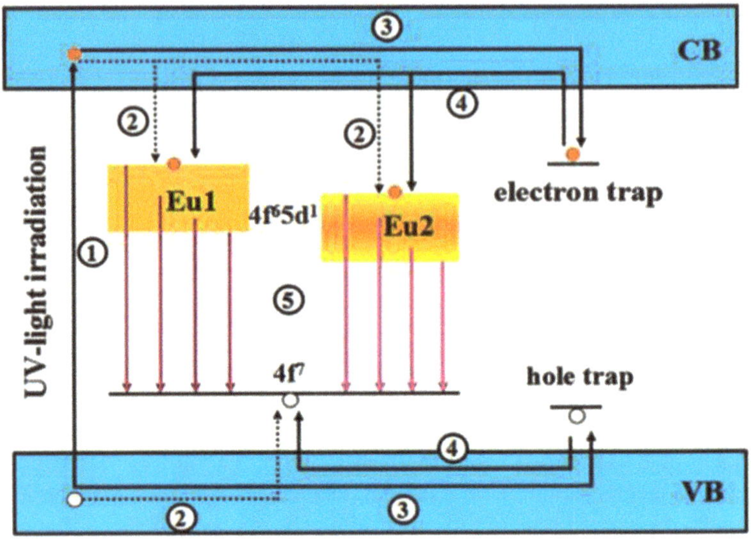

Figure 4.1.27 The schematic diagram of phosphorescence mechanism for $Ca_6BaP_4O_{17}$ doped with Eu^{2+} and Ho^{3+} ions [53].

efficient hosts and shown phosphorescence [57]. The defects generated on this material on excitation by a suitable source are responsible for the delayed luminescence. On cutting off the excitation light, $SrAl_4O_7$ doped with Eu^{2+}, Dy^{3+}, and Ce^{3+} ions show very bright green afterglow [58]. This indicates many trap formations due to the introduction of Dy^{3+} and Ce^{3+} into $SrAl_4O_7$: Eu^{3+} (SAE) phosphors. The LLP decay analysis shows a slow decaying process, which confirms the increase in defect/trap levels after incorporating lanthanide ions (Dy^{3+}, Ce^{3+}). $Ca_6BaP_4O_{13}$ is also an excellent host that shows persistent luminescence on doping with Eu^{2+} and Ho^{3+} ions. Eu^{2+} ions, which occupy Ca_1 and Ca_2 sites, act as an emission center and Ho^{3+} ions, which occupy Ca sites, act as a trap center in the codoped samples. Therefore, this is the possible mechanism of the introduction of Ho^{3+} ions into the $Ca_6BaP_4O_{13}$ matrix given as follows [59].

$$2Ho^{3+} + 3Ca^{2+} \rightarrow 2Ho^{3+}_{Ca2+} + VCa^{2+''}$$

This creates two positive defects $Ho_{Ca}^{\cdot}$ and one negative defect VCa'' to get the charge compensation. A possible afterglow mechanism of Eu^{2+} and Ho^{3+} ions codoped $Ca_6BaP_4O_{17}$ is represented in Fig. 4.1.27. Eu^{2+} ions are the excellent activator for long persistent phosphors since their 5d electron state is usually near the host's CB, making the trapping of electrons possible. With UV irradiation, electrons in the VB are excited directly through the host CB (process 1). Some of them can shift freely in the CB and transfer to $4f^65d^1$ level of Eu^{2+} ions leaving behind some holes that can move in VB randomly. Hereafter, electron transition happens from the $4f^65d^1$

level to lower levels $4f^7$, and Eu^{2+} emission generates (process 2). After switching off the UV source, the residual electrons in the CB can be trapped by $Ho_{Ca}^{\cdot}$, whereas holes in the VB can be trapped by VCa″ instead of returning to the ground state (process 3). Because of thermal agitation, the carriers (electrons and holes) get unconfined from the trapping centers and transfer via the host to the Eu^{2+} ions after a series of nonradiative transitions (process 4) followed by the recombination and Eu^{2+} persistent emission generates (process 5). Here the trap levels formed because of the codoping contributes to generating persistent luminescence.

6. Conclusions

This chapter broadly presents the significant defects that can be introduced in a matrix during synthesis or because of doping. The influences of these defects on the photoluminescence behavior of oxide-based materials are discussed in detail. The main emphasis has been given to analysis and explaining the fundamental knowledge about their creation and the constructive or destructive effects on the photoluminescence properties of various oxide-based phosphors. This might be very useful for the researchers to get a direction for improvising the photoluminescence intensity via creating and tuning various defects.

References

[1] L.Z. Liu, et al., Optical identification of oxygen vacancy types in SnO_2 nanocrystals, Appl. Phys. Lett. 102 (2013) 031916.
[2] Q.G. Zeng, Z.J. Ding, Z.M. Zhang, Synthesis, structure and optical properties of Eu^{3+}/TiO_2 nanocrystals at room temperature, J. Lumin. 118 (2006) 301–307.
[3] W.F. Lim, K. Yew Cheong, Oxygen vacancy formation and annihilation in lanthanum cerium oxide as a metal reactive oxide on 4H-silicon carbide, Phys. Chem. Chem. Phys. 16 (2014) 7015.
[4] C. Zhang, J. Lin, Defect-related luminescent materials: synthesis, emission properties and applications, Chem. Soc. Rev. 41 (2012) 7938–7961.
[5] C. Feldmann, T. Justel, C.R. Ronda, P.J. Schmidt, Inorganic luminescent materials: 100 years of research and application, Adv. Funct. Mater. 13 (2003) 511–516.
[6] R.C. Ropp, Luminescence and the Solid State, vol. 12, Elsevier, Amsterdam, Netherlands, 1991, p. 283.
[7] T. Hayakawa, A. Hiramitsu, M. Nogami, White light emission from radical carbonyl-terminations in Al_2O_3–$SiO_2Al_2O_3$–SiO_2 porous glasses with high luminescence quantum efficiencies, Appl. Phys. Lett. 82 (2003) 2975–2977.
[8] E. Cordoncillo, F.J. Guaita, P. Escribano, C. Philippe, B. Viana, C. Sanchez, Blue emitting hybrid organic–inorganic materials, Opt. Mater. 18 (2001) 309–320.
[9] M.E. Gimon-Kinsel, K. Groothuis, K.J. Balkus Jr., Photoluminescent properties of MCM-41 molecular sieves, Microporous Mesoporous Mater. 20 (1998) 67–76.
[10] F. Wang, X. Liu, Comprehensive Nanoscience and Nanotechnology, second ed., 2011.

[11] Z. Wang, et al., Single-phased chromaticity-tunable phosphor of $Sr_4Al_{14}O_{25}$: $Eu^{2+/3+}$ co-doped with Tb^{3+} for white-light-emitting diodes, Mater. Res. Express 6 (2019) 115903.
[12] M.G. Brik, A.M. Srivastava, On the optical properties of the Mn^{4+} ion in solids, J. Lumin. 133 (2013) 69–72.
[13] K. Vanheusden, et al., Mechanisms behind green photoluminescence in ZnO phosphor powders, J. Appl. Phys. 79 (1996) 7983.
[14] R. Raji, et al., ZnO nanostructures with tunable visible luminescence: effects of kinetics of chemical reduction and annealing, J. Sci.-Adv. Mater. Dev. 2 (2017) 51–58.
[15] W. Fuhs, I. Ulber, G. Weiser, M.S. Bresler, O.B. Gusev, A.N. Kuznetsov, V. Kh Kudoyarova, E.I. Terukov, I.N. Yassievich, Excitation and temperature quenching of Er-induced luminescence in a-Si:H(Er), Phys. Rev. B 56 (1997) 9545–9551.
[16] J.A.D. Allan, J. Waldorf, B.T. Sullivan, L.M. Plante, Optical coatings deposited by reactive ion plating, Appl. Opt. 32 (1993) 5583–5592.
[17] T.G.I. Tsimperidis, H.H.P. Th Bekman, C.J.G.M. Langerak, Direct observation of the two-stage excitation mechanism of Er in Si, Phys. Rev. Lett. 81 (1998) 4748–4751.
[18] F.G.J. Palm, B. Zheng, J. Michel, L.C. Kimerling, Electroluminescence of erbium doped silicon, Phys. Rev. B 54 (1996) 17603–17615.
[19] P. Dang, et al., Mixing the valence control of Eu^{2+}/Eu^{3+} and energy transfer construction of Eu^{2+}/Mn^{2+} in the solid solution $(1-x)Ca_3(PO_4)_2-xCa_9Y(PO_4)_7$ for multichannel photoluminescence tuning, Inorg. Chem. Front. 6 (2019) 2837–2849.
[20] A. Kumar Kunti, S. Kumar Sharma, R. Janay Choudhary, H. Christoffel Swart, Structural and luminescence properties of laser assisted Eu^{3+} doped $BaZrO_3$ thin films, J. Alloys Compd. 801 (2019) 99–111.
[21] J.P. Arago, E. Cordoncillo, R.A.S. Ferreira, L.D. Carlos, P. Escribano, Synthesis, characterization and optical studies on lanthanide-doped CdS quantum dots: new insights on CdS/lanthanide energy transfer mechanisms, J. Mater. Chem. 21 (2011) 1162–1170.
[22] P.M. Tanet al., New insights on the energy transfer mechanisms of Eu-doped CdS quantum dots, Phys. Chem. Chem. Phys. 22 (2020) 6266–6274.
[23] L.X. Liu, Z.W. Ma, Y.Z. Xie, Y.R. Su, H.T. Zhao, M. Zhou, J.Y. Zhou, J. Li, E.Q. Xie, Photoluminescence of rare earth 3+ doped uniaxially aligned HfO_2 nanotubes prepared by sputtering with electro spun polyvinylpyrrolidone nanofibers as templates, J. Appl. Phys. 107 (2010) 024309.
[24] Y. Yu, et al., Enhanced emissions of Eu^{3+} by energy transfer from ZnO quantum dots embedded in SiO_2 glass, Nanotechnology 19 (2008) 055711.
[25] P. Kofstad, Defect chemistry in metal oxides, Phase Transitions 58 (1996) 75–93.
[26] T. Tanaka, A. Sumiya, H. Sawada, Y. Kondo, K. Takayanagi, Direct observation of interstitial titanium ions in TiO_2 substrate with gold nanoparticle, Surf. Sci. 619 (2014) 39–43.
[27] M.K. Nowotny, T. Bak, J. Nowotny, C.C. Sorrell, Titanium vacancies in nonstoichiometric TiO_2 single crystal, Phys. Status Solidi B 242 (2005) R88–R90.
[28] D. Li, K. Lee, B.Y. Wang, M. Osada, S. Crossley, H.R. Lee, Y. Cui, Y. Hikita, H.Y. Hwang, Superconductivity in an infinite-layer nickelate, Nature 572 (2019) 624.
[29] J. Garcia-Barriocanal, A. Rivera-Calzada, M. Varela, Z. Sefrioui, E. Iborra, C. Leon, S.J. Pennycook, J. Santamaria, Colossal ionic conductivity at interfaces of epitaxial ZrO_2: $Y_2O_3/SrTiO_3$ heterostructures, Science 321 (2008) 676.
[30] A. Grimaud, O. Diaz-Morales, B. Han, W.T. Hong, Y.-L. Lee, L. Giordano, K.A. Stoerzinger, M.T.M. Koper, Y. Shao-Horn, Activating lattice oxygen redox reactions in metal oxides to catalyse oxygen evolution, Nat. Chem. 9 (2017) 457.

[31] R.A. De Souza, Oxygen diffusion in $SrTiO_3$ and related perovskite oxides, Adv. Funct. Mater. 25 (2015) 6326.
[32] C. Baeumer, et al., Verification of redox-processes as switching and retention failure mechanisms in Nb:$SrTiO_3$/metal devices, Nanoscale 8 (2016) 13967–13975.
[33] D. Christensen, Y. Frenkel, Y. Xie, Z. Chen, Y. Hikita, A. Smith, Y. Chen, L. Klein, H. Hwang, N. Pryds, B. Kalisky, Strain-tunable magnetism at oxide domain walls, Nat. Phys. 15 (2019) 269.
[34] J. Chen, et al., Ab initio study of a charged vacancy in yttrium aluminium garnet, J. Phys. Condens. Matter 20 (2008) 325212.
[35] C. Linderalv, et al., Luminescence quenching via deep defect states: a recombination pathway via oxygen vacancies in Ce-doped YAG, Chem. Mater. 33 (2021) 1 73–80.
[36] D. Dash, N.R. Panda, D. Sahu, Photoluminescence and photocatalytic properties of europium doped ZnO nanoparticles, Appl. Surf. Sci. 494 (2019) 666–674.
[37] X.Y. Chen, G.K. Liu, The standard and anomalous crystal-field spectra of Eu^{3+}, J. Solid State Chem. 178 (2005) 419–428.
[38] H.F. Brito, et al., Defects and charge compensation in $CdSiO_3$: a DFT and synchrotron study, Phys. Procedia 44 (2013) 1–9.
[39] B. Cojocaru, et al., Nanoscale insights into doping behaviour, particle size and surface effects in trivalent metal doped SnO_2, Sci. Rep. 7 (2017) 9598.
[40] T.T. Le, Giang, et al., Effect of the Eu^{3+}-Sn^{4+} substitution on the structural and optical properties of SnO_2:Eu^{3+}, J. Lumin. 228 (2020) 117624.
[41] I.E. Kolesnikov, et al., Morphology and doping concentration effect on the luminescence properties of SnO_2:Eu^{3+} nanoparticles, J. Alloys Compd. 822 (2020) 153640.
[42] A. Chakraborty, et al., Identifying the correct host - guest combination to sensitize trivalent lanthanide (guest) luminescence: titanium dioxide nanoparticles as a model host system, J. Phys. Chem. C 120 (2016) 41 23870–23882.
[43] M. Naseem Siddique, P. Tripathi, Lattice defects formulated ferromagnetism in nonmagnetic La (III) ion doped NiO nanostructures: role of oxygen vacancy, J. Alloys Compd. 825 (2020) 154071.
[44] B. Choudhury, R. Verma, A. Choudhury, Oxygen defect assisted paramagnetic to ferromagnetic conversion in Fe doped TiO_2 nanoparticles, RSC Adv. 4 (2014) 29314.
[45] X. Zhu, Y. Pu, X. Pu, X. Xie, M. Liu, Q. Li, Correlation of oxygen defects variations to band gap changes in luminescence properties of $SrZnO_2$: 0.02 Eu^{3+} phosphors, Mater. Lett. 254 (2019) 21–23.
[46] P. Kumari, J. Manam, Structural, optical and special spectral changes of Dy^{3+} emissions in orthovanadates, RSC Adv. 5 (2015) 107575.
[47] V.R. Akshay, B. Arun, G. Mandal, G.R. Mutta, A. Chanda, M. Vasundhara, Observation of optical band-gap narrowing and enhanced magnetic moment in Co-doped sol–gel-derived anatase TiO_2 nanocrystals, J. Phys. Chem. C (2018), https://doi.org/10.1021/acs.jpcc.8b06646.
[48] P. Dash, A. Manna, N.C. Mishra, S. Varma, Synthesis and characterization of aligned ZnO nanorods for visible light photocatalysis,2019 Phys. E Low-dimens. Syst. Nanostruct., https://doi.org/10.1016/j.physe.2018.11.007.
[49] S. Balaji, A.R. Allu, M.K. Pandey, P. Kumari, S. Das, Down Converted Photoluminescence of Trivalent Rare-Earth Activated Glasses for Lighting Applications, Book: Luminescent Materials in Display and Biomedical Applications, CRC Press, 2020, pp. 73–104. Taylor and Francis, ISBN- 9780367112127.

[50] G.B. Kadri, D. Das, S.K. Gupta, K. Sudarshan, Influence of Li^{+} co-doping on the luminescence of $MgO:Eu^{3+}$ nanocrystals: probing asymmetry, energy transfer and defects, Solid State Sci. 105 (2020) 106286.
[51] S.K. Gupta, K. Sudarshan, P.S. Ghosh, A.P. Srivastava, S. Bevara, P.K. Pujaria, R.M. Kadam, Role of various defects in the photoluminescence characteristics of nanocrystalline $Nd_2Zr_2O_7$: an investigation through spectroscopic and DFT calculations, J. Mater. Chem. C 4 (2016) 4988–5000.
[52] P.P. Pal, J. Manam, Enhanced luminescence of $ZnO:RE^{3+}$ (RE¼ Eu, Tb) nanorods by Li^{+} doping and calculations of kinetic parameters, J. Lumin. 145 (2014) 340–350.
[53] N. Shehata, K. Meehan, I. Ashry, I. Kandas, Y. Xu, Lanthanide-doped ceria nanoparticles as fluorescence-quenching probes for dissolved oxygen, Sensor. Actuator. B Chem. 183 (2013) 179–186.
[54] D. Das, S.K. Gupta, C.S. Datrik, P. Nandi, K. Sudarshan, Role of alkali charge compensation in the luminescence of $CaWO_4:Nd^{3+}$ and $SrWO_4:Nd^{3+}$Scheelites, New J. Chem. (2020), https://doi.org/10.1039/d0nj00651c.
[55] P. Kumari, J. Manam, Enhanced red emission on co-doping of divalent ions (M^{2+} = Ca^{2+}, Sr^{2+}, Ba^{2+}) in $YVO_4:Eu^{3+}$ phosphor and spectroscopic analysis for its application in display devices, Spectrochim. Acta A Mol. Biomol. Spectrosc. 152 (2016) 109–118.
[56] S. Basu, P.S. Devi, N.R. Bandyopadhyay, Sintering and densification behaviour of pure and alkaline earth (Ba^{2+}, Sr^{2+} and Ca^{2+}) substituted $La_2Mo_2O_9$, J. Eur. Ceram. Soc. 33 (2013) 79.
[57] R.E. Rojas-Hernandez, F. Rubio-Marcos, M. Ángel Rodriguez, J. Francisco Fernandez, Long lasting phosphors: $SrAl_2O_4$: Eu, Dy as the most studied material, Renew. Sustain. Energy Rev. 81 (2018) 2759–2770.
[58] A.N. Yerpudea, S.J. Dhoble, Combustion synthesis of rare earth activated and co-activated $SrAl_4O_7$ green long-lasting phosphors, Optik (2016), https://doi.org/10.1016/j.ijleo.2016.01.040.
[59] H. Guo, W. Chen, W. Zeng, G. Li, Y. Wang, Y. Li, Y. Li, X. Ding, Structure and luminescent properties of a novel yellow super long lasting phosphate phosphor $Ca_6BaP_4O_{17}$: Eu^{2+}, Ho^{3+}, J. Mat. Chem. C (2015) https://doi.org/10.1039/c0xx00000x.

4.2 Defects in rare-earth-doped inorganic materials

Anju Singh [1] *and Raunak Kumar Tamrakar* [2]
[1]Central Institute of Plastics Engineering and Technology, Korba, Chhattisgarh, India;
[2]Department of Applied Physics, Bhilai Institute of Technology (Seth Balkrishan Memorial), Durg, Chhattisgarh, India

1. Introduction

Rare-earth-doped inorganic materials are well known for applications such as solid-state lighting, screens, lasers, scintillators, afterglow materials, or sensors [1–4]. Advances in these fields required the development of suited materials and understanding the relations between chemical, structural and optical properties. A well-known example is the rare-earth-doped compounds, which are nowadays widely used in the field of white-light-emitting diodes [5,6]. Initially, Schnick and co-workers [7–11] discovered the chemically and thermally stable Eu^{2+}doped nitridosilicates and oxonitridosilicates. This discovery then led to great improvements in the multiphosphor approach [12] for generating white light. And despite the progress already made, further efforts are made to find new narrow-band-emitting red phosphors [13], but also further phosphors for special applications such as horticultural lighting [14].

Many such kinds of developments are possible due to the contributions of inorganic and materials chemists and their close collaborations with researchers in optical spectroscopy. In another field of applications for rare-earth-doped inorganic materials are quantum information processing, it is also expected that progress in the development of tailormade materials will lead to great improvements. In the beginning of the technology, optical studies were often limited to the available single-crystalline materials and commercial optical fibers. With the rapidly increasing need for quantum technologies, for example for secure communication, important progress on measurement techniques was made. Yet, the development of these techniques critically depends on the type and quality of the available materials. As we shall explain later, prerequisites are for example a combination of long optical and spin coherences times, which require the control of spectral diffusion or the possibility to control inhomogeneous and homogeneous line broadening. Thus, there is a need to develop materials beyond single crystals and to understand and control spectral diffusion in these materials to overcome the significant limitations and the lack of suited materials.

Luminescent materials (or phosphors) mostly are solid inorganic materials consisting of a host lattice, usually intentionally doped (Fig. 4.2.1). The absorption of energy takes place via either the host lattice or on impurities. In addition, the transfer of energy through the lattice can take place. In almost all cases, the emission takes place on

Figure 4.2.1 Luminescent material, excited by UV photons or electrons, containing activator ions and sensitizing ions (on which UV absorption takes place).

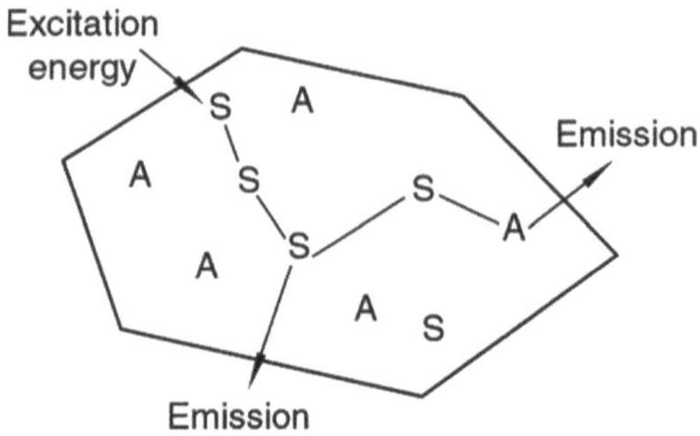

intentionally doped impurities, like rare-earth ions, which are present in relatively low concentrations (a few mole percent or less). Luminescent materials are applied widely. Major applications are in emissive displays, fluorescent lamps and LEDs and systems to detect X-rays or g-rays, for example, used in medical imaging. In the latter kind of applications, luminescent materials are excited by photons with high energy, and part of the excitation energy is used to create visible light. In the last decades, a large number of luminescent materials based on rare-earth ions or rare-earth-based host lattices have been invented. Quite a few of these materials have found their way into applications. In many cases, the application of rare-earth phosphors dramatically improved the performance of the devices in which they are applied.

The field of luminescence from rare-earth ions has been one of steady growth during the past decade, principally due to the ever-increasing demand for optical sources and amplifiers operating at wavelengths compatible with fiber communications technology. In particular, there has been enormous demand for sources that can be readily integrated with existing fiber and/or silicon microelectronics technology, and in this context rare-earth doping of silica and silicon have been areas of particular importance. However, the technological interest in rare-earth luminescence reaches beyond telecommunications: displays, laser materials, data storage, radiation detection, and medical applications are all areas in which luminescent rare earths are playing an increasingly important role. A recent article also demonstrates that not only is rare-earth luminescence in itself technologically important, but it can also be used as a valuable diagnostic tool to probe the physical and optical properties of a range of optoelectronic materials [15].

Lanthanides or rare-earth-element (REE) ions have an essential role in technological applications. Since their discovery, started at the end of the XVII century, they attracted the attention of chemists and physicists because of their unique electronic properties. To date, their use ranges from catalysis to the production of glasses, magnets, and emitting devices. By the special electronic structure, the lanthanides ions show radiative transition spanning from the UV to infrared region and, in recent years, great efforts have been made to explore the optical properties of lanthanides in transparent matrices and the effect of crystal field and defects on these transitions.

The emission of rare earth ions is generally due to optical transitions within the f-manifold. In REE the 4f electrons are well shielded from the chemical surrounding environment, thus allowing the REE transitions to almost retain their atomic character. As a consequence, the f–f emission spectra typically consist of sharp lines. These radiative recombinations are generally slow (from microsecond to millisecond decay time) due to the recombination selection rules (partially or totally spin forbidden). On the other hand, in some important cases, such as Eu and Ce, when the recombination is related to 5d–4f optical transitions the emission bands are relatively broad, such as, for example, the transition to 4f7 level in Eu^{2+} and the one to 4f1 level in Ce^{3+} In addition, since the d–f optical transitions are allowed, the recombination is much faster (in the nanosecond regime in case of Ce^{3+}).

These unique electronic properties made REE ions the leading choice for lightning devices. Solid-state lighting is the illumination standard of the future, and REE finds large applications both in light-emitting diodes (LED) or compact fluorescent lamps. Indeed LEDs performances guarantee brighter light than their incandescent counterparts, either cold or warm white light, and have a considerably longer lifespan, thus promoting them as "greener" more environmentally friendly devices. The unique electronic configuration of REE allows to retain their spectral properties irrespective of the host matrix. Indeed, contrary to some active transition metal ions like Ti and Cr, which may exhibit broad absorption and emission bands and, consequently, wide wavelength tunability albeit low gain and short upper-level lifetimes, REE emissions are weakly tunable because of shielding effects. One of the main advantages of REE compounds is the high optical gain of the phosphors, an essential feature for lighting applications.

An interesting effect due to the interaction of REE with the host matrix and with other dopant ions is not directly considered in the already mentioned interactions is related to the role of the lanthanides at the surface of the material. This is the common situation of rare-earth-doped nanoparticles, where the specific surface area is relatively high, and the interaction with the surface defects should be taken into account. Besides nano-dimensional compounds, the REE-defect interaction is also observed when the RE ion is not able to enter in the crystalline network of the matrix during the doping process. Generally, this situation happens when the radius of the cation of the matrix is much lower compared with the dopant REE and the crystalline structure is highly stable [16–20]. The interaction of lanthanide ions with defect states generally produces nonradiative depletion processes from the excited states; in these cases, beyond the decrease of the external quantum efficiency, an abrupt decrease of the time decay constant is observed. The gathered time trend is related to the radiative and nonradiative lifetime contribution.

For technological applications such as laser materials, scintillators, or phosphors, great attention has been devoted to the crystalline imperfections and to the consequent trapping and detrapping mechanism of the excited electrical charges during the electromagnetic radiation-solid state matrix interaction, to improve optical emission efficiency. Trapping sites are represented by defects in the crystalline structure such as vacancies and impurities. On the one hand, the presence of defects may be deleterious for devices' performances, determining nonradiative sources of energy losses. On the other hand, the presence of defects can be exploited in specific applications, such as for imaging and dosimetry [21,22].

To deep understand the role of specific impurities or vacancy defects, the energy level scheme of a selected ion within a certain host matrix must be known in detail. It is worth to note that the same species of Ln ions, once incorporated in a specific matrix can play as a phosphor while in a different matrix it acts as a trapping site, losing its emitting features. When it works as the optical emitting center, the Ln excited and the ground levels are placed within the energy bandgap of the matrix making the rare earth a good recombination center of the electron-hole pairs generated in the excitation process. On the other hand, when the excited states (5d levels) relapse into the conduction band of the host matrix, the Ln site is no longer suitable to capture the excited electron and the optical recombination (5d → 4f) does not take place, thus quenching the optical response. The last situation is typical when the ground state of the 2^+ ion falls into the bandgap of the matrix: an excited electron may be trapped in that site, decreasing the optical quantum yield of the overall optical process.

Stronger intensity, longer time, and lower cost with flexible wavelengths in persistent luminescence are increasingly needed as demands for biochemical and materials engineering rapidly expand. Persistent luminescence, which is based on the long decay-time phosphorescence concept [23–25], is challenged to downscale to smaller nanosized particle synthesis with lower cost and longer time for in vivo imaging. The rare-earth metal ions assisted phosphor luminescence technique has aroused tremendous interest in biological, chemical, and physical applications [26–32], which play a leading role in modulating luminescence properties. To modulate the emission wavelength, doping is a popular technology. RE metal ions as dopants have fine-structure levels of f-orbitals (4f for lanthanides and 5f actinides) to emit the specific wavelength light we need. As illustrated in the right panel of Fig. 4.2.2, the RE dopant contributes 4f and 5d levels for electron–hole recombination, whose emission state is just below the conduction band. The second type of RE ion dopant has deeper trap levels than the intrinsic defect levels of host materials, which prolongs the duration of the luminescence and extends the electron relaxation time from deeper levels.

2. Brief description of defects

Defects play an important part in both the chemical and physical behavior of solids and much of modern science and technology centers upon the exploitation or suppression of the properties that defects confer upon a material. The simplest notion of a defect is a mistake at an atom site in a crystal and these are called *point defects*. Not long after the recognition of point defects, the concept of *linear defects*, dislocations, was invoked to explain the mechanical properties of metals. In later years, it became apparent that *planar defects*, which include grain boundaries and domain walls in ferromagnetic and ferroelectric crystals, and *volume defects*, such as precipitates, also have important roles to play in influencing the physical and chemical properties of the host matrix.

The chemical and physical aspects of a defect are intimately connected. Thus, defects such as dislocations, which have a profound effect on the mechanical properties of a solid, also influence chemical properties such as rates of dissolution and reaction.

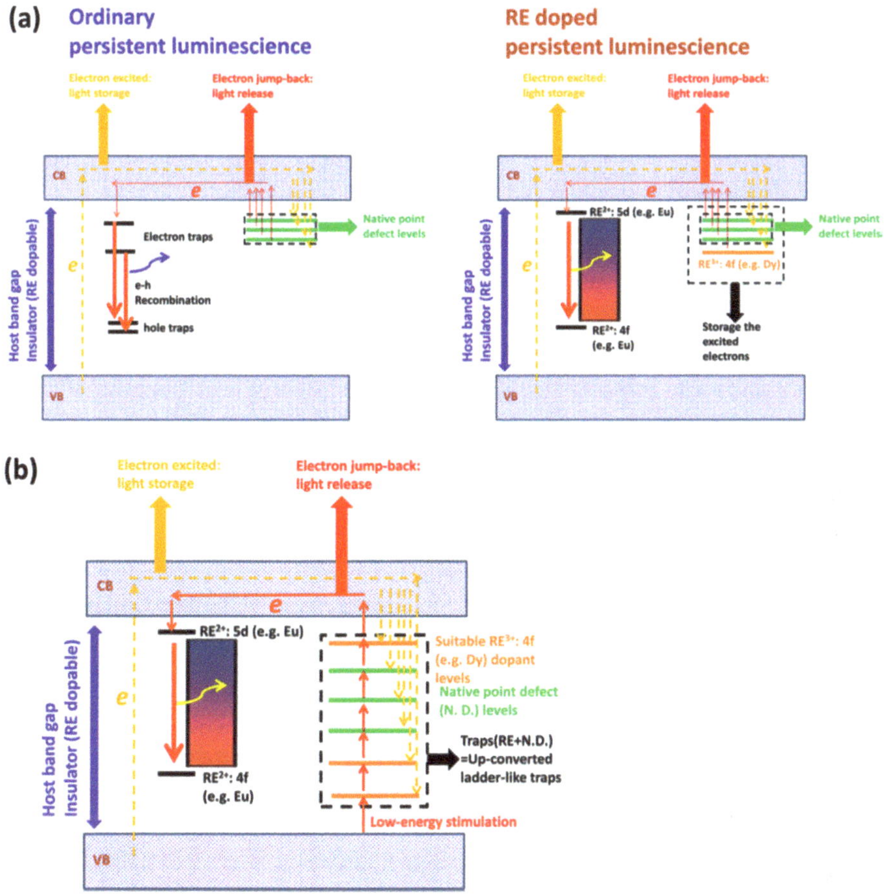

Figure 4.2.2 (A) Schematic processes of ordinary (undoped) and RE-doped insulators for persistent luminescence (both light storage and release) related to the dopant levels and native point defect levels. (B) Schematic process of upconverted persistent luminescence with the combined energy trap levels from the native point defects and suitably chosen RE dopants.

Similarly, defects that can be considered as more chemical in nature, such as impurity atoms, have profound effects on the physical properties of materials, as witnessed by doped semiconductors or solid-state lasers.

Besides the multiplicity of defects that can be envisaged, there is a wide range of host solid phases within which such defects can reside. The differences among an alloy, a metallic sulfide, a crystalline fluoride, and a silicate glass are significant from both chemical and physical viewpoints. By default, defects have come to be associated with crystalline solids. This is because a crystal has a regular repetition of atoms throughout its volume. A disturbance of this regularity then constitutes the defect. In this sense, the nature of a defect in, for example, a glass, is difficult to picture.

2.1 Point defects

Point defects are, in principle, isolated mistakes at single atom positions in a crystal. In a pure monatomic crystal, these can consist of an atom missing from a normally occupied position, a *vacancy*, or an atom in a position not normally occupied in the crystal, a *self-interstitial*. In a polyatomic crystal, the same defects can be found on each of the different atomic sublattices, together with some of the component atoms on incorrect sites, forming *antisite defects*. A population of some or all of these point effects will exist even in a completely pure crystal, as it lowers the Gibbs energy of the system, mainly owing to the entropy contribution of the defect populations. They are thus called *intrinsic defects*.

Point defects can also arise because of impurities, and these are called *extrinsic defects*, as, in principle, they can be removed by careful chemical or physical treatment. An impurity on a normally occupied site forms a *substitutional defect* and on a normally unoccupied position, an *interstitial defect*. In more complex structures, these unoccupied sites can exist in tunnels or between layers. The occupation of these sites is then called *interpolation* or *intercalation*.

Interactions between defects are of considerable importance, and the simple model of isolated point defects is often inadequate with respect to the structural, chemical, or physical properties of solids. Defects interrupt the regularity of the surrounding crystal structure and may give rise to electrical and mechanical stress fields, resulting in local electronic interactions and strain. When defect density is small, a population of isolated point defects may provide a sound structural model for the interpretation of properties. However, as the defect density increases, energetically unfavorable interactions can often be reduced by the association of defects, so that isolated point defects, as such, become replaced with clusters or extended defects. In many cases, the interaction energies are of the same order of magnitude as thermal energy. Thus, at low temperatures, defects tend to order. This ordering, particularly when promoted by long annealing times, can result in the formation of new structures in which the defects are eliminated. Intermediate stages of ordering are also common, in which clusters or aggregates of defects exist. In both of these cases, the ordering may be lost at high temperatures. In such cases, the low temperature ordered structure, which may be defect free, is transformed initially into one containing more extended defects before evolving into a population of noninteracting point defects. In the sections that follow, the temperature variation will not be repetitively stressed, but one should always be aware that defect structures investigated at one temperature may not occur in materials at another.

2.2 Defect nomenclature

To describe the chemical and physical consequences of point defect populations in a crystal, a notation is required analogous to that used in chemical equations. The most widely employed system is the Kröger–Vink notation [33] the principal features of which follow. *Vacancies* are indicated by the italic symbol *V*X, where the subscript designates the normal occupant of the site. (Note that the atom vanadium is written with nonitalic V). *Interstitial atoms* are denoted by a subscript i attached to the normal symbol for the element. *Impurity atoms* are given their normal chemical symbol and the site occupied is written as a subscript. *Associated defects*, that is, point defects that are grouped or clustered, are indicated by enclosing the components of such a group in parenthesis.

Charges on defects: The Kroger-Vink defect notation is given in Table 4.2.1. The Kröger–Vink notation bypasses the problem of deciding on the real charges on defects by considering only *effective charges* on defects (Table 4.2.1). The effective charge is the charge that the defect has with respect to that at the same position in a defect-free crystal structure. To distinguish effective charges from real charges, the superscript $'$ is used for each unit of negative charge and the superscript $^{\bullet}$ is used for each unit of positive charge. An effective charge relative to the host lattice is possible with any defect. These are added as superscripts to the appropriate defect symbol. When a defect bears no effective charge and the situation needs to be emphasized, the superscript x is used. Multiple effective charges can exist and are written using superscript n'. Multiple effective positive charges can exist and are written using superscript $^{n\bullet}$. The effective charges on an ionic defect can be considered to be linked to the defect by an imaginary bond. If the bond is weak, the effective charge can be liberated, say by thermal energy, so that it becomes free to move in an applied electric field and may contribute to the electronic conductivity of the material. Such mobile electrons in a crystal are denoted by the symbol e' and mobile holes by the symbol$^{h\bullet}$.

2.3 Defect formation

The chemical accounting system that equations represent can be modified to take into account defect populations using the following rules.

Table 4.2.1 Summary of the Kröger–Vink notation defect notation.

Defect	Notation
Nonmetal vacancy at nonmetal site	V_X
Metal vacancy at metal site	V_M
Interstitial metal	M_i
Interstitial nonmetal	X_i
Impurity metal (A) at metal site	A_M
Impurity nonmetal (Y) at nonmetal site	Y_X
Neutral vacancies	V_M^x, V_X^x
Nonmetal vacancy with effective positive charge	$V^{\bullet}_X$
Metal vacancy with effective negative charge	V'^{X}
Metal interstitial with effective positive charge	$M_i^{M\cdot}$
Nonmetal interstitial with effective negative charge	X'^{i}
Mobile electron[a]	e'
Mobile positive hole[a]	$h^{\bullet}$

[a]Concentrations of these defects are frequently designated by n and p, respectively.

1. The number of metal atom sites must always be in the correct proportion to the number of nonmetal atom sites in the crystal. As long as this proportion is maintained, the total number of sites can vary, as this simply corresponds to more or less substance present. If the crystal contains vacancies, these must also be counted in the total number of sites. Interstitial atoms do not occupy normal sites and so are not relevant when this rule is being applied.
2. The total number of atoms on both sides of the equation must be equal.
3. The crystal must always be electrically neutral. This means not only that the total charge on one side of the equation must be equal to the total charge on the other side, but also that the sum of the charges on each side of the equation must be equal to zero. In this assessment, both effective and real charges must be counted if both sorts are present.
4. Only neutral atoms are involved in chemical reactions. After the reaction, neutral atoms can dissociate into charged species if this is thought to represent the real situation in the crystal.

2.4 Point defects in stoichiometric crystals

2.4.1 Monatomic crystals

A pure monatomic crystal such as diamond can contain vacancies or self-interstitials. This defect population, in both cases, can be treated as a chemical equilibrium. Application of the law of mass action in its simplest form, assuming that there is no interaction between the defects, leads to expressions for the equilibrium number of defects present.

2.4.2 Polyatomic crystals

Populations of intrinsic point defects in other than monatomic materials must not alter the overall composition of the crystal and must maintain charge neutrality. Calculations similar to those for monatomic solids show that the free energy of crystals containing such balanced populations of point defects is lower than that for a defect-free crystal, and these defects occur in pure crystals under equilibrium conditions at all temperatures above 0 K.

2.4.3 Schottky defects

Balanced populations of vacancies on both the cation and anion sites in a crystal are known as *Schottky defects*. For example, if the crystal has a formula MX, then the number of cation vacancies must be equal to the number of anion vacancies, to maintain both composition and electrical neutrality. In such a crystal, a Schottky defect consists of one cation vacancy and one anion vacancy, although these need not be geometrically close together. In a crystal of formula MX_2, there need to be twice as many anion vacancies as cation vacancies. Similar considerations apply to compounds with other compositions. As with monatomic solids, this defect population can be treated as a chemical equilibrium.

2.4.4 Frenkel defects

A population of vacancies on one subset of atoms created by displacing some atoms into normally unoccupied interstitial sites constitutes an arrangement of paired point defects termed *Frenkel defects.* Because one species of atom or ion is simply being redistributed in the crystal, charge balance is not an issue. A Frenkel defect in a crystal of formula MX consists of one interstitial cation plus one cation vacancy, or one interstitial anion plus one anion vacancy. Equally, a Frenkel defect in a crystal of formula MX_2 can consist of one interstitial cation plus one cation vacancy, or one interstitial anion plus one anion vacancy. As with the other point defects, it is found that the free energy of a crystal is lowered by the presence of Frenkel defects and so a population of these intrinsic defects is to be expected at temperatures above 0 K. The calculation of the number of Frenkel defects in a crystal can proceed along lines parallel to those for Schottky defects.

2.4.5 Antisite defects

An antisite defect is an atom on a site normally occupied by a different chemical species that exists in the compound. For example, in a compound of formula AB, the antisite defects that can occur are an A atom on a site normally occupied by a B atom, or a B atom on a site normally occupied by an A atom. Antisite defects are not important in binary ionic compounds, as the resultant increase in electrostatic energy is generally prohibitive.

2.5 Electronic and ionic defects

The presence of defects in a solid can profoundly modify its properties. Moreover, the defect population can be changed by interaction with a surrounding gaseous atmosphere, making controlled manipulation of the properties possible. Accounting for these changes is important in applications such as fuel cells. The simplest way to assess the effects of such changes, particularly on electronic conductivity, is to assume that the point defects are noninteracting species, distributed at random in the crystal and that the defect energies are constant and not a function either of concentration or temperature. In this case, defect reaction equations can be used to deduce how defect populations and hence electronic conductivity or ionic diffusion may respond to changes in external variables.

2.6 Color centers

Point defects in a crystal often carry an effective charge. Electrons or holes can be trapped at such defects and if these are then able to interact with electromagnetic radiation so that the electronic ground state of the defect is excited to higher energy states, the optical properties may be modified to cause a colorless host crystal to become colored. Color centers that owe their properties to trapped electrons are generally known as electron-excess centers and those with trapped holes are called hole-excess centers, although electron-deficit centers are perhaps a more understandable descriptive. Note that transition metal and lanthanoid ion dopants that engender color in an otherwise colorless matrix are frequently called color centers. These are described elsewhere.

Color centers can be formed in many ways. Heating a crystal in an atmosphere that will force the introduction of point defects, electrolysis, or simple doping may suffice, but the principal way in which many crystals develop color centers, especially naturally occurring minerals, is through exposure to high-intensity ionizing radiation. Regardless of the method of formation, the color produced in any particular crystal by a specific center is constant, regardless of the method of formation.

2.6.1 Electron-excess centers

The best-known electron-excess center, the F-center, consists of an anion vacancy plus a trapped electron. These defects were originally characterized in the alkali halides. They arise in crystals that are slightly nonstoichiometric and contain a small excess of alkali metal. This imbalance creates an equal number of anion vacancies, and the outer s electron on each additional metal atom diffuses to a vacancy, which has an effective positive charge, where it is trapped. The color of the F-centers is indicative of the host structure and not the excess metal.

2.6.2 Hole-excess centers

The color of the minerals smoky quartz and amethyst is owing to hole-excess centers. Smoky quartz is a form of silica-containing aluminum as an impurity. The aluminum has only three outer bonding electrons instead of the four found on silicon. These three are used in fulfilling the bonding requirements of the structure, but one bond of the four is incomplete and lacks an electron, making the defect a hole-excess center. To preserve charge neutrality, equal amounts of H^+ are incorporated into the crystal.

2.7 Interpolation and intercalation

The difference between interstitial defects and interpolated or intercalated species is largely one of scale. Interstitial sites are thought of as normally unoccupied positions in relatively closely packed crystal structures while interpolation tends to refer to foreign atoms located in normally unoccupied cages or tunnels in a rather open crystal structure. Similarly, intercalation is mainly associated with the insertion of atoms or molecules between the layers of a crystal structure in such a way that the layers maintain their integrity.

2.8 Extended defects

Point defects can, for the sake of cataloging, be considered to be zero-dimensional. Point defect clusters, although somewhat more extensive, still occupy very small volumes in the crystal lattice. In this section, defects that can span the whole width of a crystal are described: linear (one dimensional) defects and planar (two-dimensional) faults. Close to these defects, coordination polyhedral and nearest neighbor geometry differ from those in the bulk. This may provide sites for unusual chemical reactions, introduce significant changes in physical properties, or cause a change in composition. In particular, planar boundaries, if ordered, give rise to a homologous series of phases, each with a definite composition and chemical formula.

2.8.1 Dislocations

Dislocations are defects that characterize a line of disruption in the crystal. This implies that the defect must either end on the surface of a crystal, on another dislocation or else form a closed loop. Dislocations strongly influence the mechanical properties of a solid and can have a negative influence on the electronic and optical properties of crystals, so much so that much effort is placed on the production of dislocation-free thin films and crystals for many applications. The disruption to the crystal introduced by a dislocation is characterized by the Burgers vector, **b**, and the width of a dislocation, w, which is equated to the distance from the dislocation line over which the atom displacements are greater than $b/4$, where b is the length of the dislocation's Burgers vector. Although there are many different types of dislocations, they can all be thought of as combinations of two fundamental types, *edge dislocations*, which have a Burgers vector perpendicular to the dislocation line, and *screw dislocations*, with a Burgers vector parallel to the dislocation line. Edge dislocations consist of an extra half plane of atoms inserted into the crystal. The dislocation can be conceptually formed by cutting halfway through a crystal and sliding the regions on each side of the cut parallel to the cut, to create spiraling atom planes. The dislocation line is the central axis of the spiral. The Burgers vector of most dislocations is neither perpendicular nor parallel to the dislocation line. Such dislocations have an intermediate character and are called *mixed dislocations*. In this case, the atom displacements in the region of the dislocation are a complicated combination of edge and screw components.

Dislocation loops can be formed by the aggregation of vacancies or interstitials on a plane in a crystal. Populations of vacancies or interstitials are often high in some crystals, especially at high temperatures. Slow cooling can allow these defects to order, not in three-dimensional structures, as described earlier, but as two-dimensional sheets. Planar aggregations of vacancies may be large enough to allow for lattice collapse to form a dislocation loop. In the case of interstitials, if the spread of the array is large enough, the surrounding crystal lattice distorts to form a dislocation loop.

A dislocation is surrounded by displaced atoms and the shifts give rise to a surrounding strain field. The strain field will depend on the type of dislocation, its width, and its Burgers vector. The strain field contributes to enhanced chemical reactivity, and reaction rates are increased near to dislocations. For this reason, dislocations that intersect the surface of a crystal can be revealed by placing the crystal in a solvent. The regions where the dislocations emerge dissolve preferentially to produce small pits called etch figures, on the surface. The disruption that occurs near to a dislocation line also facilitates atomic movement, and diffusion of atoms along a dislocation is normally much higher than diffusion through the bulk solid. The strain field of a dislocation interacts with point defects and can act to "attract" them into the dislocation core. If a sufficiently high concentration of point defects occurs in this way, then they can amalgamate to form a precipitate and the dislocation is said to be decorated. This technique was used to reveal dislocations before the widespread use of transmission electron microscopy and served as one of the first proofs that dislocations did indeed exist. In an ionic material such as sodium chloride, the dislocation can bear a charge, which will depend on the core dislocation structure. If the core has an excess of cations, it will be positively charged, whereas if it has an excess of anions, it will be negatively charged. Similarly, when the dislocation reaches the surface, a charge may be present, which may enhance chemical reactivity at this site.

References

[1] S. Schmiechen, P. Pust, P.J. Schmidt, W. Schnick, Nachr. Chem. 62 (2014) 847–851.
[2] G. Blasse, B.C. Grabmaier, In Luminescent Materials, Springer, Berlin, 1994.
[3] T. Justel, H. Nikol, C. Ronda, Angew. Chem. Int. Ed. 37 (1998) 3084–3103.
[4] H.A. Hoppe, Angew. Chem. Int. Ed. 48 (2009) 3572–3582.
[5] W. Schnick, Phys. Status Solidi RRL 3 (2009) A113–A114.
[6] P. Pust, P.J. Schmidt, W. Schnick, Nat. Mat. 14 (2015) 454–458.
[7] R. Mueller-Mach, G. Mueller, M.R. Krames, H.A. Hoppe, F. Stadler, W. Schnick, T. Juestel, P. Schmidt, Phys. Status Solidi 202 (2005) 1727–1732.
[8] T. Schlieper, W. Milius, W. Schnick, Z. Anorg. Allg. Chem. 621 (1995) 1380–1384.
[9] H. Huppertz, W. Schnick, Acta Crystallogr. C 53 (1997) 1751–1753.
[10] W. Schnick, H. Huppertz, R. Lauterbach, J. Mater. Chem. 9 (1999) 289–296.
[11] H.A. Hoppe, H. Lutz, P. Morys, W. Schnick, A. Seilmeier, J. Phys. Chem. Solid. 61 (2000) 2001–2006.
[12] M. Zeuner, S. Pagano, W. Schnick, Angew. Chem. Int. Ed. 50 (2011) 7754–7775.
[13] P. Pust, V. Weiler, C. Hecht, A. Tucks, A.S. Wochnik, A.-K. Hens, D. Wiechert, C. Scheu, P.J. Schmidt, W. Schnick, Nat. Mat. 13 (2014) 891–896.
[14] C. Poesl, W. Schnick, Chem. Mater. 29 (2017) 3778–3784.
[15] A. Polman, Physica B 300 (2001) 78.
[16] D.W. Cooke, K.J. McClellan, B.L. Bennett, J.M. Roper, M.T. Whittaker, R.E. Muenchausen, et al., J. Appl. Phys. 88 (2000) 7360–7362.
[17] P.C. Ricci, C.M. Carbonaro, A. Geddo Lehmann, F. Congiu, B. Puxeddu, G. Cappelletti, et al., J. Alloys Compd. 561 (2013) 109–113.
[18] J.H. Campbell, T.I. Suratwala, J. Non-Cryst. Solids 263–264 (2000) 318–341.
[19] P.R. Ehnnann, J.H. Campbell, J. Am. Ceram. Soc. 85 (2002) 1061–1069.
[20] A. Ishii, M. Hasegawa, Sci. Rep. 5 (2015) 11714.
[21] P.C. Ricci, C.M. Carbonaro, D. Chiriu, A. Anedda, J. Mater. Chem. 21 (2011) 18425.
[22] P.C. Ricci, A. Casu, M. Salis, E. Fortin, A. Anedda, Phys. Status Solidi A 207 (2010) 1751–1755.
[23] J. Hölsä, Electrochem. Soc. Interface 18 (2009) 42–45.
[24] T. Matsuzawa, Y. Aoki, N. Takeuchi, Y. Murayama, J. Electrochem. Soc. 143 (1996) 2670–2673.
[25] Z. Pan, Y.-Y. Lu, F. Liu, Nat. Mater. 11 (2012) 58–63.

[26] T. Maldiney, A. Bessiere, J. Seguin, E. Teston, S.K. Sharma, B. Viana, A.J.J. Bos, P. Dorenbos, M. Bessodes, D. Gourier, D. Scherman, C. Richard, Nat. Mater. 13 (2014) 418–426.
[27] M. Allix, S. Chenu, E. Véron, T. Poumeyrol, E.A. Kouadri- Boudjelthia, S. Alahraché, F. Porcher, D. Massiot, F. Fayon, Chem. Mater. 25 (2013) 1600–1606.
[28] A. Lecointre, A. Bessiere, A.J.J. Bos, P. Dorenbos, B. Viana, S. Jacquart, J. Phys. Chem. C 115 (2011) 4217–4227.
[29] D.C. Rodríguez Burbano, S.K. Sharma, P. Dorenbos, B. Viana, J.A. Capobianco, Adv. Opt. Mater. 3 (2015) 551–557.
[30] F. Liu, Y. Liang, Z. Pan, Phys. Rev. Lett. 113 (2014) 177401.
[31] X. Liu, C.-H. Yan, J.A. Capobianco, Chem. Soc. Rev. 44 (2015) 1299–1301.
[32] K. Van den Eeckhout, D. Poelman, P. Smet, Materials 6 (2013) 2789–2818.
[33] N.G. Connelly, T. Damhus, R.M. Hartshorn, A.T. Hutton, Nomenclature of Inorganic Chemistry, RSC Publishing, Cambridge, 2005. IUPAC Recommendations (Red Book).

Synthesis of rare-earth-activated phosphors

5

Ruby Priya [1], *Smriti Bhatia* [2], *Sandeep Kaur* [2], *Vikas Dubey* [3] *and O.P. Pandey* [2]
[1]Department of Physics, Chandigarh University, Mohali, Punjab, India; [2]Functional Material Lab, School of Physics and Materials Science, Thapar Institute of Engineering & Technology, Patiala, Punjab, India; [3]Department of Physics, Bhilai Institute of Technology Raipur, Naya Raipur, Chhattisgarh, India

1. Introduction

In the present era, rare-earth (RE)-doped phosphors hold a prestigious place in optoelectronic devices due to their tremendous properties. The RE-actived phosphors exhibit efficient luminescent properties due to the presence of f-f- and f-d transitions. The doping, codoping, or tridoping in the host material leads to desirable emission properties according to the application required. Moreover, emission efficiency also depends significantly on the synthesis route followed. There are many synthesis routes such as high-temperature solid-state method, wet chemical routes such as hydrothermal method, solvothermal methods, sol−gel method, green synthesis route, and melt quench method for glass phosphors are some examples of the synthesis routes, which are opted to develop these phosphors. The different engineering processes involved in the synthesis route determine the phase formation, crystallinity, and morphology of the end products. All these aforementioned factors affect the luminescent characteristics. So, the choice of synthesis route to synthesize these phosphors plays a vital role. In the following sections, synthesis of rare-earth activated phosphors via various routes has been discussed in detail.

2. Types of synthesis routes

2.1 *Solid-state method*

Solid-state synthesis method also popularly known as the Ceramic method, is a high-temperature synthesis technique used for the production of materials at a larger scale. This synthesis route is relatively inexpensive, has a short reaction time, and does not require complex apparatus. Moreover, the large yield of desired materials can be prepared in a relatively simple manner. The products are generally microcrystalline with pure phase. The oxides and carbonates are used as initial precursors and are mixed in mortar and pestle or via ball milling technique. The high synthesis temperature can be lowered by using fluxing agents such as NH_4Cl, NaOH, KOH, and boric acid.

Rare-Earth-Activated Phosphors. https://doi.org/10.1016/B978-0-323-89856-0.00001-8

These well-grounded materials are then subjected to high-temperature furnaces for calcination at various temperatures for the synthesis of phosphors. Table 5.1 shows the phosphors synthesized using the solid-state method.

2.2 Hydro(solvo)thermal method

Hydrothermal synthesis is a low-temperature synthesis route for the fabrication of nanomaterials with varying shapes and sizes. In this method, the hydrolysis and condensation of reactants take place in a stainless steel autoclave with a solvent. If the solvent is any other than water medium, then this method is known as the solvothermal method. The autoclave is placed in a furnace or an oven for varying temperatures for different holding durations. To vary the morphology, pH is varied and different types of solvents, surfactants, and templates are used. The reaction conditions and various outcomes of phosphors synthesized by hydro(solvo)thermal method are summarized in Table 5.2.

2.3 Combustion method

Combustion method is a facile low-temperature synthesis route. It results in highly crystalline, pure, and uniform end products. This synthesis route is quite economical and gives a high yield of nanoparticles. Various oxidizing and reducing agents also known as fuels are used during the synthesis. Metal nitrates are used as oxidizing agents, and urea glycine, citric acid, and hydrazine-based compounds are the most commonly used reducing agents. These fuels provide conditions for the highly exothermic reaction to take place. The stoichiometric amounts of fuel and oxidizing agents are calculated based on propellant chemistry. When the valency of reducing agents balances the valency of oxidizing agents and this ratio is 1 then the reaction is considered to be stoichiometrically balanced. The precursors and fuels are then added to an aqueous solution and are heated at room temperature (if needed). This aqueous solution is put in a preheated furnace where the liberation of gases such as N_2 and CO_2 takes place with a flame. The temperature of the flame is sufficiently high that it results in the formation of desired products. If there is a mixed phase or some impurity in the as-synthesized products, they are further calcined to obtain the pure phase nanoparticles. Table 5.3 shows the phosphors synthesized using the combustion method.

2.4 Sol–gel method

Sol–gel synthesis is a wet chemical route known for its yield of uniform size, high purity, and crystalline nanoparticles. This method uses a chemical solution referred to as sol to produce an integrated network called gel. This method uses a chelating or complexing agent to form gel via hydroxyl and carboxyl groups. Generally, citric acid is used as a chelating agent, so this method is also called the citrate-gel method. To maintain the pH and increase the polymerization, ethylene glycol is used as the medium. All the precursors are mixed with a chelating agent in an aqueous solution

Table 5.1 List of phosphors synthesized using solid-state synthesis.

S. no.	Host: dopant	Precursors	Fluxing agent	Calcination temp/time	Morphology	Ref.
1.	Cs_3CuI_5	CsI, CuI	–	350°C/6 h	Aggregated particles	[1]
2.	$CaTiO_3:Eu^{3+}/Li$	$CaCO_3$, TiO_2, Li_2CO_3, Eu_2O_3	–	900–1400°C/4 h	Spherical at 1000–1200°C and irregular at 1300°C	[2]
3.	$NaCaBO_3:Eu^{2+}Mn^{2+}$	$NaCO_3$, Eu_2O_3, $MnCO_3$	H_3BO_3	850°C/3 h	–	[3]
4.	$LiSrPO_4:Eu^{2+}/Mn^{2+}$, Eu^{2+}/Tb^{3+}, Eu^{2+}, Mn^{2+}, Tb^{3+}	$SrCO_3$, Eu_2O_3, $MnCO_3$, Tb_4O_7, LiH_2PO_4	–	1300°C/4 h	–	[4]
5.	$SrSnO_3:Tb^{3+}$	$SrCO_3$, Tb_2O_3	H_3BO_3	1350°C/2 h	Irregular grains	[5]
6.	$Sr_3B_2O_6:Ce^{3+}$ and Dy^{3+}	$SrCO_3$, CeO_2, Dy_2O_3	H_3BO_3	1100°C/3 h	–	[6]
7.	$GdAlO_3:Eu^{3+}$	$Gd(NO_3)_3$, $Al(NO_3)_3$, $Eu(NO_3)_3$	H_3BO_3	1000°C/2 h	Regular shape with several voids and pores	[7]
8.	$Y_3Al_{15}O_{12}$: Eu^{3+}	Y_2O_3, Eu_2O_3, Al_2O_3	–	1400°C/2 h	Spherical	[8]
9.	$Li_2SnO_3:Mn^{4+}$	Li_2CO_3, $MnCO_3$, SnO_2	–	1100°C/10 h	–	[9]
10.	$BaMoO_4:Sm^{3+}$	$BaCO_3$, Sm_2O_3, Li_2CO_3, $(NH_4)_6Mo_7O_{24}.4H_2O$	–	850°C/4 h	Relatively dispersed and uniform particles	[10]

Table 5.2 The reaction conditions and end products of phosphors synthesized via hydro(solvo)thermal method.

S. no.	Host: dopant	Precursor	Surfactant/solvent/ precipitating agent	Autoclave Temp/time	Morphology	Ref.
1.	$Gd_2O_3:Eu^{3+}$	$Gd(NO_3)_3$, $Eu(NO_3)_3$	NaOH	180°C/24 h	Smooth rods	[11]
2.	$ZnGa_2O_4$	$GaO(OH)$, $Ga_2(SO_4)_3 \cdot 16H_2O$, Ga_2O_3	NaOH	180°C/ 8–24 h	Cubic (pH = 8) Hexagonal (pH = 12)	[12]
3.	$BaTiF_6:Mn^{4+}$	HF, $KMnO_4$, $Ba(OH)_2$, H_2TiF_6	–	180°C/8 h	Rod like (10% HF), petal like (10% HF)	[13]
4.	$GdVO_4:Ln^{3+}$ (Ln = Eu, Sm, Dy)	Gd_2O_3, Eu_2O_3, Dy_2O_3, Sm_2O_3, $NaVO_3$	Ammonia	180°C/24 h	Rhombic (pH = 4.2) Spherical and cubic (pH = 4.2, 7.2) Nanorods (pH = 9.3)	[14]
5.	$GdBO_3:Eu^{3+}$	Gd_2O_3, Eu_2O_3, H_3BO_3	NH_4OH	200, 240°C/ 3–10 h	Spherical particles	[15]
6.	Nd_2O_3	–	NaOH	200°C/24 h	Rod-like	[16]
7.	$Ca_2V_2O_7:Eu^{3+}$	CaO, Eu_2O_3, NH_4VO_3	CTAB	170°C/ 1,2,3 days	Bulk, nanorod, and flake-like	[17]
8.	$K_2NaAlF_6:Mn^{4+}$	KF, NaF, AlF_3, HF, KHF_2, $KMnO_4$	H_2O_2	180°C/6 h	Octahedron	[18]
9.	$NaYF_4:Yb^{3+}$, Tm^{3+}	YCl_3, $TmCl_3$, $YbCl_3$, NH_4F, NaCl	Sodium tartrate	180°C/–	Irregular hexagonal prisms	[19]
10.	$Gd_2Zr_2O_7:Tb^{3+}$	$Gd(NO_3)_3.6H_2O$, $Tb(NO_3)_3 \cdot 6H_2O$, $ZrOCl_2 \cdot 8H_2O$	Ammonia solution	200°C/20 h	Uniform short rods	[20]
11.	$LaPO_4:Eu^{3+}$	La_2O_3, Eu_2O_3, $(NH_4)_2HPO_4$, HNO_3	NH_4OH	140°C/3 h	Nanorods and nanospheres	[21]
12.	$Sr_3Al_2O_6:Eu^{3+}$	$Sr(NO_3)_2$, $Al(NO_3)_3 \cdot 9H_2O$, $Eu(NO_3)_3 \cdot 6H_2O$	$NH_3.H_2O$	250°C/10 h	Regular cube	[22]

13.	$Y_2Zr_2O_7:Tb^{3+}$	$Y(NO_3)_3 \cdot 6H_2O$, $Tb(NO_3)_3 \cdot 6H_2O$, $ZrOCl_2 \cdot 8H_2O$	Ammonia solution	200°C/20 h	Straight rods	[23]
14.	$YVO_4:Eu^{3+}/In^{3+}$	Eu_2O_3, $Y(NO_3)_3 \cdot 6H_2O$, $In(NO_3)_3 \cdot 4.5H_2O$, $Na_3VO_4 \cdot 12H_2O$	NaOH	180°C/24 h	Spherical particles	[24]
15.	$BaMoO_4:Sm^{3+}$	$Ba(NO_3)_3$, Sm_2O_3, HNO_3, Na_2MoO_4	—	180°C/24 h	Octahedron	[25]

Table 5.3 Phosphors synthesized using combustion synthesis.

S. No.	Host: dopant	Fuel	Furnacing temp	Annealing temp/time	Morphology	Ref.
1.	$ZnAl_2O_4:Eu^{3+}$	Urea	550°C	800°C/2 h	Irregular spheres	[26]
2.	$SrMg_2Al_{16}O_{27}:Eu^{2+}$	Urea	550°C	–	Foam-like agglomerated crystallites	[27]
3.	$SrAl_2O_4:Ce^{3+}$	Urea	550°C	–	Plate-like structures	[28]
4.	$Gd_2O_3:Dy^{3+}$	Glycine	500°C	1000, 1100, 1200, 1300 and 1400°C/1 h	Featureless square	[29]
5.	$Y_3Al_5O_{12}:Eu^{3+}$	Urea	–	1200°C/2 h	Irregularly spherical or elliptical	[30]
6.	$LiCaBO_3$: Tb^{3+}, Dy^{3+}	Urea	550°C	700°C/2 h	–	[31]
7.	$NaCaPO_4$: Ce^{3+}, Dy^{3+}, Eu^{3+}	Urea	550–650°C	–	Foam-like agglomerated crystallites	[32]
8.	$LnAlO_3$: Eu^{3+} (Ln = Y, Gd)	Glycine	500°C	1000°C/3 h	Regular shape particles	[33]
9.	$Ca_2BO_3Cl:Eu^{2+}$	Urea	500°C	600°C/1 h	Irregular flake-like	[34]
10.	$LiBaPO_4:Tb^{3+}$	Urea	550°C	950°C/2 h	Regular stick shaped agglomerated particles	[35]
11.	$BaAl_2O_4:Eu^{2+}$, Nd^{3+}	Urea	500°C	–	Regular hexagonal platelets	[36]
12.	$Gd_2O_3:Bi^{3+}$	Glycine	–	1100°C/1 h	Aggregates of nanospheres	[37]

and are subjected to heating at 80–100°C (depending on vaporization), and this results in gel formation. This gel is dried and subjected to various temperatures to obtain the desired end product. Table 5.4 shows the phosphors, their synthesis conditions, morphology, etc. synthesized by sol–gel method.

2.5 Coprecipitation method

The coprecipitation synthesis method is a highly preferred synthesis method due to its ease of synthesis on a large-scale production, monetary benefits, and environment-friendly approach. Compared to other well-known synthesis methods that require sophisticated instruments and synthesis conditions, this is a facile and economic route that yields mono-disperse, well-shaped, and pure phase crystalline nanomaterials. In this method, the initial precursors such as chlorides, nitrates, and acetates are mixed in stoichiometric amounts in an aqueous solution. The precipitating agents such as NaOH, KOH, NH_4OH, urea, and $NaHCO_3$ are added under mild or room temperature conditions. The addition of these precipitating agents results in the formation of oxalates, hydroxides, or some other intermediate compounds that are then washed using ethanol or distilled water. This is followed by drying of the as-synthesized precipitates. The reaction conditions and various outcomes of phosphors synthesized by coprecipitation method are summarized in Table 5.5.

2.6 Spray pyrolysis method

Spray pyrolysis is a simple and low-cost route for the fabrication of materials in the form of thin films and powders. This method is well known because of its simple apparatus and large-scale production. Also, a clean or high vacuum is not a necessary condition for this technique. This method atomizes the solution and heats the droplets to form solid particles. The solution is sprayed on the heated surface, where the reaction of constituents takes place to form desired compounds. This method has been widely used by researchers for the synthesis of phosphors. Table 5.6 shows the phosphors synthesized using the spray pyrolysis technique.

2.7 Laser ablation method

In this synthesis technique, a high-power laser removes a material from its surface. The nanoparticles produced by this technique possess unique properties and such materials cannot be produced by any other synthesis techniques. It is a simple method as there is no need for a surfactant or any chemical addition for the fabrication of nanomaterials. The important parameters that need to be considered while following this method are ablation time, repetition rate of laser, wavelength of laser, and absorption of an aqueous solution. This method has been widely used for the synthesis of phosphors. Cho et al. [70] prepared YAG:Ce nanophosphors using laser ablation of micron size commercial YAG:Ce target phosphor. The emission intensity of the as-prepared nanophosphors was almost equal to that of the target phosphor. The emission spectra consisted of a broadband at around 550 nm. The as-prepared phosphors can be used as

Table 5.4 Phosphors synthesized using sol–gel synthesis.

S. No.	Host: dopant	Chelating agent/ complexant	Medium	Stirring temp./ drying ttemp.	Calcination temp.	Morphology	Ref.
1.	$Y_3Al_5O_{12}$:Ce^{3+}	Citric acid/ ethylene glycol	Water	–/120°C	1000°C–4 h	–	[38]
2.	$BaLn_2ZnO_5$: Eu^{3+} (Ln = La, Gd, and Y)	Citric acid	Water	70°C/120°C	700–1200°C–5 h	Spherical	[39]
3.	$ZnTiO_3$:Eu^{3+}	–	Water	5 h/80°C–24 h	500–1000°C–24 h	Hexagonal with certain agglomeration	[40]
4.	$SrAl_2O_4$: Eu^{3+}, Dy^{3+}	Citric acid	Water	80°C–3 h/–	900°C	–	[41]
5.	$ZnAl_2O_4$: Mn	–	Water	90°C–4 h/–	600°C	–	[42]
6.	ZnO:Mg	–	Water	80°C–1 h/ 120°C–24 h	–	Hexagonal	[43]
7.	Y_2O_3:Eu^{3+}	Methoxyethanol and citric acid	Water	–/100°C	700–1400°C–2 h	Crystalline and smooth	[44]
8.	Gd_2O_3:Er^{3+}, Yb^{3+}	Citric acid	Water	–/130°C–24 h	800–900°C–2 h	Spherical	[45]
9.	Zn_2SiO_4:Mn^{2+}	NH_4OH	Water and ethanol	–	800–1100°C–2 h	Heavily agglomerated particles	[46]
10.	$Gd_2Ti_2O_7$:Eu^{3+}, V^{4+}	NH_4OH	Water	2 h/–	800°C–5 h	–	[47]
11.	$CaNb_2O_6$:	Citric acid/ ethylene glycol		–/120°C–10 h	600–700°C–3 h	–	[48]
12.	$InVO_4$:Eu^{3+}	Citric acid/ ethylene glycol	Water	–	950°C–6 h	Pebble-like	[49]

Table 5.5 List of phosphors synthesized via coprecipitation route.

S. No.	Host: dopant	Precursor	Precipitating agent	Drying temp/ time	Morphology	Ref.
1.	$Y_3Al_5O_{12}$:Ce	Y_2O_3, $Ce(NO_3)_3 \cdot 6H_2O$, $Al(NO_3)_3 \cdot 9H_2O$	Ammonia water and ammonium hydrogen carbonate	120°C/12 h	Sharply grown particles and aggregated	[50]
2.	$Lu_2O_3:Eu^{3+}$	Lu_2O_3, Eu_2O_3	$NH_3.H_2O$ and NH_4HCO_3	100°C/24 h	Agglomerated particles	[51]
3.	$Ba_2MgSi_2O_7$: Eu^{2+}	Eu_2O_3, $(C_2H_5O)_4Si$, $Ba(NO_3)_2 \cdot 4H_2O$, $Mg(NO_3)_2 \cdot 6H_2O$	NH_4OH	200°C/–	Homogeneous aggregates	[52]
4.	$MgWO_4:Eu^{3+}$	$MgCl_2$, Eu_2O_3, Na_2WO_4, HCl	$NH_3.H_2O$ and NH_4HCO_3	–	Irregular particles	[53]
5.	$Lu_3Al_5O_{12}:Re^{3+}$ (Ce, Eu)	Lu_2O_3, Eu_2O_3, $Ce(NO_3)_3$, $Al(NO_3)_3 \cdot 9H_2O$	NH_4HCO_3	–	Spherical	[54]
6.	$Sr_2MgSi_2O_7$: Eu^{2+}, Dy^{3+}	$Sr(NO_3)_2$, $Mg(NO_3)_2$, $Eu(NO_3)_3$, $Dy(NO_3)_3$	Aminopropyl triethoxysilane	90°C/4 h	Aggregated particles	[55]
7.	$Na_3GaF_6:Mn^{4+}$	NaF, Ga_2O_3, HF, NaCl, NaAc, Na_2SO_4, $KMnO_4$	Ethylene glycol	70°C/10 h	Irregular polyhedron	[56]
8.	$CdMoO_4:Dy^{3+}$	$Cd(NO_3)_2$, $DyCl_3$, Na_2MoO_4	–	70°C/24 h	Spherical (with increase in annealing temp aggregation of grains takes place)	[57]
9.	$Gd_2O_3:Er^{3+}$	Gd_2O_3, Er_2O_3	LiOH	120°C/-	Smooth surface particles	[58]
10.	$SrMoO_4:Eu^{3+}$	Eu_2O_3, $Sr(NO_3)_2$, $(NH_4)_6Mo_7O_{24}$	Polyethylene glycol	140°C/12 h	Uniform roundish shaped particles	[59]
11.	$LaPO_4$:Ce, Tb	La_2O_3, Tb_4O_7, HNO_3, HCl	Citric acid	120°C/–	Ball shaped	[60]
12.	$Gd_6WO_{12}:Eu^{3+}$	Gd_2O_3, Eu_2O_3, $Na_2WO_4 \cdot 2H_2O$, HCl	NaOH	80°C/10 h	Spherical	[61]

Table 5.6 Phosphors synthesized using spray pyrolysis method.

S. No.	Host: dopant	Precursors	Surfactant/solvent/fuel	Particle/crystallite size	Morphology	Ref.
1.	$LaPO_4$:Ce, Tb	$La(NO_3)_3$, $Tb(NO_3)_3$, $Ce(NO_3)_3$	Distilled water	0.7 μm/76.9 nm	Spherical	[62]
2.	ZrO_2:Eu^{3+}	Eu_2O_3, $ZrOCl_2 \cdot 8H_2O$	Citric acid/HNO_3	–	Spherical	[63]
3.	$SrAl_2O_4$:Eu^{3+}/Dy^{3+}	$Sr(NO_3)_2$, $Al(NO_3)_3 \cdot 8H_2O$, $Eu(NO_3)_3 \cdot 5H_2O$, $Dy(NO_3)_3 \cdot xH_2O$	Distilled water and methanol	–/43 nm	Spherical	[64]
4.	Sr_2CeO_4	$Ce(NO_3)_3 \cdot 6H_2O$, $Sr(NO_3)_2$	Citric acid, PEG/Distilled water	–	Spherical	[65]
5.	Gd_2O_3:Eu	$Gd(NO_3)_3 \cdot 6H_2O$, $Eu(NO_3)_3 \cdot 6H_2O$	Distilled water	0.65 μm/–	Spherical	[66]
6.	Y_2O_3:Eu^{3+}	Y_2O_3, Eu_2O_3	HNO_3/Distilled water	1.1 μm/–	Spherical	[67]
7.	Zn_2SiO_4:Mn^{2+}	$Zn(NO_3)_2 \cdot 6H_2O$, zinc acetate, zinc-2-ethylhexanoate, $Si(OC_2H_5)_4$, $Mn(CH_3COO)_2$. $4H_2O$	Distilled water or ethanol	100–300 nm/–	Agglomerated particles	[68]
8.	$MnGa_2O_4$: Mn^{2+}	$Mg(NO_3)_2$, Ga_2O_3, manganese acetate	Distilled water	–	Spherical	[69]

yellow-emitting phosphors in high-power white LEDs. Yi et al. [71] synthesized Eu-doped $GdVO_4$ thin film phosphors on Si substrates using this technique. It was noted that the morphology, luminescence intensity, and crystallinity of the thin film were strongly dependent on the synthesis conditions, in particular, substrate temperature and oxygen pressure. These phosphors were found to be a potential candidate in display applications. Shi et al. [72] prepared chromium-doped $ZnGa_2O_4$ thin films on quartz substrates by using laser ablation. It was found that the grain size decreased, band gap increased, crystallinity improved, and the intensity of the red emission band enhanced with an increase in oxygen pressure from 0 to 1 Pa.

2.8 Other methods

Several other methods are also being used for the synthesis of phosphors. Ishigaki et al. [73] prepared $SrAl_2O_4:Eu^{3+}$, Dy^{3+} using microwave synthesis technique using domestic microwave oven operated at 2.45 GHz. Carbon was used as a microwave absorber and was used to generate a reductive atmosphere. The SEM images revealed that the obtained sample grain size was uniform about 2 μm. Photoluminescence results revealed that under excitation at 364 nm, an intense emission peak was found at 516 nm. Another method known as the Polyol method was used by Bedekar et al. [74] to synthesize Dy-doped Gd_2O_3 nanophosphors. The synthesized particles possessed spherical morphology, and characteristic blue and yellow emissions were observed. The luminescence properties were observed with the prepared phosphors in solid form, an aqueous solution, and when coated on a quartz substrate. Aldo keto synthesis method is another soft synthesis root that requires low temperature and less time for synthesis of phosphors. Koparkar et al. [75] synthesized $YPO_4:Eu^{3+}$ nanophosphors via this route. The particles formed were spherical with slight agglomeration and particle size less than 100 nm. Green synthesis is defined as a method that uses environmentally friendly materials for the synthesis of nanoparticles. It is an alternative to chemical-based synthesis routes, and no toxic by-products are produced. This method is an energy-saving method as no high temperature or pressure conditions are necessary while synthesis. Malleshappa et al. [76], Vidya et al. [77], and Ravikumar et al. [78] synthesized $CeO_2:Eu^{3+}$, $Gd_2O_3:Eu^{3+}$, and $ZnAl_2O_4:Dy^{3+}$ phosphors, respectively, using *Euphorbia tirucalli* plant latex. Eu-doped CeO_2 and Dy-doped $ZnAl_2O_4$ had flaky type morphology, while Eu-doped Gd_2O_3 showed rose-like morphology. Photoluminescent properties of these nanophosphors were analyzed. The high intense photoluminescence (PL) peak was found at 613 nm for $CeO_2:Eu$, 612 nm for $Gd_2O_3:Eu$, and at 476 and 573 nm for $ZnAl_2O_4:Dy$.

3. Factors affecting the synthesis routes

The fabrication of phosphors is influenced by many factors such as the use of surfactants, fluxing agents, and templates. Also, the factors like temperature, dopants, and precipitating agents make the difference. These factors are discussed in brief in the following section.

3.1 Use of surfactants/polymers

RE compounds are promising functional materials that have a wide range of applications in the field of optoelectronic devices, biosensors, bio-imaging, etc. [79–83]. These applications depend strongly on the composition and structure of the materials [80,81]. It is worth noting that surfactants are known for their amphipathic character and exhibit a double affinity [81]. Surfactants help to retain the homogeneity of the product and reduce the particles' surface tension hence facilitating the formation of the new phase [81]. Priya and Pandey [79] reported that the addition of the surfactants cetyl trimethyl ammonium bromide (CTAB) and trioctylphosphine oxide (TOPO) altered the morphological features but not the crystal structure of Gd_2O_3:Eu phosphors. The crystallite size also decreased with the use of surfactants as compared to the bare sample. The morphology results showed that CTAB-assisted samples were agglomerated and nonuniform in shape as compared to TOPO-assisted samples that were almost in a rod shape. It is worth noting that the use of surfactants decreased the rate of nonradiative transitions and increased the luminescence intensity and lifetime values of the samples. Similar results were reported by Yang et al. [82] who studied the effect of polyethylene glycol (PEG), Tween-80, sodium dodecyl sulfonate, and CTAB surfactants on $CaMoO_4:Eu^{3+}$ phosphors. The introduction of surfactants did not change the crystal structure of $CaMoO_4:Eu^{3+}$ phosphors. However, the samples showed uniform sphere-like morphology with diameters of 300 nm, and the strongest red luminescent intensity for PEG20000 is 5 wt%. Tan et al. [83] compared different surfactants for their effectiveness to enhance infrared-to-visible upconversion emissions of $NaYF_4$:Yb–Er. The use of different surfactants was found to significantly change the optical efficiency of as-synthesized NaYF4:Yb–Er particles. The brightness and efficiency ranking was polyvinylpyrrolidone > PEG monooleate > trioctylphosphine > unmodified particles.

3.2 Use of templates

The nano-materials are being extensively used in various technological fields, because of their unique optical and chemical properties [84]. Hence, their fabrication methods are of paramount significance. In comparison with conventional fabrication routes, the template methods are preferred as they offer control on the morphology, particle size, and structure during the nano-materials formation [84]. The different templates are chosen according to their structures, which are divided into hard template and soft template. Walsh et al. [85] used dextran as a sacrificial template. They prepared dextran-templated silver sponges in the form of macroscopic wires, 1 cm^3 cubes, or cylindrical monoliths approximately 10 cm in diameter. Authors also stated that Dextran has benefits like the metal-ion-loaded dextran gels are readily shaped and patterned through the use of appropriate molds with a range of macroscopic morphologies. Also, dextran contains reductive aldehyde groups that can be exploited for in situ reduction of metal ions and nucleation of monodisperse metallic clusters. Finally, the use of dextran did not require stringent conditions, so could be readily extended to the fabrication of macroporous frameworks of gold, copper oxide, and composites of silver/copper oxide or silver/titania. Koczkur et al. [86] explained that colloidal synthesis offers a route to nanoparticles with controlled composition and structural features. They emphasized the use of polyvinylpyrrolidone (PVP) to obtain such nanostructures as PVP can serve as a surface stabilizer, growth modifier, nanoparticle dispersant, and reducing agent depending upon the synthetic conditions. Vidya et al. [77] studied an eco-friendly, cost effective, and bio-template route for the preparation of cubic $Gd_2O_3:Eu^{3+}$ nanophosphors using *Euphorbia tirucalli* (*E. tirucalli*) plant latex as fuel. Authors found that the increasing latex concentration favors the transformation from nanoplates to rose-like nanoflowers with sizes varying from 20 to 30 nm. It also showed that hexagonal $Gd(OH)_3:Eu^{3+}$ phase converted to the pure cubic phase of $Gd_2O_3:Eu^{3+}$ on calcination at 600°C for 3 h. Whereas the decolorization of dye was found to decrease from 83% to 54% with an increase in the plant latex concentration.

3.3 Use of fluxing agents

It is a well-known fact that a good fluxing agent provides uniform particle size, increases PL efficiency, and crystalline character as compared to the phosphors synthesized without using it. In a recent study, Tao et al. [87] studied a series of $YAG:Ce^{3+}$ prepared by the conventional solid-state reaction with different fluxing agents like fluorides (NaF, AlF_3, SrF_2, KF, LiF, CaF_2, MgF_2, CeF_3, NH_4F), chloride ($SrCl_2$, NaCl, KCl, $CaCl_2$, $BaCl_2$, $MgCl_2$, NH_4Cl, LiC), oxides (SiO_2, Li_2O, Bi_2O_3, B_2O_3, CaO, BaO), phosphates (KH_2PO_4, K_3PO_4, Na_3PO_4, $(NH_4)_2HPO_4$, and carbonates (Li_2CO_3, Na_2CO_3, $NaHCO_3$). It was observed that the emission spectra and the crystallinity of $YAG:Ce^{3+}$ were influenced by the different fluxing agents. YAG: Ce^{3+} prepared with fluxes presented a more uniform particles size compared with the phosphor prepared without the fluxing agents. The results also indicated that the chloride fluxing agent reduced the synthesis temperature of $YAG:Ce^{3+}$. Also, most of the carbonate fluxing agents resulted in a decrease in emission intensity. Whereas, oxide fluxing agents increased the uniformity of $YAG:Ce^{3+}$ but did not have a significant effect on crystallinity. The addition of Na^+ ions reduced the particle size of YAG: Ce^{3+} on contrary to ammonium salt. Liu et al. [88] also revealed that the use of fluxing agents of H_3BO_3 or $LiBO_2$ in the materials synthesis can play a significant role in improving the PL properties of phosphors and the surface morphology. Similar results were obtained by Sivakumar et al. [89] during their study on flux materials like H_3BO_3, NH_4Cl, NH_4F, NH_4Br, $BaCl_2$, $BaBr_2$, CaF_2, and BaF_2. The various studies reveal that an appropriate fluxing agent can not only improve the crystallinity and morphology but also increase the PL emission of the phosphor.

3.4 Effect of temperature

The emission intensity in the phosphors is strongly dependent on the annealing temperature. Several studies [90–93] reported the dependence of the emission intensity on the temperature. Tseng et al. [90] reported that phosphors with the compositions of x% Eu_2O_3-doped $SrAl_2O_4$ ($x = 0.25$, 0.5, 0.75, and 1.0 mol%) were synthesized at different temperatures. It was observed that the volume of the nano-cotton-filament-structured crystals decreased with raising synthesis temperature. The emission intensities of the characteristic peaks increased as the temperature was increased to 1300°C from 1000°C in steps of 100°C. In another study, Liu et al. [91] described the effect of synthesis temperature on $Ca_{1.975}Eu_{0.025}MgSi_2O_7$ powder in the temperature range of 1150−1350°C. It was found that the reduction was infused during the synthesis process, and it was removed as the temperature went to 800°C. At 1150°C, the $Ca_2MgSi_2O_7$, $CaSiO_3$, and $CaMgSiO_4$ phases coexisted, and the PL spectrum had only one broad emission band with a centered wavelength of 475 nm (blue light), which was dominated by Eu^{2+}-doped $CaSiO_3$ and $CaMgSiO_4$ phases. As the temperature was increased to 1350°C only $Ca_2MgSi_2O_7$ phase was observed, and Eu^{2+}-doped $Ca_2MgSi_2O_7$ phase dominated emission band with centered wavelength 529 nm (green light). In the samples synthesized at 1200−1300°C, the diffraction intensities of $CaSiO_3$ and $CaMgSiO_4$ phases decreased, and that of $Ca_2MgSi_2O_7$ phase increased. It was observed that when the reduction atmosphere removing temperature increased, because the concentration of Eu^{2+} ions decreased, both the emission intensity and decay time decreased. In different studies, Shinde et al. [92] and Kaewnuam et al. [93] reported that the PL intensity of the phosphors increased with increasing synthesizing temperatures.

3.5 Effect of precipitating agents

The morphology of the materials can also be controlled with the help of precipitating agents such as NaOH, KOH, and NH_4OH. Ayse Dulda [94] studied the effects of alkali addition rate and the type of alkali on the growth rates of the crystal facets in α-GaO(OH) particles via precipitation. The author analyzed that the single-phase α-Ga_2O_3 was obtained after thermal treatment of NaOH precipitated precursor, while β-Ga_2O_3 was formed when NH_4OH-precipitated precursor was calcinated. The author also observed that similar spheroidal morphology was obtained by uniform and slow release of OH^{-1} ions from decomposing urea, and spherical morphology was observed with fast addition of NH_4OH or NaOH. The crystallite size also decreased when NaOH was replaced by NH_4OH, which indicates that type of prectipating agent has a significant effect on the morphology. Zhang et al. [95] also prepared nanostructured scandium hydrous at 180°C, using NaOH, NH_4OH, and KOH as the bases. It was found that the nature and concentration of the bases remarkably influenced the phase composition, the texture behavior (shape and size), and the surface state of the hydrothermal products. The usage of a weak base (NH_4OH) within pH 10−12, resulted in nanosized ç-ScOOH lozenge-like plates as the main products. When using a strong base (NaOH or KOH), we obtained R−ScOOH nanorods, nanosized hexagonal-like plates, and cubic $Sc(OH)_3$ cubes/cuboids in turn across the basicity value varied from pH 10 to 5 mol L^{-1}.

3.6 Effect of dopants

The structure, morphology, and optical properties of phosphors also depend upon dopants. The doping with transition and RE elements can not only improve the optical properties of the phosphors but also bring changes to the morphology. Kumar et al. [96] found that the as-synthesized $Gd_2O_3:Ho^{3+}/Yb^{3+}$ sample had a flaky structure that became rod-like on annealing. On codoping with Ca^{2+}, the as-synthesized sample showed elongated particles that became spherical after annealing. On the other hand, Zn^{2+} codoped as-synthesized sample exhibited spherical particles that grew in the same shape with connecting particles after annealing. The upconversion characteristics showed that Ca^{2+} doping did not enhance the upconversion intensity, whereas Zn^{2+} enhanced the upconversion intensity up to 15 mol% codoping. Whereas, the Ca^{2+} and Zn^{2+} codoping increased the downconversion intensity that was in contrast to the upconversion process. One of the possible explanations for this can be related to the effective sizes of the ions. The effective ionic radius of Zn^{2+} (0.088 nm) is smaller than Gd^{3+} (0.1078 nm) thereby compensating all defects of the host lattice and enhancing the emission intensity. On the other hand, the effective ionic radii of Ca^{2+} (0.114 nm) is slightly higher than Gd^{3+} ions. Hence, it can reside at interstitial positions and create a problem in the upconversion emission. But in the downshifting process, the enhancement in emission intensity due to Ca^{2+} codoping has probably happened due to absorption of the 449 nm excitation and cooperative energy transfer from Ca^{2+} ion to the Ho^{3+} activator center [96]. Vinoditha et al. [97] studied Eu^{2+}-doped ZnO nanoparticles and found that the length of the ZnO rods shortened and became thinner with increasing Eu^{2+}-doping concentration. Correspondingly, the PL intensity increased with increasing Eu^{2+} concentration up to 0.20 at%.

4. Different types of structures

4.1 Core–shell structure

To improve the luminescence intensity of the phosphors, the synthesis of core–shell nanostructures has come into the picture. The core-shell structures are developed via chemical routes. The morphology and the size of the core–shell materials can be customized by varying the shape, size, or shell thickness. Some of the core–shell-structured RE-activated phosphors are given in Table 5.7.

4.2 Quantum dots

Quantum dots (QDs) have attracted great interest from scientists as an approach to further advancement in LEDs because of their size-dependent emission. QDs offer high quantum efficiencies, and due to the quantum confinement effect, their emissions are readily tuneable between 500 and 650 nm [103]. A few examples of RE-activated QDs are tabulated in Table 5.8.

4.3 1D/2D/3D structures

In a recent study, Mejia et al. [111] proposed the euclidian spatial dimensionality on the luminescence produced by upconversion energy transfer among randomly

Table 5.7 List of core–shell-structured rare-earth-activated phosphors.

Core–shell structure	Synthesis route	Dopant	Morphology	Ref.
$Y_3Al_5O_{12}$@SiS	Solvothermal method	Eu^{3+}	Spherical	[98]
$Y_3Al_5O_{12}$@SiS	Solvothermal method	Ce^{3+}	Spherical	[98]
YAG:Ce/YAG:Cr	Urea homogeneous precipitation method	Eu^{3+}	Spherical	[99]
RE:$LaPO_4$ (muti core–shell)	Hydrothermal method	Tb^{3+}	–	[100]
RE:$LaPO_4$	Hydrothermal method	Ce^{3+}	–	[100]
RE: $LaPO_4$	Hydrothermal method	Dy^{3+}	–	[100]
Y_2O_3@ SiO_2	Chemical route	Eu^{3+}	Spherical	[101]
$BaGeF_6$@PPG	Coprecipitation method	Mn^{2+} + Dy^{3+}	Flower like morphology (rod shape)	[102]
$BaGeF_6$: @PPG-$NaGdF_4$	Hydrothermal method	Mn^{2+} + Dy^{3+}	Spherical	[102]

Table 5.8 List of quantum dots doped in different phosphors.

Quantum dot	Synthesis method	Dopant	Ref.
ZnSe	Chemical route using ethylene glycol as stabilizing agent	Gadolinium (Gd^{3+})	[104]
ZnS	Sol–gel	Europium (Eu^{3+})	[105]
ZnO	Sol–gel	Terbium (Tb^{3+}) Holmium (Ho^{3+}) Lanthanum (La^{3+})	[106]
CdTe	Colloidal method	Gadolinium (Gd^{3+})	[107]
ZnS	Chemical route	Cerium	[108]
$CaMoO_4:Na^+$	High gravity rotating packed bed (RPB) technology	Eu^{3+} Tb^{3+}	[109]
Gd_2O_3	Sucrose combustion method	Eu^{3+}	[110]

distributed optical centers with the help of an analytical model. Its analysis indicated that under similar conditions, the rate of the upconversion follows the 1D > 2D > 3D tendency at initial times, but this conduct reverses at some (concentration and pump power independent) time point, resulting in a higher total upconversion luminescence

Table 5.9 List of phosphors with different structures.

Name of phosphor	Synthesis method	Structure (1D/2D/3D)	Ref.
$K_2SiF_6:Mn^{4+}$	Liquid−solid-solution route	Nanorods (1D)	[113]
$SiO_2:Tb^{3+}$	Vapor−liquid−solid mechanism	Nanowires (1D)	[114]
$BiOCl:Er^{3+}$	Hydrothermal method	Nanosheets (2D)	[115]
$Eu_{0.56}Ta_2O_7$	Soft chemical exfoliation method	Nanosheets (2D)	[116]
$Rb_2GeF_6:Mn^{4+}$	Coprecipitation method	Bulk single crystallite phosphor (3D)	[117]

for larger dimensionalities. Finally, the study shows that upconversion luminescence is very sensitive to the value of the minimum possible distance of optical centers mutual approach, especially in the low-dimensional medium. The physical origin for this trend (1D > 2D > 3D) allows predicting that in steady state the intensity of the upconversion will be higher as the dimensionality of the medium is lower. Paulraj et al. [112] compared the photoluminescence efficiencies of nanocrystalline versus bulk $Eu:Y_2O$ phosphor and found that the nanocrystalline phosphor material has lower PL efficiency compared with the sintering aid grown material, due to higher diffused reflectance. Further, phosphor material with spherical morphology, optimum particle size (0.5−1.0 mm), and higher crystallite size has the highest PL efficiency [112]. Table 5.9 summarizes different types of structures of phosphors synthesized by opting different synthesis routes.

4.4 Glass phosphors

RE-doped glasses possess an increasing number of applications in optoelectronics because of flexibility in size, shape, and composition. Glass phosphors mostly use RE^{3+} ions as active ions, because of low coupling of the RE^{3+} with the host vibrations, which reduces the nonradiative transitions and hence increases the quantum efficiency of the emitting level. Glasses can be prepared by various methods such as melt-quenching, sol−gel, and vapour-deposition method. Although melt-quenching is the conventional way to synthesize glasses, in some special cases, sol−gel method is preferred over the former [118,119]. This is mainly due to the high melting temperatures required to prepare the glasses or due to high loss during the emission. The different glass phosphors synthesized by different synthesis routes are enlisted in Table 5.10.

Table 5.10 Emission characteristics of some glass phosphors.

Glass composition (mol%)	Synthesis route	Dopant	Excitation wavelength	Emission transitions	Ref.
$95SiO_2{-}5Pr(NO_3)_3$	Sol−gel method	Praseodymium (Pr)	450 nm	$^3P_0{-}^3H_4$, $^3P_1{-}^3H_5$, $^1D_2{-}^3H_4$, and $^3P_0{-}^3F_2$ (495, 540, 615, and 650 nm)	[118]
$57ZrF{-}36BaF_2{-}1LaF_3{-}3EuF_3{-}3AlF_3$	Sol−gel method	Europium (Eu)	394 nm	$^5D_0{-}^7F_{J\,(J=0,\,1,\,2,\,3\text{ and }4)}$ (579, 591, 612, 655, and 702 nm)	[119]
$5La_2O_3{-}95TeO_2$	Melt-quenching technique	Lanthanum (La)	325 nm	325, 405, 447, 538, and 582 nm in the blue/green region due to intrinsic defects	[120]
$25Bi_2O_3{-}5B_2O_3{-}(70\text{-}x)P_2O_5{-}xCeF_3$	Melt-quenching technique	Cerium (Ce)	317 nm	5D − 4F (360 nm)	[121]
$x(Pr_6O_{11}){-}100\text{-}x$(SLS−ZnO)	Melt-quenching technique	Praseodymium (Pr)	444 nm	$^3P_o{-}^3H_{4,5,6}$, $^3F2$ (530, 556, 613, and 650 nm)	[122]
$40Bi_2O_3{-}20BaO{-}(40{-}x)B_2O_3$	Melt-quenching technique	Neodymium (Nd)	585 nm	$^4F_{3/2}{-}^4I_{9/2,11/2,13/2}$ (900, 1060, 1345 nm)	[123]
$20PbO{-}5CaO{-}5ZnO{-}10NaF{-}59B_2O_3{-}1Sm_2O_3$	Melt-quenching technique	Samarium (Sm)	402 nm	$^4G_{5/2}{-}^6H_{5/2}$, $^6H_{7/2}$, $^6H_{9/2}$, and $^6H_{11/2}$ (565, 602, 648, and 710 nm)	[124]
$20PbO{-}5CaO{-}5ZnO{-}10NaF{-}59B_2O_3{-}1Tb_2O_3$	Melt-quenching technique	Terbium (Tb)	378 nm	$^4F_7{-}^5D_1$ $^7F_6{-}^5D_{3,4}$ (418, 442, and 458 nm)	[124]

$(70-x)TeO_2-20B_2O_3-xDy_2O_3$	Melt-quenching technique	Dysprosium (Dy)	383 nm	$^4I_{13/2}-^4F_{9/2}-^6H_{15/2}$, $^6H_{13/2}$, and $^6H_{11/2}$ (484, 576, and 664 nm)	[125]
$(Sb_2O_3)_{90-x}-(Na_2O)_{10}-(ZnO)_x$	Melt-quenching technique	Erbium (Er)	514.5 nm	$^4S_{3/2}$, $^4F_{9/2}-^4I_{15/2}$, $^4S_{3/2}-^4I_{13/2}$, $^4S_{3/2}-^4I_{13/2}$, $I_{11/2}-^4I_{15/2}$, $^4S_{3/2}-^4I_{11/2}$, $^4I_{13/2}-^4I_{15/2}$ (550, 670, 850, 980, 1230, and 1530 nm)	[126]
$Bi_2O_3-GeO_2-$	Melt-quenching technique	Thulium (Tm)	808 nm	$^3P_1-^3P_0$ (broadband 100–1500 nm) (Due to Bi^+) $^3H_4-^3F_4$ (broadband in 1350–1600 nm)	[127]
$60TeO_2-30ZnF_2-10NaF$	Physical and chemical dehydration technique	Holmium (Ho)	1163 nm	2.85 mm $^5I_6-^5I_7$	[128]

5. Summary

In conclusion, it can be observed that different synthesis routes lead to different structural, morphological, and optical features of the phosphors. For large-scale production, the solid-state method is an effective route, but the synthesized phosphors are of microcrystalline with no defined morphology. Wet chemical routes (hydrothermal, coprecipitation, sol−gel), laser ablation, and spray pyrolysis methods result in phosphors in nanosize dimensions. For glass phosphors, sol−gel and melt-quench methods are popularly used. Melt-quench results in phosphors that do not have morphology and have amorphous nature, while the sol−gel results in the glass phosphors with nanosize particle formation. From a biomedical application point of view, the sol−gel method is effective over the melt quench method. Further, the end products in various synthesis routes depend on various parameters such as choice of precipitating agents, templates, surfactants, and dopants. The size and shape of the phosphors depend on the type of synthesis route followed and the application to be trailed.

References

[1] X. Huang, Q. Sun, B. Devakumar, Facile low-temperature solid-state synthesis of efficient blue-emitting $Cs_3Cu_2I_5$ powder phosphors for solid-state lighting, Mater. Today Chem. 17 (2020) 100288, https://doi.org/10.1016/j.mtchem.2020.100288.

[2] Y. Wu, Z. Sun, K. Ruan, et al., Enhancing photoluminescence with Li-doped $CaTiO_3$: Eu^{3+} red phosphors prepared by solid state synthesis, J. Lumin. 155 (2014) 269−274, https://doi.org/10.1016/j.jlumin.2014.06.051.

[3] Y. Wang, H. Zhang, Q. Wei, et al., Solid state synthesis, tunable luminescence and thermal stability of $NaCaBO_3$: Eu^{2+}/Mn^{2+} phosphors, Ceram. Int. 42 (2016) 12422−12426, https://doi.org/10.1016/j.ceramint.2016.04.182.

[4] Y. Cao, N. Liu, J. Tian, X. Zhang, Solid state synthesis and tunable luminescence of $LiSrPO_4$: $Eu^{2+}/Mn^{2+}/Tb^{3+}$ phosphors, Polyhedron 107 (2016) 78−82, https://doi.org/10.1016/j.poly.2016.01.016.

[5] Z. Kotan, M. Ayvacikli, Y. Karabulut, et al., Solid state synthesis, characterization and optical properties of Tb doped $SrSnO_3$ phosphor, J. Alloys Compd. 581 (2013) 101−108, https://doi.org/10.1016/j.jallcom.2013.07.048.

[6] H. Liu, Z. Guo, Ce^{3+} and Dy^{3+} doped $Sr_3B_2O_6$: solid state synthesis and tunable luminescence, J. Lumin. 187 (2017) 181−185, https://doi.org/10.1016/j.jlumin.2017.03.008.

[7] K. Upadhyay, R.K. Tamrakar, V. Dubey, High temperature solid state synthesis and photoluminescence behavior of Eu^{3+} doped $GdAlO_3$ nanophosphor, Superlattice. Microst. 78 (2015) 116−124, https://doi.org/10.1016/j.spmi.2014.11.030.

[8] H.K. Yang, J.H. Jeong, Synthesis, crystal growth, and photoluminescence properties of YAG: Eu^{3+} phosphors by high-energy ball milling and solid-state reaction, J. Phys. Chem. C 114 (2010) 226−230, https://doi.org/10.1021/jp908903t.

[9] R. Cao, W. Wang, J. Zhang, et al., Synthesis and luminescence properties of Li_2SnO_3: Mn^{4+} red-emitting phosphor for solid-state lighting, J. Alloys Compd. 704 (2017) 124−130, https://doi.org/10.1016/j.jallcom.2017.02.079.

[10] Z. Xia, D. Chen, Synthesis and luminescence properties of $BaMoO_4$:Sm^{3+} phosphors, J. Am. Ceram. Soc. 93 (2010) 1397–1401, https://doi.org/10.1111/j.1551-2916.2009.03574.x.

[11] R. Priya, O.P. Pandey, Hydrothermal synthesis of Eu^{3+}- doped Gd_2O_3 nanophosphors and its Judd-Ofelt analysis, Vacuum 156 (2018) 283–290, https://doi.org/10.1016/j.vacuum.2018.07.038.

[12] L. Chen, Y. Liu, Z. Lu, K. Huang, Hydrothermal synthesis and characterization of $ZnGa_2O_4$ phosphors, Mater. Chem. Phys. 97 (2006) 247–251, https://doi.org/10.1016/j.matchemphys.2005.08.024.

[13] Y. Zhou, Q. Zhou, Y. Liu, et al., Hydrothermal synthesis and luminescent properties of $BaTiF_6$:Mn^{4+} red phosphor for LED backlighting, Mater. Res. Bull. 73 (2016) 14–20, https://doi.org/10.1016/j.materresbull.2015.08.022.

[14] S. Tang, M. Huang, J. Wang, et al., Hydrothermal synthesis and luminescence properties of $GdVO_4$: Ln^{3+} (Ln = Eu, Sm, Dy) phosphors, J. Alloys Compd. 513 (2012) 474–480, https://doi.org/10.1016/j.jallcom.2011.10.093.

[15] T. Kim, S. Kang, Hydrothermal synthesis and photoluminescence properties of nano-crystalline $GdBO_3$: Eu3+ phosphor, Mater. Res. Bull. 40 (2005) 1945–1954, https://doi.org/10.1016/j.materresbull.2005.06.001.

[16] B. Umesh, B. Eraiah, H. Nagabhushana, et al., Synthesis and characterization of spherical and rod like nanocrystalline Nd_2O_3 phosphors, J. Alloys Compd. 509 (2011) 1146–1151, https://doi.org/10.1016/j.jallcom.2010.09.143.

[17] J. Gu, B. Yan, Hydrothermal synthesis and luminescent properties of $Ca_2V_2O_7$: Eu^{3+} phosphors, J. Alloys Compd. 476 (2009) 619–623, https://doi.org/10.1016/j.jallcom.2008.09.084.

[18] H. Cheng, Y. Song, G. Liu, et al., Hydrothermal synthesis of narrow-band red emitting K_2NaAlF_6: Mn^{4+} phosphor for warm-white LED applications, RSC Adv. 7 (2017) 45834–45842, https://doi.org/10.1039/c7ra09671b.

[19] W. Shan, R. Li, J. Feng, et al., Hydrothermal synthesis and up-conversion luminescence properties of $NaYF_4$: Yb^{3+}, Tm^{3+} phosphors, Mater. Chem. Phys. 162 (2015) 617–627, https://doi.org/10.1016/j.matchemphys.2015.06.035.

[20] L. Gao, H. Zhu, L. Wang, G. Ou, Hydrothermal synthesis and photoluminescence properties of $Gd_2Zr_2O_7$: Tb^{3+} phosphors, Mater. Lett. 65 (2011) 1360–1362, https://doi.org/10.1016/j.matlet.2011.02.004.

[21] S. Ray, G.B. Nair, P. Tadge, et al., Size and shape-tailored hydrothermal synthesis and characterization of nanocrystalline LaPO4: Eu^{3+} phosphor, J. Lumin. 194 (2018) 64–71, https://doi.org/10.1016/j.jlumin.2017.10.015.

[22] X. Li, H. Pan, A. Tang, et al., Hydrothermal synthesis and luminescent properties of Eu^{3+} doped $Sr_3Al_2O_6$ phosphor for white LED, J. Nanosci. Nanotechnol. 16 (2016) 3474–3479, https://doi.org/10.1166/jnn.2016.11847.

[23] L. Gao, Y. An, H. Zhu, et al., Hydrothermal synthesis and photoluminescence properties of $Y_2Zr_2O_7$: Tb^{3+} phosphors 7, 2011, pp. 1337–1340. https://doi.org/10.1007/s10853-010-4924-3.

[24] G. Zhao, W. Liu, Z. Lou, et al., Hydrothermal synthesis and photoluminescence properties of in 3+ co-doped YVO_4: Eu^{3+} phosphors, Adv. Powder Technol. 27 (2016) 1225–1232, https://doi.org/10.1016/j.apt.2016.04.008.

[25] Z. Li, Y. Wang, J. Cao, et al., Hydrothermal synthesis and luminescent properties of $BaMoO_4$: Sm^{3+} red phosphor, J. Rare Earths 34 (2016) 143–147, https://doi.org/10.1016/S1002-0721(16)60006-6.

[26] R. Priya, A. Negi, S. Singla, O.P. Pandey, Luminescent studies of Eu doped $ZnAl_2O_4$ spinels synthesized by low-temperature combustion route, Optik 204 (2020) 164173, https://doi.org/10.1016/j.ijleo.2020.164173.
[27] V.B. Pawade, S.J. Dhoble, Novel blue-emitting $SrMg_2Al_{16}O_{27}$: Eu^{2+} phosphor for solid-state lighting, Luminescence 26 (2011) 722–727, https://doi.org/10.1002/bio.1304.
[28] K.E. Foka, F.B. Dejene, H.C. Swart, Photoluminescence properties of Ce^{3+}-doped $SrAl_2O_4$ prepared using the solution combustion method, Phys. B Condens. Matter 439 (2014) 177–180, https://doi.org/10.1016/j.physb.2013.11.031.
[29] M. Jayasimhadri, B.V. Ratnam, K. Jang, et al., Combustion synthesis and luminescent properties of nano and submicrometer-size Gd_2O_3: Dy^{3+} phosphors for white LEDs, Int. J. Appl. Ceram. Technol. 8 (2011) 709–717, https://doi.org/10.1111/j.1744-7402.2010.02499.x.
[30] S. Shikao, W. Jiye, Combustion synthesis of Eu^{3+} activated $Y_3Al_5O_{12}$ phosphor nanoparticles, J. Alloys Compd. 327 (2001) 82–86, https://doi.org/10.1016/S0925-8388(01)01399-8.
[31] N.S. Bajaj, S.K. Omanwar, Combustion synthesis and luminescence characteristic of rare earth activated $LiCaBO_3$, J. Rare Earths 30 (2012) 1005–1008, https://doi.org/10.1016/S1002-0721(12)60169-0.
[32] K.N. Shinde, S.J. Dhoble, A. Kumar, Combustion synthesis of Ce^{3+}, Eu^{3+} and Dy^{3+} activated $NaCaPO_4$ phosphors, J. Rare Earths 29 (2011) 527–535, https://doi.org/10.1016/S1002-0721(10)60492-9.
[33] S.D. Han, S.P. Khatkar, V.B. Taxak, et al., Combustion synthesis and luminescent properties of Eu^{3+}- doped $LnAlO_3$ (Ln = Y and Gd) phosphors, Mater. Sci. Eng. B 127 (2006) 272–275, https://doi.org/10.1016/j.mseb.2005.10.016.
[34] Z. Xia, L. Liao, Z. Zhang, Y. Wang, Combustion synthesis and luminescence properties of yellow-emitting phosphors Ca_2BO_3Cl: Eu^{2+}, Mater. Res. Bull. 47 (2012) 405–408, https://doi.org/10.1016/j.materresbull.2011.11.006.
[35] C.B. Palan, N.S. Bajaj, A. Soni, et al., Combustion synthesis and preliminary luminescence studies of $LiBaPO_4$: Tb^{3+} phosphor, Bull. Mater. Sci. 38 (2015) 1527–1531, https://doi.org/10.1007/s12034-015-0964-2.
[36] A.H. Wako, F.B. Dejene, H.C. Swart, Combustion synthesis, characterization and luminescence properties of barium aluminate phosphor, J. Rare Earths 32 (2014) 806–811, https://doi.org/10.1016/S1002-0721(14)60145-9.
[37] Y. Zou, L. Tang, J.L. Cai, et al., Combustion synthesis and luminescence of monoclinic Gd_2O_3: Bi phosphors, J. Lumin. 153 (2014) 210–214, https://doi.org/10.1016/j.jlumin.2014.03.026.
[38] A. Boukerika, L. Guerbous, N. Brihi, Ce-doped YAG phosphors prepared via sol-gel method: effect of some modular parameters, J. Alloys Compd. 614 (2014) 383–388, https://doi.org/10.1016/j.jallcom.2014.06.133.
[39] C. Guo, X. Ding, Y. Xu, Luminescent properties of Eu^{3+}- doped $BaLn_2ZnO_5$ (Ln=La, Gd, and Y) phosphors by the sol-gel method, J. Am. Ceram. Soc. 93 (2010) 1708–1713, https://doi.org/10.1111/j.1551-2916.2010.03640.x.
[40] M. Li, B. Jiao, Synthesis and photoluminescence properties of $ZnTiO_3$: Eu^{3+} red phosphors via sol-gel method, J. Rare Earths 33 (2015) 231–238, https://doi.org/10.1016/S1002-0721(14)60408-7.
[41] T. Peng, L. Huajun, H. Yang, C. Yan, Synthesis of $SrAl_2O_4$: Eu, Dy phosphor nanometer powders by sol-gel processes and its optical properties, Mater. Chem. Phys. 85 (2004) 68–72, https://doi.org/10.1016/j.matchemphys.2003.12.001.

[42] D. Zhang, C. Wang, Y. Liu, et al., Green and red photoluminescence from $ZnAl_2O_4$: Mn phosphors prepared by solgel method, J. Lumin. 132 (2012) 1529–1531, https://doi.org/10.1016/j.jlumin.2012.01.025.
[43] Q. Shi, J. Zhang, D. Zhang, et al., Red luminescent and structural properties of Mg-doped ZnO phosphors prepared by sol-gel method, Mater. Sci. Eng. B 177 (2012) 689–693, https://doi.org/10.1016/j.mseb.2012.03.045.
[44] J.Y. Cho, K.Y. Ko, Y.R. Do, Optical properties of sol-gel derived Y_2O_3: Eu^{3+} thin-film phosphors for display applications, Thin Solid Films 515 (2007) 3373–3379, https://doi.org/10.1016/j.tsf.2006.09.029.
[45] H. Hao, X. Zhang, Y. Wang, L. Liang, Color modulation and temperature sensing investigation of Gd_2O_3: 1 mol% Er^{3+}, 1 mol% Yb^{3+} phosphors under different excitation condition, J. Lumin. 215 (2019) 116556, https://doi.org/10.1016/j.jlumin.2019.116556.
[46] J.M. Zhang, D.L. Zhao, Z.M. Shen, Synthesis and luminescent properties of Mn_{2+}-doped zinc silicate phosphors by sol-gel method, Key Eng. Mater. 434–435 (2010) 214–216, https://doi.org/10.4028/www.scientific.net/KEM.434-435.214.
[47] K.M. Lin, C.C. Lin, C.Y. Hsiao, Y.Y. Li, Synthesis of $Gd_2Ti_2O_7$: Eu^{3+}, V^{4+} phosphors by sol-gel process and its luminescent properties, J. Lumin. 127 (2007) 561–567, https://doi.org/10.1016/j.jlumin.2007.03.010.
[48] Y.J. Hsiao, C.W. Liu, B.T. Dai, Y.H. Chang, Sol-gel synthesis and the luminescent properties of $CaNb_2O_6$ phosphor powders, J. Alloys Compd. 475 (2009) 698–701, https://doi.org/10.1016/j.jallcom.2008.07.142.
[49] Y.S. Chang, Z.R. Shi, Y.Y. Tsai, et al., The effect of Eu_{3+}-activated $InVO_4$ phosphors prepared by sol-gel method, Opt. Mater. 33 (2011) 375–380, https://doi.org/10.1016/j.optmat.2010.09.018.
[50] K. Zhang, H.Z. Liu, Y.T. Wu, W.B. Hu, Co-precipitation synthesis and luminescence behavior of Ce-doped yttrium aluminum garnet (YAG:Ce) phosphor: the effect of precipitant, J. Alloys Compd. 453 (2008) 265–270, https://doi.org/10.1016/j.jallcom.2006.11.101.
[51] Q. Chen, Y. Shi, S. Wang, et al., A novel co-precipitation synthesis of a new phosphor Lu_2O_3: Eu^{3+}, J. Eur. Ceram. Soc. 27 (2007) 191–197, https://doi.org/10.1016/j.jeurceramsoc.2006.01.022.
[52] T.T.H. Tam, N.V. Du, N.D.T. Kien, et al., Co-precipitation synthesis and optical properties of green-emitting $Ba_2MgSi_2O_7$: Eu^{2+} phosphor, J. Lumin. 147 (2014) 358–362, https://doi.org/10.1016/j.jlumin.2013.11.066.
[53] X. Feng, W. Feng, M. Xia, et al., Co-precipitation synthesis, photoluminescence properties and theoretical calculations of $MgWO_4$: Eu^{3+} phosphors, RSC Adv. 6 (2016) 14826–14831, https://doi.org/10.1039/c5ra22631g.
[54] Z. Wang, M. Xu, W. Zhang, M. Yin, Synthesis and luminescent properties of nano-scale LuAG: RE^{3+} (Ce, Eu) phosphors prepared by co-precipitation method, J. Lumin. 122–123 (2007) 437–439, https://doi.org/10.1016/j.jlumin.2006.01.199.
[55] W. Pan, G. Ning, X. Zhang, et al., Enhanced luminescent properties of long-persistent $Sr_2MgSi_2O_7$: Eu^{2+}, Dy^{3+} phosphor prepared by the co-precipitation method, J. Lumin. 128 (2008) 1975–1979, https://doi.org/10.1016/j.jlumin.2008.06.009.
[56] H. Xu, F. Hong, G. Pang, et al., Co-precipitation synthesis, luminescent properties and application in warm WLEDs of Na_3GaF_6: Mn^{4+} red phosphor, J. Lumin. 219 (2020) 116960, https://doi.org/10.1016/j.jlumin.2019.116960.

[57] N.P. Singh, N.R. Singh, Y.R. Devi, et al., Effects of annealing temperature on structural and luminescence properties of $CdMoO_4$:Dy^{3+} phosphor synthesized at room temperature by co-precipitation method, Solid State Sci. 102 (2020) 106172, https://doi.org/10.1016/j.solidstatesciences.2020.106172.

[58] P. Liu, F. Wang, B. Yang, Upconversion/downconversion luminescence of color-tunable Gd_2O_3: Er^{3+} phosphors under ultraviolet to near-infrared excitation, Solid State Sci. 102 (2020) 106165, https://doi.org/10.1016/j.solidstatesciences.2020.106165.

[59] Y. Tan, X. Luo, M. Mao, et al., Optimization red emission of $SrMoO_4$: Eu^{3+} via hydrothermal co-precipitation synthesis using orthogonal experiment, Curr. Appl. Phys. 18 (2018) 1403−1409, https://doi.org/10.1016/j.cap.2018.08.005.

[60] L. Zhiqi, R. Le, Z. Zhaowu, et al., Synthesis of $LaPO_4$: Ce, terbium by co-precipitation method, J. Rare Earths 24 (2006) 137−140, https://doi.org/10.1016/S1002-0721(07)60343-3.

[61] Y. Tian, B. Chen, R. Hua, et al., Synthesis and characterization of novel red emitting nanocrystal Gd_6WO_{12}: Eu^{3+} phosphors, Phys. B Condens. Matter 404 (2009) 3598−3601, https://doi.org/10.1016/j.physb.2009.06.007.

[62] I. Wuled Lenggoro, B. Xia, H. Mizushima, et al., Synthesis of $LaPO_4$: Ce, Tb phosphor particles by spray pyrolysis, Mater. Lett. 50 (2001) 92−96, https://doi.org/10.1016/S0167-577X(00)00422-5.

[63] Z.W. Quan, L.S. Wang, J. Lin, Synthesis and characterization of spherical ZrO_2: Eu^{3+} phosphors by spray pyrolysis process, Mater. Res. Bull. 40 (2005) 810−820, https://doi.org/10.1016/j.materresbull.2005.02.006.

[64] S. Gültekin, S. Yıldırım, O. Yılmaz, et al., Structural and optical properties of $SrAl_2O_4$: Eu^{2+}/Dy^{3+} phosphors synthesized by flame spray pyrolysis technique, J. Lumin. 206 (2019) 59−69, https://doi.org/10.1016/j.jlumin.2018.10.011.

[65] X. Liu, Y. Luo, J. Lin, Synthesis and characterization of spherical Sr_2CeO_4 phosphors by spray pyrolysis for field emission displays, J. Cryst. Growth 290 (2006) 266−271, https://doi.org/10.1016/j.jcrysgro.2005.12.091.

[66] Y. Iwako, Y. Akimoto, M. Omiya, et al., Photoluminescence of cubic and monoclinic Gd_2O_3: Eu phosphors prepared by flame spray pyrolysis, J. Lumin. 130 (2010) 1470−1474, https://doi.org/10.1016/j.jlumin.2010.03.014.

[67] Y. Shimomura, N. Kijima, High-temperature spray pyrolysis of Y_2O_3: EU^{3+} red phosphor, Electrochem. Solid State Lett. 7 (2004) 4−7, https://doi.org/10.1149/1.1633692.

[68] J.S. Lee, M.H. Oh, P. Kumar, et al., Mn-doped Zn_2SiO_4 phosphors synthesis using flame spray pyrolysis, J. Therm. Spray Technol. 20 (2011) 1001−1008, https://doi.org/10.1007/s11666-011-9630-4.

[69] S. Choi, K. Kim, Y.M. Moon, et al., Rapid synthesis of spherical-shaped green-emitting $MgGa_2O_4$: Mn^{2+} phosphor via spray pyrolysis, Mater. Res. Bull. 45 (2010) 979−981, https://doi.org/10.1016/j.materresbull.2010.04.014.

[70] Y.S. Cho, Y.D. Huh, C.R. Park, Y.R. Do, Preparation with laser ablation and photoluminescence of $Y_3Al_5O_{12}$: Ce nanophosphors, Electron. Mater. Lett. 10 (2014) 461−465, https://doi.org/10.1007/s13391-014-4024-7.

[71] S.S. Yi, J.S. Bae, K.S. Shim, et al., Photoluminescence behaviors of Eu-doped $GdVO_4$ thin film phosphors grown by pulsed laser ablation, J. Alloys Compd. 408 (412) (2006) 890−893, https://doi.org/10.1016/j.jallcom.2004.12.107.

[72] Q. Shi, C. Wang, D. Zhang, et al., Luminescence of Cr^{3+}-doped $ZnGa_2O_4$ thin films deposited by pulsed laser ablation, Thin Solid Films 520 (2012) 6845−6849, https://doi.org/10.1016/j.tsf.2012.07.088.

[73] T. Ishigaki, H. Mizushina, K. Uematsu, et al., Microwave synthesis technique for long phosphorescence phosphor $SrAl_2O_4$: Eu^{2+}, Dy^{3+} using carbon reduction, Mater. Sci. Eng. B 173 (2010) 109–112, https://doi.org/10.1016/j.mseb.2009.11.004.

[74] V. Bedekar, D.P. Dutta, A.K. Tyagi, White light emission from spin coated Gd_2O_3: Dy nano phosphors synthesized using polyol technique, J. Nanosci. Nanotechnol. 10 (2010) 8234–8238, https://doi.org/10.1166/jnn.2010.3008.

[75] K.A. Koparkar, N.S. Bajaj, S.K. Omanwar, Photoluminescence in a novel aldo-keto synthesized YPO_4: Eu^{3+} nanophosphor, Solid State Phenom. 222 (2015) 179–185, https://doi.org/10.4028/www.scientific.net/SSP.222.179.

[76] J. Malleshappa, H. Nagabhushana, S.C. Prashantha, et al., Eco-friendly green synthesis, structural and photoluminescent studies of CeO_2: Eu^{3+} nanophosphors using E. tirucalli plant latex, J. Alloys Compd. 612 (2014) 425–434, https://doi.org/10.1016/j.jallcom.2014.05.101.

[77] Y.S. Vidya, K.S. Anantharaju, H. Nagabhushana, S.C. Sharma, Euphorbia tirucalli mediated green synthesis of rose like morphology of Gd_2O_3: Eu^{3+} red phosphor: structural, photoluminescence and photocatalytic studies, J. Alloys Compd. 619 (2015) 760–770, https://doi.org/10.1016/j.jallcom.2014.09.050.

[78] B.S. Ravikumar, H. Nagabhushana, D.V. Sunitha, et al., Plant latex mediated green synthesis of $ZnAl_2O_4$: Dy^{3+} (1–9 mol%) nanophosphor for white light generation, J. Alloys Compd. 585 (2014) 561–571, https://doi.org/10.1016/j.jallcom.2013.09.080.

[79] R. Priya, O.P. Pandey, Structural, morphological, luminescent and magnetic studies of CTAB and TOPO assisted Gd_2O_3: Eu phosphors synthesized via co-precipitation route, J. Alloys Compd. 847 (2020) 156388, https://doi.org/10.1016/j.jallcom.2020.156388.

[80] C. Wu, W. Qin, G. Qin, et al., Photoluminescence from surfactant-assembled Y_2O_3: Eu nanotubes, Appl. Phys. Lett. 82 (2003) 520–522, https://doi.org/10.1063/1.1542685.

[81] X. Zhou, X. Wang, Comparison of the effects of cationic and nonionic surfactants on the properties of Y_2O_3: Eu^{3+} phosphors synthesized by a co-precipitation-molten salt method, Optik 126 (2015) 4800–4803, https://doi.org/10.1016/j.ijleo.2015.09.112.

[82] Y. Yang, X. Li, W. Feng, et al., Effect of surfactants on morphology and luminescent properties of $CaMoO_4$: Eu^{3+} red phosphors, J. Alloys Compd. 509 (2011) 845–848, https://doi.org/10.1016/j.jallcom.2010.09.106.

[83] M.C. Tan, L. Al-Baroudi, R.E. Riman, Surfactant effects on efficiency enhancement of infrared-to-visible upconversion emissions of $NaYF_4$: Yb-Er, ACS Appl. Mater. Interfaces 3 (2011) 3910–3915, https://doi.org/10.1021/am200768u.

[84] Y. Xie, D. Kocaefe, C. Chen, Y. Kocaefe, Review of research on template methods in preparation of nanomaterials, J. Nanomater. 2016 (2016), https://doi.org/10.1155/2016/2302595.

[85] D. Walsh, L. Arcelli, T. Ikoma, et al., Dextran templating for the synthesis of metallic and metal oxide sponges, Nat. Mater. 2 (2003) 386–390, https://doi.org/10.1038/nmat903.

[86] K.M. Koczkur, S. Mourdikoudis, L. Polavarapu, S.E. Skrabalak, Polyvinylpyrrolidone (PVP) in nanoparticle synthesis, Dalton Trans. 44 (2015) 17883–17905, https://doi.org/10.1039/c5dt02964c.

[87] C. Tao, P. Li, N. Zhang, et al., Improvement the luminescent property of $Y_3Al_5O_{12}$: Ce^{3+} by adding the different fluxing agents for white LEDs, Optik 179 (2019) 632–640, https://doi.org/10.1016/j.ijleo.2018.10.169.

[88] W.R. Liu, B. Lei, Flux-Assisted preparation and photoluminescence of emission-Tunable (Sr, Eu) $Al_2Si_2O_8$ phosphors, J. Nanosci. Nanotechnol. 18 (2018) 374–380, https://doi.org/10.1166/jnn.2018.14565.

[89] V. Sivakumar, A. Lakshmanan, Pyrolysis synthesis of Zn_2SiO_4: Mn^{2+} phosphors—effect of fuel, flux and co-dopants, J. Lumin. 145 (2014) 420–424, https://doi.org/10.1016/j.jlumin.2013.08.016.
[90] H.-W. Tseng, W.-C. Tzou, S. Wei, et al., Effects of synthesis temperature and Eu_2O_3 concentration on the crystalline phases and photoluminescence properties of $SrAl_2O_4$ phosphors, J. Mater. Res. Technol. 9 (2020) 14051–14060, https://doi.org/10.1016/j.jmrt.2020.10.003.
[91] J. Liu, Y.T. Wang, C.Y. Lin, C.F. Yang, Effect of synthesis temperature on the crystalline structures and photoluminescence properties of the green-light $Ca_{1.975}Eu_{0.025}MgSi_2O_7$ phosphors, Cryst. Growth Des. 20 (2020) 3154–3162, https://doi.org/10.1021/acs.cgd.0c00027.
[92] K.N. Shinde, V.B. Pawade, S.J. Dhoble, A. Hakeem, Impact of temperature on $Na_2Sr(PO_4)F$: Eu^{3+} phosphor, Luminescence 29 (2014) 352–356, https://doi.org/10.1002/bio.2551.
[93] E. Kaewnuam, J. Kaewkhao, Synthesis-temperature effect on the luminescence under light and UV excitation of Eu^{3+} doped lithium lanthanum borate phosphor, Mater. Today Proc. 5 (2018) 15086–15091, https://doi.org/10.1016/j.matpr.2018.04.062.
[94] A. Dulda, Morphology controlled synthesis of α-GaO(OH) nanoparticles: thermal conversion to Ga_2O_3 and photocatalytic properties, Adv. Mater. Sci. Eng. 2016 (2016), https://doi.org/10.1155/2016/3905625.
[95] Y.W. Zhang, J.H. Liu, R. Si, et al., Phase evolution, texture behavior, and surface chemistry of hydrothermally derived scandium (hydrous) oxide nanostructures, J. Phys. Chem. B 109 (2005) 18324–18331, https://doi.org/10.1021/jp051870b.
[96] A. Kumar, J.C.G.E. da Silva, K. Kumar, et al., Improvement in upconversion/down-shifting luminescence of Gd_2O_3: Ho^{3+}/Yb^{3+} phosphor through Ca^{2+}/Zn^{2+} incorporation and optical thermometry studies, Mater. Res. Bull. 112 (2019) 28–37, https://doi.org/10.1016/j.materresbull.2018.11.031.
[97] U. Vinoditha, K.M. Balakrishna, B.K. Sarojini, et al., Effect of Eu^{3+} doping on the structural, morphological and luminescence properties ZnO nanostructures, AIP Conf. Proc. 1953 (2018), https://doi.org/10.1063/1.5032567.
[98] K. Korthout, P.F. Smet, D. Poelman, Rare earth doped core-shell particles as phosphor for warm-white light-emitting diodes, Appl. Phys. Lett. 98 (2011) 1–4, https://doi.org/10.1063/1.3606540.
[99] H. Ding, H. Qin, S. Feng, et al., Full spectrum core-shell phosphors under ultraviolet excitation, Chem. Commun. 55 (2019) 12188–12191, https://doi.org/10.1039/c9cc04827h.
[100] C. Phosphors, J.A. Dorman, J.H. Choi, et al., High-quality white light using core–shell RE^{3+}: $LaPO_4$ (RE = Eu, Tb, Dy, Ce) phosphors, J. Phys. Chem. 116 (23) (2012) 12854–12860, 4.
[101] Y.X. Fu, Y.H. Sun, Comparative study of synthesis and characterization of mono-dispersed SiO_2 @ Y_2O_3: Eu^{3+} and SiO_2 @ Y_2O_3: Eu^{3+} @ SiO_2 core-shell structure phosphor particles, J. Alloys Compd. 471 (2009) 190–196, https://doi.org/10.1016/j.jallcom.2008.03.055.
[102] F. Hong, H. Xu, G. Pang, et al., Moisture resistance, luminescence enhancement, energy transfer and tunable color of novel core-shell structure $BaGeF_6$: Mn^{4+} phosphor, Chem. Eng. J. 390 (2020) 124579, https://doi.org/10.1016/j.cej.2020.124579.

[103] H.S. Jang, H. Yang, S.W. Kim, et al., White light-emitting diodes with excellent color rendering based on organically capped CdSe quantum dots and Sr_3SiO_5: Ce^{3+}, Li^+ phosphors, Adv. Mater. 20 (2008) 2696–2702, https://doi.org/10.1002/adma.200702846.
[104] S. Sureshkumar, B. Jothimani, T.M. Sridhar, B. Venkatachalapathy, Synthesis and characterization of gadolinium-doped ZnSe quantum dots for fluorescence imaging of cancer cells, RSC Adv. 6 (2016) 16081–16086, https://doi.org/10.1039/c5ra18773g.
[105] J. Planelles-Aragó, B. Julián-López, E. Cordoncillo, et al., Lanthanide doped ZnS quantum dots dispersed in silica glasses: an easy one pot sol-gel synthesis for obtaining novel photonic materials, J. Mater. Chem. 18 (2008) 5193–5199, https://doi.org/10.1039/b809254k.
[106] J. Sowik, M. Miodyńska, B. Bajorowicz, et al., Optical and photocatalytic properties of rare earth metal-modified ZnO quantum dots, Appl. Surf. Sci. 464 (2019) 651–663, https://doi.org/10.1016/j.apsusc.2018.09.104.
[107] A. Arivarasan, S. Bharathi, S. Ezhil Arasi, et al., Investigations of rare earth doped CdTe QDs as sensitizers for quantum dots sensitized solar cells, J. Lumin. 219 (2020) 116881, https://doi.org/10.1016/j.jlumin.2019.116881.
[108] H. Chu, D. Yao, J. Chen, et al., Double-emission ratiometric fluorescent sensors composed of rare-earth-doped ZnS quantum dots for Hg^{2+} detection, ACS Omega (2020), https://doi.org/10.1021/acsomega.0c00861.
[109] Y. Pu, L. Lin, J. Liu, et al., High-gravity-assisted green synthesis of rare-earth doped calcium molybdate colloidal nanophosphors, Chin. J. Chem. Eng. 28 (2020) 1744–1751, https://doi.org/10.1016/j.cjche.2020.03.023.
[110] A. Jain, P.G.J. Fournier, V. Mendoza-Lavaniegos, et al., Functionalized rare earth-doped nanoparticles for breast cancer nanodiagnostic using fluorescence and CT imaging, J. Nanobiotechnol. 16 (2018) 1–18, https://doi.org/10.1186/s12951-018-0359-9.
[111] L. Mejía, C.Z. Hadad, Effect of the Euclidean dimensionality on the energy transfer up-conversion luminescence, J. Photochem. Photobiol. Chem. 382 (2019) 111908, https://doi.org/10.1016/j.jphotochem.2019.111908.
[112] A. Paulraj, P. Natarajan, K. Munnisamy, et al., Photoluminescence efficiencies of nanocrystalline versus bulk Y_2O_3: Eu phosphor-revisited, J. Am. Ceram. Soc. 94 (2011) 1627–1633, https://doi.org/10.1111/j.1551-2916.2010.04246.x.
[113] F. Paquin, J. Rivnay, A. Salleo, et al., Multi-phase semicrystalline microstructures drive exciton dissociation in neat plastic semiconductors, J. Mater. Chem. C 3 (2015) 10715–10722, https://doi.org/10.1039/b000000x.
[114] J. Lin, Y. Huang, J. Zhang, et al., Characterization and photoluminescence properties of Tb-doped SiO_2 nanowires as a novel green-emitting phosphor, Chem. Mater. 19 (2007) 2585–2588, https://doi.org/10.1021/cm070271+.
[115] Y. Li, T. Wang, W. Ren, et al., BiOCl: Er^{3+} nanosheets with tunable thickness for photon avalanche phosphors, ACS Appl. Nano Mater. 2 (2019) 7652–7660, https://doi.org/10.1021/acsanm.9b01735.
[116] T.C. Ozawa, K. Fukuda, K. Akatsuka, et al., $Eu_{0.56}Ta_2O_7$: a new nanosheet phosphor with the high intrananosheet site photoactivator concentration, J. Phys. Chem. C 112 (2008) 1312–1315, https://doi.org/10.1021/jp711699c.
[117] S. Sakurai, T. Nakamura, S. Adachi, Related content Synthesis and properties of Rb_2GeF_6: Mn emitting phosphors, Jpn. J. Appl. Phys. 57 (2018) 022601.

[118] A.L. Fanai, U. Khan, S. Rai, Luminescence enhancement of Pr^{3+} doped sol-gel silica glass as a result of Al^{3+} co-doping, J. Non-Cryst. Solids 503 (504) (2019) 89–93, https://doi.org/10.1016/j.jnoncrysol.2018.09.027.

[119] M. Dejneka, E. Snitzer, R.E. Riman, Spectroscopic characterization of Eu^{3+}-doped inorganic and alkoxide sol-gel derived fluorozirconate glass and zirconium fluoride gels, J. Non-Cryst. Solids 202 (1996) 23–34, https://doi.org/10.1016/0022-3093(96)00170-6.

[120] R. Kaur, A. Khanna, Photoluminescence and thermal properties of trivalent ion-doped lanthanum tellurite anti-glass and glass composite samples, J. Lumin. 225 (2020) 117375, https://doi.org/10.1016/j.jlumin.2020.117375.

[121] N. Chanthima, T. Sareein, Y. Tariwong, et al., Investigations on luminescence properties of Ce^{3+} ion doped bismuth borophosphate glasses, Appl. Mech. Mater. 879 (2018) 22–26, https://doi.org/10.4028/www.scientific.net/amm.879.22.

[122] N. Mohamed, J. Hassan, K.A. Matori, et al., Influence of Pr doping on the thermal, structural and optical properties of novel SLS-ZnO glasses for red phosphor, Results Phys. 7 (2017) 1202–1206, https://doi.org/10.1016/j.rinp.2017.03.018.

[123] R. Ruamnikhom, P. Limsuwan, M. Horprathum, et al., Up- and downconversion luminescence properties of Nd^{3+} ions doped in $Bi_2O_3-BaO-B_2O_3$ glass system R, Adv. Mater. Sci. Eng. 13 (2019), https://doi.org/10.1016/j.rinp.2019.102165.

[124] C. Madhukar Reddy, B. Deva Prasad Raju, N. John Sushma, et al., A review on optical and photoluminescence studies of RE^{3+} (RE=Sm, Dy, Eu, Tb and Nd) ions doped LCZSFB glasses, Renew. Sustain. Energy Rev. 51 (2015) 566–584, https://doi.org/10.1016/j.rser.2015.06.025.

[125] S. Kaur, O.P. Pandey, C.K. Jayasankar, N. Chopra, Spectroscopic, thermal and structural investigations of Dy^{3+} activated zinc borotellurite glasses and nano-glass-ceramics for white light generation, J. Non-Cryst. Solids 521 (2019), https://doi.org/10.1016/j.jnoncrysol.2019.119472.

[126] J. Zavadil, Z.G. Ivanova, P. Kostka, et al., Photoluminescence study of Er-doped zinc – sodium – antimonite glasses, J. Alloys Compd. 611 (2014) 111–116.

[127] M.M. Martins, L.R.P. Kassab, D.M. da Silva, C.B. de Araújo, Tm^{3+} doped Bi_2O_3-GeO_2 glasses with silver nanoparticles for optical amplifiers in the short-wave-infrared-region, J. Alloys Compd. 772 (2019) 58–63, https://doi.org/10.1016/j.jallcom.2018.08.146.

[128] J. He, Z. Zhou, H. Zhan, et al., 2.85 μm fluorescence of Ho-doped water-free fluorotellurite glasses, J. Lumin. 145 (2014) 507–511, https://doi.org/10.1016/j.jlumin.2013.08.020.

Degradation of rare-earth-activated phosphors

6

Neeraj Verma [1], *Teerath Ram* [1], *Vikas Dubey* [2], *Neha Dubey* [1] *and Jagjeet Kaur* [1]
[1]Department of Physics, Govt. V.Y.T. PG. Autonomous College Durg, Durg, Chhattisgarh, India; [2]Department of Physics, Bhilai Institute of Technology Raipur, Naya Raipur, Chhattisgarh, India

1. Introduction

To date, a number of new and promising LED phosphors have been invented, developed, or modified that have demonstrated their suitability in white LEDs. These phosphors include oxides, sulfides, phosphates, and nitrides. Among those, nitride phosphors are considered a novel class of luminescent materials and are getting much more attention in recent years. Nitride phosphors show significantly red-shifted excitation and emission spectra, high chemical stability, and very small thermal degradation; thus they are very suitable for use in white LEDs, typically warm white LEDs, and high-power LEDs [1].

1.1 Small thermal quenching

Thermal quenching is a measure of thermal stability of phosphors, which is one of the most important technical parameters in their practical uses. Thermal quenching is one of the reasons for the degradation of LED lamps, having a great influence in LED's lifetime and chromaticity. To attain high efficiency and long lifetime of LED lamps, it is therefore required that the thermal quenching of a phosphor be as small as possible. Thermal quenching relies significantly on the crystal structure and chemical composition of the host lattice [1].

However, a number of lanthanide activators display emission spectra with spectral intensities and positions that are easily affected by environmental temperature. Heat is generally detrimental to phosphors, and phosphor efficiency decreases through nonradiative relaxation as device temperature increases. This phenomenon indicates TQ, and, consequently, phosphors shift to emission peak wavelengths, which decrease luminescent intensity. Excited electrons can relax through radiative (Fig. 6.1B [*dashed green line*]) and nonradiative (Fig. 6.1B [*gray line*]) processes such as photon emission and collisional quenching, respectively. The TQ property can also be elucidated by using the configurational coordinate diagram, in which thermal energy assists excited electrons across the intersection point between the excited and ground-state parabola curves, thereby resulting in nonradiative decay. Fig. 6.1C shows the relative positions between the localized 5d electron states of the activators and the delocalized conduction band states of the hosts. First, auto-ionization spontaneously occurs, and no 5d−4f

Rare-Earth-Activated Phosphors. https://doi.org/10.1016/B978-0-323-89856-0.00002-X

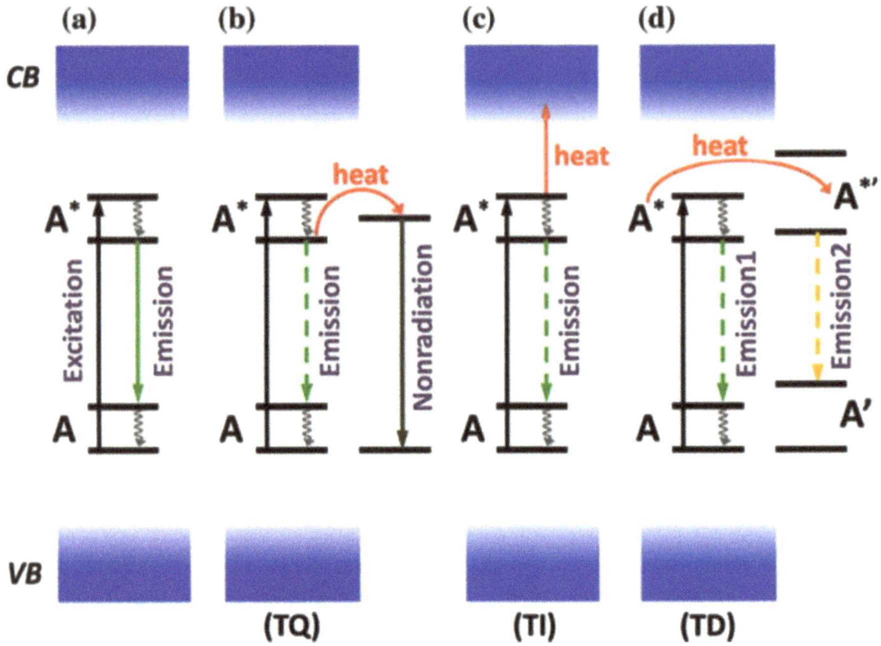

Figure 6.1 Luminescent mechanisms of thermal effects on inorganic solids. (A) Emission from a luminescent activator on excitation. (B) TQ results in a nonradiative pathway associated with heat. (C) TI excites electrons toward the conduction band through heat. (D) TD can lead to other emissions as a result of heat. A and A* represent the ground and excited states of the activator, respectively. A′ and A*′ represent the ground and excited states of the activator with different charges, respectively. VB and CB represent the valence and conduction bands of the host, respectively.
Reprinted from Ref. [2], Open Access.

Ref. No.	Authors and year	Study	Remarks
[9]	Wang et al. (2020)	$CaAlSiN_3:Eu^{2+}$ phosphor-in-glass	Photoluminescence properties and thermal degradation
[10]	Takeda et al. (2019)	Degradation of phosphor excitation efficiency	Short wavelength vacuum ultraviolet radiation in plasma discharge devices
[11]	Kong et al. (2019)	Micro-structural corrosion mechanism	Thermal Degradation of $BaMgAl_{10}O_{17}:Eu^{2+}$ phosphor
[12]	Jaffer et al. (2019)	Cathodoluminescence degradation	Bi-doped La_2O_3 and La_2O_2S phosphor powders

emission is observed when the lowest 5d state is above the bottom of the conduction band. Such cases include $Ba_{10}(PO_4)_4(SiO_4)_2:Eu^{2+}$ [3], $Ln_2O_3:Ce^{3+}$ [4], $LaAlO_3:Ce^{3+}$ [5], and the Eu^{2+} on trivalent RE sites in oxide compounds [6]. Second, the 5d states of the activators are below the conduction band of the hosts in most 5d−4f emission situations. The 5d electrons are ionized to the conduction band through thermal ionization, which depends on the energy EdC between the 5d state (d) of the activator and the bottom of the conduction band (C) [7,8]. The activator Eu^{2+}, located in the fluffy structure, is easily oxidized to the trivalent species at high temperatures. Therefore, the existence of Eu^{3+} can be observed in the PL and X-ray absorption spectra. This phenomenon is called the "thermal-degradation effect" (Fig. 6.1D).

2. Comparative study of some rare-earth-activated phosphor for degradation

Red component is crucial to achieve high-quality white light with excellent color-rendering property and color uniformity. Thus, red-emitting PiG composed of red phosphor and glass matrix is a crucial component for high-quality white LEDs [12,13].

Recently, PiG materials prepared by cosintering of glass powders and phosphor powders under low temperature (less than 800°C) have been attracted a booming attention owing to its merits of high thermal conductivity and low thermal expansion coefficient [14−18]. TeO_2-based and SiO_2-based phosphor-in-glass (PiG)s have been applied successfully to disperse $Ce^{3+}:Y_3Al_5O_{12}$ (Ce:YAG) yellow phosphors [2,19]. In our previous work, $Lu_3Al_5O_{12}:Ce^{3+}$ phosphors were successfully incorporated into the tellurite glass to obtain green-emitting PiG composite by the cosintering method [20]. Among them, the red-emitting PiG materials in favor of improving the color index in wLED industry are eager to demand, and a robust red-emitting PiG has been fabricated by distributing the $CaAlSiN_3:Eu^{2+}$ phosphor powders into $ZnO-B_2O_3-BaO-Al_2O_3$ glass matrix under the temperature of 620−680°C [21]. Until now, the luminescent intensity of the red-emitting PiG is still limited by the high cosintering temperature owing to the thermal quenching and decomposition of the phosphor at >600°C [22−24]. It is highly desirable to investigate the thermal degradation and dispersing of red phosphor in an ultra-low melting glass. The tin fluorophosphate (Sn−P−F−O) glass is an ideal encapsulated material in view of significant advantages of low melting temperature, good chemical stability, and facile synthesis [25]. Further, the thermal stability of red phosphors is vital to the optical properties of PiGs, which will be greatly influenced by the composition, crystal structure, and particle morphology during the different preparation conditions. $CaAlSiN_3:Eu^{2+}$ phosphor has been regarded as a promising red component in white LEDs due to its high quantum efficiency, excellent thermal quenching characteristics, and high chemical stability [26−30].

3. Characterization of some rare-earth-activated phosphor (degradation study)

Fig. 6.2 shows the projection of the crystal structure of $CaAlSiN_3$ crystal viewed along [001] and the local coordination of Si/Al and Ca.

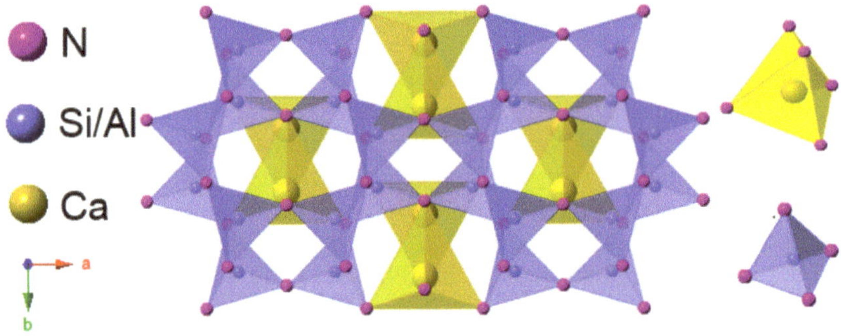

Figure 6.2 Projection of the crystal structure of $CaAlSiN_3$ crystal viewed along [001] and the local coordination of Si/Al and Ca. Wang et al. [9].

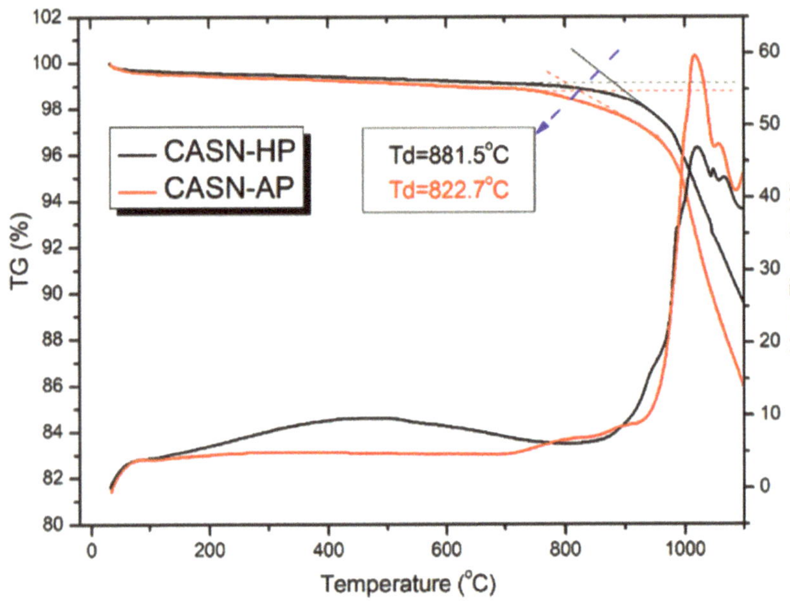

Figure 6.3 TG and DSC curves of CASN-HP and CASN-AP phosphors [9].

Fig. 6.3, the TG curves for both CASN-HP and CASN-AP phosphors exhibit similar trends of weight loss when heated up to 650°C, which can be ascribed to the release of an absorbed water molecule. When further heated to 1200°C, CASN-HP and CASN-AP phosphors present the weight loss of 10.6% and 13.8%, respectively, which can be ascribed to the decomposition of the phosphors. Compared to the CASN-HP phosphor, CASN-AP phosphor undergoes a lower initial decomposition temperature (822.7°C) and more weight loss. Although the same decomposition heat flow peak at 1017°C can be found for CASN-HP and CASN-AP phosphors, the initial decomposition temperature of CASN-AP phosphor determined by the DSC curve is corresponding with the results of the thermogravimetric analysis.

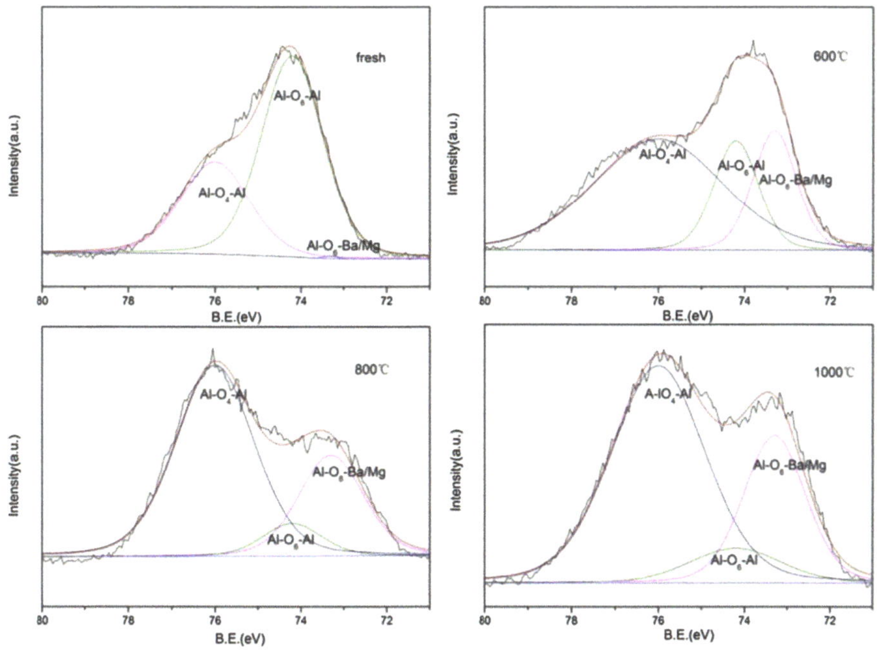

Figure 6.4 Al element XPS spectra of BAM phosphors with different heat treatment temperatures. Kong et al. [11].

Therefore, it can be concluded from the results of FESEM observation and thermal analysis that the thermal stability of the CASN-HP phosphor is superior to that of CASN-AP phosphor owing to the uniform prism-like morphology in a high-pressure preparation condition [9].

In Fig. 6.4, the XPS spectra of Al^{2+} band of the BAM phosphors prepared at different heat-treatment temperatures are shown including the fitting curves. It could be seen that the XPS spectra of Al elements were obviously different from other elements, and the peaks were obvious asymmetry [11].

4. Thermal degradation mechanism

BAM phosphors inevitably have a lot of crystal defects due to a physical crushing and mechanical grinding process applied after a high-temperature solid-state reaction. The charge imbalance and surface energy increase can be provided by the process. Therefore, it is necessary to restore electrical neutrality and decrease surface energy through recombination or adjustment of surface structures and adsorption of foreign molecules and ions. That will form a structural difference between phosphor surface and bulk. These defects of the matrix material will reduce the excitation light absorption in the luminescence process, and stimulate the nonradiative transitions to consume a part of the absorbed energy, thereby reducing the luminous efficiency. The concentration of defect gradually increases with increasing temperature according to the Boltzmann distribution. The higher temperature rises, the more lattice distortion and defect concentration increase. Therefore, the high temperature would aggravate the distortion of microstructure in this experiment. Oppositely, the oxidation of Eu^{2+} ion to Eu^{3+} state may be considered as a crystal structure transformation to an equilibrium state with lower free energy. In this process, the defect concentration decreases. The elimination of defects was attributed to the adjustment of microstructures. According to the evidence about obvious microstructural corrosion researched by TL and XPS, we gave a microstructural corrosion mechanism on the thermal degradation of BAM phosphors-based microstructural changes. Fig. 6.5 shows the graph of microstructural modification (Fig. 6.6).

Fig. 6.7 presents the fitting of the Bi 4f spectra for the sputter cleaned samples. These were fitted by two doublets, each having the expected spin-orbit separation of 5.3 eV [32]. The large energy separation of the doublets (~2 eV) is similar to that reported by Eom et al. [31] for pulsed laser deposited Bi_2O_3 thin films. The lower binding energy doublet corresponds to metallic Bi. As, in general, elemental Bi is known to oxidize when exposed to air, the existence of metallic Bi on the surface of the samples in this study was unexpected and it may be that metallic Bi nanoparticles were instead embedded just below the surface inside the host. The higher binding energy doublet

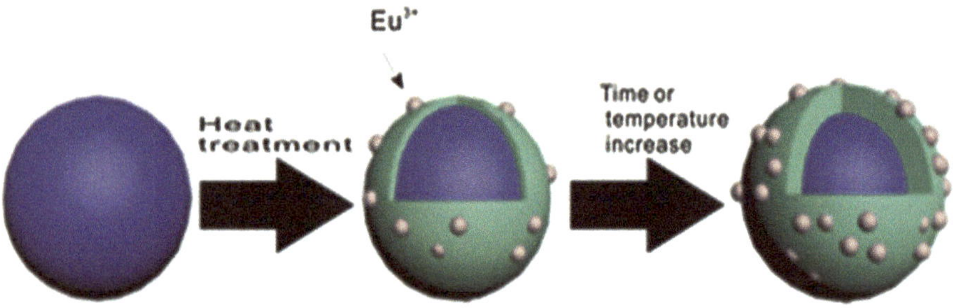

Figure 6.5 Model diagram of the thermal degradation process. Kong et al. [11].

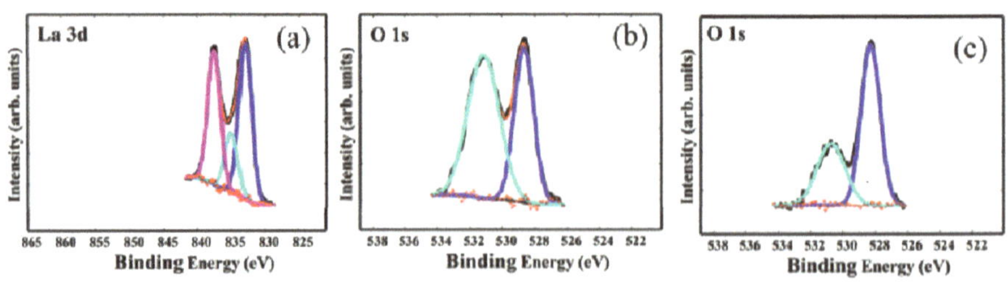

Figure 6.6 Fitting of selected XPS spectra of degraded samples given in the figure reported by Jaffar et al. [12]. (A) La 3d, after sputtering. (B) O 1s, before sputtering. (C) O 1s, after sputtering [12].

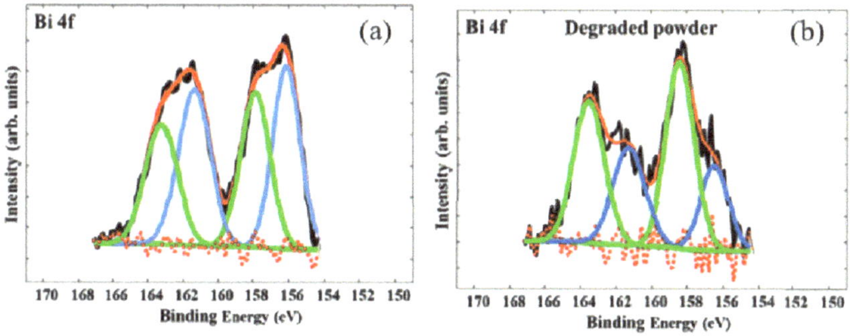

Figure 6.7 Fitting of the Bi 4f XPS spectra obtained after sputter cleaning of the (A) fresh sample and (B) degraded sample [12].

corresponds to Bi^{3+} ions bonded to oxygen. Although the simplest explanation for this peak is that it originates from Bi^{3+} ions substituting La^{3+} ions in the host, it is also possible that there is a contribution from Bi^{3+} ions in Bi_2O_3 formed by atmospheric oxidation of Bi metal, which itself was formed by segregation of Bi^{3+} ions to the surface and their reduction during annealing [12].

5. Conclusion

It is concluded by the study of degradation of rare-earth-activated phosphors most of the phosphors are degraded by thermal quenching mechanism. Our chapter represents a few of the samples phosphors for the degradation.

References

[1] R.-J. Xie, Y. Qiang Li, H. Yamamoto, Nitride Phosphor for Solid State Lighting Book, CRC press, 2011.

[2] Ru-Shi Liu Springer Phosphors, Up Conversion Nano Particles, Quantum Dots and Their Applications, 2017.

[3] Z.J. Yu, X.W. Huang, W.D. Zhuang, X.Z. Cui, H.W. Li, Crystal structure transformation and luminescent behavior of the red phosphor for plasma display panels, J. Alloys Compd. 390 (2005) 220.

[4] Z.J. Yu, X.W. Huang, W.D. Zhuang, X.Z. Cui, H.Q. He, H.W. Li, Co-precipitation preparation and luminescent behavior of (Y, Gd) BO3:Eu phosphor, J. Rare Earths 22 (2004) 829.

[5] H.B. Xu, W.D. Zhuang, X.F. Wen, R.H. Liu, Y.S. Hu, T. Xia, Effect of Li + Ions doping on structure and luminescence of (Y, Gd)BO3:Tb3+, J. Rare Earths 28 (2010) 701.

[6] H.B. Xu, W.D. Zhuang, T. Xia, R.H. Liu, Y. Tao, Y. Huang, Synthesis and properties of PDP green phosphor (Y, Gd)BO3 Tb3+, Chin. J. Rare Met. 34 (2010) 887.

[7] H.B. Xu, W.D. Zhuang, Y.S. Hu, T. Xia, X.F. Wen, R.H. Liu, Influence of alkaline metal ions doping on properties of (Y, Gd)BO3:Tb, J. Chin. Rare Earths Soc. 28 (2010) 268.
[8] S.S. Zhang, W.D. Zhuang, C.L. Zhao, Y.S. Hu, H.Q. He, X.W. Huang, Study on (Y, Gd) 3(Al, Ga)5O12:Ce^{3+} phosphor, J. Rare Earths 22 (2004) 118.
[9] R. Wang, M.Y. Wang, G. Li, J.H. Zhang, Y.J. Zhang, H. Lin, E.Y.B. Pun, D.S. Li, Red-emitting improvement of $CaAlSiN_3$:Eu^{2+} phosphor-in-glass: insight into the effect of atmospheric pressure preparation on photoluminescence properties and thermal degradation, J. Lumin. 225 (2020) 117390.
[10] E. Takeda, T. Zukawa, T. Ishibashi, K. Yoshino, Y. Morita, M. Fujii, Mechanisms for the degradation of phosphor excitation efficiency by short wavelength vacuum ultraviolet radiation in plasma discharge devices, J. Phys. Chem. Solid. 124 (2019) 274−280.
[11] S. Konga, J. Xiaa, Z. Zhanga, W. Penga, T. Qia, Z. Huang, A micro-structural corrosion mechanism on the thermal degradation of BaMgAl10O17: Eu2+ phosphor, Optik. - Int. J. Light Electr. Optics 181 (2019) 202−208.
[12] B.M. Jaffar, H.C. Swart, H.A.A. Seed Ahmed, A. Yousif, R.E. Kroon, Cathodoluminescence degradation of Bi doped La_2O_3 and La_2O_2S phosphor powders, Phys. B Condens. Matter 574 (2019) 411659.
[13] P. Zheng, S.X. Li, L. Wang, T.L. Zhou, S.H. You, T. Takeda, N. Hirosaki, R.J. Xie, Unique color converter architecture enabling phosphor-in-Glass (PiG) films suitable for high-power and high-luminance laser driven white lighting, ACS Appl. Mater. Interfaces 10 (2018) 14930−14940.
[14] H. Lin, T. Hu, Y. Cheng, M.X. Chen, Y.S. Wang, Glass Ceramic Phosphors: towards long-lifetime high-power white light-emitting-diode applications-a review, Laser Photon. Rev. 12 (2018) 1700344.
[15] D.Q. Chen, W.D. Xiang, X.J. Liang, J.S. Zhong, H. Yu, M.Y. Ding, H.W. Lu, Z.G. Ji, Advances in transparent glass-ceramic phosphors for white light-emitting diodes-a review, J. Eur. Ceram. Soc. 35 (2015) 859−869.
[16] J.K. Deng, W. Li, H.R. Zhang, Y.L. Liu, B.F. Lei, H.M. Zhang, L.S. Liu, X. Bai, H.Y. Luo, H.Z. Liu, W.R. Liu, J. Wang, Eu3þ-doped phosphor-in-glass: a route toward tunable multicolor materials for near-UV high-power warm-white LEDs, Adv. Opt. Mater. 5 (2017) 1600910.
[17] Y.H. Kim, N.S.M. Viswanath, S. Unithrattil, H.J. Kim, W.B. Im, Review—phosphor plates for high-power LED applications: challenges and opportunities toward perfect lighting, ECS J. Solid State Sci. Technol. 7 (2018) R3134−R3147.
[18] H. Lin, B. Wang, J. Xu, R. Zhang, H. Chen, Y. Yu, Y. Wang, Phosphor-in-glass for high-powered remote-type white AC-LED, ACS Appl. Mater. Interfaces 6 (2014) 21264−21269.
[19] R. Zhang, H. Lin, Y.L. Yu, D.Q. Chen, J. Xu, Y.S. Wang, A new-generation color converter for high-power white LED: transparent Ce3þ: YAG phosphor-in-glass, Laser Photon. Rev. 8 (2014) 158.
[20] D. Han, D.S. Li, Y.J. Zhang, H. Lin, E.Y.B. Pun, Crystalline-phase depuration in Ce3 þ activated Lu3Al5O12-tellurite green emitting phosphor-in-glass, J. Lumin. 214 (2019) 116531.
[21] Q.Q. Zhu, X. Xu, L. Wang, Z.F. Tian, Y.Z. Xu, N. Hirosaki, R.J. Xie, A robust red emitting phosphor-in-glass (PiG) for use in white lighting sources pumped by blue laser diodes, J. Alloys Compd. 702 (2017) 193−198.
[22] Z. Lin, H. Lin, J. Xu, F. Huang, H. Chen, B. Wang, Y. Wang, Highly thermal-stable warm w-LED based on Ce: YAG PiG stacked with a red phosphor layer, J. Alloys Compd. 649 (2015) 661−665.

[23] J. Zhong, D. Chen, Y. Zhou, Z. Wan, M. Ding, Z. Ji, Stable and chromaticity-tunable phosphor-in-glass inorganic color converter for high-power warm white light-emitting diode, J. Eur. Ceram. Soc. 36 (2016) 1705–1713.

[24] Z. Lin, H. Lin, J. Xu, F. Huang, H. Chen, B. Wang, Y. Wang, A chromaticity-tunable garnet-based phosphor-in-glass color converter applicable in w-LED, J. Eur. Ceram. Soc. 36 (2016) 1723–1729.

[25] W. Mao, M.Z. Cai, W.Q. Xie, P.P. Li, W.Y. Zhen, S.Q. Xu, J.J. Zhang, Tunable white light in trivalent europium single doped tin fluorophosphates ultra-low melting glass, J. Alloys Compd. 805 (2019) 205–210.

[26] W.Q. Liu, D. Wu, H. Chang, R.X. Duan, W.J. Wu, G. Amu, K.F. Chao, F.Q. Bao, O. Tegus, The enhanced red emission and improved thermal stability of $CaAlSiN_3$:Eu2þ phosphors by using nano-EuB6 as raw material, Nanomaterials 8 (2018) 66.

[27] S.X. Li, X.J. Liu, R.H. Mao, Z.R. Huang, R.J. Xie, Red-emission enhancement of the $CaAlSiN_3$:Eu2þ phosphor by partial substitution for Ca_3N_2 by $CaCO_3$ and excessive calcium source addition, RSC Adv. 93 (2015).

[28] Q.Q. Zhu, X.J. Wang, L. Wang, N. Hirosaki, T. Nishimura, Z.F. Tian, Q. Li, Y.Z. Xu, X. Xu, R.J. Xie, β-sialon: Eu phosphor-in-glass: a robust green color converter for high power blue laser lighting, J. Mater. Chem. C 3 (2015) 10761–10766.

[29] K. Uheda, N. Hirosaki, Y. Yamamoto, A. Naito, T. Nakajima, H. Yamamoto, Luminescence properties of a red phosphor, $CaAlSiN_3$:Eu2þ for white light-emitting diodes, Electrochem. Solid State Lett. 9 (2006) H22–H25.

[30] T. Suehiro, R.J. Xie, N. Hirosaki, Gas-Reduction Nitridation synthesis of $CaAlSiN_3$: Eu^{2+} fine powder phosphors for solid-state lighting, Ind. Eng. Chem. Res. 53 (2014) 2713–2717.

[31] J.-ho Eom, H.-june Jung, J.-hee Han, J.-yong Lee, S.-gil Yoon, formation of bismuth nanocrystals in Bi_2O_3 thin films grown at 300 K by pulsed laser deposition for thermoelectric applications, ECS J. Solid State Sci. Technol. 3 (2014) 315–319.

[32] J.F. Moulder, W.F. Stickle, P.E. Sobol, K.D. Bomben, J. Chastain (Eds.), Handbook of X-Ray Photoelectron Spectroscopy, Revised Ed., Perkin Elmer Corporation, Eden Prairie, Minnetsoa, USA, 1992, 978-0-9627026-2-4.

Further reading

[1] G.T. Bauer, Kinetics of X-ray irradiation-caused degradation of photoluminescence of some phosphors, J. Electrochem. Soc. 123 (1976) 79.

Spectroscopy techniques for rare-earth-activated phosphors

7

Sadhana Agrawal
Physics, National Institute of Technology Raipur, Raipur, Chhattisgarh, India

1. Introduction

Rare-earth-activated phosphors have attracted researchers because of their enhanced energy efficiency, color rendering index (CRI), lumen output, and greater radiation stability [1]. In the trivalent rare-earth ions, the emissions take place due to 4f electrons that remain unperturbed in the host lattice [2]. Because of this, rare-earth-activated phosphors have high luminescence intensity, greater quantum yield, high luminescence efficiency, high stability, etc. [3] and hence resulting in their application in X-ray photography, phosphor thermometry, color TV picture tubes, fluorescent lamps, light-emitting diodes, etc. [4].

1.1 Rare-earth elements

Rare-earth elements along with yttrium and scandium constitute the lanthanide series given by International Union of Pure and Applied Chemistry. Rare-earth elements are further subdivided into two categories: light rare-earth elements and heavy rare-earth elements depending upon their atomic number [5].

Rare earths are elements with interesting characteristics. These are brittle, hard, malleable, ductile elements and are stable in the atmosphere [6]. Rare-earth elements react very slowly with water and are soluble in different acids. The salts of rare earths are generally white, pink, green, and yellow in color. They are extracted from different ores such as monazite, bastnasite, and gadolinite [7]. The partially filled 4f electrons surrounded by $4p^6$ and $5s^2$ electrons in the rare-earth elements contribute to the luminesce phenomena [8]. They give various emissive wavelengths in ultraviolet, visible, and near-infrared regions, which found rare-earth elements, suitable for different practical applications. Some of the transitions of different rare-earth elements are described here.

- **Europium**—The trivalent europium (Eu^{3+}) ions give emission in the red region corresponding to $^5D_0 - {}^7F_J$ ($J = 0, 1, 2, 3, 4, 5, 6$) transitions [9]. These hypersensitive transitions occur due to electric dipole and magnetic dipole transitions (Fig. 7.1).
- **Terbium**—The trivalent terbium (Tb^{3+}) ions give emission in the green, blue, and yellow regions corresponding to $^5D_3$–^7F_J, $^5D_4$–^7F_J ($J = 6, 5, 4, 3, \ldots$), respectively [10] (Fig. 7.2).
- **Ytterbium**—The ytterbium (Yb^{3+}) ions give emission in the red region corresponding to $4f^{14}$–$4f^{13}5d^1$ transitions at 350 and 550 nm, respectively [11]. The transitions occur due to the crystal field splitting of 5d levels (Fig. 7.3).

Rare-Earth-Activated Phosphors. https://doi.org/10.1016/B978-0-323-89856-0.00009-2

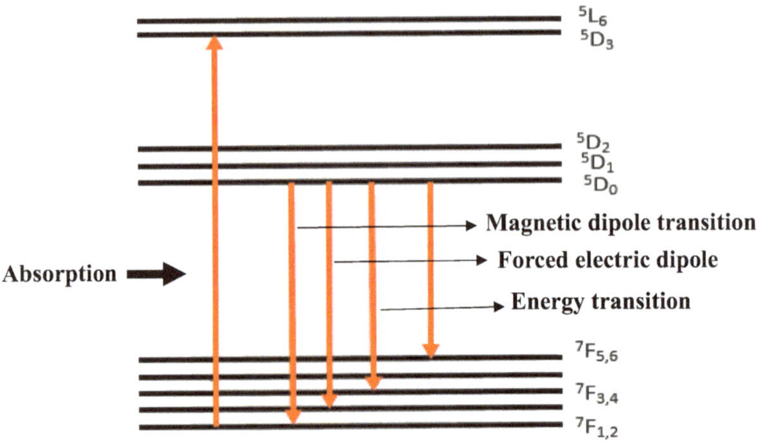

Figure 7.1 Energy level diagram of Eu^{3+}.

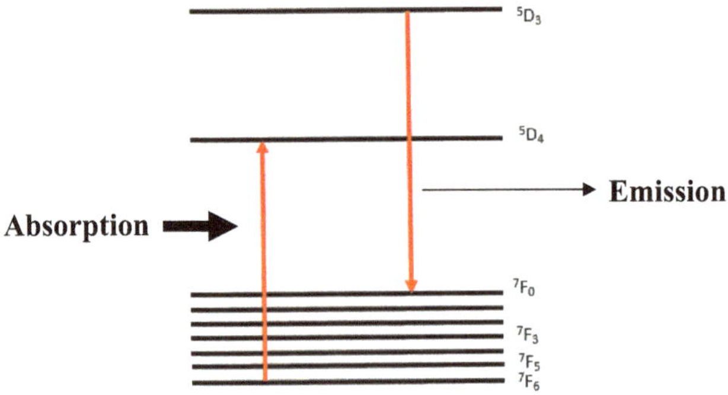

Figure 7.2 Energy level diagram of Tb^{3+}.

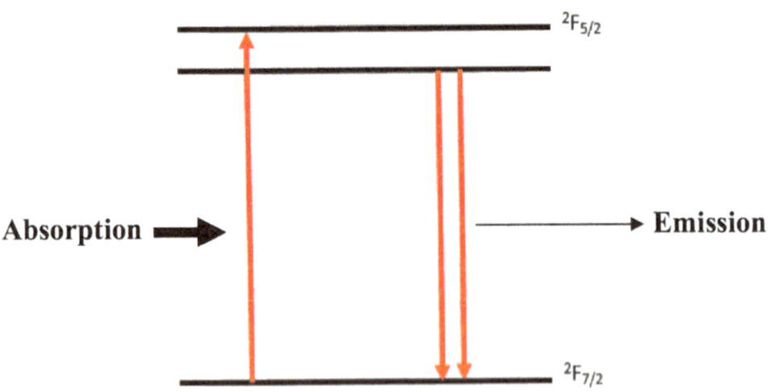

Figure 7.3 Energy level diagram of Yb^{3+}.

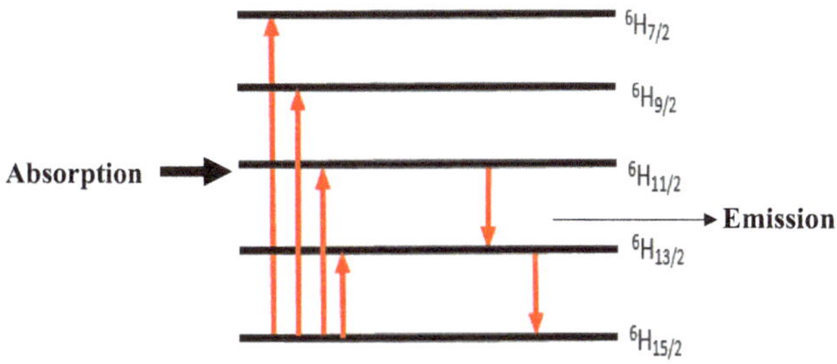

Figure 7.4 Energy level diagram of Dy^{3+}.

- **Dysprosium**—The dysprosium (Dy^{3+}) ions give hypersensitive transition at $^6H_{13/2}$ (570 nm) and $^6H_{15/2}$ (470 nm), respectively, with white emission and in general white emission is being turned into yellow [12] (Fig. 7.4).
- **Samarium**—The trivalent samarium (Sm^{3+}) ions give emission in an orange-red region corresponding to $^4G_{5/2}$–^6H_J ($J > 5/2$) transition of $4f^5$ levels [13] (Fig. 7.5).
- **Gadolinium**—The gadolinium (Gd^{3+}) ions give transitions at $^8S_{7/2}$–^6P_J ($J > 7/2$) in the blue and green region at 460 and 550 nm, respectively [14] (Fig. 7.6).
- **Praseodymium**—The praseodymium (Pr^{3+}) ion gives transition due to the participation of $4f^2$ electrons. If the transition occurs by $3P^0$ level it gives green emission; however, if the transition occurs due to $^3P_0$–$^3H_6$ and $^3P_0$–$^3F_2$, it gives red emission [15] (Fig. 7.7).

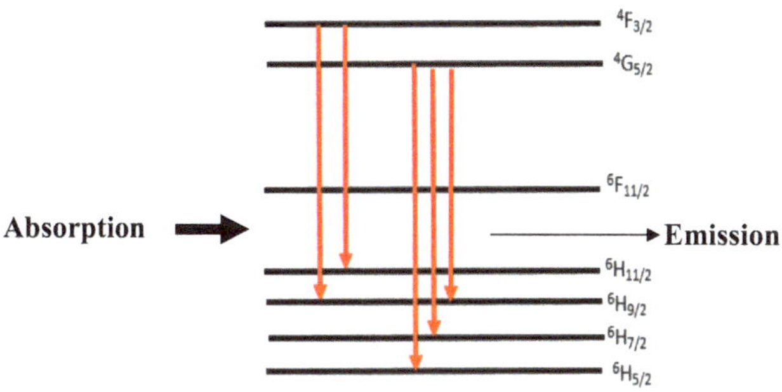

Figure 7.5 Energy level diagram of Sm^{3+}.

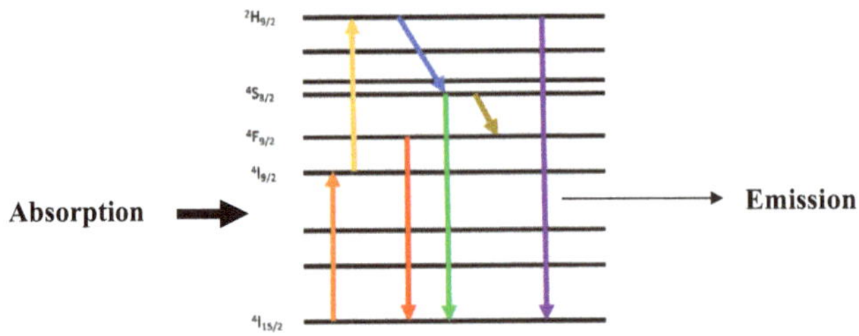

Figure 7.6 Energy level diagram of Gd^{3+}.

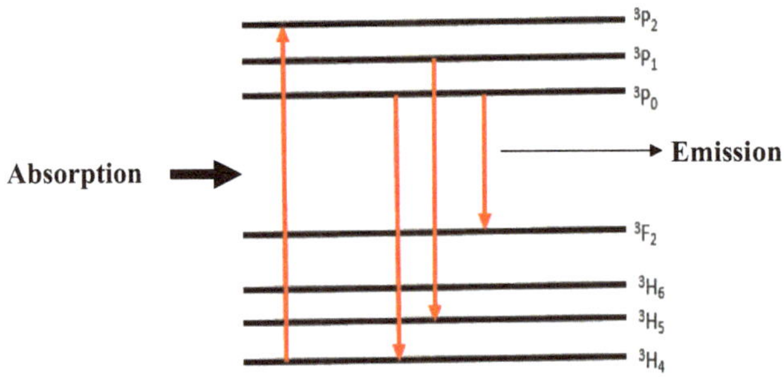

Figure 7.7 Energy level diagram of Pr^{3+}.

1.2 Rare-earth-activated phosphors and their synthesis

Substitution of rare-earth elements in various host lattices is done for their applications in LEDs, laser hosts, lithium-ion batteries, sensors, actuators, chips, amplifiers, catalytic converters, etc. [16]. Oxides, borates, ferrites, apatites, etc. are some of the hosts in which rare-earth ions can easily be substituted to enhance the luminescent properties of phosphors.

Aka et al. reported that calcium oxide doped with Dy^{3+} ions shows the application in laser materials, light-emitting diodes, scintillators, dosimeters, etc. [17]. Petit et al. reported about $YCa_4O(BO_3)_3$ phosphor as good laser material operating at the wavelength 1.06 μm [18]. Som et al. in the luminescence study of gadolinium-doped yttrium oxide phosphors reported an intense peak in the red region corresponding to $5D_0{-}7F_2$ transition [19]. Silver et al. reported that the erbium-doped and ytterbium-codoped yttrium oxide phosphors show emission in blue, blue-green, and green-yellow regions [20]. Tessari et al. reported that europium-doped yttrium oxide phosphors show an increase in luminescence intensity with increasing crystallite size [21]. Lu and Jagannathan et al. reported that $Y_3Al_5O_{12}$:Ce phosphors show an increase in luminescence intensity when crystallite size changes from bulk to microcrystalline range [22]. Vijyaprasath et al. reported that ZnO:Gd nanoparticles show band edge emission in visible range [23]. Mathan Kumar et al. reported the change in bandgap by doping of Gd ions in WO_3/TiO_2 composites [24]. Kiisk et al. reported about the rare-earth-based oxide materials for application in photoluminescence-based gas sensors and reported that the sensing mechanism is mainly due to fluorescence [25]. Kumar et al. reported about the luminescence properties of rare-earth-doped oxide materials that show application in light-emitting diodes [26]. Xia reported about the rare-earth-doped magnetic oxide and shows the application in magnetic sensors, spintronic devices, solid oxide fuel cells, magnetic refrigeration, biomedicine, and catalysts [27]. Muthulselvi reported about the rare-earth-doped reduced graphene oxide and shows the application in capacitors [28]. The synthesis of phosphors plays a vital role in the structural, morphological, and spectroscopic properties. Among various synthesis methods, the conventional solid-state reaction technique has been selected here to synthesize rare-earth-activated phosphors in which starting materials are taken in powder form. The reactants do not react in normal time scale and temperature, so at sufficiently high temperature and over a certain time scale, along with manual grinding the synthesis process is carried out [29]. To aid homogenization and proper mixing, acetone is added during the grinding process. The powder is then calcined and sintered at a high temperature depending upon the material properties. A schematic diagram of the solid-state reaction method is shown in Fig. 7.8.

Substitution of rare earths in the host lattice results in enhanced luminescence properties. These enhanced properties are investigated using various spectroscopic techniques such as ultraviolet spectroscopy (UV–Vis), photoluminescence (PL) spectroscopy, and thermoluminescence spectroscopy.

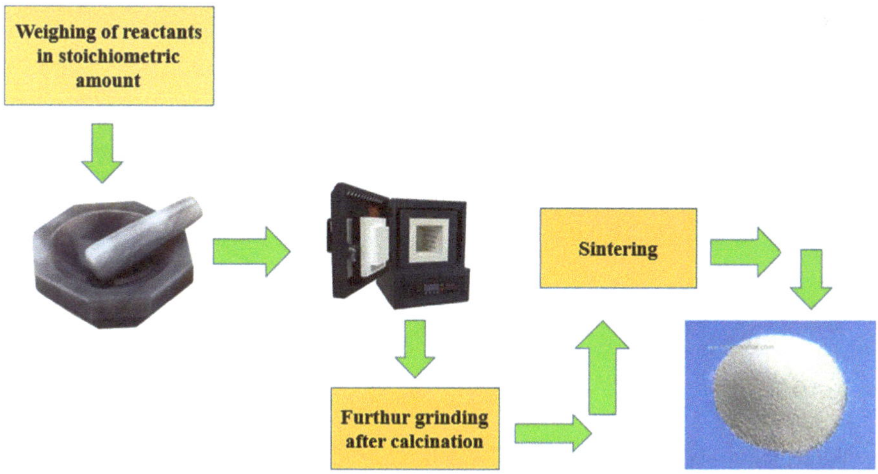

Figure 7.8 Flow chart of solid-state reaction method.

2. Spectroscopic techniques

Spectroscopic techniques have been widely used because of their versatility in the field of optical, electrical, biomedical, and astronomical science. Spectroscopy techniques, however, have an importance that makes them ideal for use in different kinds of analytical measurements. Spectroscopic techniques are important because of the following reasons:

- Fast and simple qualitative and quantitative measurements are possible
- No use of polluting solvents
- Direct and nondestructive analyses are possible
- Inexpensive with respect to analysis time and reagents
- Determination of chemical reactions among a functional group

Rare-earth-activated phosphors synthesized by solid-state reaction method were characterized by X-ray diffraction, scanning electron microscopy, transmission electron microscopy, Fourier transform infrared spectroscopy, Raman spectroscopy, UV−visible spectroscopy, photoluminescence spectroscopy, and thermoluminescence spectroscopy.

2.1 X-ray diffraction analysis

X-ray diffraction is an analytical tool to find out the crystallographic structure of the material. It is a fruitful technique used to determine the crystallographic parameters of different types of samples. It functions in two modes, that is, spectroscopic mode and photographic mode [30]. In this technique, a collimated monochromatic beam of X-rays is allowed to fall on the specimen and the diffracted beam, received by the detector, constitutes a diffraction pattern that obeys Bragg's law (Fig. 7.9) [31]. The nature of peaks in the diffraction pattern gives information about the lattice constants, phase, degree of crystallinity, and crystallite size of the material. More intense and individual XRD peaks indicate the crystalline nature of the material [32].

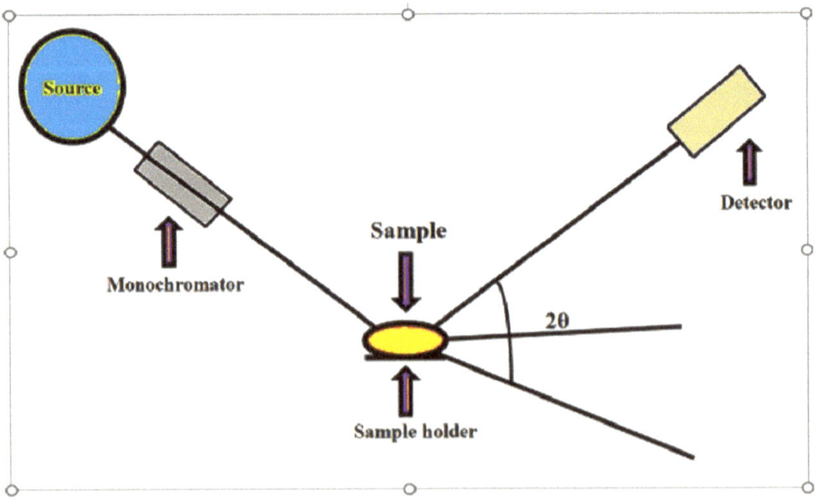

Figure 7.9 X-ray diffraction technique.

The crystallite size is obtained using Scherrer's formula [33].

$$D = \frac{k\lambda}{\beta \cos \theta}$$

where λ is X-ray wavelength, β is full-width half maxima, θ is Bragg's angle, and k is the shape factor.

X-ray diffraction spectra using Cu Kα powder diffractometer for gadolinium-doped calcium yttrium oxide phosphors are shown in Fig. 7.2. XRD analysis confirms the formation of a single-phase hexagonal crystal structure with space group R_3C [30]. XRD spectra show no additional peaks due to the substitution of Gd^{3+} ion, and this indicates that the host lattice was not affected by the dopant ions [34] (Fig. 7.10).

The Rietveld refinement analysis was done by using material analysis by diffraction software as shown in Fig. 7.11. The Rietveld parameters were calculated by using the relation

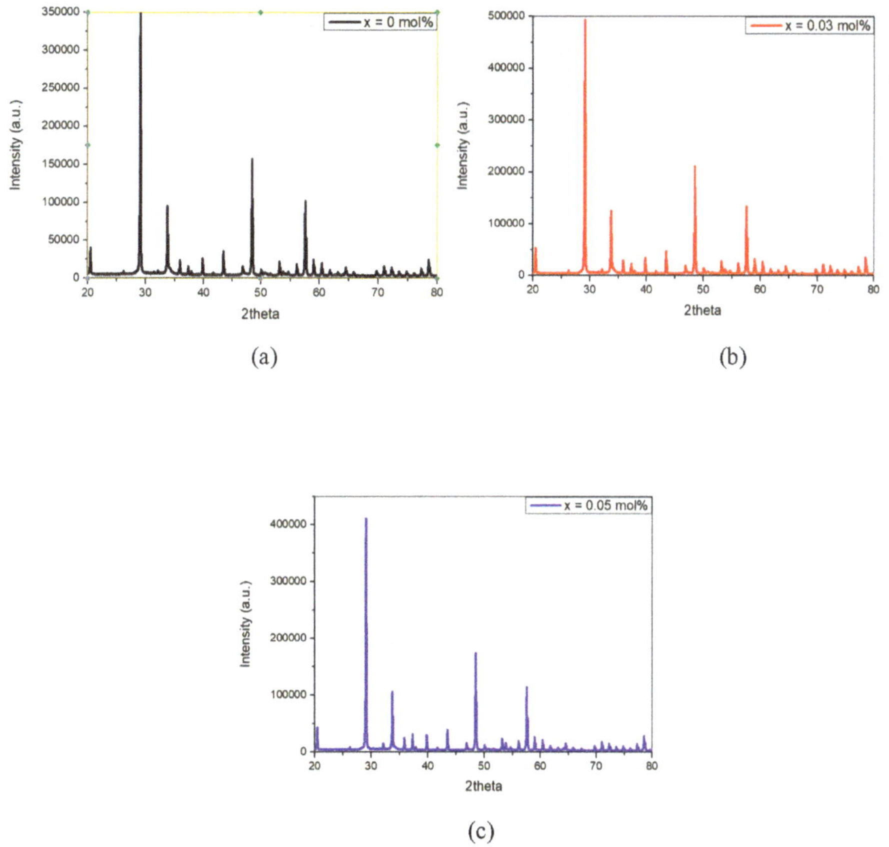

Figure 7.10 XRD pattern of $Ca_{2-x}Y_2O_5$:$x$$Gd^{3+}$: (A) $x = 0$ mol%; (B) $x = 0.03$ mol%; (C) $x = 0.05$ mol%.

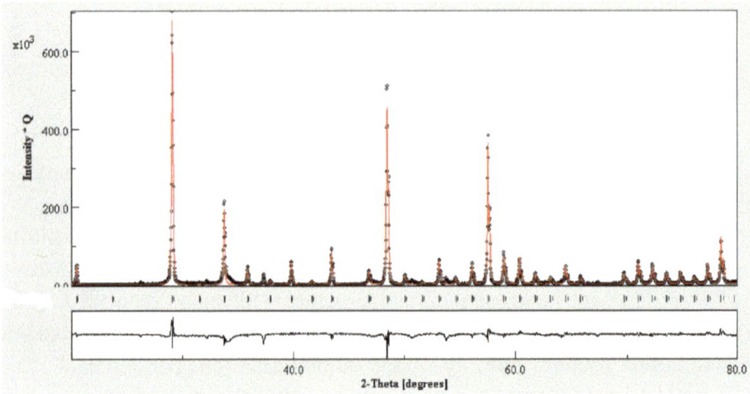

(a)

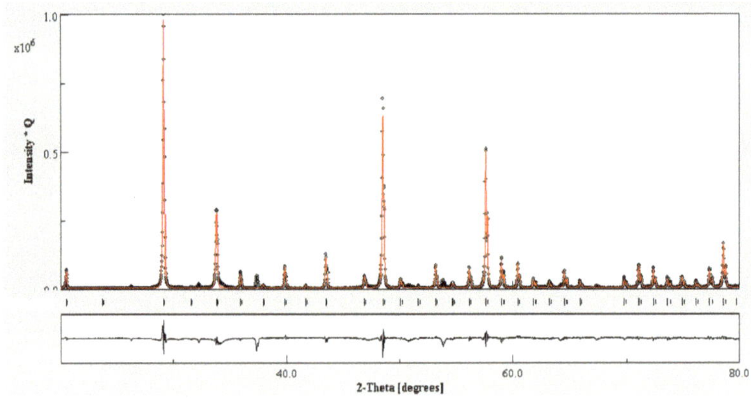

(b)

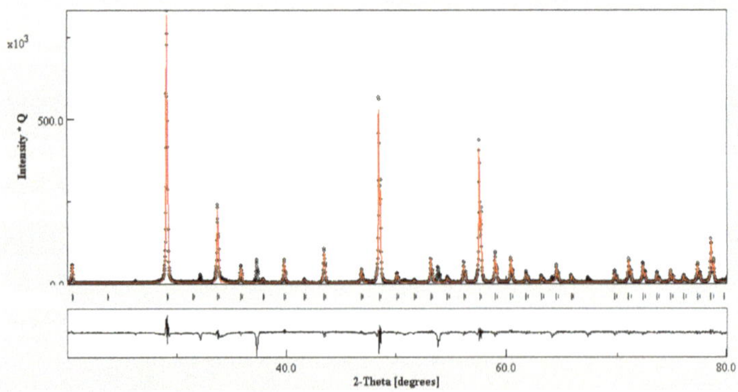

(c)

Figure 7.11 Reitveld refined pattern of $Ca_{2-x}Y_2O_5$:$x$$Gd^{3+}$: (A) $x = 0$ mol%; (B) $x = 0.03$ mol%; (C) $x = 0.05$ mol%.

$$R_{wp} = \left(\frac{\sum_{i=1}^{n} y_i w_i}{\sum_{i=1}^{n} w_i} \right)^{1/2} \times 100$$

where $w_i = \frac{1}{y_i}$

$$R_{exp} = \left(\frac{N - p}{\sum w_i y_i^2} \right)^{1/2} \times 100$$

where N is the total number of parameters and P is the number of refined parameters

$$\text{Goodness of fit(GoF)} = \frac{R_{wp}}{R_{exp}}$$

The Rietveld parameters are shown in Table 7.1.

2.2 Scanning electron microscopy

The high energetic beam of electrons is used for the morphological investigation of the material in scanning electron microscopy (SEM). The electrons generated from the electron gun are focused and made to fall on the specimen. After hitting the specimen, signals are generated due to the scattering of electrons and give information about the external morphology and texture of the material as shown in Fig. 7.12 [35].

Electron beam produces mainly three signals in the form of secondary electron beam, backscattered electron beam, and energy dispersive X-rays. The secondary electron beam produces two-dimensional intensity distribution, backscattered beam is used to detect contrast between areas with different chemical compositions and energy dispersive X-rays provide information about the elemental composition of the material [36].

The morphological investigation for gadolinium-doped calcium yttrium oxide phosphors is done by using JEOL-JSM-5200 with 20 kV operating voltage. The SEM images (Fig. 7.13) show the agglomerated and porous grains with irregular shape and size [37].

Table 7.1 Rietveld refined parameters for $Ca_{2-x}Y_2O_5{:}xGd^{3+}$: (a) $x = 0$ mol%; (b) $x = 0.03$ mol%; (c) $x = 0.05$ mol%.

Molar concentration (x mol%)	Phase	R_{wp}	R_{exp}	R_p
$x = 0$	Cubic	0.1345	1.1165	0.099
$x = 0.03$	Cubic	0.1438	1.0450	0.098
$x = 0.05$	Cubic	0.1489	1.0978	0.095

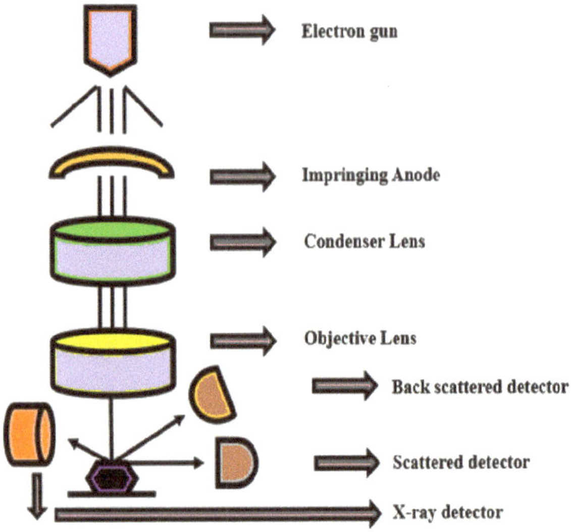

Figure 7.12 Scanning electron microscope.

Figure 7.13 SEM images of $Ca_{2-x}Y_2O_5$:$x$$Gd^{3+}$ ($x = 0.05$ mol%).

2.3 Energy-dispersive X-ray spectroscopy

Energy-dispersive X-ray spectroscopy (EDXS) is a microanalysis technique used for identifying the elements present in the sample. This technique also quantifies the elemental composition of the sample. In the SEM instrument, the atoms on the surface are excited by the electron beam, resulting in the emission of specific wavelengths of X-rays that are characteristic of the atomic structure of the elements [38]. An energy-dispersive detector analyzes these X-rays emission and represents elements along with their compositions that are present in the sample [39]. The EDX spectra depict X-ray energy versus counts that are used to find the elemental composition of the specimen [40].

Energy-dispersive X-ray spectroscopy for gadolinium-doped calcium yttrium oxide phosphors was done by INCA 250 EDS ZEISS EVO (Fig. 7.14). The determination of elemental composition was done at energy resolution of 20 eV, probe current of 5 μA, beam current of 1000 A, detection limit of 0.005%, and electron high tension value of 20 kV. The elemental analysis shows the presence of Ca, Y, O, and Gd atoms with no trace of other elements.

2.4 Transmission electron microscopy

Transmission electron microscopy (TEM) is a unique technique in which electrons are ejected thermionically from LaB_6 filament when heated above 2000 K. The schematic diagram of the transmission electron microscope is shown in Fig. 7.15.

In a TEM, a beam of electrons is accelerated through the potential from 40 to 100 kV (kV) and passed through a strong magnetic field [41]. The resolution of TEM is about 0.2 nm. The highly energetic beam of electrons gets diffracted from the different planes, and the TEM image that can be seen on the screen is formed. TEM images give information about the morphological, compositional, and crystallographic parameters of the specimen [42].

Fig. 7.16 shows the TEM image obtained by using TEM analyzer Hitachi H-600 for 0.4 mol% terbium (Tb^{3+}) ions-doped $Y_6Ba_4(SiO_4)_6F_2$. It shows the rod-like structure of the prepared phosphors with an average length of 1.157 μm.

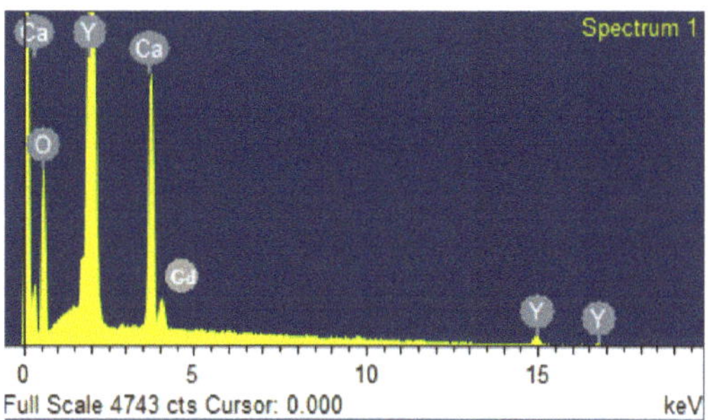

Figure 7.14 EDXS image of $Ca_{2-x}Y_2O_5$:$x$$Gd^{3+}$ ($x = 0.05$ mol%).

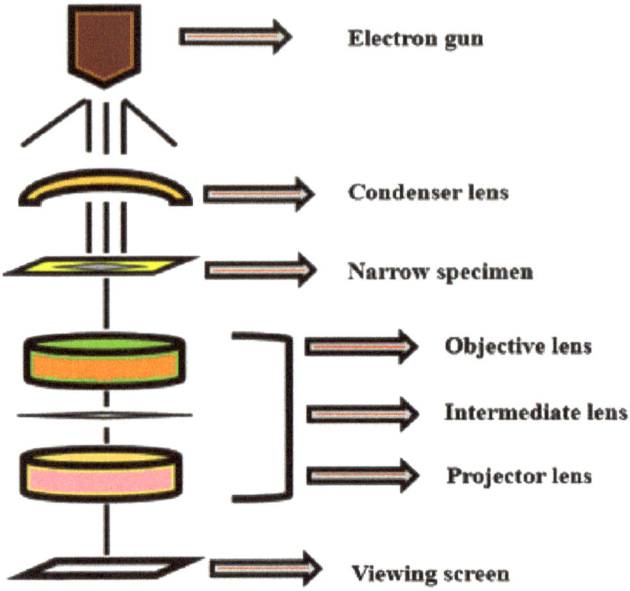

Figure 7.15 Transmission electron microscope.

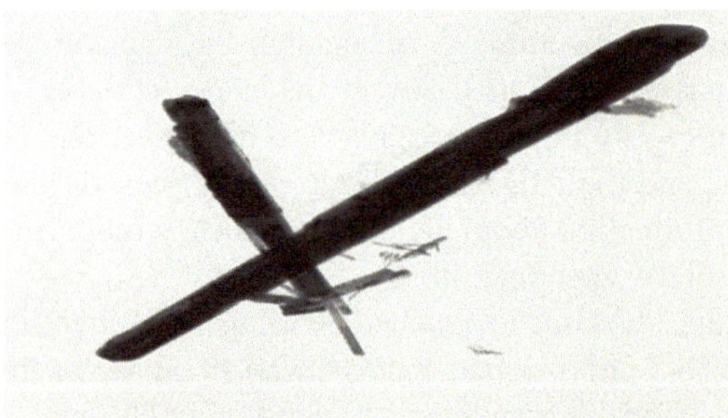

Figure 7.16 TEM image of 0.4 mol% Tb^{3+}-doped $Y_6Ba_4(SiO_4)_6F_2$.

2.5 Selected area electron diffraction pattern

Selected area electron diffraction (SAED) is a crystallographic experimental technique that is performed by using a transmission electron microscope. In the SAED pattern, each obtained spot corresponds to a satisfied diffraction condition [43]. SAED of 0.4 mol% terbium (Tb^{3+}) ions-doped $Y_6Ba_4(SiO_4)_6F_2$ phosphor is shown in Fig. 7.17. The concentric ring-like structure reveals that the phosphor is of polycrystalline nature. Interplanar distances are calculated from the SAED pattern and were found to be 0.31, 0.248, 0.200, and 0.135 nm corresponding to the different reflection planes.

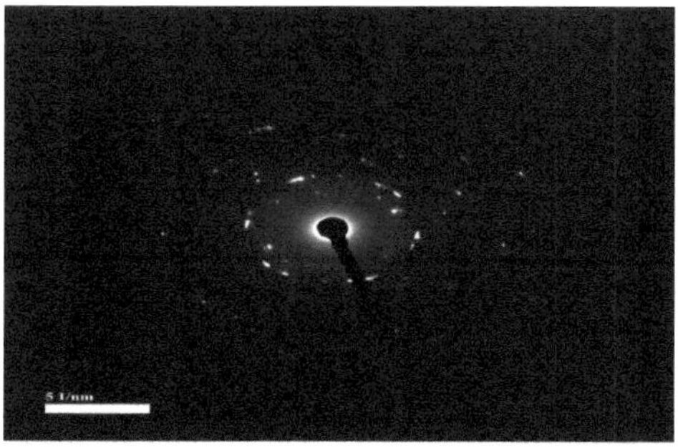

Figure 7.17 SAED image of 0.4 mol% Tb^{3+}-doped $Y_6Ba_4(SiO_4)_6F_2$.

2.6 HRTEM

The HRTEM stands for high-resolution transmission electron microscopy. High-resolution transmission electron microscopy is based on the use of both the transmitted and the scattered beams to create an interference image. It is a phase contrast image that contains a size as small as that of the unit cell dimension [44]. At lower magnification, the shape and size of the particle can be seen and if the magnification is high then along with the shape, lattice planes can also be observed as shown in Fig. 7.18. The HRTEM image of 0.4 mol% terbium (Tb^{3+}) ions-doped $Y_6Ba_4(SiO_4)_6F_2$ phosphor has the resolution around 0.05 nm. The interplanar distance calculated from the HRTEM image was found to be 0.256 nm.

2.7 FTIR spectroscopy

FTIR stands for Fourier transform infrared spectroscopy. This spectroscopy is used to detect the functional groups present in the compound. FTIR instrument consists of a set of sources, sample stage, detector, amplifier, A/D convertor, and screen [45]. The radiation that falls on the specimen is transmitted as well as reflected while some rays are absorbed by the specimen. After passing through the interferometer, the signal is amplified and converted into the digital signal and FTIR spectra are obtained [46]. The spectra are the molecular fingerprint of the specimen and can be used to identify unknown materials, decomposition, and oxidation (Fig. 7.19).

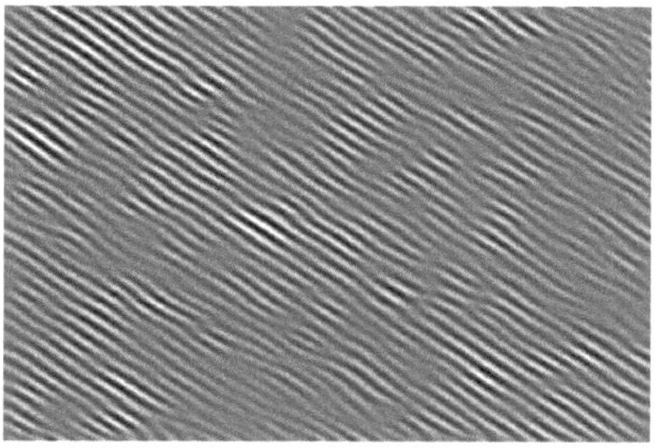

Figure 7.18 HRTEM image of 0.4 mol% Tb^{3+}-doped $Y_6Ba_4(SiO_4)_6F_2$.

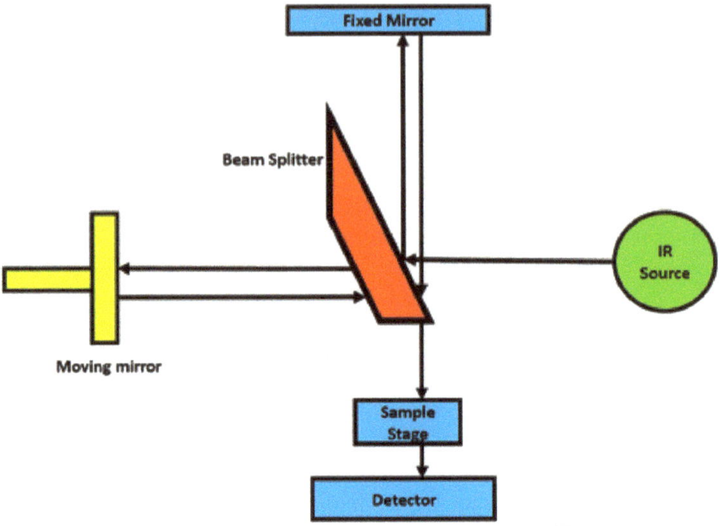

Figure 7.19 Schematic diagram of Fourier transform infrared spectroscopy.

Fourier transform infrared spectroscopy was done by using Bruker FTIR spectrometer. FTIR spectra of gadolinium (Gd^{3+})-doped $Ca_2Y_2O_5$ phosphors are shown in Fig. 7.20. It shows five broad peaks at 873 cm^{-1}, 1417 cm^{-1}, 1107 cm^{-1}, 1021 cm^{-1}, and 1259 cm^{-1}. The peaks show the presence of functional group and symmetric stretching and bending modes of Ca–O and Y–O [47]. The peak obtained at 873 cm^{-1} is due to the symmetric bending between O–Ca–O bonds [48]. Peaks at 1417 cm^{-1} are due to the symmetric stretching of C–O bonds [49]. The split peaks

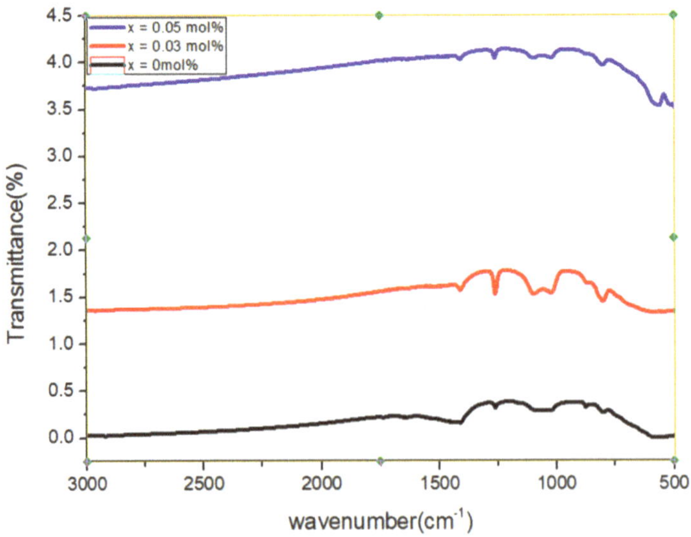

Figure 7.20 FTIR spectra of Gd^{3+}-doped $Ca_{2-x}Y_2O_5$.

obtained at 1107 cm^{-1} and 1021 cm^{-1} may be due to the incorporation of the CO_3^{2-} group at the time of synthesis of phosphor [50]. The peaks at 1259 cm^{-1} are due to the stretching of Y—O bonds [51].

2.8 Raman spectroscopy

Raman spectroscopy is a technique where inelastic scattering takes place when the monochromatic light is made incident on a specimen. The incident light interacts with the excitations present in the specimen and gets scattered at different wavelengths greater than the incident light which is known as Raman scattering [52]. The obtained Raman spectra are the chemical fingerprint of material and give information about the chemical structure, phase, molecular interaction, chemical bonds, contamination, etc. (Fig. 7.21).

Raman spectra of gadolinium (Gd^{3+})-doped $Ca_2Y_2O_5$ phosphors were taken by using RIAR-0403 spectrometer and shown in Fig. 7.22. Two peaks obtained at 249 cm^{-1} and 380 cm^{-1} are due to the lattice model of Ca^{2+} and Y^{3+} ions [53]. Sharp peaks were obtained at 653 cm^{-1} and 741 cm^{-1} due to the asymmetric bending between O—Ca—O bond [54]. Peaks at 566 cm^{-1} are due to the symmetric bending of O—Y—O bonds [55]. Sharp peaks at 824 cm^{-1} are due to the symmetric stretching of the Y—O bond [56]. Peaks found at 980 cm^{-1} and 1063 cm^{-1} are due to the asymmetric stretching of the Ca—O bond [57]. The broadband at 2429 cm^{-1} is due to the presence of $(OH)^-$ group [58].

2.9 UV—visible spectroscopy

Ultraviolet (UV) visible spectroscopy is a well-known technique in which light absorbed by the specimen is measured. The absorbance is taken over the UV and visible ranges of electromagnetic radiation [59].

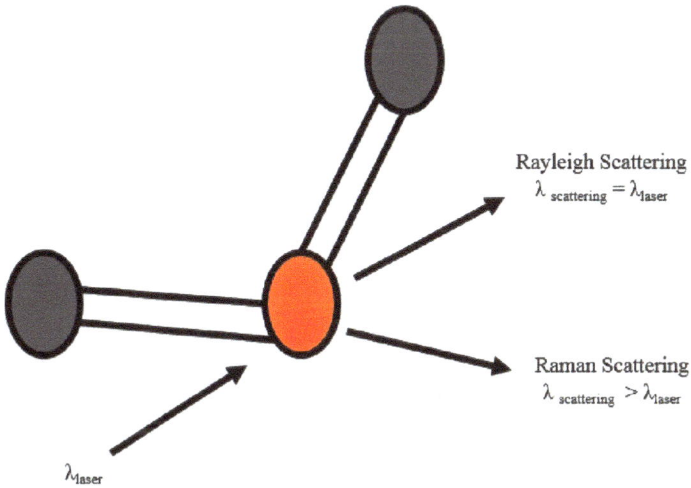

Figure 7.21 Schematic diagram of Raman spectroscopy.

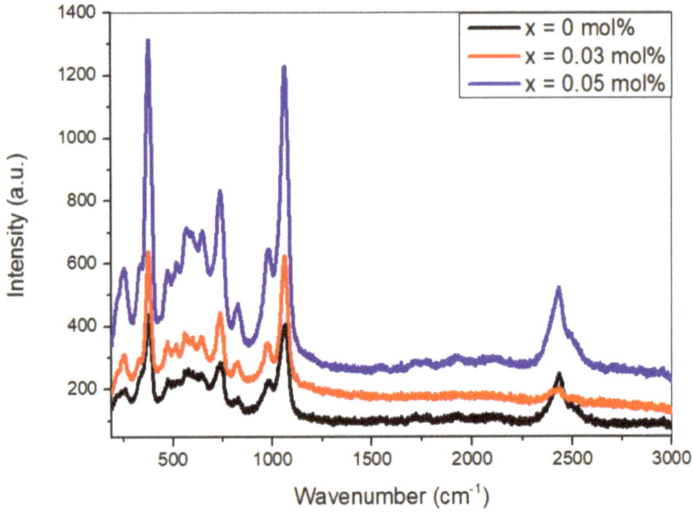

Figure 7.22 Raman spectra of Gd^{3+}-doped $Ca_{2-x}Y_2O_5$.

After absorbing the radiation in UV−visible region, the atoms get excited and move from a lower energy state to an excited energy state. Materials with smaller bandgaps absorb the shorter wavelength of electromagnetic radiation while materials with larger bandgaps absorb a larger wavelength of electromagnetic radiation to change the atomic states. The energy required for these transitions is electrochemically specific [60] (Fig. 7.23).

The spectra drawn between $(\alpha h\upsilon)^2$ and $h\upsilon$ are known as Tauc plots that give the information about the bandgap of the materials in solid and liquid states [61]. Tauc plots for $Ca_{2-x}Y_2O_5$ phosphor doped with different concentrations of Gd^{3+} ions are shown in Fig. 7.24 For undoped $Ca_2Y_2O_5$ phosphor, the bandgap was found to be 2.43 eV. However, by doping Gd^{3+} ions, the value of the bandgap reduces to 1.61 eV. The

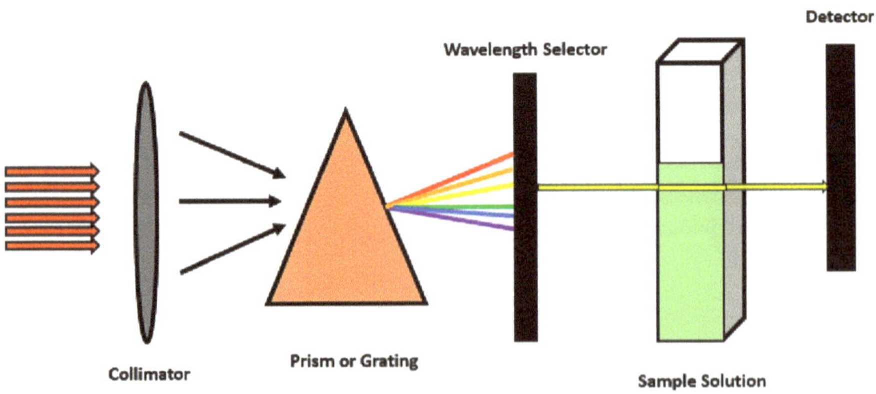

Figure 7.23 Schematic diagram of UV−visible spectroscopy.

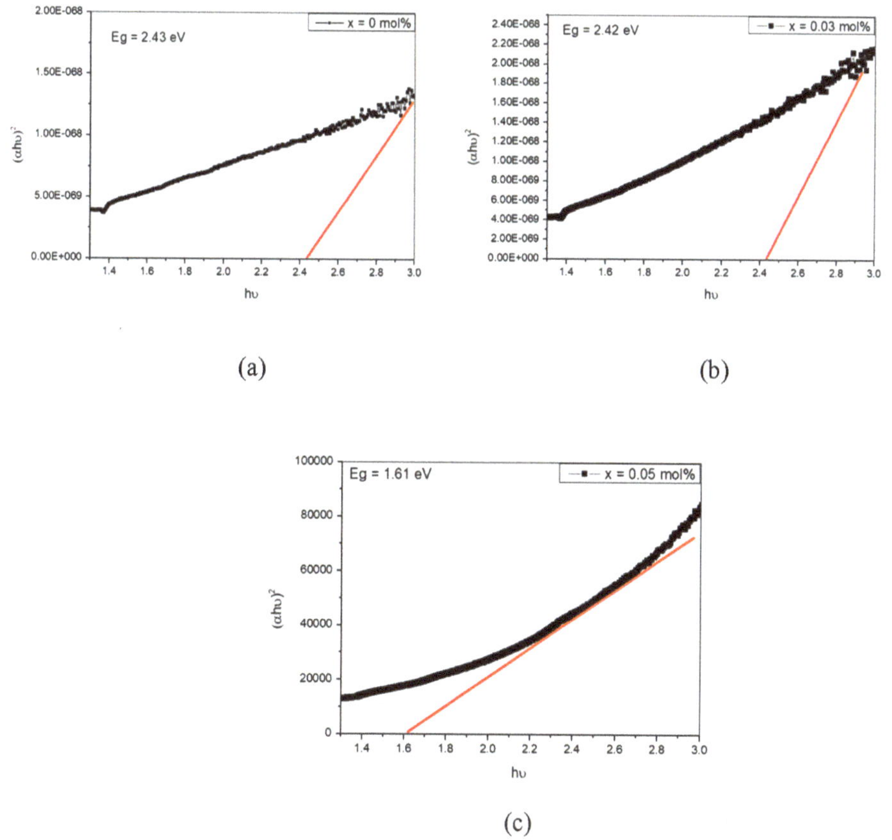

Figure 7.24 UV–visible spectra of $Ca_{-x}Y_2O_5{:}xGd^{3+}$: (A) $x = 0$ mol%; (B) $x = 0.03$ mol%; (C) $x = 0.05$ mol%.

reduction in the bandgap of Gd^{3+}-doped phosphors is due to the formation of metastable states by the dopant ions [62].

2.10 Photoluminescence spectroscopy

PL spectroscopy is an optical phenomenon in which UV radiation of about 254 nm from a Xenon lamp is allowed to fall on the material and thereby excites the electrons to a higher energy level [63]. These highly energetic electrons after losing energy come back to the ground state by releasing photons in the visible region as detected by the photomultiplier tube (Fig. 7.25).

CRI is the measurement of how colors look under a light source when compared with sunlight. The index is measured from 0 to 100, with a perfect 100 indicating that colors under the light source appear the same as they would under natural sunlight. Lights with a CRI that is measured greater than 80 are considered to be more than acceptable for most applications.

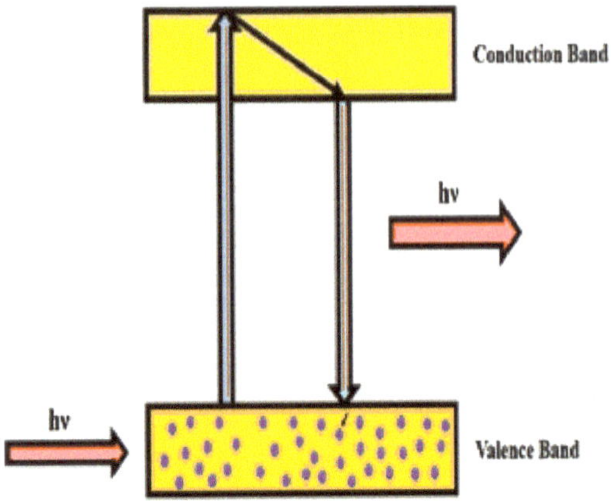

Figure 7.25 Photoluminescence (PL) spectroscopy.

Lights with a CRI that is measured greater than 90 are generally considered "High CRI" lights.

Correlated color temperature (CCT) is used to find the application of light sources in warm or cool light-emitting diodes. It is measured in Kelvin (K) and it basically lies between 2200 and 6500 K, respectively. Warm light sources such as incandescent bulbs have a CCT value between 2200 and 3000 K and feature more lights in red, orange, and yellow regions of the electromagnetic spectrum. However cool light sources, such as some fluorescent lamps have a CCT value > 4000 K and feature lights in the blue region. The CCT temperature was calculated by using the equation

$$CCT = -449\,n^3 + 525\,n^2 - 6823.3\,n + 5520.32$$

where $n = \dfrac{X - 0.3320}{Y - 0.1858}$ and X, Y are Commission Internationale de I'Eclairage (CIE) coordinates.

The chromaticity diagram is expressed in the form of a 2-D diagram and shows the value perceivable by the emitted color on a set of coordinates along x and y axes that are known as chromaticity coordinates. The color of light is expressed by using chromaticity coordinates (x, y) as a function of tristimulus values (X, Y, and Z):

$$x = \frac{X}{X + Y + Z}$$

$$y = \frac{Y}{X + Y + Z}$$

The Judd–Ofelt theory was discovered in 1962 by Brian R. Judd and George S. Ofelt, who describes the transitions of 4f electron of rare-earth elements present in the liquid and solid form of materials. It is based on three approximations, that is, static, free ion, and single configuration. In the static model, the centralized ion is affected by the surrounding ions in the presence of a static crystal or ligand field. In the free ion model, the surrounding ions produce a crystal field that is treated as a perturbation on the free-ion Hamiltonian. In the single configuration model, the electron interaction within the configuration is neglected. The utility of the Judd–Ofelt theory is that it provides a theoretical expression for the line strength, given by

$$\sum_{\lambda=2,4,6} \Omega_k \left(f''[SL]J \| U^\lambda \| f''[S'L']J'\right)^2$$

where Ω_k are the Judd–Ofelt parameters. The terms in brackets are doubly reduced matrix elements for intermediate coupling.

Fig. 7.26shows the PL excitation spectrum of $Ca_{2-x}Y_2O_5{:}xGd^{3+}$ ($x = 0$, 0.03, 0.05 mol%) phosphors. The broad peak obtained at 231 nm shows that the phosphor can easily absorb UV radiation [64].

Fig. 7.27 shows the PL emission spectra of $Ca_{2-x}Y_2O_5{:}xGd^{3+}$phosphor ($x = 0$, 0.03, 0.05 mol%). The peaks obtained at 690 nm in the red region is corresponding to $^6G_{7/2}$ to $^6P_{5/2}$ transition under 231 nm UV excitation [65]. The transition occurred due to the defects created by doping of Gd^{3+}ions in the host lattice [66]. With the increase in dopant concentration, the increase in PL intensity is observed.

The CIE coordinates are suggested under the convention of CIE. The chromaticity gamut chart shows the area enclosed within the coordinate axes [67]. Every color is specified by X, Y, and Z values corresponding to the lightness, hue, and saturation of a color. The chromaticity coordinates are given as

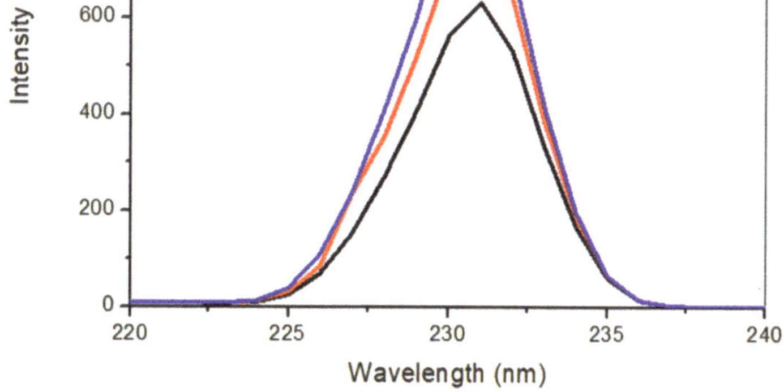

Figure 7.26 PL excitation spectra of Gd^{3+}-doped $Ca_{2-x}Y_2O_5$.

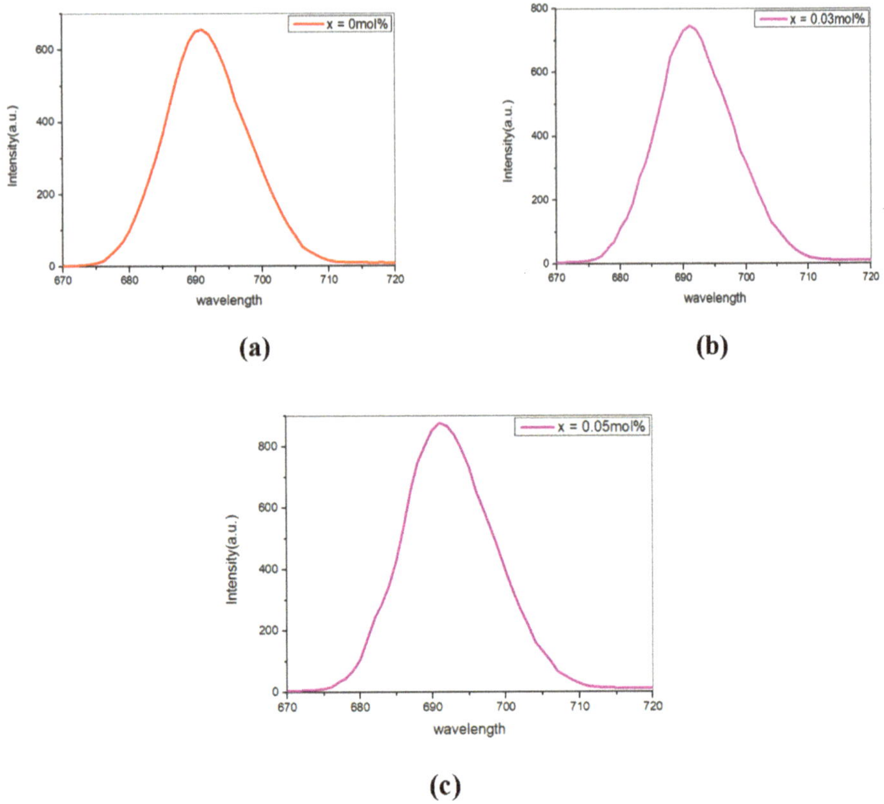

Figure 7.27 PL emission spectra of $Ca_{2-x}Y_2O_5$:$x$$Gd^{3+}$: (A) $x = 0$ mol%; (B) $x = 0.03$ mol%; (C) $x = 0.05$ mol%.

$$X = \frac{X}{X + Y + Z}$$

$$Y = \frac{Y}{X + Y + Z}$$

$$Z = \frac{X}{X + Y + Z}$$

where $X + Y + Z = 1$

The CIE coordinates values of 0.05 mol% of Gd^{3+}-doped $Ca_2Y_2O_5$ phosphor obtained at CIE $X = 0.680$ and CIE $Y = 0.280$ confirm that the phosphor gives cool red emission as shown in Fig. 7.28. The chromaticity coordinates values show a good agreement with national television standard committee (NTSC) values [68].

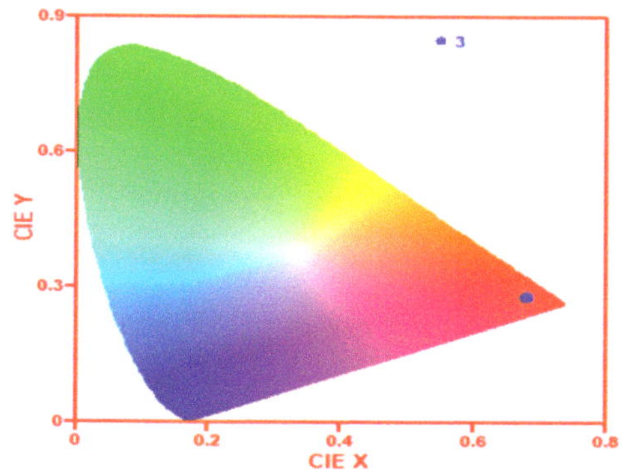

Figure 7.28 CIE chromaticity diagram of $Ca_{2-x}Y_2O_5{:}xGd^{3+}$ ($x = 0.05$ mol%).

2.11 Thermoluminescence spectroscopy technique

Thermoluminescence (TL) spectroscopy is a technique in which the material is preexposed to radiation, that is, UV, α, β, γ, etc. The material after exposure is heated due to which the light is emitted [9]. The thermoluminescence mechanism is shown in the figure. Phenomena of thermoluminescence have applications in radiation dose measurement, TL mapping, dating, etc. (Fig. 7.29).

The mechanism of thermoluminescence can be explained by shallow and deep trap levels. A trap is created due to the presence of impurity/imperfection in the semiconductor. The imperfection gives rise to various electronic states within the energy gap of the material. Depending upon the localized position of these traps, they are basically categorized as shallow/deep traps. If a trap is present nearer to the band edge, they are termed as shallow traps; however, if they are present farther from the band edge they are termed as deep traps. In the deep level trapping, the charges moving in the localized states are trapped by deep traps and released back to the energy gap by gaining sufficient thermal energy (Fig. 7.30).

The thermoluminescence spectra (thermograms) of $Ca_{2-x}Y_2O_5$ phosphor doped with different concentrations of Gd^{3+} ions were obtained by using TL/OSL Reader

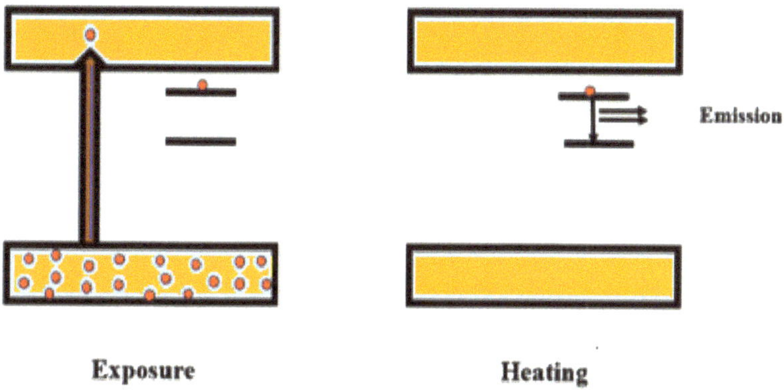

Figure 7.29 Thermoluminescence (TL) spectroscopy.

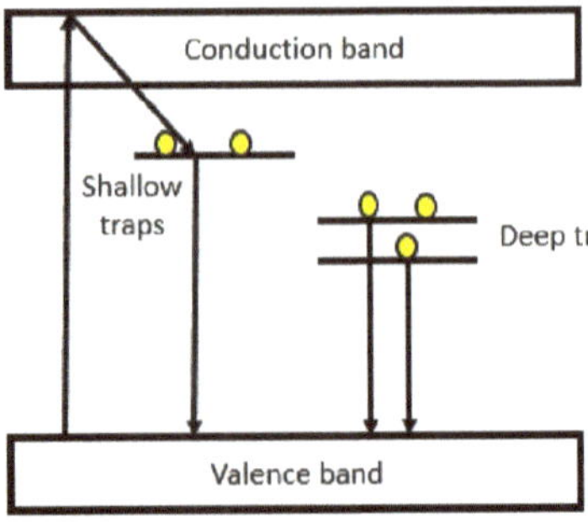

Figure 7.30 Energy level diagram for deep trapping phenomena.

supplied by Nucleonix System Private limited. These phosphors were irradiated with UV radiation for 5 min. The thermograms obtained by plotting a graph between temperature and intensity give information about the kinetic parameters like activation energy (*E*), order of kinetics (*s*), and the figure of merit Fig. 7.31 [69]. With the increase in dopant concentration, the increase in lumen count is observed.

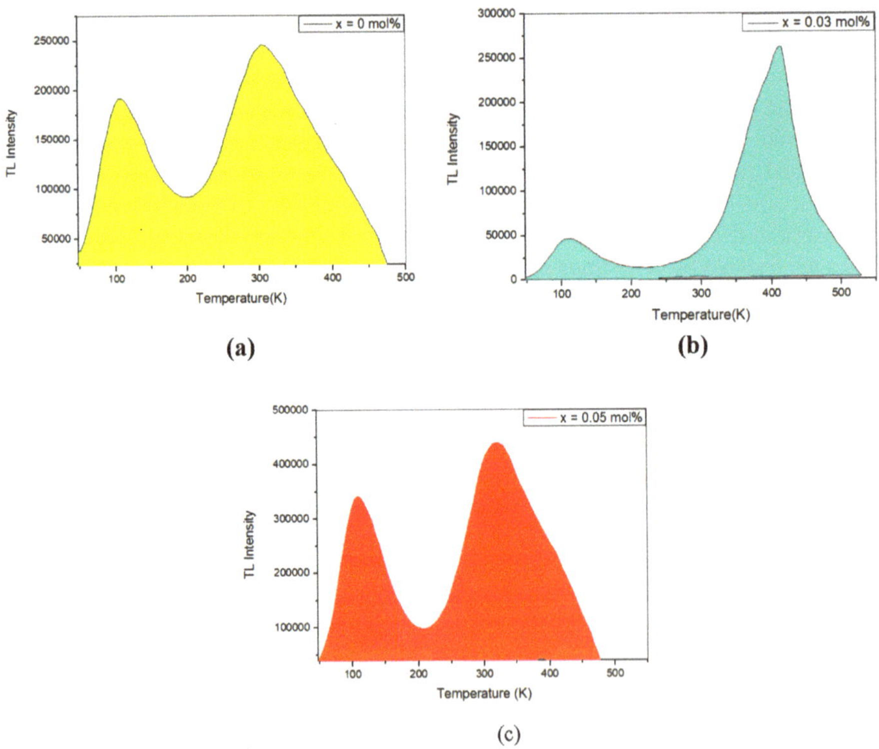

Figure 7.31 TL spectra of $Ca_{2-x}Y_2O_5{:}xGd^{3+}$: (A) $x = 0$ mol%; (B) $x = 0.03$ mol%; (C) $x = 0.05$ mol%.

Fig. 7.32 shows the thermoluminescence fitted curve of 0.05 mol% of Gd^{3+}-doped $Ca_{2-x}Y_2O_5$ phosphor. The thermogram was deconvoluted by computerized glow curve deconvolution technique (CGCD) using Glowfit. The CGCD technique is used for the evaluation of TL kinetic parameters for a given peak in the glow curve. The kinetic parameters such as activation energy (E), order of kinetics (b), and frequency factor (s) were calculated using the peak shape method employed by Grosswener, Lushchik and Halperin, and Braner. Nowadays, the TL analysis is also employed in radiation dosimetry, and several methods, models, programs were investigated for TL glow curve analysis. The CGCD glow curve analysis is based on first-order kinetics given by Randall and Wilkins. The TL intensity of the glow curve is given by

$$I = I_m \exp\left(\frac{E}{kT_m} - \frac{E}{kT}\right)\exp\left(-\frac{E}{kT_m^2}\int_{T_m}^{T}\left(\frac{E}{kT_m} - \frac{E}{kT}dT\right)\right)$$

As shown in the earlier equation, the glow curve is the nonlinear function of I_m, T_m, and E. To find the fitted value of TL parameters, the iterative method is used. CGCD is capable of deconvoluting as many as 10 glow peaks from the TL glow curves. The figure of merit is calculated by

$$\mathrm{FOM} = \frac{\Sigma \mathrm{TL_{exp}}}{\Sigma \mathrm{TL_{Theo}}}$$

where $\mathrm{TL_{exp}}$ and $\mathrm{TL_{Theo}}$ are the experimental and theoretical values of TL glow curves.

The glow peak parameters were calculated by following relations [70]

$\boldsymbol{\delta = T_2 - T_{max}}$

$\boldsymbol{\tau = T_{max} - T_1}$

$\boldsymbol{\omega = T_2 - T_1}$

$\boldsymbol{\mu_g = \delta/\omega}$.

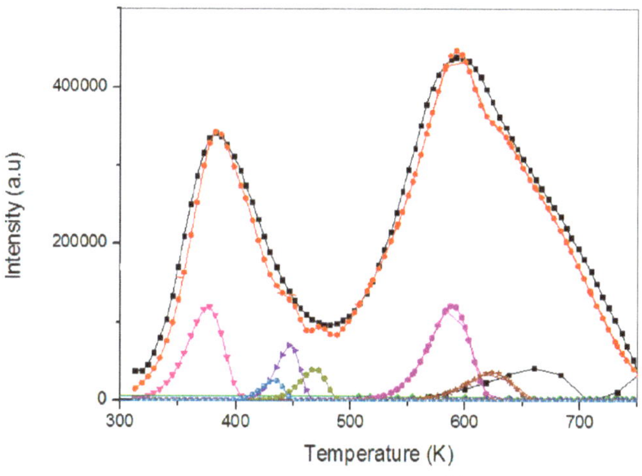

Figure 7.32 Deconvoluted TL spectrum of $Ca_{2-x}Y_2O_5{:}xGd^{3+}$ ($x = 0.05$ mol%).

Here form factor (μ_g) is independent of activation energy (E), but it depends on the order of kinetics (b). Form factor ranges from 0.42 to 0.52. When μ_g is close to 0.42, it represents first-order kinetics; and when μ_g is close to 0.52, it represents second-order kinetics. The modified Halperin and Braner equation by Chen [71] for calculation of activation energy (E) is given as follows:

$$E_t = \left[1.51 + 3\left(\mu_g - 0.42\right)\right] \frac{KT_{\text{max}}^2}{\tau} - \left[1.58 + 4.2\left(\mu_g - 0.42\right)\right] 2kT_{\text{max}}$$

$$E_\delta = \left[0.976 + 7.3\left(\mu_g - 0.42\right)\right] \frac{KT_{\text{max}}^2}{\tau}$$

$$E_\omega = \left[2.52 + 10.2\left(\mu_g - 0.42\right)\right] \frac{KT_{\text{max}}^2}{\tau} - 2kT_{\text{max}}$$

Preexponential factor or frequency factor (s) for first and general order kinetics is given as follows:

$$s = \frac{\beta E}{kT_{\text{max}}^2} \exp\left[E / kT_{\text{max}}\right]$$

$$s = \frac{\beta E}{kT_{\text{max}}^2} \exp\left[\left(\frac{-E}{kT_{\text{max}}}\right)(1 + (b-1)) \frac{2kT_{\text{max}}}{E}\right]^{b/b-1}$$

When an isolated TL peak is considered without taking into account the retrapping phenomena (first-order kinetics), then the rate of release of electrons is given by

$$\frac{dn}{dt} = -p.n$$

where p is the escape probability. By integrating the previous equation, we get

$$n = n_0 \exp(-p.t)$$

where n_0 is the initial number of trapped charges.

We know that the TL intensity is proportional to the release of trapped charges

$$I(t) \prec -\frac{dn}{dt}$$

By substituting the values, we obtain

$$I(t) = I_0 \exp(-p.t)$$

Table 7.2 Calculated TL parameters.

I_m	T_m	E
2.673×10^5	397.2	0.5002
1.204×10^5	376	0.7721
1.2925×10^5	566.4	0.5
2.7114×10^5	643.8	0.5246
7.157×10^4	447	1.7
1.222×10^5	589	1.7
3.666×10^4	624	1.7
4.096×10^4	469	1.7
2.671×10^4	433	1.7

where I_0 is the TL intensity at time $t = 0$.

The calculated TL parameters such as maximum intensity (I_m), maximum temperature (T_m), and activation energy (E) for 0.05 mol% Gd^{3+}-doped $Ca_2Y_2O_5$ phosphor are shown in Table 7.2.

The phosphors exhibit a deep trapping mechanism and second-order kinetics with activation energy ranging from 0.5 to 1.7 eV. The second-order kinetics confirms the trapping and detrapping repeatedly. These rare-earth-activated phosphors can be used in TL mapping, sensing, and monitoring applications because of enhanced lumen count and deep trapping mechanisms [72].

3. Conclusions

This chapter provides brief information on various spectroscopic techniques and explains the structural, morphological, and luminescence behavior of rare-earth-activated phosphors. Here, we have synthesized gadolinium doped calcium yttrium oxide by solid-state reaction method. The XRD analysis confirms the formation of the cubic phase with space group R_3C. Morphological analysis shows the formation of agglomerated and porous grains with irregular shapes and sizes. The FTIR spectroscopy shows five broad peaks at 873 cm^{-1}, 1417 cm^{-1}, 1107 cm^{-1}, 1021 cm^{-1}, and 1259 cm^{-1}. The peaks show the presence of functional group and symmetric stretching and bending modes of Ca—O and Y—O. The Raman spectroscopy depicts the symmetric and asymmetric bending and stretching between Ca—O and Y—O bonds. UV—Visible spectroscopy shows that $Ca_2Y_2O_5$ phosphor in the bandgap was found to be 2.43 eV. However, by doping Gd^{3+} ions, the value of the bandgap reduces to 1.61 eV. The PL emission peak obtained at 690 nm in the red region is corresponding to $^6G_{7/2}$ to $^6P_{5/2}$ transition under 231 nm UV excitation. Thermoluminescence spectroscopy depicts that the phosphors exhibit a deep trapping mechanism and second-order kinetics with activation energy ranging from 0.5 to 1.7 eV. These rare-earth-activated phosphors can be used in TL mapping, sensing, and monitoring applications.

References

[1] S. Ekambaram, K.C. Patil, Synthesis and properties of rare earth doped lamp phosphors, Bull. Mater. Sci. 18 (1995) 921−930.

[2] M. Nazarov, D.Y. Noh, Rare earth double activated phosphors for different applications, J. Rare Earths 28 (2010) 1−11.

[3] H. Zhu, et al., Highly efficient non-rare-earth red emitting phosphor for warm white light-emitting diodes, Nat. Commun. 5 (2014) 4312−4322.

[4] T. Minakova, et al., High efficient $YVPO_4$ luminescent materials activated by europium, Crystals 9 (2019) 1−14.

[5] V. Balaram, Rare earth elements: a review of applications, occurrence, exploration, analysis, recycling, and environmental impact, Geosci. Front. 10 (2019) 1285−1303.

[6] C.D.B. Amaral, Determination of rare earth elements in geological and agricultural samples by ICP-OES, Spectroscopy 32 (2017) 32−36.

[7] S.K. Bhushan, Geology of the kamthai rare earth deposit, J. Geol. Soc. India 85 (2015) 537−546.

[8] V. Balaram, Recent trends in the instrumental analysis of rare earth elements in geological and industrial materials, Trends Anal. Chem. 15 (1996) 475−486.

[9] Z. wang, et al., Investigating the luminescence behaviors and temperature sensing properties of rare-earth-doped $Ba_2In_2O_5$ phosphors, Inorg. Chem. 57 (2018) 8841−8849.

[10] R. Marin, et al., Europium-doped ZnO nanosponges − controlling optical properties and photocatalytic activity, J. Mater. Chem. C 7 (2019) 3909−3919.

[11] Z. Wang, et al., Rapid preparation of terbium-doped titanium dioxide nanoparticles and their enhanced photocatalytic performance, R. Soc. Open Sci. 6 (2019) 191077−191090.

[12] X. Jiang, et al., The synthesis and characterization of ytterbium-doped TiO_2 hollow spheres with enhanced visible-light photocatalytic activity, RSC Adv. 7 (2017) 24598−24606.

[13] N. Aggarwal, et al., Nanoscale properties of dysprosium-doped zinc oxide nanoparticles, J. Supercond. Nov. Magnetism 33 (2020) 883−888.

[14] S.R. Yashodha, Synthesis and photoluminescence properties of Sm^{3+} doped LaOCl phosphor with reddish orange emission and it's Judd- Ofelt analysis, Mater. Res. Express 7 (2020) 015003−015020.

[15] M. Mazhdi, et al., Gadolinium doping affecting on structural, magnetic and dielectric properties of ZnO nanoparticles, Appl. Phys. A 126 (2020) 272−280.

[16] F. Bondioli, et al., Synthesis and characterization of praseodymium-doped ceria powders by a microwave-assisted hydrothermal (MH) route, J. Mater. Chem. 15 (2005) 1061−1066.

[17] T. Aitasalo, et al., Persistent luminescence phenomena in materials doped with rare earth ions, J. Solid State Chem. 171 (2003) 114−122.

[18] G. Aka, et al., Self-frequency conversion in nonlinear laser crystals, Opt. Mater. 22 (2013) 89−94.

[19] S. Som, et al., Luminescence studies of rare earth doped yttrium gadolinium mixed oxide phosphor, Phys. B Condens. Matter 407 (2012) 3515.

[20] Y. Tian, et al., Upconversion luminescence properties of Y2O3:Yb, Er and Y2O2S:Yb, Er nanoparticles prepared by complex precipitation, J. Nanomater. 5 (2015) 1−7.

[21] G. Tessari, et al., Synthesis and optical properties of nanosized powders: lanthanide-doped Y_2O_3, Appl. Surf. Sci. 144 (1999) 686−689.

[22] J. McKittrick, et al., The influence of processing parameters on luminescent oxides produced by combustion synthesis, Displays 19 (1999) 169–172.
[23] C.-H. Lu, R. Jagannathan, Cerium-ion-doped yttrium aluminum garnet nanophosphors prepared through sol–gel pyrolysis for luminescent lighting, Appl. Phys. Lett. 80 (2002) 3608–3610.
[24] G. Vijayaprasath, et al., Optical and magnetic studies on Gd doped ZnO nanoparticles synthesized by co-precipitation method, J. Lumin. 178 (2016) 1–8.
[25] V. Kiisk, et al., Oxygen sensitive photoluminescence of rare earth ions in TiO_2 thin films, J. Phys. Chem. C 123 (2019) 17908–17914.
[26] V. Kumar, et al., Luminescence Properties of Rare Earth Doped Oxide Materials, Elsevier, 2020.
[27] W. Xia, Z. Pei, K. Leng, X. Zhu, Research progress in rare earth-doped perovskite manganite oxide nanostructures, Nanoscale Res. Lett. 9 (2020) 130–137.
[28] M. Muthuselvi, et al., Rare earth metal oxide-doped reduced graphene-oxide nanocomposite as binder-free hybrid electrode material for supercapacitor application, Int. J. Energy Res. 45 (2021) 8255–8266.
[29] M.K. Ganesan, et al., Synthesis and functional properties of nanostructured Gd-doped WO_3/TiO_2 composites for sensing applications, Mater. Sci. Semicond. Process. 105 (2020) 104732–104741.
[30] Y. Zhu, Y. Zhou, Preparation of pure ZnO nanoparticles by a simple solid-state reaction method, Appl. Phys. A 92 (2008) 275–278.
[31] R. Elilarassi, G. Chandrasekaran, Structural, optical and magnetic characterization of Cu-doped ZnO nanoparticles synthesized using solid state reaction method, J. Mater. Sci. Mater. Electron. 21 (2010) 1168–1173.
[32] K. He, et al., Method for determining crystal grain size by X-ray diffraction, Cryst. Res. Technol. 53 (2018) 1–6.
[33] H.A. Foner, N. Adan, The characterization of papers by X-ray diffraction (XRD): measurement of cellulose crystallinity and determination of mineral composition, J. Forensic Sci. Soc. 23 (1983) 313–321.
[34] V. Causin, et al., Forensic differentiation of paper by X-ray diffraction and infrared spectroscopy, Forensic Sci. Int. 197 (2010) 70–74.
[35] A.C. Mohan, B. Renjanadevi, Preparation of zinc oxide nanoparticles and its characterization using scanning electron microscopy (SEM) and X-ray diffraction(XRD), Procedia Technol. 24 (2016) 761–766.
[36] M.G. Ruiz, et al., Characterization of calcium carbonate, calcium oxide, and calcium hydroxide as starting point to the improvement of lime for their use in construction, J. Mater. Civ. Eng. 21 (2009) 694–698.
[37] M. Kouzu, et al., Solid base catalysis of calcium oxide for a reaction to convert vegetable oil into biodiesel, Adv. Powder Technol. 21 (2010) 488–494.
[38] K.D. Vernon-Parry, Scanning electron microscopy: an introduction, III-vs Rev. 13 (2000) 40–44.
[39] A.J. Wilkinson, P.B. Hirsch, Electron diffraction based techniques in scanning electron microscopy of bulk materials, Micron 28 (1997) 279–308.
[40] C. Bhavya, et al., Synthesis of calcium oxide nanoparticles and its mortality study on fresh water fish *Cyprinus carpio*, J. Environ. Sci. 10 (2016) 55–60.
[41] M. Manso, Investigation of the composition of historical and modern Italian papers by energy dispersive X-ray fluorescence (EDXRF), X-ray diffraction (XRD), and scanning electron microscopy energy dispersive spectrometry (SEM-EDS), Appl. Spectrosc. 65 (2011) 52–59.

[42] R.P. Williams, et al., Quantification of the extent of reaction of metakaolin-based geopolymers using X-ray diffraction, scanning electron microscopy, and energy-dispersive spectroscopy, J. Am. Ceram. Soc. 94 (2011) 2663–2670.
[43] B. Cerqueira, Time of flight secondary ion mass spectrometry and high-resolution transmission electron microscopy/energy dispersive spectroscopy: a preliminary study of the distribution of Cu^{2+} and Cu^{2+}/Pb^{2+} on a Bt horizon surfaces, J. Hazard Mater. 195 (2011) 422–431.
[44] O.L. Krivanek, et al., High-energy-resolution monochromator for aberration-corrected scanning transmission electron microscopy/electron energy-loss spectroscopy, Phil. Trans. R. Soc. A 367 (2009) 3683–3697.
[45] S. Chenna, P.A. Crozier, Operando transmission electron microscopy: a technique for detection of catalysis using electron energy-loss spectroscopy in the transmission electron microscope, ACS Catal. 2 (2012) 2395–2402.
[46] T.E. Weirich, et al., Structures of nanometre-size crystals determined from selected-area electron diffraction data, Acta Cryst. A 56 (2000) 29–35.
[47] A. Thust, High-resolution transmission electron microscopy on an absolute contrast scale, Phys. Rev. Lett. 102 (2009) 100–109.
[48] R. Kumar, et al., Fourier transform infrared spectroscopy and chemometrics for the characterization and discrimination of writing/photocopier paper types: application in forensic document examinations, Spectrochim. Acta Mol. Biomol. Spectrosc. 170 (2017) 19–28.
[49] K.K. Pandey, A study of chemical structure of soft and hardwood and wood polymers by FTIR spectroscopy, J. Appl. Polym. Sci. 71 (1999) 1969–1975.
[50] P.S. Araújo, et al., Thermal degradation of calcium and sodium alginate: a greener synthesis towards calcium oxide micro/nanoparticles, Int. J. Biol. Macromol. 140 (2019) 749–760.
[51] S. Pramanik, K.K. Kar, Nanohydroxyapatite synthesized from calcium oxide and its characterization, Int. J. Adv. Manuf. Technol. 66 (2013) 1181–1189.
[52] R. Srinivasan, et al., Structural and optical studies of yttrium oxide nanoparticles synthesized by co-precipitation method, Mater. Res. Bull. 45 (2010) 1165–1170.
[53] R. Srinivasan, et al., Structural and optical properties of europium doped yttrium oxide nanoparticles for phosphor applications, J. Alloys Compd. 496 (2010) 472–477.
[54] R. Jones, et al., Raman techniques: fundamentals and frontiers, Nanoscale Res. Lett. 14 (2019) 1–34.
[55] P. McMillan, N. Ross, The Raman spectra of several orthorhombic calcium oxide perovskites, Phys. Chem. Miner. 16 (1988) 21–28.
[56] L. L Boyer, J. R Hardy, Theoretical study of the structural phase transition in RbCaF3, Phys. Rev. B 24 (1981) 2577–2591.
[57] W. Cochran, A. Zia, Structure and dynamics of perovskite type crystals, Phys. Status Solidi 25 (1968) 273–283.
[58] R.A. Cowley, Lattice dynamics and phase transitions of strontium titanate, Phys. Rev. A 134 (1964) 981–997.
[59] F.S. Rocha, et al., Experimental methods in chemical engineering: ultraviolet visible spectroscopy—UV-Vis, the Canadian, J. Chem. Eng. 96 (2018) 2512–2517.
[60] S. Biyada, et al., Spectroscopic characterization of organic matter transformation during composting of textile solid waste using UV–visible spectroscopy, Infrared spectroscopy and X-ray diffraction (XRD), Microchem. J. 159 (2020) 105314–105410.
[61] S. Singh, et al., A phosphate-dependent shift in redox state of cerium oxide nanoparticles and its effects on catalytic properties, Biomaterials 32 (2011) 6745–6753.

[62] C.M. Magdalane, et al., Facile synthesis of heterostructured cerium oxide/yttrium oxide nanocomposite in UV light induced photocatalytic degradation and catalytic reduction: synergistic effect of antimicrobial studies, J. Photochem. Photobiol. 173 (2017) 23−34.
[63] N.L. Ignjatovic, et al., Rare-earth (Gd^{3+},Yb^{3+}/Tm^{3+}, Eu^{3+}) co-doped hydroxyapatite as magnetic, up-conversion and down-conversion materials for multimodal imaging, Sci. Rep. 9 (2019) 16305−16320.
[64] R. Rajaramakrishna, et al., Structural analysis and luminescence studies of Ce^{3+}: Dy^{3+} co-doped calcium zinc gadolinium borate glasses using EXAFS, Radiat. Phys. Chem. 171 (2020) 108695−108700.
[65] V.Dubey et al., Photoluminescence and thermoluminescence behavior of Gd doped Y_2O_3 phosphor, Optik Int. J. Light Electr. Opt. 126(1) 1-15.
[66] M.M. Obeid, et al., Unraveling the effect of Gd doping on the structural, optical, and magnetic properties of ZnO based diluted magnetic semiconductor nanorods, RSC Adv. 9 (2019) 33207−33221.
[67] V. Paterlini, et al., Characterization and luminescence of Eu^{3+}- and Gd^{3+}-doped hydroxyapatite $Ca_{10}(PO_4)_6(OH)_2$, Crystals 10 (2020) 806−820.
[68] M.S. Pathak, et al., Narrowband ultraviolet B emitting Gd doped $CaAl_2O_4$ phosphors − an ESR and photoluminescence study, J. Lumin. 204 (2018) 75−80.
[69] A. Pandey, et al., Thermoluminescence and photoluminescence characteristics of nanocrystalline $LiNaSO_4$: Eu phosphor, J. Phys. D Appl. Phys. 36 (2003) 1−19.
[70] A. Ozdemir, et al., Optically Stimulated Luminescence (OSL) dosimetric properties of $Li_2B_4O_7$:Ag,Gd and its relationship with thermoluminescence (TL) glow-curves, J. Alloys Compos. 751 (2018) 159−169.
[71] J.R. Jayaramaiah, et al., Luminescence studies of europium doped yttrium oxide nano phosphor, Sensor. Actuator. B Chem. 173 (2012) 234−238.
[72] A.S. Rajashekharaiah, et al., Photoluminescence, thermoluminescence and photocatalytic studies of sonochemical synthesis of $Bi_2Zr_2O_7$:Sm^{3+} nanomaterials, J. Mater. Sci. 31 (2020) 15627−15643.

Section II

Applications of rare-earth-activated phosphors

Rare-earth-activated phosphors for LED applications

8

Irfan Ayoub [1], Umer Mushtaq [1], Nisar Hussain [1], Seemin Rubab [1], Rakesh Sehgal [2], Hendrik C. Swart [3] and Vijay Kumar [1,3]

[1]Department of Physics, National Institute of Technology Srinagar, Srinagar, Jammu and Kashmir, India; [2]Department of Mechanical Engineering, National Institute of Technology Srinagar, Srinagar, Jammu and Kashmir, India; [3]Department of Physics, University of the Free State, Bloemfontein, South Africa

1. Introduction

Phosphors are luminescent materials that emit photons once excited with the help of an excitation source [1]. These excitation sources can be of many types such as electron beam, UV light, or high-intensity electric field. Based on the excitation source used, this property of spontaneous emission of light upon excitation (luminescence) is divided into many categories such as photoluminescence (excitation with photons of higher energy), cathodoluminescence (excitation with the help of electron beam), or electroluminescence (excitation with the help of strong electric field). Phosphor is composed of two main components: one is a host and the second is an activator. The host used must be inert while the activator should be rare-earth (RE) ion or 3d transition metal ion. In this chapter, the main focus is targeted on the 4f RE-activated nanophosphors as they have been found to be very useful for many applications such as flat panel displays, optoelectronic devices, and light-emitting diodes. The RE-activated nanophosphors have generated huge attention due to their use in the field of biomedical imaging, labeling, and radiation detection [2–5]. Based on the type of transition, RE-activated nanophosphors are of two types: one group is where the d–f transition takes place, which generates a broad emission band; the second is the narrow emission band generating transition that takes place within f levels. The 4f ions used generally for light-emitting diode (LED) including Eu^{3+}, Eu^{2+}, Tb^{3+}, Gd^{3+}, Ce^{3+}, Yb^{3+}, Dy^{3+}, and Sm^{3+} are used as dopants as well as codopants in different nanophosphor hosts [6]. The luminescent properties of these RE-activated nanophosphors mainly depend on the type of host used, changes in its chemical composition, activators–host lattice coupling, and energy scheme [7,8]. Some RE ions such as Ce^{3+} and Eu^{2+} when used with particular hosts emit photons of lower frequency; thus, it is very important to look at the structural model of such nanophosphors to ensure color tunability and better luminescence properties [9]. To synthesize a nanophosphor having the property of color-tunable emission, one must keep track of some factors such as the host must have a broad transparency range, low cut-off frequency, tunable crystal phase, high refractivity, and increased thermal stability [10].

Rare-Earth-Activated Phosphors. https://doi.org/10.1016/B978-0-323-89856-0.00013-4

The property of the emitted photon also depends on the chemical composition of the host as well as an activator–host coupling and mechanism of energy transfer [11–13]. The RE-activated nanophosphors have been widely used in tricolor fluorescent lamps, bioimaging equipment, and in the case of persistent luminescent pigments. These types of nanophosphors have been seen to be exceptionally stable with better optical characteristics and environment friendliness. The quantum efficiency (QE) of RE-activated nanophosphors mainly relies on the absorption cross-section of RE ion and their integration inside the host lattices. While doping, the host with an RE ion replaces a cation in it. Thus, the dopant needs to have its atomic radii comparable with that of the cation that it must replace. The doping concentration is generally below 10% [14,15]. In this chapter, different types of lighting and types of solid-state lightings are discussed. Further, the RE-activated nanophosphor, different types of electronic transition responsible for the generation of photons, and the use of such nanophosphors for the construction of efficient LEDs are also discussed.

2. Types of lighting

For the energy sustainable strategy, it has become very important to develop energy-saving lighting devices, as one-fifth of global power consumption is used for lightning [16,17]. This necessitates the importance of designing energy-saving lightning, which is sustainable for the long run. Keeping in view the necessity and need, new devices are continuously being fabricated due to advancements and achievements in material and lightning technology. Lightning devices can be broadly categorized into two types as given in Fig. 8.1 [16,17].

General lighting refers to the direct application of light and apparatus need to give illumination whereas specialized lightning refers to the indirect application of light [17]. Light-emitting devices, for example, fluorescent lamps were initially fabricated for common lighting along with for industrial lightning [18,19]. They had been used continuously for the past so many years because of their less power consumption and longer life along with their efficiency [17]. The efficiency of the fluorescent lamps in comparison with the previously available system is given in Table 8.1.

The main part of the fluorescent lamps is mercury vapor along with a coating of phosphor inside the tube [21–24]. The purpose of mercury is the conversion of energy from electrical to radiant in the UV region, then into the visible region with the help of phosphor [25]. The quantity of light emitted from the fluorescent lamp depends on the

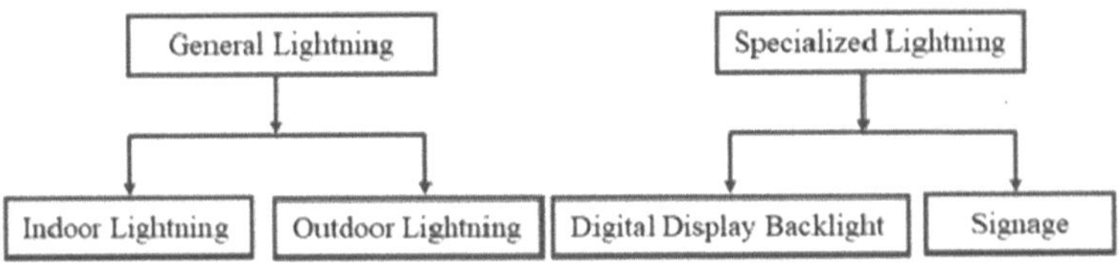

Figure 8.1 Types of lightning.

Table 8.1 Efficiency values of lightning devices.

Light source	Efficiency [lm/W]	Reference
Incandescent (no blast)	10–18	[18]
Halogens (no blast)	15–20	[18]
Compact fluorescent	35–60	[18,20]
Linear fluorescent	60–100	[18]
Metal halide	50–90	[18]

phosphor utilized as different chemicals have different wavelengths. Mostly, the phosphors used in fluorescent lamps are halo-phosphates (calcium, antimony, chlorine, etc.) [16]. Despite many innovations, only 63% of energy can be transformed to UV radiation by an efficient fluorescent lamp. After the conversion of UV radiation to visible in the phosphor, the overall efficiency cannot exceed 28% [16,26]. Due to the environmental hazards, such as emission of UV light, environmental contamination by mercury and phosphor when disposed of, and because of their shapes, structure, collection, and recycling, fluorescent lamps are rapidly being replaced by solid-state light (SSL)-emitting devices [16,21,22,24,27]. These devices have gained broad recognition as the familiar lights that give information about frequently used devices such as monitor, cell phone, and oven [28]. It refers to the phenomenon of light generated by solid-state electroluminescence (the phenomenon in which light is emitted in response to the flow of electric current or electric field) unlike incandescent bulbs or fluorescent tubes. In comparison with the others, SSL produces light with less heat generation and less energy dissipation. In 1999, the most important question was how SSL will get the victory over the entrenched lightning technology. In this regard, five parameters need to be satisfied, which are performance, quality of light, lifetime system cost, availability of investment capital cost, and constraints on the ability of the existing industry to react [29]. By now, SSL has almost met our expectations. The typical varieties of SSL devices are given in Fig. 8.2 [16].

Research in the field of SSL is growing day by day, because of its various advantages over the previously available system such as being mercury-free, low energy consumption, high durability, long life, and so on [30–32]. All these newly

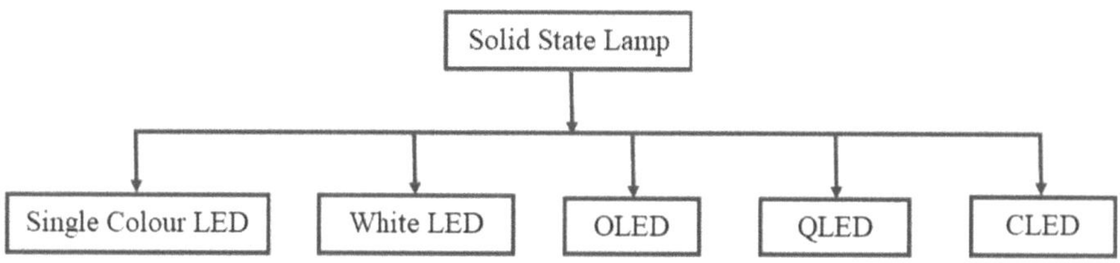

Figure 8.2 Types of solid-state lamps. OLED stands for organic light-emitting diodes, QLED stands for quantum-dot light-emitting diodes, and CLED stands for carbon-based light-emitting diodes.

developed lightning devices are considered to be prominent technologies. In this chapter, a general introduction of all the above-mentioned types will be given, but the main focus will be devoted to RE-activated phosphors along with organic LEDs.

3. Light-emitting diodes

After the discovery of electroluminescence, the first practical p−n junction diode was developed almost after 60 years. This was a p−n junction emitter of the red color [33]. White light emitters came into being after the expansion of light emitters from GaN and related materials in 1994 [34−36]. LED is a semiconductor light source that is close to a semiconductor chip rather than a lightning device. The basic structure of LED is similar to that of semiconductor devices, which is the formation of junction between p and n-type semiconductors. The passage of electric current across the junction only in one direction like ordinary diodes, as the current flows the LED is forward biased and thus emits light. Generally, AlGaInP, GaAlAs, and AlGaInN emit monochromatic light corresponding to the bandgap of their semiconductor compounds used [16]. These LEDs do not emit white light (polychromatic). White LEDs are employed as general lightning sources and are also capable of generating full-color displays [16,37]. In the first case, a three-color chip is used and white light is obtained by mixing different colors. As the device does not radiate any energy in some wavelength bands, the design may generate some odd colors during illumination. It is generally used for displays in backlighting. In the second case, the RGB phosphor is excited by using an LED with a wavelength shorter than blue. In this case, the phosphor emission shows very little change with the variation in LED wavelength, and also in this case the color can be controlled. In the last case, a yellow phosphor is used to convert monochromatic blue light to white light [26,37]. Monochromatic LEDs have been used in a large number of applications including information display panels. With the growth of GaN LEDs, full color LED displays have been developed, which was impossible due to the inefficiency of blue and green LEDs [34,35]. The LEDs based on GaN have become the driving force for the current technologies. These LEDs not only improved the efficiency but also made possible the production of monochromatic multichip-based full-color LEDs and phosphor-based white LEDs [16]. The use of LEDs as backlighting devices has brought faster response, broader color spectrum, longer lifetimes, thinner structure, and no mercury [38,39]. Latest LED lighting devices include downlights [40], road lights [37], base lamps [37], retrofit LED lamps [37], and so on.

4. Rare-earth-activated phosphor

The discovery of light emission from RE ions dates back to the 19th century. The rare-earth ions have picked up a lot of attention because of their excellent luminescence properties with sharp emission lines and high lumen counterpart, trivalent lanthanide ions have a unique and attractive luminescence. Among the large number of the elements available so far in the universe, the total of all the RE elements is 17. These include scandium [41], yttrium [13], and lanthanide series [42–56], which play the role of dopants and codopants in almost all types of luminescent materials [57,58]. The fundamental reason for the luminescence in RE (3+) ions is a parity-restricted electronic transformation within 4f-shells [59]. Also, the remarkable feature revealed from the studies predicted that the optical properties of the 13 RE elements from Ce^{3+}–Yb^{3+} are regulated by the screening of the partially filled 4f-shells shielded by the outer filled shells (5s and 5p) [60,61]. All the elements mentioned earlier possess their own set of characteristic energy levels. Among them, the 13 RE elements from Ce^{3+}–Yb^{3+} show the phenomenon of luminescence near the visible portion of the electromagnetic spectrum [62,63]. In comparison, ions that do not possess the 4f energy levels that include Sc(III), Y(III), La(III), and Lu(III), show the phenomenon of the absorption and emission within the visible range of the electromagnetic spectrum [64]. The strategy for controlling the photoluminescence properties that apply to lightning applications is mainly determined by the crystal host, possible chemical composition adjustments, activator host lattice coupling, and mechanism of the energy transfer, which are depicted in Fig. 8.3 [7,8]. The primary prerequisite for fabricating optimal phosphor is the selection of the host lattice with the appropriate structural arrangements for a better understanding of the reliable excitation and emission spectra according to the existence of activators doped [65]. In a few host lattices, the dopants such as Ce^{3+} and Eu^{2+} possess a longer wavelength in their emission spectrum. As a result, a thorough understanding and analysis of suitable structural models are needed to monitor photoluminescence properties [9]. The essential criteria for an exemplary host matrix include wide clarity radius, high refractivity, tunability, low photon intensity, significant photochemical and thermal stability, and so on [10]. The modification in the chemical structure is the second most significant criterion for improving luminescence and color tunability [11–13].

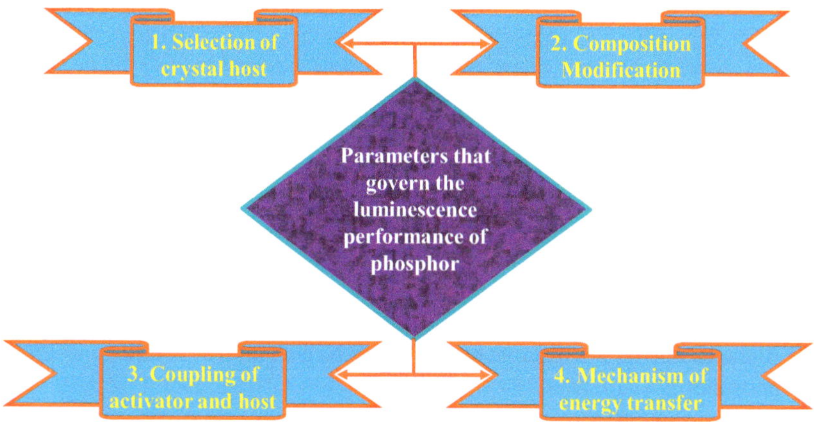

Figure 8.3 Parameters for designing the energy-efficient phosphor material.

Besides these two criteria, the activator–host pairing and the mechanism of the energy transfer are also important factors affecting the phosphors' luminescent action. Thus, for the fabrication and layout of the phosphors with tunable emission for commercial and technical applications, the previously mentioned parameters should be taken into consideration. The RE-ion family has received a lot of recognition in recent years for their different uses such as, in tricolor fluorescent lamps (La, Y, Ce, Tb, Eu), for surgical devices (Lu, Gd, Ce) in colored televisions (Y, Ce, Tb, Eu), in radiation detectors along with the effective emitter in a variety of solid-state host lattices. Due to their extraordinary stability, optical and electric properties, and eco-friendliness, an enormous review has been granted on different categories of light-emitting RE-activated phosphors. The transitions associated with the RE-doped LEDs have been classified into two categories, which are narrow spectral lines that arise because of f–f transitions and broad emission bands that arise due to f–d transitions [57,58,64]. As the ligand field affects the spectral location of permitted f–d transitions, they are incompatible with f–f electronic transitions [66]. Among the REs, the most commonly utilized activators include Ce(III), Eu(III)/Eu(II), Gd(III), Yb(III), Dy(III), Sm(III), Tm(III), Er(III), and Nd(III). Their QE is determined not only by the RE ions absorption cross-section but by the way they remain integrated into the host lattices. The vast majority of RE-activated phosphors products are inorganic, with just a few organic compounds. In general, the phosphors are made up of a host lattice in which an RE ion replaces any of the cations. In most cases, the dopant is added in such amounts to make sure that the substitution ratio is kept below 10%. To achieve this replacement, the host matrix should be made up of cations whose ionic radii should be comparable/equal to that of lanthanides. In the case of the divalent RE ions, inactive host cations such as Ba(II) and Sr(II) remain present in the host lattice. Throughout most trivalent rear-earth phosphors, La^{3+} and Y^{3+} are primarily used as inactive host cations [14,15]. The maximum of the RE ions accumulate the largest of coordination numbers from octahedral to hexahedral based on the ionic radii. Lu(III) possesses the lowest coordination number that is octahedral while La(III) possesses the highest coordination number that is hexahedral. As a consequence, the lattice has a very diverse appearance. Different host matrices, such as oxides [67–69], aluminates [70–74], silicates [47–49], phosphates [53,75], vanadates [56,76], and phosphors, utilizing various other host lattices may be used to investigate phosphor compounds.

4.1 Representations and quantum efficiency in phosphors

Phosphors are typically made up of two parts: a lattice and an activator that is carefully applied in a regulated quantity ranging from ppm to a few mL percent of the matrix. The activator ion acts as the main luminescent core of the host lattice. As a result, for a substance to be significantly useful, the charge on the dopant cation must be identical to that of the charge on the inactive host cation. The general way for the representation of the phosphor is as follows:

$$X_{1-a-b} : Y_a \ Z_b$$

Here, the host is represented by X, the activator cation is Y, and the coactivator cation is Z in this situation. In general, the phosphor that possesses a QE of at least 80% is considered to be the efficient one. To get a reliable phosphor content, the host lattice and the activator must work together effectively such that the majority of the photons incident on the luminescent center gets absorbed. As most of the absorbed photons are deactivated during the relaxation period, thus the activated phosphor does not release all the absorbed photons. The QE is stated to assess the efficiency of phosphor, which is defined as the ratio of the number of photons emitted to the number of photons absorbed.

4.2 Comparative study of the emission properties of RE ions and other emitting centers

The emission in the RE ions is due to the optical transformation within the f-manifold. As the f-electrons are well protected from the chemical world, they have nearly maintained their atomic nature. As a result, the f−f emissions spectra are characterized by sharp lines. Furthermore, the f−f transitions are partly prohibited, and some of them are also spin prohibited. As a result, these optical transitions are very long, thus taking microseconds to the millisecond to complete [77]. Besides these, wide emission bands have also been noted for a variety of RE ions, with important examples as Eu^{2+} ($4f^7$) and Ce^{3+} ($4f^1$). In these situations, the responsible transitions for the emission are 4d−5f optical transitions. The d-electron engages in chemical bonding, because of which the d−f emission spectra possess strong broad bands. Besides that, the d−f optical transitions are permitted and as a result are extremely fast, and it takes place within a few seconds [66]. It is fascinating to have a comparison between the RE-phosphors emission characteristics to those of other luminescent centers. Most of the transition metal ions such as Mn (2+) are well known for acting as a luminescent in most of the hosts such as in $ZnSiO_4$:Mn. In this host, $^4T_1$ (t_{2g}, $^4e_g^1$)−$^6A_1$ (t_{2g}, $^3e_g^2$) transition results in the green emission that finally results in shifting of the chemical bonding. This shift occurs due to the transfer of an electron from t_{2g}-orbitals to e_g orbitals. It is because of this shift that the emission band formed is broad [78]. Also due to the spin−orbit interactions, the emission spectrum of Mn (4+) in Mg_4GeO_5.5F:Mn possesses a relatively thin emission line with a certain structure. This emission is caused due to spin-flip transitions in between the t_{2g}^3 manifolds, which means that the chemical bonding is barely altered. This latter example demonstrates that sharp emission bands can be found in optical transitions among states that engage in chemical bonding where the chemical bond intensity in the ground and excited states is not substantially different [79]. The earlier approach demonstrates that line emission is not a unique characteristic of RE ions and that the broad emission spectra can be derived from RE ions as well, relying on the optical transitions involved. Thus, the RE is not peculiar in this regard. In comparison to line emission produced by the transition metal ions, the spectral location of the emission lines formed by the RE ions is almost autonomous of the host lattice. Furthermore, some line emitting ions such as

Tb (3+), Eu (3+) emit at spectral positions thereby resulting in high lumen efficacies and excellent light efficiency, also the spectra of these ions are well suited to the human eye.

5. Rare-earth-doped phosphors for light-emitting diodes

The RE-doped LEDs possess an enormous number of benefits over other traditional light sources, much on luminous efficiency, power-efficient, reliability, environmental friendliness, and long operating life [80,81]. These LEDs mainly consist of a phosphor deposited on the LED chip, which is being activated by the different types of relevant light sources. The mainstay of these LEDs is inorganic materials mostly activated with RE ions. The RE or lanthanide ions are well known for their ability to be activated by either absorbing the UV or near-UV light, this results in the emission of visible light. RE ions are members of the f-block family elements in the periodic table. Although actinides are also included in this sequence, they are conveniently excluded from the RE-doped LEDs. Actinides are never favored in phosphors because of their radioactive origin. Apart from the La and Lu, rest of the elements in the series can emit light. Lanthanides are very much effective in emitting light in the +3 oxidation state. However, Eu has the unique property of emitting light in both the +2 and +3 oxidation states [82–89]. Besides these, certain transition metal elements have also been found to act as possible luminescence dopants. The well known among them are Cu, Mn, and Cr, which act as luminescent centers with oxidation states of +1, +2, +3, and +4. The dopants serve as activators in any of the inert host matrices. An efficient phosphor considered has a bandgap difference between 3 and 5 eV. In the case of the absence of the luminescent ions in the host matrix, the majority of the host lattices are inert and thus specifically do not show luminescence. To enable the luminescence process in such hosts, an activator ion is added to them. Besides the activator ions, in some situations are unable to capture the energy and transform it to luminescence emission. In such cases, the sensitizer ion plays a supportive role, by effectively absorbing energy and transferring a portion or all of it to the activator, thus resulting in a noticeable increase in their luminescence emission [90–95]. The key need for a phosphor to be used in the rare-earth-doped LEDs is that it should have a high UV or near-UV absorption, along with the effective emission in the visible light field. Depending on the characteristics of the dopants, different color emitting RE-doped phosphors came into existence. The well-known LEDs are red-emitting, yellow-emitting, green-emitting, blue-emitting, and white-emitting diodes based on RE-doped phosphors. A brief discussion about these is discussed in the next subsections.

5.1 Red light-emitting phosphor

Owing to the strong emission from the 593–650 nm, Eu^{3+}-doped phosphor acts as an efficient and desired for the generation of the red-emitting phosphor [96].

The transition that generates the perfect red portion of absolute color purity is $^5D_0-^7F_2$, which corresponds to an emission wavelength of 610–625 nm [97–99]. It is widely accepted that if the RE ion is even doped in earth/sand, they will emit the color of their characteristic emission. When it comes to phosphor for illumination, importance should be given to those materials that possess wide excitation bands. Besides the fact that Eu^{3+} are well known for their excitation and emission peaks that are due to the prohibited f–f transitions, they also possess valuable charge transfer bands (CTBs) in the UV field that is being used for the habituation of the red emission [100,101]. The CTBs function as sensitizers by absorbing UV light and transferring it to Eu^{3+} ions thus converting them to Eu^{2+}. It has also been observed that the CTBs are more extreme than the Eu^{3+} excitation peaks [102,103]. However, in the near-ultraviolet (NUV), the charge transfer property of the red-emitting LEDs has little benefits as the Eu^{3+} ion CTB mostly covers the near-UV spectrum [104]. Therefore, the reason for using the Eu^{3+} ions in the red-emitting NUV LEDs is to select a host wherein the said dopant possesses a low symmetry site [105,106]. As it has been observed that when Eu^{3+} ion occupies the noncentrosymmetric site, there occurs a relaxation in the parity forbidden 4f–4f transitions that result in an enhancement in the performance. Few examples of the Eu^{3+}-doped red-emitting phosphors identified so far are $Y_2O_3:Eu^{3+}$ [107–109], $YPO_4:Eu^{3+}$ [110], $CaTiO_3:Eu^{3+}$ [106], $La_2CaZnO_5:Eu^{3+}$, etc. [111]. The phosphor-converted assembly of red $CaLa_2ZnO_5:5\%Eu^{3+}$ phosphor and a blue LED torch is shown in Fig. 8.4. Besides this, searching for the Eu^{2+} doped nitrates or oxynitride phosphors is another way applicable for the development of an efficient red-emitting phosphor [112]. Nitrate-based phosphors are considered to possess good physical and chemical stabilities, as well as efficient luminescence properties [113,114]. Furthermore, these hosts prove to have a strongly covalent environment for 2+ doped Eu ions, thereby allowing them to undergo a heavy crystal field effect and thus resulting in shifting of their emission bands to longer wavelengths. The potential of Sialon-style hosts for their strong absorption in the UV to blue regions of the electromagnetic spectrum has drawn a lot of interest [115,116]. Besides the RE ions, the transition metals are also being used for the development of the red-emitting phosphor [117–119].

5.2 *Yellow light-emitting phosphor*

Whenever one thinks about the yellow-emitting phosphor, the first thing that should come to one's mind is cerium-doped aluminum garnet phosphor. The mentioned phosphor, that is, YAG:Ce is the well-known consistently active yellow-emitting phosphor that has been used for the production of white light by intermixing with the blue LEDs. For a long time, YAG:Ce has overshadowed the rest of the available yellow phosphors in the visible spectrum, as one may wonder whether there is any need to search for the mentioned color-emitting phosphor in the glorious presence of YAG:Ce. It possesses appropriate phonon energy and other important parameters that govern the transition, thus allowing Ce^{3+} emission to occur efficiently

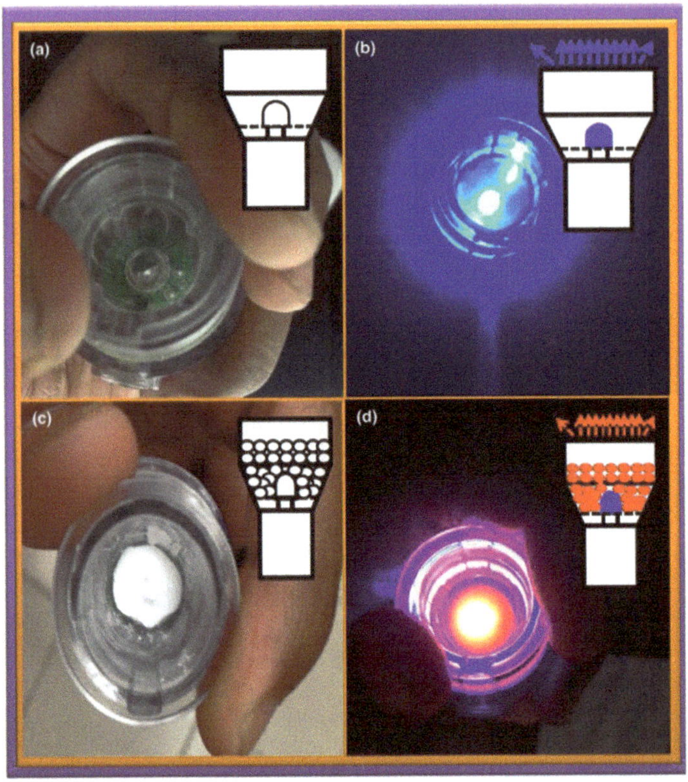

Figure 8.4 Assembly of the red $CaLa_2ZnO_5$:5%Eu^{3+} phosphor and a blue LED torch, showing the red light emission.
Reproduced with permission from Ref. [111].

[120–122]. Choosing an appropriate reduction atmosphere, the rate of the Ce^{3+} emissions can be increased by transferring the remaining Ce^{4+} ions in YAG to Ce^{3+}. Also, by optimizing the calcination temperature and time, the luminescence efficiency rate can be increased, which in turn results in the uniform distribution of Ce^{3+} ions in the YAG-host [123]. Besides Ce^{3+} ions, YAG has also been doped with Ga^{3+} ions for attaining better thermal stability [124]. Despite this, there is a desperate need for the investigation of the new yellow phosphor that possesses better color rendering index (CRI) and correlated color temperature (CCT) values [125,126]. Their applicability in high-power LEDs is restricted due to their inefficiency in providing a similar spectral distribution at higher temperatures [127]. Many investigators have also attempted to replace the Y^{3+} ion with Gd^{3+}, La^{3+}, or Lu^{3+} ions, but their color emissions did not improve to the target amount [128,129]. The yellow-emitting nitride phosphors that are triggered with Ce^{3+} and Eu^{2+} ions include $Ba_2Si_5N_8$:Eu^{2+} [130], $SrSi_2O_2N_2$:Eu^{2+} [131], $CaAlSiN_3$:Ce^{3+} [132], Eu^{2+}-doped Ca–SiAlON [133,134], and La_3Si_6N11:Ce^{3+} [127].

5.3 Green light-emitting phosphor

The green-emitting phosphor is the initiate in its nature regarding which least literature is available as compared to other tricolor phosphors. However, the current literature is so vast that a thorough discussion and analysis of the advancements so far achieved in the green-emitting phosphors in the LED technology can be made. Generally, till now, the green emission is produced by the doping of the Tb^{3+} and Mn^{2+} ions in a variety of host materials [135−137]. The sensitization of the Mn^{2+} ions with the Ce^{3+} or Eu^{2+} ions results in luminescence that is very much intensified in nature. The emission spectrum of Mn^{2+} ranges from green to red depending on the crystal field strength. However, owing to the forbidden $^4T_1-^6A_1$ transition, their poor UV absorption is a confounding consideration from the standpoint of LED manufacturers. On providing a favorable atmosphere to the Tb^{3+} ions, they also emit only green luminescence. Till now, other than green, these ions have not yet been observed to emit any other color. Thus, it is the nature of the crystal atmosphere of the dopant that governs whether Tb^{3+} ion will emit green light emission or no emission at all. However, there also exists a possibility that doped Tb ion will transit into a +4 charge state in which no luminescence will be witnessed. In this case, postheating of the phosphor is done, which reverts the Tb^{4+} ions to Tb^{3+} ions. The peak positions of the Tb^{3+} ions are situated at 490, 540, and 620 nm matching the transitions $^5D_4-^7F_6$, $^5D_4-^7F_5$, $^5D_4-^7F_4$, and $^5D_4-^7F_3$, respectively, [138]. Due to the spin-forbidden 4f−4f transitions, the emission peaks belonging to these are narrow, thus their peak locations can only vary by ±5 nm from host to host. Among the previously mentioned transitions, the only transition, that is, $^5D_4-^7F_5$ possesses the highest intensity and thus leads to the phosphors' overall green emission. Consequently, despite their clear green luminescence, most Tb^{3+}-doped phosphors struggle to achieve the LED condition [138−140]. Few examples of the green-emitting phosphors are $BaY_2Si_3O_{10}:Ce^{3+}$, Tb^{3+} [141], $Sr_3Y\ (PO_4)_3:Eu^{2+}$, Tb^{3+} [142], $SrBaSiO_4:Eu^{2+}$ [143], etc.

5.4 Blue light-emitting phosphor

Since the introduction of InGaN LED chips in 1994, many advances in the field of illumination have been made [144]. White light is known to be produced by combining a blue LED chip with a yellow phosphor coating, specifically a yellow-emitting YAG: Ce^{3+} phosphor. A new combination has been proposed, where a near-UV LED chip is used to excite the tricolor red, green, and blue phosphors. This method is thought to be one of the best for producing white light with a high CRI and high luminous efficiency [145]. $BaMgAl_{10}O_{17}:Eu^{2+}$ phosphor (BAM) is one of the most widely used blue phosphors [146,147]. The success met by the $BAM:Eu^{2+}$ phosphor has prompted many investigators to dig deeper into its optical and structural properties, besides focusing on the development of the synthesizing methods [148−150]. Its excitation spectrum spans a wide range of wavelengths, from 225 to 410 nm. However, the excitation band displays a significant absorption in the 360−410 nm regions and little absorption in the 360−410 nm range, thus making it unsuitable for LED fabrication. As a result, researchers start looking for some alternative blue-emitting phosphors that would meet the LED specifications. Under near-UV excitation, many of Eu^{2+} doped phosphors have exhibited improved blue luminescence. Phosphates are one of the familiar classes of phosphors that allow Eu^{2+} ions to luminescence in the blue region of the electromagnetic spectrum. They possess a broad bandgap, low phonon capacity, and excellent chemical and thermal stability. The well-known phosphate-based phosphors that produce blue luminescence include $RbBaPO_4:Eu^{2+}$ [151], $SrCaP_2O_7:Eu^{2+}$ [152], $Ca_3Mg_3(PO_4)_4:Eu^{2+}$ [153], $NaMgPO_4:Eu^{2+}$ [154], and others. In addition, Ce^{3+}-triggered phosphors such as $Li_4SrCa(SiO_4)_2:Ce^{3+}$ [155] and $Gd_5Si_3O_{12}N:Ce^{3+}$ [156] have also shown blue luminescence.

5.5 White-light emitting phosphor

Owing to the beneficial biocompatibility, high luminous power, energy efficiency, long life cycle, and being free from hazardous mercury, white LEDs are indisputably considered as the next-generation solid-state lighting sources. White light-emitting diodes can be produced using three different methods. In the first method, the white light is achieved by an optimal mixing of red, green, and blue phosphors that can be excited by near-UV LEDs. But due to the different lifetime possessed by each color, there occurs a gradual decay in the phosphor, and the final color of the discharge will get changed over time. The inconsistency of the color temperature and the unnecessary energy loss encountered in this method compelled the researchers to devise a new method. It is convenient to make white light by combining the red, green, and blue-emitting LEDs, but the color instability issue still exists. In the current situation, the second and more common solution for the attainment of the white light is to combine a blue-emitting LED with a yellow-emitting phosphor [80,126]. In this regard, YAG: Ce^{3+} phosphor pumped by the application of blue-emitting LED is used in all commercial white LEDs [145,157,158]. While this method yielded a lower energy loss than the tricolor phosphor blending method, it has certain imperfections that must be addressed. One such problem is the emitted white light's poor CRI, which is produced by the dearth of a red portion in the blend. To overcome the shortcomings of these two methods, a third method for producing white light from a single-phase host lattice has been introduced. Large research has been done in support of the single-phase host materials, but to date parameters satisfying the commercial stage have not been attained. Thus, there is a need to devote time to attain better results. Few examples in this regard are $Ca_8MgBi(PO_4)_7:Dy^{3+}$ [159], $CaZr_4(PO_4)_6:Dy^{3+}$ [160,161]. The basic mechanism involved in the emission of these host matrices is briefly discussed in the next section.

6. Luminescence mechanism of RE-activated nanophosphors

RE-activated nanophosphors are crystalline substances that have a very small amount of RE ions present in them. They have been extensively used for different types of

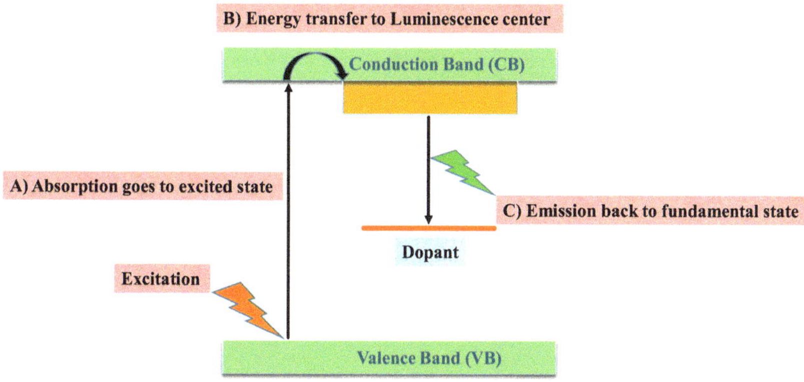

Figure 8.5 Schematic representation of energy level diagram in case of a host + dopant phosphor materials.
Reproduced with permission from Ref. [60].

applications like display panels, etc. [162,163]. The mechanism of such nanophosphors primarily depends on four basic steps viz. excitation of atoms or molecules from the ground state to a higher excited state, deexcitation to the ground state, decay time, and emission of the photons after deexcitation as represented in Fig. 8.5 [60].

The excitation of these RE-activated nanophosphors is brought about by the incidence of highly energetic electron beam (cathode rays) X-rays or UV rays. The decay mechanism is more or less the same for all inorganic nanophosphors, and it is independent of the type of excitation source used [164]. The process of excitation takes place when either an RE ion absorbs this energy directly or the host lattice absorbs energy that in turn transfers that energy to the dopant. The absorption efficiency of the activator ions can be increased by adding a codopant to the host lattice. Codoping is done when in the case of activator ions, the optical transition is not allowed by selection rules. After the excitation of the charge carriers with the help of an external excitation source, they come to the corresponding ground state after the emission of a photon or a phonon. The probability of emission of the photon increases considerably when the energy gap between the consecutive levels is very large. The process of energy transfer from host lattice to activator ions can take place via four processes that include radiative energy transfer from sensitizer (a chemical species or an active center that is inserted into the host lattice together with the activator ion such that the excitation of activator ions must be improved [165]), nonradiative resonant energy transfer from sensitizer to RE-activator ion, multiphoton-assisted nonradiative energy transfer, and cross-relaxation among activator ions as represented in Fig. 8.6 [60].

For the multiphoton-assisted nonradiative case the transfer of energy takes place when the energy of excited states is different for one another but if activator ions and sensitizer are having the same energy then cross-relaxation takes place [166,167]. In the case of RE-activated nanophosphors, color tunability can be made possible by adding suitable dopants to the host material. Due to the presence of a partly filled 4f subshell, sharp line spectra are obtained in RE-activated nanophosphors.

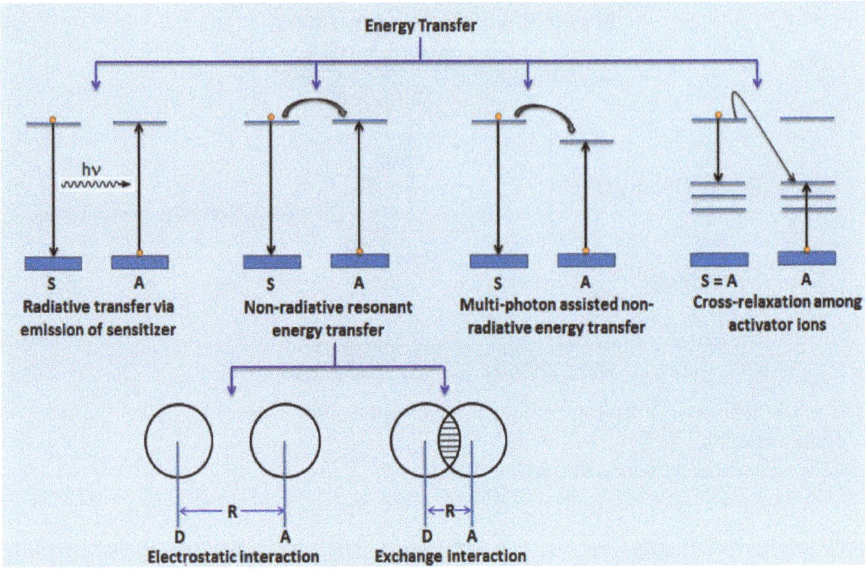

Figure 8.6 The mechanism of energy transfer involving various processes. Reproduced with permission from Ref. [60].

As the electrons of the 4f subshell are shielded by 5s and 5p orbitals, many distinct energy levels are generated [14,15]. The luminescence of RE-activated nanophosphors can be understood by exploring the transition shown by RE-doped nanophosphor viz. f–f transition, f–d transition, and charge transfer electronic transition.

6.1 f–f transition

In the case of RE trivalent ions, a very large number of distinct 4f energy states are available in visible and IR wavelength range except for Ce^{3+} and Yb^{3+}. The electric field due to the surrounding atoms causes the splitting of energy levels up to the magnitude of 100 cm^{-1}. This splitting is still not very large as the 4f energy levels are not very sensitive. Among these levels, only those energy levels that correspond to the PL emission are considered. These energy states arising because of the 4f electronic configuration in rare-earth ions were introduced by Dieke [168] and then distinguished by Carnall and his coworkers [169].

6.2 f–d transition

The f–d electronic transition involves the transfer of an electron from the 4f shell to a higher energy 5d energy state. This 4f to 5d transition is permitted by the Laporte selection rule, and such transitions are accessible by rare-earth ions that have ease of oxidation such as Eu^{2+}, Sm^{2+}, Yb^{2+}, Tb^{3+}, Pr^{3+}, and Ce^{3+}. In the case of such ions firstly, the emission band is broad as in the case of 5d orbital there is huge splitting because of the crystalline field as this orbital is increasingly exposed to the surrounding atoms. Secondly, these transitions are dipole allowed and thus are highly probable transition with the strength of 10^6 times as that of f−f transition [170]. When an electron shifts to the 4f shell from surrounding atoms or ions, then the process is called a charge-transfer transition. Such transitions are fully allowed by the selection rules and thus produce a wide excitation band which is known as the CTB. An important property of rare-earth ions is that they are easily reducible and thus generate such types of excitation bands as in the case of Sm^{3+}, Yb^{3+}, Ce^{4+}, Eu^{3+}, Tb^{4+}, and Pr^{4+} [171].

7. Organic light-emitting diodes

The organic light-emitting diodes (OLEDs) had appeared to be one of the most promising lighting technologies, because of their potential application for the fabrication of full color [172,173] and multicolor displays [174,175], and multilayer structures [176,177]. These are the LEDs that possess a thin emissive film of the organic electroluminescent compound that emits light in response to electric current. After the discovery of electroluminescence in organic materials, the first successful attempt in the field of organic displays and lightning technology took place in 1987 after the work of Tang and Vanslyke [172]. Another breakthrough to the technology was the discovery of electroluminescence from polymers at Cambridge University [178]. OLEDs have generated significant interest because of their wide range of applications. The key features of OLEDs for flat panel display include their self-emitting property, high luminous efficiency, full-color capability, wide viewing angle, high contrast, low power consumption, low weight, potentially large area color display, and flexibility [179−182]. To date, OLEDs of different colors such as the development of efficient green OLEDs [183−185], red OLEDs [186−188], blue OLEDs [189−191], and white OLEDs [192−194] have been reported. Out of the previously mentioned OLEDs, the white OLEDs have generated great interest, because of their application in backlight units for liquid crystal displays and OLED displays [177,192]. A large number of methods had been developed for the fabrication that includes a single emitting layer (EML) with doping from fluorophores or phosphors [195−202], different color emitting doped multi-EML [177,203−210], tandem structured OLEDs [211−213], OLEDs developed by incorporating polymers [214−216], coupling of blue-emitting OLEDs with yellow color emitting phosphor layer [217,218], and lastly utilizing the effect of microcavity from emissive layer [216]. However, more work is to be done on blue OLEDs. The structural history of OLEDs is depicted in Fig. 8.7. An OLED consists of a thin film of organic electroluminescent material sandwiched between two electrodes as shown in Fig. 8.8.

The luminescent material generally consists of a hole transporting layer, a light-emitting layer, and an electron transporting layer. The anode is generally made of indium tin oxide (ITO) because of its high work function and good transparency in the visible range, which is transparent while the cathode made of metal is reflective. The thickness of the organic layer varies between 100 and 150 nm [174]. When the

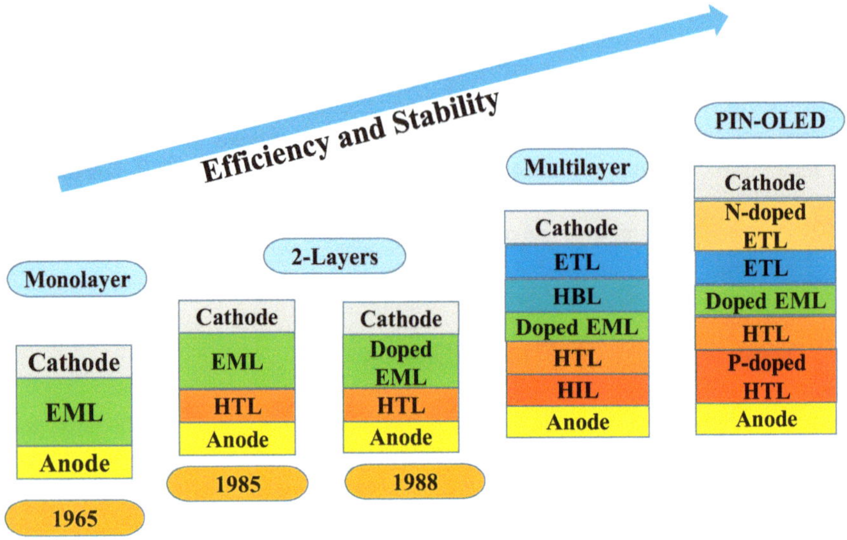

Figure 8.7 Evolution of OLEDs structures.

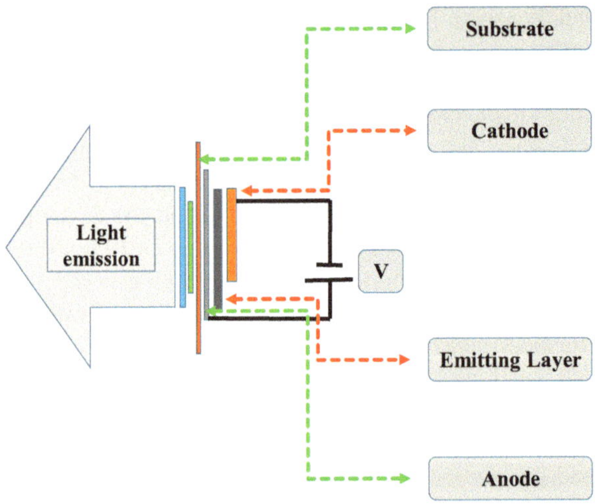

Figure 8.8 Basic structures of OLED.

voltage is switched on in between the electrodes, the motion of charges takes place and gets injected onto the organic material. By the process of hopping the charges move inside the thin layer of organic material and recombine to form excitons. These excitons recombine after diffusion takes place and results in the emission of a photon. The color of the photon is governed by the energy difference between the highest occupied molecular orbital (HOMO) and the lowest unoccupied molecular orbital (LUMO). Generally, a low barrier of the order of 5–6 eV is required about the HOMO level of the organic material for efficient hole injection. Similarly, a low barrier for electrons of the order of 2–3 eV is required concerning the LUMO level of the organic material [174]. On the behalf of the materials used, the OLEDs are categorized as small molecule organic light-emitting diodes (SM-OLED) and polymer light-emitting diodes. The light-emitting process for both is the same but the difference lies in the deposition process. Narrowing of the emission spectrum and shifting of emission peak from lower energy to higher energy has been observed in the multilayer structure of organic light-emitting diode [219]. On the basis of the emission of light, the OLEDs have been classified as top-emitting and bottom-emitting OLEDs. The device is referred to as a bottom emission device if the light is emitted from the bottom electrode through the substrate, and the device is referred to as a top emission device if the light is emitted from the top or lid of the device. The demand for top emission OLEDs has shown an enormous growth rate because of their large number of applications. Different methods were used for their development such as the vacuum deposition method was used for the development of surface-emitting organic light-emitting devices [220]. The top-emitting OLED possesses an easier integration with backplane electronics but requires a transparent top electrode instead of an opaque metal cathode. Current technology is based on thin semitransparent metal electrodes instead of ITO deposition, which is difficult for implementation in large-area devices. The OLED based on an inverted organic layer has also been fabricated [221]. The primary criteria for choosing a material for OLEDs are high luminescence quantum yield, good carrier mobility and film-forming properties, good thermal and oxidative stability, and good color purity [177]. The first-generation OLEDs were mainly based on fluorescent materials [172]. In these OLEDs, the emission of light was as a result of the recombination of the single excitons but the quantum efficiency remained confined within 25% [174]. This limits the efficiency arising due to only singlet relaxation and thus nonparticipation of the triplet exciton population. The second generation of OLEDs is based on phosphorescent materials. In these OLEDs, all excitons are converted into emissive triplet states through intercrossing. These materials prove more efficient than fluorescent ones. The first successful attempt of using phosphorescent materials in OLEDs was done in1998 which shows more efficiency (up to 90%) than fluorescent materials [222]. Efficiencies up to 100% have also been reported in the case of phosphorescent OLEDs [223,224]. The stability of the device is very important for OLEDs, especially the differential aging of the three primary colors. Device stability is defined as the meantime to half brightness. A lifetime of about 20,000 h with at least 100 cd/m^{-2} brightness is considered to be necessary. For the achievement of such a lifetime, the luminance for each pixel takes higher values [174]. It has been observed that for high luminance values the lifetime of the OLEDs shows degradation. The degradation phenomenon mainly occurs in three different ways.

Dark spot degradation: As suggested by the name itself, it refers to the decrease in the luminescence of OLED due to the formation of dark spots on the emissive area. The formation of dark spots takes place in both operative and nonoperative modes. Often, it is suggested that the mechanism for the formation of the dark spots is either electrochemical or thermally activated as a result of current flow (Joules heating). As the OLED is exposed to air, the formation of the dark spots increases rapidly and becomes visible to the naked eye [225]. It was seen that OLED operated in air lost its electroluminescence efficiency up to 99% within a period of 150 min [226]. The main cause behind the formation of dark spots is cathode delamination, change of organic materials at the cathode interface. Also, it was investigated that hole transport layer and anode do not play role in the formation of dark spots [227]. The other methods include coating with a thick layer of calcium (1000 Å) [228] and heating the substrate before deposition of the organic layer [229].

Catastrophic failure: It refers to the total loss of luminescence as a result of electrical shorts. These electrical shorts arise due to the loss of a large leakage current, because of the defects present in the organic layer that attains heat when the device is operated. This effect is more prominent in small molecule-based OLEDs wherein pixels short circuit under continuous operation, hence referred to as progressive electrical shorts [230]. This defect can be removed by using materials with better filming properties and stable morphology [225], by using an AC driving system that inhibits the accumulation of charge in the organic layer [231], and by applying the preoxidative method to the device [230].

Intrinsic degradation: It refers to the decrease in the luminance of the device during its operation. This change is not associated with any change in the device appearance, it thus refers to a decrease in electro luminance quantum efficiency of OLED. Several models such as morphological instability model, unstable cationic ALQ3 model, indium migration model, mobile ionic impurities model, and immobile positive charge accumulation model for the internal degradation [225] were suggested. The lifetime of about 20,000 h has been achieved in advanced RGB materials with a luminance up to 100 cd m^{-2}. Also, the lifetime shows a drastic change with the increase in temperature. Typical lifetime values of second generation are given in Table 8.2 [174].

Table 8.2 Performance of phosphorescent materials.

Color	Current efficiency (CdA^{-1})	Lifetime (hours)
Red	15 (at 500 cd m^{-2})	22,000 (at 500 cd m^{-2})
Green	29 (at 1000 cd m^{-2})	20,000 (at 1000 cd m^{-2})
Blue	19 (at 100 cd m^{-2})	$<$1000 (at 200 cd m^{-2})

8. Quantum-dot LED

In the present scenario, more attention is being diverted toward semiconductor quantum dots because of their significant optical properties viz. narrow emission bands, emission tunability, high quantum yield, and optical stability. On the basis of these properties, they are being widely used for LED applications [16]. The first design of the QLED was prepared from CdSe nanocrystals. Initially, these nanocrystals were dissolved in toluene and then exposed to ITO or ITO/poly(p-phenylenevinylene) (PPV) plates. These nanocrystals get attached to the surface because of the pretreatment of the plate's hexane dithiol. The variability of the band-gap of the quantum dot with size has also been used for the generation of white light or other emission spectra by integrating them with organic semiconductor materials or OLEDs [153]. In QLEDs the generation of light takes place by the recombination of electrons and holes injected into the light-emitting layer via cathode and anode boundaries. QD is being used in displays and for lightning in a number of ways such as QD can be used as a color-changing medium for LEDs such as the use of green and red CdSe/ZnS/CdSZnS quantum dots and polymer composites of quantum dots for the production of full-color emission spectrum [232,233]. QD incorporated with light-emitting polymers is used better electroluminescent spectra such as utilizing the combination of CdSe/ZnS (red emission) QDs and organic material (green/blue emission) [234]. On application of the voltage in forward bias, the transportation of electrons and holes toward the QD layer takes place wherein they recombine and generate photon. The performance of the QLED depends on the material of the QD and the structure of the device. In comparison with available OLEDs or LEDs, the QLEDs possess more stability, solution processability, excellent color purity, and narrow emission spectra for the quantum dots. But the degradation of the organic material, oxidation of reactive metals, and presence of heavy metals in QDs which proves harmful limit their application [16]. The different electroluminescent QD materials, their luminescence, and external QE are studied by various investigators [235–240]. The quality of light generated by an RE-activated nanophosphor can be determined by the way it is perceived by a naked eye. The physical properties of light emitted from a phosphor are characterized via two ways viz radiometry and photometry. Radiometry is where we are concerned with photon energy, the intensity of the emission, and radiant flux while photometry deals with how the human eye characterizes the light and how is its color sensation [241]. CRI value denotes the ability of an LED to generate a true reflection of colors from an object. The measurement of the ability of the light source to express each color is called as CRI. The CRI value is measured in comparison with broad daylight for which the value of CRI is 100%. The higher value of CRI means that the light emitted from a light source is closer to daylight. For the construction of a better LED besides greater efficiency and higher brightness, a higher value of CRI must be ensured. The use of RE-doped LED is generally costly thus while fabricating phosphor-converted LED generally low-quality phosphors are used, which has adverse effects on the CRI of the emitted light [242].

For explaining the appearances of the color emitted by an LED, we use the color temperature by relating it with the temperature of an ideal black body. The color temperature is used in terms of kelvin (K). This term can only be defined for light sources that generate some heat like incandescent bulbs that generate thermal radiations. In the case of light sources where light is generated using luminescence, the temperature of the light source does not provide the correct value of color temperature. For such sources where light is generated using the phenomenon of luminescence, the color temperature is measured in terms of CCT. Depending on the appearance of emission color, the CCT is rated how warm or cool the light source is [145]. Cool-colored light is nearer to blue-colored radiation and warm-colored light is near to infrared radiation while the color temperature of cool light is over 5000 K while the warm light has a color temperature in the range of 2500–3500 K [243,244].

The efficiency of RE-activated nanophosphor is generally expressed in terms of QE, and it is defined as the ratio of the number of photons emitted to the number of photons absorbed by a phosphor. The case of LED efficiency is defined in terms of external quantum efficiency (EQE), which measures the ability of an LED to convert electrons into photons. The EQE is given by the ratio of the number of photons emitted to the number of electrons passing through the LED. Its value depends on internal quantum efficiency, injection efficiency, and extraction efficiency. The value of EQE is calculated by using the following relation:

$$\text{EQE} = \text{internal quantum efficiency} \times \text{injection efficiency} \times \text{extraction efficiency}$$

The term injection efficiency is given by the ratio of the total number of electrons passing into the device to the number of electrons injected into the active region. Not all the recombinations produce photons, and the ones that do are called radiative recombinations. The fraction of all the radiative transitions is called internal quantum efficiency. The extraction efficiency is defined as the fraction of photons produced because of radiative transitions that can escape the device.

9. Conclusions

In this chapter, the main focus is on the luminescent properties of RE-activated nanophosphors and their use in the field of light-emitting diodes. Nowadays, LEDs are being used for numerous applications other than lighting and thus a number of modifications have been done in their configurations. The use of nanophosphors has enabled us to achieve high CRI and CCT values. A lot of progress has been made during the last decade in achieving better efficiency, cost-effectiveness, sustainable, and clean energy technology but still a lot more work needs to be done in this field. As phosphors are the prospective candidates of the future thus the fabrication of novel, improved and highly efficient phosphors is the need of the hour.

References

[1] V.B. Pawade, H.C. Swart, S.J. Dhoble, Review of rare earth activated blue emission phosphors prepared by combustion synthesis, Renew. Sustain. Energy Rev. 52 (2015) 596–612, https://doi.org/10.1016/j.rser.2015.07.170.

[2] V. Pawade, N. Dhoble, S. Dhoble, Promising blue emitting Ca3.5Mg0.5Si3O8Cl4: Eu2+ nano-phosphor for near UV-excited white-LEDs, Optoelectron. Adv. Mater. Rapid Commun. (2011) 208–210.

[3] N. Yeh, J.P. Chung, High-brightness LEDs-Energy efficient lighting sources and their potential in indoor plant cultivation, Renew. Sustain. Energy Rev. 13 (2009) 2175–2180, https://doi.org/10.1016/j.rser.2009.01.027.

[4] N.G. Yeh, C.-H. Wu, T. Chih Cheng, Light-emitting diodes-their potential in biomedical applications, Renew. Sustain. Energy Rev. 14 (n.d.) 2161–2166. https://doi.org/10.1016/j.rser.2010.02.015.
[5] V.B. Pawade, S.J. Dhoble, Spin-orbit splitting difference and Stokes shift in cerium 3+-activated aluminate phosphors, Spectrosc. Lett. 46 (2013) 472–475, https://doi.org/10.1080/00387010.2012.731025.
[6] V.B. Pawade, S.J. Dhoble, Novel blue-emitting SrMg 2Al 16O 27:Eu 2+ phosphor for solid-state lighting, Luminescence 26 (2011) 722–727, https://doi.org/10.1002/bio.1304.
[7] Z. Xia, Z. Xu, M. Chen, Q. Liu, Recent developments in the new inorganic solid-state LED phosphors, Dalton Trans. 45 (2016) 11214–11232, https://doi.org/10.1039/c6dt01230b.
[8] Z. Mao, Z. Lu, J. Chen, B.D. Fahlman, D. Wang, Tunable luminescent Eu2+-doped dicalcium silicate polymorphs regulated by crystal engineering, J. Mater. Chem. C. 3 (2015) 9454–9460, https://doi.org/10.1039/c5tc02049b.
[9] J. Zheng, Q. Cheng, S. Wu, Z. Guo, Y. Zhuang, Y. Lu, Y. Li, C. Chen, An efficient blue-emitting Sr5(PO4)3Cl:Eu2+ phosphor for application in near-UV white light-emitting diodes, J. Mater. Chem. C. 3 (2015) 11219–11227, https://doi.org/10.1039/c5tc02482j.
[10] S. Zhang, Y. Nakai, T. Tsuboi, Y. Huang, H.J. Seo, The thermal stabilities of luminescence and microstructures of Eu 2+-doped KBaPO4 and NaSrPO4 with β-K 2SO4 type structure, Inorg. Chem. 50 (2011) 2897–2904, https://doi.org/10.1021/ic102504x.
[11] J.M. Song, J.S. Park, S. Nahm, Luminescence properties of Eu2+ activated Ba2Si 5N8 red phosphors with various Eu2+ contents, Ceram. Int. 39 (2013) 2845–2850, https://doi.org/10.1016/j.ceramint.2012.09.056.
[12] V.D. Luong, W. Zhang, H.R. Lee, Preparation of Sr2Si5N8:Eu2+ for white light-emitting diodes by multi-step heat treatment, J. Alloys Compd. 509 (2011) 7525–7528, https://doi.org/10.1016/j.jallcom.2011.04.109.
[13] B. Lee, S. Lee, H.G. Jeong, K.S. Sohn, Solid-state combinatorial screening of (Sr, Ca, Ba, Mg)2Si 5N8:Eu2+ phosphors, ACS Comb. Sci. 13 (2011) 154–158, https://doi.org/10.1021/co100046u.
[14] J. Lucas, P. Lucas, T. Le Mercier, A. Rollat, W. Davenport, Introduction to rare earth luminescent materials, in: Rare Earths, Elsevier, 2015, pp. 251–280, https://doi.org/10.1016/b978-0-444-62735-3.00014-0.
[15] R. Withnall, J. Silver, Physics of light emission from rare-earth doped phosphors, in: Handbook of Visual Display Technology, Springer, Berlin Heidelberg, 2012, pp. 1019–1028, https://doi.org/10.1007/978-3-540-79567-4_68.
[16] T.H. Kim, W. Wang, Q. Li, Advancement in materials for energy-saving lighting devices, Front. Chem. Sci. Eng. 6 (2012) 13–26, https://doi.org/10.1007/s11705-011-1168-y.
[17] H. Mubarak, S.B.A. Kashem, Comparison of different energy saving lights using solar panel, Front. Energy 10 (2016) 466–472, https://doi.org/10.1007/s11708-016-0417-7.
[18] M. Arik, A. Setlur, Environmental and economical impact of LED lighting systems and effect of thermal management, Int. J. Energy Res. 34 (2010) 1195–1204, https://doi.org/10.1002/er.1639.
[19] B. Jacob, Lamps for improving the energy efficiency of domestic lighting, Light. Res. Technol. 41 (2009) 219–228, https://doi.org/10.1177/1477153509339610.
[20] H.Y. Shin, J.P. Yang, Design and implementation of a digital control IC for driving multiple cold cathode fluorescent lamps, Int. J. Circ. Theor. Appl. 38 (2010) 343–358, https://doi.org/10.1002/cta.561.

[21] E.G. Thaler, R.H. Wilson, D.A. Doughty, W.W. Beersb, Measurement of mercury bound in the glass envelope during operation of fluorescent lamps, J. Electrochem. Soc. 142 (1995) 1968–1970, https://doi.org/10.1149/1.2044224.
[22] C. Raposo, C. Carvalhinho Windmöller, W.A. Durão Júnior, Mercury speciation in fluorescent lamps by thermal release analysis, Waste Manag. 23 (2003) 879–886, https://doi.org/10.1016/S0956-053X(03)00089-8.
[23] H.-S. Koo, C.-H.J. Chang, N.-K. Cho, J.-H. Lee, Development and application of less-mercury flat fluorescent lamps for backlights and general lighting, J. Soc. Inf. Disp. 16 (2008) 759, https://doi.org/10.1889/1.2953483.
[24] T.C. Chang, S.J. You, B.S. Yu, C.M. Chen, Y.C. Chiu, Treating high-mercury-containing lamps using full-scale thermal desorption technology, J. Hazard Mater. 162 (2009) 967–972, https://doi.org/10.1016/j.jhazmat.2008.05.129.
[25] M. Aucott, M. McLinden, M. Winka, Release of mercury from broken fluorescent bulbs, J. Air Waste Manag. Assoc. 53 (2003) 143–151, https://doi.org/10.1080/10473289.2003.10466132.
[26] M.S. Shur, A. Žukauskas, Solid-state lighting: toward superior illumination, Proc. IEEE 93 (2005) 1691–1703, https://doi.org/10.1109/JPROC.2005.853537.
[27] A.B. Mukherjee, R. Zevenhoven, J. Brodersen, L.D. Hylander, P. Bhattacharya, Mercury in waste in the European Union: sources, disposal methods and risks, Resour. Conserv. Recycl. 42 (2004) 155–182, https://doi.org/10.1016/j.resconrec.2004.02.009.
[28] D.A. Steigerwald, J.C. Bhat, D. Collins, R.M. Fletcher, M.O. Holcomb, M.J. Ludowise, P.S. Martin, S.L. Rudaz, Illumination with solid state lighting technology, IEEE J. Sel. Top. Quant. Electron. 8 (2002) 310–320, https://doi.org/10.1109/2944.999186.
[29] R. Haitz, J. Tsao, Solid-state lighting: why it will succeed, and why it won't be overtaken, Opt. Photonik. 6 (2011) 26–30, https://doi.org/10.1002/opph.201190325.
[30] D.K. Yim, I.S. Cho, C.W. Lee, J.H. Noh, H.S. Roh, K.S. Hong, Preparation and photoluminescence properties of γ-KCaPO4: Eu2+ phosphors for near UV-based white LEDs, Opt. Mater. 33 (2011) 1036–1040, https://doi.org/10.1016/j.optmat.2011.02.031.
[31] T. Murata, T. Tanoue, M. Iwasaki, K. Morinaga, T. Hase, Fluorescence properties of Mn4+ in CaAl12O 19 compounds as red-emitting phosphor for white LED, J. Lumin. 114 (2005) 207–212, https://doi.org/10.1016/j.jlumin.2005.01.003.
[32] T. Nishida, T. Ban, N. Kobayashi, High-color-rendering light sources consisting of a 350-nm ultraviolet light-emitting diode and three-basal-color phosphors, Appl. Phys. Lett. 82 (2003) 3817–3819, https://doi.org/10.1063/1.1580649.
[33] N. Holonyak, S.F. Bevacqua, Coherent (visible) light emission from Ga(As1-xPx) junctions, Appl. Phys. Lett. 1 (1962) 82–83, https://doi.org/10.1063/1.1753706.
[34] S. Nakamura, M. Senoh, N. Iwasa, S.I. Nagahama, High-brightness InGaN blue, green and yellow light-emitting diodes with quantum well structures, Jpn. J. Appl. Phys. 34 (1995) L797–L799, https://doi.org/10.1143/JJAP.34.L797.
[35] S. Nakamura, III-V nitride based light-emitting devices, Solid State Commun. 102 (1997) 237–243, https://doi.org/10.1016/S0038-1098(96)00722-3.
[36] S. Nakamura, JJAP91 Nakamura GaN growth using GaN buffer layer.pdf, Jpn. J. Appl. Phys. 30 (1991) L1705–L1707.
[37] M. Bessho, K. Shimizu, Latest trends in LED lighting, Electron. Commun. Jpn. Part I Commun. 95 (2012) 1–7, https://doi.org/10.1002/ecj.10394.
[38] S.Y. Lee, J.W. Kwon, H.S. Kim, M.S. Choi, K.S. Byun, New design and application of high efficiency LED driving system for RGB-LED backlight in LCD display, PESC Rec. IEEE Annu. Power Electron. Spec. Conf. (2006) 1–5, https://doi.org/10.1109/PESC.2006.1711925.

[39] H.J. Chiu, S.J. Cheng, LED backlight driving system for large-scale LCD panels, IEEE Trans. Ind. Electron. 54 (2007) 2751–2760, https://doi.org/10.1109/TIE.2007.899938.
[40] R.J. Lin, M.-S. Tsai, C.-C. Sun, Novel optical lens design with a light scattering freeform inner surface for LED down light illumination, Opt. Express 23 (2015) 16715, https://doi.org/10.1364/oe.23.016715.
[41] G. Zhu, S. Xin, A novel blue light pumped yellow-emitting phosphor RbZnPO4:Dy3+ with satisfactory color tuning and thermal properties for high-power warm white light emitting diodes, RSC Adv. 5 (2015) 106795–106799, https://doi.org/10.1039/c5ra24286j.
[42] D. Singh, V. Tanwar, A.P. Simantilleke, B. Mari, P.S. Kadyan, I. Singh, Rapid synthesis and enhancement in down conversion emission properties of BaAl2O4:Eu2+,RE3+ (RE3+=Y, Pr) nanophosphors, J. Mater. Sci. Mater. Electron. 27 (2016) 2260–2266, https://doi.org/10.1007/s10854-015-4020-1.
[43] S. Kadyan, S. Singh, A.P. Simantilleke, D. Singh, Synthesis and optical studies of nanocrystalline Eu2+-doped and RE3+(Nd3+, Dy3+)-codoped Ba4Al14O25 materials for UV-LEDs, Optik 212 (2020) 164671, https://doi.org/10.1016/j.ijleo.2020.164671.
[44] W. Liang, L. Niu, Y. Gao, Photon cascade emission of Pr3+-doped K2YZr (PO4) 3 phosphor, Adv. Mater. Res. 815 (2013) 677–681. Trans Tech Publ. (n.d.).
[45] W. Liang, Y. Wang, Energy transfer between Pr3+ and Mn2+ in K 2YZr(PO4)3: Pr, Mn phosphor, Mater. Chem. Phys. 127 (2011) 170–173, https://doi.org/10.1016/j.matchemphys.2011.01.052.
[46] C.H. Huang, W.R. Liu, T.M. Chen, Single-phased white-light phosphors Ca9Gd(PO4) 7: Eu2+, Mn2+ under near-ultraviolet excitation, J. Phys. Chem. C 114 (2010) 18698–18701, https://doi.org/10.1021/jp106693z.
[47] N. Guo, Y. Huang, H. You, M. Yang, Y. Song, K. Liu, Y. Zheng, Ca9Lu(PO4)7: Eu2+,Mn 2+: a potential single-phased white-light-emitting phosphor suitable for white-light-emitting diodes, Inorg. Chem. 49 (2010) 10907–10913, https://doi.org/10.1021/ic101749g.
[48] G.S. Ningombam, T.S. David, N.R. Singh, Enhancement of Eu3+ emission in YVO4: Eu3+ nanocrystals by Li+ codoping: an oxidant-resistant dispersion and polymer film, ACS Omega 4 (2019) 13762–13771, https://doi.org/10.1021/acsomega.9b01265.
[49] J. Zhang, Z. Zhang, Z. Tang, Y. Tao, X. Long, Luminescent properties of the BaMgAl10O17:Eu2+,M3+ (M = Nd, Er) phosphor in the VUV region, Chem. Mater. 14 (2002) 3005–3008, https://doi.org/10.1021/cm011744u.
[50] B. Liu, Y. Wang, J. Zhou, F. Zhang, Z. Wang, The reduction of Eu3+ to Eu2+ in BaMgAl10 O17:Eu and the photoluminescence properties of BaMgAl10 O17: Eu2+ phosphor, J. Appl. Phys. 106 (2009) 053102, https://doi.org/10.1063/1.3211301.
[51] N. Pradal, G. Chadeyron, S. Thérias, A. Potdevin, C.V. Santilli, R. Mahiou, Investigation on combustion derived BaMgAl10O 17:Eu2+ phosphor powder and its corresponding PVP/BaMgAl10O17:Eu2+ nanocomposite, Dalton Trans. 43 (2014) 1072–1081, https://doi.org/10.1039/c3dt51915e.
[52] Y. Parganiha, J. Kaur, N. Dubey, V. Dubey, R. Shrivastava, S.J. Dhoble, H.C. Swart, Luminescence and structural properties of Gd2SiO5:Eu3+ phosphors synthesized from the modified solid state method, Ceram. Int. 43 (2017) 9084–9091, https://doi.org/10.1016/j.ceramint.2017.04.054.
[53] Y. Chen, B. Liu, C. Shi, M. Kirm, M. True, S. Vielhauer, G. Zimmerer, Luminescent properties of Gd2SiO5 powder doped with Eu3+ under VUV-UV excitation, J. Phys. Condens. Matter 17 (2005) 1217–1224, https://doi.org/10.1088/0953-8984/17/7/014.

[54] S. Singh, D. Singh, Synthesis and spectroscopic investigations of trivalent europium-doped M2SiO5 (M = Y and Gd) nanophosphor for display applications, J. Mater. Sci. Mater. Electron. 31 (2020) 5165–5175, https://doi.org/10.1007/s10854-020-03076-5.
[55] D. Jia, W. Jia, Y. Jia, Long persistent alkali-earth silicate phosphors doped with Eu2+, Nd3+, J. Appl. Phys. 101 (2007) 023520, https://doi.org/10.1063/1.2409767.
[56] C. Zhang, Y. Liu, J. Zhang, X. Zhang, J. Zhang, Z. Cheng, J. Jiang, H. Jiang, A single-phase Ba9Lu2Si6O24:Eu2+, Ce3+, Mn2+ phosphor with tunable full-color emission for NUV-based white LED applications, Mater. Res. Bull. 80 (2016) 288–294, https://doi.org/10.1016/j.materresbull.2016.04.009.
[57] J. Shen, L.D. Sun, C.H. Yan, Luminescent rare earth nanomaterials for bioprobe applications, Dalton Trans. (2008) 5687–5697, https://doi.org/10.1039/b805306e.
[58] J.H.L. Voncken, in: Physical and Chemical Properties of the Rare Earths, Springer, Cham, 2016, pp. 53–72, https://doi.org/10.1007/978-3-319-26809-5_3.
[59] L.U. Khan, Z.U. Khan, Rare earth luminescence: electronic spectroscopy and applications, in: Handbook of Materials Characterization, Springer International Publishing, 2018, pp. 345–404, https://doi.org/10.1007/978-3-319-92955-2_10.
[60] I. Gupta, S. Singh, S. Bhagwan, D. Singh, Rare earth (RE) doped phosphors and their emerging applications: a review, Ceram. Int. 47 (2021) 19282–19303, https://doi.org/10.1016/j.ceramint.2021.03.308.
[61] G.F. De Sá, O.L. Malta, C. De Mello Donegá, A.M. Simas, R.L. Longo, P.A. Santa-Cruz, E.F. Da Silva, Spectroscopic properties and design of highly luminescent lanthanide coordination complexes, Coord. Chem. Rev. 196 (2000) 165–195, https://doi.org/10.1016/S0010-8545(99)00054-5.
[62] J.K.A. Gschneider, L. Eyring, Handbook on the Physics and Chemistry of Rare Earths, vol. 9, 1987.
[63] H. Dong, L.D. Sun, C.H. Yan, Energy transfer in lanthanide upconversion studies for extended optical applications, Chem. Soc. Rev. 44 (2015) 1608–1634, https://doi.org/10.1039/c4cs00188e.
[64] V.S. Sastri, J.C. Bünzli, V.R. Rao, G.V.S. Rayudu, J.R. Perumareddi, Modern Aspects of Rare Earths and Their Complexes, Elsevier Inc., 2003, https://doi.org/10.1016/B978-0-444-51010-5.X5014-7.
[65] N.C. George, A.J. Pell, G. Dantelle, K. Page, A. Llobet, M. Balasubramanian, G. Pintacuda, B.F. Chmelka, R. Seshadri, Local environments of dilute activator ions in the solid-state lighting phosphor Y3-xCexAl5O12, Chem. Mater. 25 (2013) 3979–3995, https://doi.org/10.1021/cm401598n.
[66] X. Qin, X. Liu, W. Huang, M. Bettinelli, X. Liu, Lanthanide-activated phosphors based on 4f-5d optical transitions: theoretical and experimental aspects, Chem. Rev. 117 (2017) 4488–4527, https://doi.org/10.1021/acs.chemrev.6b00691.
[67] J.K. Krebs, Z.J. Barninger, Decay dynamics of Eu3+/Tb3+ in solution-processed Y2O3, J. Lumin. 130 (2010) 1305–1307, https://doi.org/10.1016/j.jlumin.2010.02.044.
[68] H. Shi, X.Y. Zhang, N.L. Wang, W.L. Dong, X.Y. Mi, Preparation of Y2O3: Eu3+, Tb3+ nanopowders with tunable luminescence by ammonium bicarbonate co-precipitation method, Funct, Mater. Lett. 8 (2015), https://doi.org/10.1142/S179360471550006X.
[69] S. Gnanam, V. Rajendran, Facile sol-gel preparation of Cd-doped cerium oxide (CeO2) nanoparticles and their photocatalytic activities, J. Alloys Compd. 735 (2018) 1854–1862, https://doi.org/10.1016/j.jallcom.2017.11.330.

[70] S. Gültekin, S. Yıldırım, O. Yılmaz, İ.Ç. Keskin, M.İ. Katı, E. Çelik, Structural and optical properties of SrAl2O4: Eu2+/Dy3+ phosphors synthesized by flame spray pyrolysis technique, J. Lumin. 206 (2019) 59−69, https://doi.org/10.1016/j.jlumin.2018.10.011.

[71] S. Singh, V. Tanwar, A.P. Simantilleke, D. Singh, Structural and photoluminescent investigations of SrAl2O4: Eu2+, RE3+ improved nanophosphors for solar cells, Nano-Struct. Nano-Objects 21 (2020) 100427, https://doi.org/10.1016/j.nanoso.2020.100427.

[72] A. Kumar, G. Kedawat, P. Kumar, J. Dwivedi, B.K. Gupta, Sunlight-activated Eu2+/Dy3+ doped SrAl2O4 water resistant phosphorescent layer for optical displays and defence applications, New J. Chem. 39 (2015) 3380−3387, https://doi.org/10.1039/c4nj02333a.

[73] R.E. Rojas-Hernandez, F. Rubio-Marcos, M.Á. Rodriguez, J.F. Fernandez, Long lasting phosphors: SrAl2O4:Eu, Dy as the most studied material, Renew. Sustain. Energy Rev. 81 (2018) 2759−2770, https://doi.org/10.1016/j.rser.2017.06.081.

[74] Y. Lu, Y. Li, Y. Xiong, D. Wang, Q. Yin, SrAl2O4: Eu2+, Dy3+ phosphors derived from a new sol-gel route, in: Microelectronics Journal, Elsevier, 2004, pp. 379−382, https://doi.org/10.1016/S0026-2692(03)00250-7.

[75] W. Liang, L.F. Niu, Y.H. Gao, Y.H. Wang, Photon cascade emission of Pr3+-doped K2YZr(PO4)3 Phosphor, Adv. Mater. Res. (2013) 677−681, https://doi.org/10.4028/www.scientific.net/AMR.815.677. Trans Tech Publications Ltd.

[76] Y.S. Chang, F.M. Huang, Y.Y. Tsai, L.G. Teoh, Synthesis and photoluminescent properties of YVO4: Eu3+ nano-crystal phosphor prepared by Pechini process, J. Lumin. 129 (2009) 1181−1185, https://doi.org/10.1016/j.jlumin.2009.05.020.

[77] A.N. Carneiro Neto, R.T. Moura, A. Shyichuk, V. Paterlini, F. Piccinelli, M. Bettinelli, O.L. Malta, Theoretical and experimental investigation of the Tb3+ → Eu3+ energy transfer mechanisms in cubic A3Tb0.90Eu0.10(PO4)3 (A = Sr, Ba) materials, J. Phys. Chem. C 124 (2020) 10105−10116, https://doi.org/10.1021/acs.jpcc.0c00759.

[78] C. Bertail, S. Maron, V. Buissette, T. Le Mercier, T. Gacoin, J.P. Boilot, Structural and photoluminescent properties of Zn2SiO 4:Mn2+ nanoparticles prepared by a protected annealing process, Chem. Mater. 23 (2011) 2961−2967, https://doi.org/10.1021/cm2005902.

[79] T. Jüstel, H. Nikol, C. Ronda, New developments in the field of luminescent materials for lighting and displays, Angew. Chem. Int. Ed. 37 (1998) 3084−3103, https://doi.org/10.1002/(SICI)1521-3773(19981204)37:22<3084::AID-ANIE3084>3.0.CO;2-W.

[80] G.B. Nair, S.J. Dhoble, A perspective perception on the applications of light-emitting diodes, Luminescence 30 (2015) 1167−1175, https://doi.org/10.1002/bio.2919.

[81] J. Liu, Z.M. Zhang, Z.C. Wu, F.F. Wang, Z.J. Li, Study on luminescence and thermal stability of blue-emitting Sr5(PO4)3F: Eu2+phosphor for application in InGaN-based LEDs, Mater. Sci. Eng. B 221 (2017) 10−16, https://doi.org/10.1016/j.mseb.2017.03.014.

[82] H. Yu, B. Yao, D. Deng, L. Chen, H. Zhao, Y. Li, Y. Hua, S. Xu, Novel tunable red to yellow emitting (Ca, Ba)4(PO4)2O: Eu2+ phosphor for blue-excited white light emitting diodes, Ceram. Int. 43 (2017) 1038−1043, https://doi.org/10.1016/j.ceramint.2016.10.036.

[83] G.B. Nair, P.D. Bhoyar, S.J. Dhoble, Exploration of electron−vibrational interaction in the 5d states of Eu2+ ions in ABaPO4 (A = Li, Na, K and Rb) phosphors, Luminescence 32 (2017) 22−29, https://doi.org/10.1002/bio.3143.

[84] N.V. Rebrova, A.Y. Grippa, A.S. Pushak, T.E. Gorbacheva, V.Y. Pedash, O.G. Viagin, V.L. Cherginets, V.A. Tarasov, V.V. Vistovskyy, A.P. Vas'kiv, S.V. Myagkota, Crystal growth and characterization of Eu2+ doped RbCaX3 (X = Cl, Br) scintillators, J. Cryst. Growth 466 (2017) 39–44, https://doi.org/10.1016/j.jcrysgro.2017.03.016.

[85] M. Dalal, V. Taxak, J. Dalal, A. Khatkar, ... S.C.-J. of A. and, Undefined 2017, Crystal Structure and Judd-Ofelt Properties of a Novel Color Tunable Blue-White-Red Ba5Zn4Y8O21: Eu3+ Nanophosphor for Near-Ultraviolet Based WLEDs, Elsevier. (n.d.).

[86] X. Sun, C. Li, X. He, L. Yu, H. Qi, Y. Gao, F. Gao, ... C.W.-J. of, Undefined 2015, Preparation and Luminescence Properties of Sr2EuxLa1− xAlO5 Phosphor, Elsevier. (n.d.).

[87] M. Guzik, E. Tomaszewicz, Y. Guyot, J. Legendziewicz, G. Boulon, Eu3+ luminescence from different sites in a scheelite-type cadmium molybdate red phosphor with vacancies, J. Mater. Chem. C. 3 (2015) 8582–8594, https://doi.org/10.1039/c5tc01109d.

[88] B. Poornaprakash, P.T. Poojitha, U. Chalapathi, S.H. Park, Achieving room temperature ferromagnetism in ZnS nanoparticles via Eu3+ doping, Mater. Lett. 181 (2016) 227–230, https://doi.org/10.1016/j.matlet.2016.06.033.

[89] S. Tamboli, P.D. Bhoyar, S.J. Dhoble, Analysis of electron-vibrational interaction in the 5d states of Eu2+ ions in A3P4O13 and A3(PO4)2 (A=Sr, Ba) phosphors, J. Lumin. 184 (2017) 23–28, https://doi.org/10.1016/j.jlumin.2016.12.003.

[90] E. Cavalli, P. Boutinaud, M. Grinberg, Luminescence dynamics in CaWO4:Pr3+ powders and single crystals, J. Lumin. 169 (2016) 450–453, https://doi.org/10.1016/j.jlumin.2014.10.069.

[91] H. Xu, D. Ying, A. Lu, X. Wang, J. Hu, Surfactant-assistant solvothermal synthesis of CaWO4:Eu3+ phosphors and luminescence, Superlattice. Microst. 83 (2015) 668–675, https://doi.org/10.1016/j.spmi.2015.04.011.

[92] H. Zhu, M. Fang, Z. Huang, G. Liu, K. Chen, X. Min, Y. Mao, M. Wang, Photoluminescence properties of Li2 Mg2 (WO4)3: Eu 3+ red phosphor with high color purity for white LEDs applications, J. Lumin. vol. 172 (2016) 180–184, https://doi.org/10.1016/j.jlumin.2015.12.021. Elsevier.

[93] R. Cao, D. Peng, H. Xu, Z. Luo, H. Ao, S. Guo, J. Fu, Synthesis and luminescence properties of Sr3(VO4)2:Eu3+ phosphor and emission enhancement by co-doping Li+ ion, Optik 127 (2016) 7896–7901, https://doi.org/10.1016/j.ijleo.2016.05.157.

[94] L. Jing, X. Liu, Y. Li, Y. Wang, Synthesis and photoluminescence properties of Ca9Y(VO4)7: Dy phosphors for white light-emitting diodes, J. Lumin. 162 (2015) 185–190, https://doi.org/10.1016/j.jlumin.2015.02.048.

[95] X. Mi, K. Du, K. Huang, P. Zhou, D. Geng, Y. Zhang, M. Shang, J. Lin, Synthesis and luminescence of Ca9Eu1-xLn x(VO4)7 (Ln = Y, La, Gd, Lu) phosphors, Mater. Res. Bull. 60 (2014) 72–78, https://doi.org/10.1016/j.materresbull.2014.08.017.

[96] G.B. Nair, A. Kumar, H.C. Swart, S.J. Dhoble, Improved steady-state photoluminescence derived from the compensation of the charge-imbalance in Ca3Mg3(PO4)4: Eu3+ phosphor, Ceram. Int. 45 (2019) 21709–21715, https://doi.org/10.1016/j.ceramint.2019.07.171.

[97] M.M. Yawalkar, G.B. Nair, G.D. Zade, S.J. Dhoble, Effect of the synthesis route on the luminescence properties of Eu3+ activated Li6M(BO3)3 (M = Y, Gd) phosphors, Mater. Chem. Phys. 189 (2017) 136–145, https://doi.org/10.1016/j.matchemphys.2016.12.006.

[98] G.B. Nair, S.J. Dhoble, Photoluminescence properties of Eu3+/Sm3+ activated CaZr4(PO4)6 phosphors, J. Fluoresc. 26 (2016) 1865–1873, https://doi.org/10.1007/s10895-016-1880-6.

[99] P. Niu, X. Liu, Y. Wang, W. Zhao, Photoluminescence properties of a novel red-emitting phosphor Ba2LaV3O11: Eu3+, J. Mater. Sci. Mater. Electron. 29 (2018) 124–129, https://doi.org/10.1007/s10854-017-7895-1.
[100] S. Ray, G.B. Nair, P. Tadge, N. Malvia, V. Rajput, V. Chopra, S.J. Dhoble, Size and shape-tailored hydrothermal synthesis and characterization of nanocrystalline LaPO4: Eu3+ phosphor, J. Lumin. 194 (2018) 64–71, https://doi.org/10.1016/j.jlumin.2017.10.015.
[101] M. Yang, H. Shi, L. Ma, Q. Gui, J. Ma, M. Lin, A. Sunna, W. Zhang, L. Dai, J. Qu, Y. Liu, Multifunctional luminescent nanofibres from Eu3+-doped La2O2SO4 with enhanced oxygen storage capability, J. Alloys Compd. 695 (2017) 202–207, https://doi.org/10.1016/j.jallcom.2016.10.164.
[102] J. Wang, Y. Cheng, Y. Huang, P. Cai, S. Il Kim, H.J. Seo, Structural and luminescent properties of red-emitting Eu3+-doped ternary rare earth antimonates R3SbO7 (R = La, Gd, Y), J. Mater. Chem. C. 2 (2014) 5559–5569, https://doi.org/10.1039/c4tc00359d.
[103] W. Jiang, R. Fu, X. Gu, P. Zhang, A. Coşgun, A red-emitting phosphor LaSr2AlO5: Eu3+/Eu2+ prepared under oxidative and reductive atmospheres, J. Lumin. 157 (2015) 46–52, https://doi.org/10.1016/j.jlumin.2014.07.018.
[104] P. Dorenbos, The Eu3+ charge transfer energy and the relation with the band gap of compounds, J. Lumin. 111 (2005) 89–104, https://doi.org/10.1016/j.jlumin.2004.07.003.
[105] N. Zhang, C. Guo, L. Yin, J. Zhang, M. Wu, Red emitting phosphors of Eu3+ doped Na2Ln2Ti3O10 (Ln = Gd, Y) for white light emitting diodes, J. Alloys Compd. 635 (2015) 66–72, https://doi.org/10.1016/j.jallcom.2015.02.014.
[106] S. Som, A.K. Kunti, V. Kumar, V. Kumar, S. Dutta, M. Chowdhury, S.K. Sharma, J.J. Terblans, H.C. Swart, Defect correlated fluorescent quenching and electron phonon coupling in the spectral transition of Eu3+ in CaTiO3 for red emission in display application, J. Appl. Phys. 115 (2014) 193101, https://doi.org/10.1063/1.4876316.
[107] K.M. Nissamudeen, S. Sankar, A.H. Bahna, K.G. Gopchandran, Studies on the influence of lithium incorporation in the photoluminescence of Y2O3: Eu3+ thin films, J. Phys. Chem. Solid. 70 (2009) 821–826, https://doi.org/10.1016/j.jpcs.2009.04.002.
[108] M.G. Ivanov, U. Kynast, M. Leznina, Eu3+ doped yttrium oxide nano-luminophores from laser synthesis, J. Lumin. 169 (2016) 744–748, https://doi.org/10.1016/j.jlumin.2015.05.036.
[109] S. Zhong, J. Chen, S. Wang, Q. Liu, Y. Wang, S. Wang, Y2O3: Eu3+ hexagonal microprisms: fast microwave synthesis and photoluminescence properties, J. Alloys Compd. 493 (2010) 322–325, https://doi.org/10.1016/j.jallcom.2009.12.092.
[110] S. Ray, A. Banerjee, P. Pramanik, Selective synthesis, characterization, and photoluminescence study of YPO4: Eu3+ nanorods and nanoparticles, Mater. Res. Bull. 45 (2010) 870–877, https://doi.org/10.1016/j.materresbull.2010.02.020.
[111] V. Kumar, S. Som, S. Dutta, S. Das, H.C. Swart, Red-light-emitting inorganic La2CaZnO5 frameworks with high photoluminescence quantum efficiency: theoretical approach, Mater. Des. 93 (2016) 203–215, https://doi.org/10.1016/j.matdes.2015.12.153.
[112] X. Teng, Y. Liu, Y. Liu, Y. Hu, H. He, W. Zhuang, Luminescence properties of Tm3+ co-doped Sr2Si5N8: Eu2+ red phosphor, J. Lumin. 130 (2010) 851–854, https://doi.org/10.1016/j.jlumin.2009.12.012.
[113] Q. Wu, J. Ding, X. Wang, Y. Li, Y. Wang, Structure modification and covalence variation induced by cation substitution in pure nitride Ca-α-sialon phosphor, Mater. Res. Bull. 83 (2016) 649–656, https://doi.org/10.1016/j.materresbull.2016.07.018.

[114] V. Bachmann, A. Meijerink, C. Ronda, Luminescence properties of SrSi2AlO2N3 doped with divalent rare-earth ions, J. Lumin. 129 (2009) 1341−1346, https://doi.org/10.1016/j.jlumin.2009.06.023.

[115] R.J. Xie, N. Hirosaki, Silicon-based oxynitride and nitride phosphors for white LEDs-A review, Sci. Technol. Adv. Mater. 8 (2007) 588−600, https://doi.org/10.1016/j.stam.2007.08.005.

[116] T. Senden, E.J. van Harten, A. Meijerink, Synthesis and narrow red luminescence of Cs2HfF6: Mn4+, a new phosphor for warm white LEDs, J. Lumin. 194 (2018) 131−138, https://doi.org/10.1016/j.jlumin.2017.10.006.

[117] A. Fu, L. Zhou, S. Wang, Y. Li, Preparation, structural and optical characteristics of a deep red-emitting Mg2Al4Si5O18: Mn4+ phosphor for warm w-LEDs, Dyes Pigm. 148 (2018) 9−15, https://doi.org/10.1016/j.dyepig.2017.08.050.

[118] S. Zhang, Y. Hu, H. Duan, Y. Fu, M. He, An efficient, broad-band red-emitting Li2MgTi3O8: Mn4+ phosphor for blue-converted white LEDs, J. Alloys Compd. 693 (2017) 315−325, https://doi.org/10.1016/j.jallcom.2016.09.203.

[119] T. Sasaki, J. Fukushima, Y. Hayashi, H. Takizawa, Synthesis and photoluminescence properties of a novel Sr2Al6O11: Mn4+ red phosphor prepared with a B2O3 flux, J. Lumin. 194 (2018) 446−451, https://doi.org/10.1016/j.jlumin.2017.10.076.

[120] G.B. Nair, S.J. Dhoble, Assessment of electron-vibrational interaction (EVI) parameters of YAG:Ce3 +, TAG: Ce3 + and LuAG:Ce3 + garnet phosphors by spectrum fitting method, Spectrochim. Acta A 173 (2017) 822−826, https://doi.org/10.1016/j.saa.2016.10.049.

[121] V. Bachmann, C. Ronda, A. Meijerink, Temperature quenching of yellow Ce3+ luminescence in YAG:Ce, Chem. Mater. 21 (2009) 2077−2084, https://doi.org/10.1021/cm8030768.

[122] D. Valiev, T. Han, V. Vaganov, S. Stepanov, The effect of Ce3+ concentration and heat treatment on the luminescence efficiency of YAG phosphor, J. Phys. Chem. Solid. 116 (2018) 1−6, https://doi.org/10.1016/j.jpcs.2018.01.007.

[123] X. He, X. Liu, R. Li, B. Yang, K. Yu, M. Zeng, R. Yu, Effects of local structure of Ce3+ ions on luminescent properties of Y3Al5O12:Ce nanoparticles, Sci. Rep. 6 (2016) 1−11, https://doi.org/10.1038/srep22238.

[124] Y. Liu, J. Zou, M. Shi, B. Yang, Y. Han, W. Li, Z. Wang, H. Zhou, M. Li, N. Jiang, Effect of gallium ion content on thermal stability and reliability of YAG: Ce phosphor films for white LEDs, Ceram. Int. 44 (2018) 1091−1098, https://doi.org/10.1016/j.ceramint.2017.10.056.

[125] W.S. Song, Y.S. Kim, H. Yang, Yellow-emitting phosphor of Sr3B2O6: Eu2+ for application to white light-emitting diodes, Mater. Chem. Phys. 117 (2009) 500−503, https://doi.org/10.1016/j.matchemphys.2009.06.042.

[126] Y. Zhao, H. Xu, X. Zhang, G. Zhu, D. Yan, A. Yu, Facile synthesis of YAG: Ce3+ thick films for phosphor converted white light emitting diodes, J. Eur. Ceram. Soc. 35 (2015) 3761−3764, https://doi.org/10.1016/j.jeurceramsoc.2015.05.017.

[127] Z. Chen, Q. Zhang, Y. Li, H. Wang, R.J. Xie, A promising orange-yellow-emitting phosphor for high power warm-light white LEDs: pure-phase synthesis and photoluminescence properties, J. Alloys Compd. 715 (2017) 184−191, https://doi.org/10.1016/j.jallcom.2017.04.270.

[128] J. Li, J.G. Li, S. Liu, X. Li, X. Sun, Y. Sakka, The development of Ce3+-activated (Gd,Lu)3Al 5O12 garnet solid solutions as efficient yellow-emitting phosphors, Sci. Technol. Adv. Mater. 14 (2013) 54201−54210, https://doi.org/10.1088/1468-6996/14/5/054201.

[129] W. Zhou, X. Ma, M. Zhang, Y. Luo, Z. Xia, Synthesis and photoluminescence properties of green-emitting Lu3(Al,Sc)5O12: Ce3+ phosphor, Ceram. Int. 41 (2015) 7140–7145, https://doi.org/10.1016/j.ceramint.2015.02.030.
[130] X. Piao, K.I. MacHida, T. Horikawa, H. Hanzawa, Self-propagating high temperature synthesis of yellow-emitting Ba 2Si5N8: Eu2+ phosphors for white light-emitting diodes, Appl. Phys. Lett. 91 (2007) 041908, https://doi.org/10.1063/1.2760038.
[131] Q.N. Fei, Y.H. Liu, T.C. Gu, D.J. Wang, Color improvement of white-light through Mn-enhancing yellow-green emission of SrSi2O2N2: Eu phosphor for white light emitting diodes, J. Lumin. 131 (2011) 960–964, https://doi.org/10.1016/j.jlumin.2011.01.001.
[132] J. Chen, Y. Zhao, G. Li, Z. Mao, D. Wang, L. Bie, Facile synthesis of yellow-emitting CaAlSiN3: Ce3+ phosphors and the enhancement of red-component by co-doping Eu2+ ions, Solid State Commun. 255–256 (2017) 1–4, https://doi.org/10.1016/j.ssc.2017.02.010.
[133] Y. Chen, J. Xu, Y. Yu, W. Chen, R. Zeng, Preparation Eu-doped ca-α-SiAlON phosphor by heterogeneous precipitation: an orange–yellow phosphor for white light-emitting diodes, Ceram. Int. 41 (2015) 11086–11090, https://doi.org/10.1016/j.ceramint.2015.05.055.
[134] Y. Ge, S. Sun, M. Zhou, Y. Chen, Z. Tian, J. Zhang, Z. Xie, Impacts of Si particle size and nitrogen pressure on combustion synthesis of Eu2 +-doped α-SiAlON yellow phosphors, Powder Technol. 305 (2017) 141–146, https://doi.org/10.1016/j.powtec.2016.09.066.
[135] R. Naik, S.C. Prashantha, H. Nagabhushana, H.P. Nagaswarupa, K.S. Anantharaju, S.C. Sharma, B.M. Nagabhushana, H.B. Premkumar, K.M. Girish, Mg2SiO4:Tb3+ nanophosphor: auto ignition route and near UV excited photoluminescence properties for WLEDs, J. Alloys Compd. 617 (2014) 69–75, https://doi.org/10.1016/j.jallcom.2014.07.100.
[136] Z.J. Zhang, O.M. Ten Kate, A. Delsing, P. Dorenbos, J.T. Zhao, H.T. Hintzen, Photoluminescence properties of Pr3+, Sm3+ and Tb3+ doped SrAlSi4N7 and energy level locations of rare-earth ions in SrAlSi4N7, J. Mater. Chem. C. 2 (2014) 7952–7959, https://doi.org/10.1039/c4tc00538d.
[137] V. Singh, R.P.S. Chakradhar, J.L. Rao, D.K. Kim, Mn2+ activated MgSrAl10O17 green-emitting phosphor-A luminescence and EPR study, J. Lumin. 128 (2008) 1474–1478, https://doi.org/10.1016/j.jlumin.2008.02.001.
[138] X. Liu, L. Yan, J. Lin, Synthesis and luminescent properties of LaAlO3: RE3+ (RE = Tm, Tb) nanocrystalline phosphors via a sol-gel process, J. Phys. Chem. C 113 (2009) 8478–8483, https://doi.org/10.1021/jp9013724.
[139] C. He, Z. Xia, Q. Liu, Microwave solid state synthesis and luminescence properties of green-emitting Gd2O2S: Tb3+ phosphor, Opt. Mater. 42 (2015) 11–16, https://doi.org/10.1016/j.optmat.2014.12.012.
[140] R.K. Tamrakar, K. Upadhyay, Combustion synthesis and luminescence behaviour of the Tb3+ doped SrY2O4 phosphor, J. Electron. Mater. 47 (2018) 651–654, https://doi.org/10.1007/s11664-017-5825-x.
[141] Z. Xia, Y. Liang, D. Yu, M. Zhang, W. Huang, M. Tong, J. Wu, J. Zhao, Photoluminescence properties and energy transfer in color tunable BaY2Si3O10: Ce, Tb phosphors, Opt Laser. Technol. 56 (2014) 387–392, https://doi.org/10.1016/J.OPTLASTEC.2013.09.014.
[142] A. Guan, Z. Lu, F. Gao, X. Zhang, H. Wang, T. Huang, L. Zhou, Synthesis and optical characterization of Eu2+,Tb3+-codoped Sr3Y(PO4)3 green phosphors, J. Rare Earths 36 (2018) 238–242, https://doi.org/10.1016/J.JRE.2017.06.013.

[143] X. Zhang, X. Tang, J. Zhang, M. Gong, An efficient and stable green phosphor SrBaSiO4: Eu2+ for light-emitting diodes, J. Lumin. 130 (2010) 2288–2292, https://doi.org/10.1016/J.JLUMIN.2010.07.006.

[144] S. Nakamura, InGaN/AlGaN blue-light-emitting diodes, J. Vac. Sci. Technol. A 13 (1995) 705–710, https://doi.org/10.1116/1.579811.

[145] S. Ye, F. Xiao, Y.X. Pan, Y.Y. Ma, Q.Y. Zhang, Phosphors in phosphor-converted white light-emitting diodes: recent advances in materials, techniques and properties, Mater. Sci. Eng. R Rep. 71 (2010) 1–34, https://doi.org/10.1016/j.mser.2010.07.001.

[146] S.S. Lee, H.J. Kim, S.H. Byeon, J.C. Park, D.K. Kim, Thermal-shock-assisted solid-state process for the production of BaMgAl10O17: Eu phosphor, Ind. Eng. Chem. Res. 44 (2005) 4300–4303, https://doi.org/10.1021/ie048953j.

[147] K.B. Kim, Y. Il Kim, H.G. Chun, T.Y. Cho, J.S. Jung, J.G. Kang, Structural and optical properties of BaMgAl10O17: Eu2+ phosphor, Chem. Mater. 14 (2002) 5045–5052, https://doi.org/10.1021/cm020592f.

[148] Y. Liu, S. Zhang, D. Pan, J. Tian, H. Liu, M. Wu, A.A. Volinsky, Mechanism and kinetics of the BaMgAl10O17: Eu2+ alkaline fusion reaction, J. Rare Earths 33 (2015) 664–670, https://doi.org/10.1016/S1002-0721(14)60468-3.

[149] C.W. Won, H.H. Nersisyan, H.I. Won, S.J. Kwon, H.Y. Kim, S.Y. Seo, Synthesis of nano-size BaMgAl10O17: Eu2+ blue phosphor by a rapid exothermic reaction, J. Lumin. 130 (2010) 678–681, https://doi.org/10.1016/j.jlumin.2009.11.017.

[150] Z.H. Zhang, Y.H. Wang, X.X. Li, Y.K. Du, W.J. Liu, Photoluminescence degradation and color shift studies of annealed BaMgAl10O17: Eu2+ phosphor, J. Lumin. 122–123 (2007) 1003–1005, https://doi.org/10.1016/j.jlumin.2006.01.351.

[151] H.J. Song, D.K. Yim, H.S. Roh, I.S. Cho, S.J. Kim, Y.H. Jin, H.W. Shim, D.W. Kim, K.S. Hong, RbBaPO4:Eu2+: a new alternative blue-emitting phosphor for UV-based white light-emitting diodes, J. Mater. Chem. C. 1 (2013) 500–505, https://doi.org/10.1039/c2tc00162d.

[152] R.L. Kohale, S.J. Dhoble, Eu2+ luminescence in SrCaP2O7 pyrophosphate phosphor, Luminescence 28 (2013) 656–661, https://doi.org/10.1002/bio.2411.

[153] G. Ju, Y. Hu, L. Chen, X. Wang, Z. Mu, Blue persistent luminescence in Eu2+ doped Ca3Mg 3(PO4)4, Opt. Mater. 36 (2014) 1183–1188, https://doi.org/10.1016/j.optmat.2014.02.024.

[154] X. Zhang, J. Song, C. Zhou, L. Zhou, M. Gong, High efficiency and broadband blue-emitting NaCaBO3: Ce 3+ phosphor for NUV light-emitting diodes, J. Lumin. 149 (2014) 69–74, https://doi.org/10.1016/j.jlumin.2014.01.012.

[155] J. Zhang, W. Zhang, Z. Qiu, W. Zhou, L. Yu, Z. Li, S. Lian, Li4SrCa(SiO4)2: Ce3+, a highly efficient near-UV and blue emitting orthosilicate phosphor, J. Alloys Compd. 646 (2015) 315–320, https://doi.org/10.1016/j.jallcom.2015.05.280.

[156] F.C. Lu, L.J. Bai, Z.P. Yang, X.N. Han, Synthesis and photoluminescence of a novel blue-emitting Gd5Si3O12N: Ce3+ phosphor, Mater. Lett. 151 (2015) 9–11, https://doi.org/10.1016/j.matlet.2015.03.031.

[157] R. Kasuya, T. Isobe, H. Kuma, J. Katano, Photoluminescence enhancement of PEG-modified YAG: Ce3+ nanocrystal phosphor prepared by glycothermal method, J. Phys. Chem. B 109 (2005) 22126–22130, https://doi.org/10.1021/jp052753j.

[158] F. Liu, Q. Liu, Y. Fang, N. Zhang, B. Yang, G. Zhao, White light emission from NaLa(PO3)4: Dy3+ single-phase phosphors for light-emitting diodes, Ceram. Int. 41 (2015) 1917–1920, https://doi.org/10.1016/j.ceramint.2014.09.078.

[159] Z.W. Zhang, A.J. Song, M.Z. Ma, X.Y. Zhang, Y. Yue, R.P. Liu, A novel white emission in Ca8MgBi(PO4) 7:Dy3+ single-phase full-color phosphor, J. Alloys Compd. 601 (2014) 231–233, https://doi.org/10.1016/j.jallcom.2014.02.165.
[160] W. Geng, G. Zhu, Y. Shi, Y. Wang, Luminescent characteristics of Dy3+ doped calcium zirconium phosphate CaZr4(PO4)6 (CZP) phosphor for warm-white LEDs, J. Lumin. 155 (2014) 205–209, https://doi.org/10.1016/j.jlumin.2014.06.044.
[161] Z.W. Zhang, L. Liu, X.F. Zhang, J.P. Zhang, W.G. Zhang, D.J. Wang, Preparation and investigation of CaZr4(PO4) 6: Dy3+ single-phase full-color phosphor, Spectrochim. Acta Part A Mol. Biomol. Spectrosc. 137 (2015) 1–6, https://doi.org/10.1016/j.saa.2014.07.052.
[162] J. Ballato, J.S. Lewis, P. Holloway, Display applications of rare-earth-doped materials, MRS Bull. 24 (1999) 51–55, https://doi.org/10.1557/s0883769400053070.
[163] V. Dubey, S. Agrawal, J. Kaur, Photoluminescence and thermoluminescence behavior of Gd doped Y2O3 phosphor, Optik 126 (2015) 1–5, https://doi.org/10.1016/j.ijleo.2014.06.175.
[164] S. Singh, S. Kadyan, S. Sheoran, B. Mari, D. Singh, Synthesis and optical investigations of Eu3+ activated MYAlO4 (M = Ca and Sr) as promising display nanomaterials, Optik 208 (2020) 164552, https://doi.org/10.1016/j.ijleo.2020.164552.
[165] N. Daldosso, L. Pavesi, Low-dimensional silicon as a photonic material, in: Nanosilicon, Elsevier Ltd, 2008, pp. 314–334, https://doi.org/10.1016/B978-008044528-1.50010-5.
[166] Investigation of Preparation and Photoluminescence Properties of Phosphors Used for Near Ultraviolet and Blue LEDs (n.d.).
[167] A.N. Belsky, J.C. Krupa, Luminescence excitation mechanisms of rare earth doped phosphors in the VUV range, Displays 19 (1999) 185–196, https://doi.org/10.1016/S0141-9382(98)00049-3.
[168] D.W. Goodwin, Spectra and energy levels of rare earth ions in crystals, Phys. Bull. 20 (1969), https://doi.org/10.1088/0031-9112/20/12/010, 525–525.
[169] W.T. Carnall, G.L. Goodman, K. Rajnak, R.S. Rana, A systematic analysis of the spectra of the lanthanides doped into single crystal LaF3, J. Chem. Phys. 90 (1989) 3443–3457, https://doi.org/10.1063/1.455853.
[170] Y.-D. Huh, Y.-S. Cho, Y.R. Do', The optical properties of(Y 1-x Gd x) 3-z (Al 1-y Ga y) 5 O 12 :Ce z phosphors for white LED, Bull. Kor. Chem. Soc. 23 (2002) 1435–1454, https://doi.org/10.5012/bkcs.2002.23.10.1435.
[171] M. Balestrieri, S. Colis, M. Gallart, G. Schmerber, M. Ziegler, P. Gilliot, A. Dinia, Photoluminescence properties of rare earth (Nd, Yb, Sm, Pr)-doped CeO2 pellets prepared by solid-state reaction, J. Mater. Chem. C. 3 (2015) 7014–7021, https://doi.org/10.1039/c5tc00075k.
[172] C.W. Tang, S.A. Vanslyke, Organic electroluminescent diodes, Appl. Phys. Lett. 51 (1987) 913–915, https://doi.org/10.1063/1.98799.
[173] S.R. Forrest, The road to high efficiency organic light emitting devices, Org. Electron. 4 (2003) 45–48, https://doi.org/10.1016/j.orgel.2003.08.014.
[174] Z. Shen, P.E. Burrows, V. Bulović, S.R. Forrest, M.E. Thompson, Three-color, tunable, organic light-emitting devices, Science 276 (1997) 2009–2011, https://doi.org/10.1126/science.276.5321.2009.
[175] M. Yoshida, A. Fujii, Y. Ohmori, K. Yoshino, Polarity-dependent multicolor organic electroluminescent device, Jpn. J. Appl. Phys. 35 (1996), https://doi.org/10.1143/jjap.35.l397.

[176] M. Era, C. Adachi, T. Tsutsui, S. Saito, Double-heterostructure electroluminescent device with cyanine-dye bimolecular layer as an emitter, Chem. Phys. Lett. 178 (1991) 488–490, https://doi.org/10.1016/0009-2614(91)87007-X.

[177] J. Kido, M. Kimura, K. Nagai, Multilayer white light-emitting organic electroluminescent device, Science 267 (1995) 1332–1334, https://doi.org/10.1126/science.267.5202.1332.

[178] J. Kido, Y. Okamoto, Organo lanthanide metal complexes for electroluminescent materials, Chem. Rev. 102 (2002) 2357–2368, https://doi.org/10.1021/cr010448y.

[179] P.E. Burrows, G. Gu, V. Bulović, Z. Shen, S.R. Forrest, M.E. Thompson, Achieving full-color organic light-emitting devices for lightweight, flat-panel displays, IEEE Trans. Electron. Dev. 44 (1997) 1188–1203, https://doi.org/10.1109/16.605453.

[180] A. Kraft, A.C. Grimsdale, A.B. Holmes, Electroluminescent conjugated polymers - seeing polymers in a new light, Angew. Chem. Int. Ed. 37 (1998) 402–428, https://doi.org/10.1002/(SICI)1521-3773(19980302)37:4<402::AID-ANIE402>3.0.CO;2-9.

[181] H. Aziz, Degradation mechanism of small molecule-based organic light-emitting devices, Science 283 (1999) 1900–1902, https://doi.org/10.1126/science.283.5409.1900.

[182] B. Geffroy, P. le Roy, C. Prat, Organic light-emitting diode (OLED) technology: materials, devices and display technologies, Polym. Int. 55 (2006) 572–582, https://doi.org/10.1002/pi.1974.

[183] J. Shi, C.W. Tang, Doped organic electroluminescent devices with improved stability, Appl. Phys. Lett. 70 (1997) 1665–1667, https://doi.org/10.1063/1.118664.

[184] V.E. Choong, S. Shi, J. Curless, C.L. Shieh, H.C. Lee, F. So, J. Shen, J. Yang, Organic light-emitting diodes with a bipolar transport layer, Appl. Phys. Lett. 75 (1999) 172–174, https://doi.org/10.1063/1.124309.

[185] X. Cai, R. Liu, H. Shi, C. Li, H. Zhu, High efficiency green phosphorescent OLEDs using double-host materials, Dyes Pigm. 143 (2017) 196–202, https://doi.org/10.1016/j.dyepig.2017.04.044.

[186] Y. Sakakibara, S. Okutsu, T. Enokida, T. Tani, Red organic electroluminescence devices with a reduced porphyrin compound, tetraphenylchlorin, Appl. Phys. Lett. 74 (1999) 2587–2589, https://doi.org/10.1063/1.123906.

[187] Y. Hamada, H. Kanno, T. Tsujioka, H. Takahashi, T. Usuki, Red organic light-emitting diodes using an emitting assist dopant, Appl. Phys. Lett. 75 (1999) 1682–1684, https://doi.org/10.1063/1.124790.

[188] P.E. Burrows, S.R. Forrest, T.X. Zhou, L. Michalski, Operating lifetime of phosphorescent organic light emitting devices, Appl. Phys. Lett. 76 (2000) 2493–2495, https://doi.org/10.1063/1.126386.

[189] L. Zhang, Z. Liu, X. Zhang, J. Chen, Y. Cao, Efficient blue light-emitting diodes based on conjugated polymers with fluorinated silole chemically doped in fluorene–carbazole main chain, J. Inorg. Organomet. Polym. Mater. 25 (2015) 64–72, https://doi.org/10.1007/s10904-014-0078-0.

[190] K.H. Lee, L.K. Kang, Y.S. Kwon, J.Y. Lee, S. Kang, G.Y. Kim, J.H. Seo, Y.K. Kim, S.S. Yoon, Blue electroluminescent materials based on 2,7-distyrylfluorene for organic light-emitting diodes, Thin Solid Films 518 (2010) 5091–5097, https://doi.org/10.1016/j.tsf.2010.02.063.

[191] X.Y. Jiang, Z.L. Zhang, X.Y. Zheng, Y.Z. Wu, S.H. Xu, A blue organic emitting diode from anthracene derivative, Thin Solid Films 401 (2001) 251–254, https://doi.org/10.1016/S0040-6090(01)01623-6.

[192] B.W. D'Andrade, S.R. Forrest, White organic light-emitting devices for solid-state lighting, Adv. Mater. 16 (2004) 1585–1595, https://doi.org/10.1002/adma.200400684.

[193] B.W. D'Andrade, J. Brooks, V. Adamovich, M.E. Thompson, S.R. Forrest, White light emission using triplet excimers in electrophosphorescent organic light-emitting devices, Adv. Mater. 14 (2002) 1032–1036, https://doi.org/10.1002/1521-4095(20020805)14:15<1032::AID-ADMA1032>3.0.CO;2-6.

[194] B.P. Yan, C.C.C. Cheung, S.C.F. Kui, H.F. Xiang, V.A.L. Roy, S.J. Xu, C.M. Che, Efficient white organic light-emitting devices based on phosphorescent platinum(II)/fluorescent dual-emitting layers, Adv. Mater. 19 (2007) 3599–3603, https://doi.org/10.1002/adma.200602683.

[195] J. Kido, K. Hongawa, K. Okuyama, K. Nagai, White light-emitting organic electroluminescent devices using the poly(N-vinylcarbazole) emitter layer doped with three fluorescent dyes, Appl. Phys. Lett. 64 (1994) 815–817, https://doi.org/10.1063/1.111023.

[196] C.H. Chuen, Y.T. Tao, Highly-bright white organic light-emitting diodes based on a single emission layer, Appl. Phys. Lett. 81 (2002) 4499–4501, https://doi.org/10.1063/1.1528736.

[197] B.W. D'Andrade, R.J. Holmes, S.R. Forrest, Efficient organic electrophosphorescent white-light-emitting device with a triple doped emissive layer, Adv. Mater. 16 (2004) 624–628, https://doi.org/10.1002/adma.200306670.

[198] C.H. Chuen, Y.T. Tao, F.I. Wu, C.F. Shu, White organic light-emitting diodes based on 2,7-bis(2,2-diphenylvinyl)-9, 9'-spirobifluorene: improvement in operational lifetime, Appl. Phys. Lett. 85 (2004) 4609–4611, https://doi.org/10.1063/1.1824178.

[199] X. Gong, S. Wang, D. Moses, G.C. Bazan, A.J. Heeger, Multilayer polymer light-emitting diodes: white-light emission with high efficiency, Adv. Mater. 17 (2005) 2053–2058, https://doi.org/10.1002/adma.200500727.

[200] J. Huang, G. Li, E. Wu, Q. Xu, Y. Yang, Achieving high-efficiency polymer white-light-emitting devices, Adv. Mater. 18 (2006) 114–117, https://doi.org/10.1002/adma.200501105.

[201] T.H. Kim, H.K. Lee, O. Ok Park, B.D. Chin, S.H. Lee, J.K. Kim, White-light-emitting diodes based on iridium complexes via efficient energy transfer from a conjugated polymer, Adv. Funct. Mater. 16 (2006) 611–617, https://doi.org/10.1002/adfm.200500621.

[202] P.I. Shin, C.F. Shu, Y.L. Tung, Y. Chi, Efficient white-light-emitting diodes based on poly(N-vinylcarbazole) doped with blue fluorescent and orange phosphorescent materials, Appl. Phys. Lett. 88 (2006) 2004–2007, https://doi.org/10.1063/1.2214141.

[203] B.W. D'Andrade, M.E. Thompson, S.R. Forrest, Controlling exciton diffusion in multilayer white phosphorescent organic light emitting devices, Adv. Mater. 14 (2002) 147–151, https://doi.org/10.1002/1521-4095(20020116)14:2<147::AID-ADMA147>3.0.CO;2-3.

[204] K.O. Cheon, J. Shinar, Bright white small molecular organic light-emitting devices based on a red-emitting guest-host layer and blue-emitting 4,4′-bis(2, 2′-diphenylvinyl)-1,1′-biphenyl, Appl. Phys. Lett. 81 (2002) 1738–1740, https://doi.org/10.1063/1.1498500.

[205] Y.S. Huang, J.H. Jou, W.K. Weng, J.M. Liu, High-efficiency white organic light-emitting devices with dual doped structure, Appl. Phys. Lett. 80 (2002) 2782–2784, https://doi.org/10.1063/1.1413220.

[206] S. Tokito, T. Iijima, T. Tsuzuki, F. Sato, High-efficiency white phosphorescent organic light-emitting devices with greenish-blue and red-emitting layers, Appl. Phys. Lett. 83 (2003) 2459–2461, https://doi.org/10.1063/1.1611620.

[207] G. Li, J. Shinar, Combinatorial fabrication and studies of bright white organic light-emitting devices based on emission from rubrene-doped 4,4′-bis(2,2′-diphenylvinyl)-1,1′-biphenyl, Appl. Phys. Lett. 83 (2003) 5359–5361, https://doi.org/10.1063/1.1635658.
[208] Y. Sun, N.C. Giebink, H. Kanno, B. Ma, M.E. Thompson, S.R. Forrest, Management of singlet and triplet excitons for efficient white organic light-emitting devices, Nature 440 (2006) 908–912, https://doi.org/10.1038/nature04645.
[209] G. Lei, L. Wang, Y. Qiu, Multilayer organic electrophosphorescent white light-emitting diodes without exciton-blocking layer, Appl. Phys. Lett. 88 (2006) 2004–2007, https://doi.org/10.1063/1.2185255.
[210] S.L. Lai, M.Y. Chan, M.K. Fung, C.S. Lee, S.T. Lee, Efficiency enhancement and voltage reduction in white organic light-emitting devices, Appl. Phys. Lett. 90 (2007) 88–91, https://doi.org/10.1063/1.2740482.
[211] F. Guo, D. Ma, White organic light-emitting diodes based on tandem structures, Appl. Phys. Lett. 87 (2005) 1–3, https://doi.org/10.1063/1.2120898.
[212] C.C. Chang, J.F. Chen, S.W. Hwang, C.H. Chen, Highly efficient white organic electroluminescent devices based on tandem architecture, Appl. Phys. Lett. 87 (2005) 1–3, https://doi.org/10.1063/1.2147730.
[213] H. Kanno, R.J. Holmes, Y. Sun, S. Kena-Cohen, S.R. Forrest, White stacked electrophosphorescent organic light-emitting devices employing MoO3 as a charge-generation layer, Adv. Mater. 18 (2006) 339–342, https://doi.org/10.1002/adma.200501915.
[214] G. Tu, C. Mei, Q. Zhou, Y. Cheng, Y. Geng, L. Wang, D. Ma, X. Jing, F. Wang, Highly efficient pure-white-light-emitting diodes from a single polymer: polyfluorene with naphthalimide moieties, Adv. Funct. Mater. 16 (2006) 101–106, https://doi.org/10.1002/adfm.200500028.
[215] J. Liu, Q. Zhou, Y. Cheng, Y. Geng, L. Wang, D. Ma, X. Jing, F. Wang, White electroluminescence from a single-polymer system with simultaneous two-color emission: polyfluorene as the blue host and a 2,1,3-benzothiadiazole derivative as the orange dopant on the main chain, Adv. Funct. Mater. 16 (2006) 957–965, https://doi.org/10.1002/adfm.200500761.
[216] J. Jiang, Y. Xu, W. Yang, R. Guan, Z. Liu, H. Zhen, Y. Cao, High-efficiency white-light-emitting devices from a single polymer by mixing singlet and triplet emission, Adv. Mater. 18 (2006) 1769–1773, https://doi.org/10.1002/adma.200502740.
[217] A.R. Duggal, J.J. Shiang, C.M. Heller, D.F. Foust, Organic light-emitting devices for illumination quality white light, Appl. Phys. Lett. 80 (2002) 3470–3472, https://doi.org/10.1063/1.1478786.
[218] B.C. Krummacher, V.E. Choong, M.K. Mathai, S.A. Choulis, F. So, F. Jermann, T. Fiedler, M. Zachau, Highly efficient white organic light-emitting diode, Appl. Phys. Lett. 88 (2006) 2004–2007, https://doi.org/10.1063/1.2186080.
[219] Y. Ohmori, A. Fujii, M. Uchida, C. Morishima, K. Yoshino, Observation of spectral narrowing and emission energy shift in organic electroluminescent diode utilizing 8-hydroxyquinoline aluminum/aromatic diamine multilayer structure, Appl. Phys. Lett. 63 (1993) 1871–1873, https://doi.org/10.1063/1.110632.
[220] V. Bulović, P. Tian, P.E. Burrows, M.R. Gokhale, S.R. Forrest, M.E. Thompson, A surface-emitting vacuum-deposited organic light emitting device, Appl. Phys. Lett. 70 (1997) 2954–2956, https://doi.org/10.1063/1.119260.

[221] T. Dobbertin, M. Kroeger, D. Heithecker, D. Schneider, D. Metzdorf, H. Neuner, E. Becker, H.H. Johannes, W. Kowalsky, Inverted top-emitting organic light-emitting diodes using sputter-deposited anodes, Appl. Phys. Lett. 82 (2003) 284–286, https://doi.org/10.1063/1.1535743.

[222] M.A. Baldo, D.F. O'Brien, Y. You, A. Shoustikov, S. Sibley, M.E. Thompson, S.R. Forrest, Highly efficient phosphorescent emission from organic electroluminescent devices, Nature 395 (1998) 151. https://www.nature.com/articles/25954.pdf.

[223] C. Adachi, M.A. Baldo, M.E. Thompson, S.R. Forrest, Nearly 100% internal phosphorescence efficiency in an organic light emitting device, J. Appl. Phys. 90 (2001) 5048–5051, https://doi.org/10.1063/1.1409582.

[224] D.F. O'Brien, M.A. Baldo, M.E. Thompson, S.R. Forrest, Improved energy transfer in electrophosphorescent devices, Appl. Phys. Lett. 74 (1999) 442–444, https://doi.org/10.1063/1.123055.

[225] H. Aziz, Z.D. Popovic, Degradation phenomena in small-molecule organic light-emitting devices, Chem. Mater. 16 (2004) 4522–4532, https://doi.org/10.1021/cm040081o.

[226] P.E. Burrows, V. Bulovic, S.R. Forrest, L.S. Sapochak, D.M. McCarty, M.E. Thompson, Reliability and degradation of organic light emitting devices, Appl. Phys. Lett. 65 (1994) 2922–2924, https://doi.org/10.1063/1.112532.

[227] Y.F. Liew, H. Aziz, N.X. Hu, H.S.O. Chan, G. Xu, Z. Popovic, Investigation of the sites of dark spots in organic light-emitting devices, Appl. Phys. Lett. 77 (2000) 2650–2652, https://doi.org/10.1063/1.1320459.

[228] M.K. Fung, Z.Q. Gao, C.S. Lee, S.T. Lee, Inhibition of dark spots growth in organic electroluminescent devices, Chem. Phys. Lett. 333 (2001) 432–436, https://doi.org/10.1016/S0009-2614(00)01407-X.

[229] M.Y. Chan, S.L. Lai, F.L. Wong, O. Lengyel, C.S. Lee, S.T. Lee, Efficiency enhancement and retarded dark-spots growth of organic light-emitting devices by high-temperature processing, Chem. Phys. Lett. 371 (2003) 700–706, https://doi.org/10.1016/S0009-2614(03)00310-5.

[230] Y. Kim, D. Choi, H. Lim, C.S. Ha, Accelerated pre-oxidation method for healing progressive electrical short in organic light-emitting devices, Appl. Phys. Lett. 82 (2003) 2200–2202, https://doi.org/10.1063/1.1564872.

[231] S.A. Van Slyke, C.H. Chen, C.W. Tang, Organic electroluminescent devices with improved stability, Appl. Phys. Lett. 69 (1996) 2160–2162, https://doi.org/10.1063/1.117151.

[232] J. Lee, V.C. Sundar, J.R. Heine, M.G. Bawendi, K.F. Jensen, Full color emission from II ± VI semiconductor, Adv. Mater. 12 (2000) 1102–1105.

[233] E. Jang, S. Jun, H. Jang, J. Lim, B. Kim, Y. Kim, White-light-emitting diodes with quantum dot color converters for display backlights, Adv. Mater. 22 (2010) 3076–3080, https://doi.org/10.1002/adma.201000525.

[234] G. Cheng, M. Mazzeo, A. Rizzo, Y. Li, Y. Duan, G. Gigli, White light-emitting devices based on the combined emission from red CdSe/ZnS quantum dots, green phosphorescent, and blue fluorescent organic molecules, Appl. Phys. Lett. 94 (2009) 23–26, https://doi.org/10.1063/1.3157131.

[235] S. Coe-Sullivan, J.S. Steckel, W.K. Woo, M.G. Bawendi, V. Bulovic, Large-area ordered quantum-dot monolayers via phase separation during spin-casting, Adv. Funct. Mater. 15 (2005) 1117–1124, https://doi.org/10.1002/adfm.200400468.

[236] J. Zhao, J.A. Bardecker, A.M. Munro, M.S. Liu, Y. Niu, I.K. Ding, J. Luo, B. Chen, A.K.Y. Jen, D.S. Ginger, Efficient CdSe/CdS quantum dot light-emitting diodes using a thermally polymerized hole transport layer, Nano Lett. 6 (2006) 463–467, https://doi.org/10.1021/nl052417e.

[237] J.M. Caruge, J.E. Halpert, V. Wood, V. Buloví, M.G. Bawendi, Colloidal quantum-dot light-emitting diodes with metal-oxide charge transport layers, Nat. Photonics 2 (2008) 247–250, https://doi.org/10.1038/nphoton.2008.34.

[238] M.C. Schlamp, X. Peng, A.P. Alivisatos, Improved efficiencies in light emitting diodes made with CdSe(CdS) core/shell type nanocrystals and a semiconducting polymer, J. Appl. Phys. 82 (1997) 5837–5842, https://doi.org/10.1063/1.366452.

[239] P.O. Anikeeva, J.E. Halpert, M.G. Bawendi, V. Bulović, Quantum dot light-emitting devices with electroluminescence tunable over the entire visible spectrum, Nano Lett. 9 (2009) 2532–2536, https://doi.org/10.1021/nl9002969.

[240] P.O. Anikeeva, J.E. Halpert, M.G. Bawendi, V. Bulović, Electroluminescence from a mixed red-green-blue colloidal quantum dot monolayer, Nano Lett. 7 (2007) 2196–2200, https://doi.org/10.1021/nl0703424.

[241] E.F. Schubert, Light-emitting Diodes, second ed., Cambridge University Press, 2006 https://doi.org/10.1017/CBO9780511790546.

[242] H. Yaguchi, A. David, T. Fuchida, ... K.H., Undefined 2017, CIE 2017 Colour Fidelity Index for Accurate Scientific Use, Lirias.Kuleuven.Be. (n.d.).

[243] C.H. Huang, T.M. Chen, W.R. Liu, Y.C. Chiu, Y.T. Yeh, S.M. Jang, A single-phased emission-tunable phosphor Ca9Y(PO 4)7:Eu2+,Mn2+ with efficient energy transfer for white-light-emitting diodes, ACS Appl. Mater. Interfaces 2 (2010) 259–264, https://doi.org/10.1021/am900668r.

[244] J. Zhang, Z. Hua, F. Zhang, Warm white-light generation in Ca9MgNa(PO4)7:Sr2+, Mn2+, Ln (Ln=Eu2+, Yb3+, Er3+, Ho3+, and Tm3+) under near-ultraviolet and near-infrared excitation, Ceram. Int. 41 (2015) 9910–9915, https://doi.org/10.1016/j.ceramint.2015.04.068.

Further reading

[1] T. Shiga, H. Fujikawa, Y. Taga, Design of multiwavelength resonant cavities for white organic light-emitting diodes, J. Appl. Phys. 93 (2003) 19–22, https://doi.org/10.1063/1.1527708.

Rare-earth-activated phosphors for forensic applications

9

Yatish R. Parauha and Sanjay J. Dhoble
Department of Physics, Rashtrasant Tukadoji Maharaj Nagpur University, Nagpur, Maharashtra, India

1. Introduction

In recent years, rare-earth-activated phosphors have great potential due to their excellent chemical, physical, and optical properties, which have become quite important materials as hosting components for various applications such as white LEDs, Thermoluminescence (TL) dosimetry, fingerprint detection, plant growth, scintillators, bio-imaging, solar cell, anticounterfeiting, and phototherapy [1–8]. Over the past few years, the use of rare-earth-activated phosphors as the labeling agents for visualization of latent fingerprints (LFPs), and this technique has been proved an important technique for the physical identification of LFPs in the field of criminology. In advanced forensic investigations, fingerprints (FP) were considered an important source for identifying individuals. It has specific features like type 1 (ridge flow), type 2 (core, delta, and bifurcation), and type 3 (sweat pores), which are characteristic of each person's identification [9]. In these cases, LFP has gained the highest publicity due to its unique and unchanging ridge characteristics. The ridge pattern at the tips of the human fingers is unique and remains uniform throughout life. When a person touches a surface at a location, the sweat of his fingers is transferred to the surface through the pores thus forming an invisible ridge pattern. Criminals can be detected by analyzing FP commonly found during crime scene investigations; it is necessary to develop efficient methods to visualize LFPs as a result. It is difficult to see the LFP with naked eyes at the crime scene, so it is not possible to identify the culprit. For this, more precise visualization techniques are required. Until now, several methods exist for fingerprint identification and development, such as ninhydrin, fluorescent dyes, multimetal deposition, and powder dusting [10]. However, most of the techniques used for LFP detection have several drawbacks, such as low sensitivity, high background barrier, complex processes, high toxicity, and poor visual image properties [10]. According to the literature, the powder dusting method is the most effective technique currently used for LFP detection. The powder dusting method has proved to be simple and can easily enable reproduction with excellent adhesion, high contrast, high sensitivity, and selectivity [11]. A variety of materials are currently used for LFP detection such as semiconductor quantum dots and magnetic nanoparticles especially Au nanoparticles and inorganic nanoparticles [2,10,12–14]. However, some technical faults, such as

Rare-Earth-Activated Phosphors. https://doi.org/10.1016/B978-0-323-89856-0.00003-1

fabricated measures, damaging, damaging character of visual photographs, and preservation time are usually delayed by their broader characteristics. Therefore, the rapid, accurate, effective, and nondestructive content requirement for the rich view of LFP is high, and the research community's search for this is continuing [11].

Currently, anticounterfeiting and fraud detection techniques are also an important part of forensic application. In the last few years, emphasis has been laid on the development of anticounterfeiting and fraud detection techniques. And emphasis has also been placed on the development of such unique materials, which allow for a dynamic fluorescence signal with higher protection than conventional anticounterfeiting counterparts. Fluorescent anticounterfeiting technologies have attracted a lot of attention in the research society due to their high concealment properties [15]. High fluorescence intensity and long fluorescence lifetime are two important properties in the development of fluorescent anticounterfeiting materials [16]. Optical storage materials (OSMs) have been thoroughly investigated for their important applications, such as biomedicine, clinical medicine, and environmental engineering. In recent years, OSMs have been successfully explored for their potential application in anticounterfeiting techniques. We have had considerable success in developing anticounterfeiting techniques, but realizing hidden multikinetic anticounterfeit material is still an important challenge due to the increasing high requirements [17]. At present, it is important to develop materials that can be applied to anticounterfeiting technologies. Many anticounterfeiting materials can also be developed by incorporating a single activator ion into a single fluorescent material, which has attracted greater attention due to lower manufacturing costs and a simplified luminescence mechanism [18]. This chapter focuses on the synthesis of rare-earth-activated phosphors for forensic application, their characterization techniques, previously investigated results, and future aspects.

2. Rare-earth-doped phosphors

Phosphor is a microcrystalline/nanocrystalline solid luminescent material, which is found to be white in color. These materials have the ability to absorb the incident light in the electromagnetic spectrum range (i.e., ultraviolet (UV), NUV, IR region) and emits photons after absorption. Phosphors are composed of an inert host lattice, which is transparent to the excitation radiation, and an activator, typically a 3d or 4f electron metal, which is excited under energy bombardment. These emission and absorption processes occur when suitable impurity elements are doped into the phosphor host to observe their luminescence properties. These phosphor materials are available in inorganic and organic forms. In inorganic, the combination of transition series metal oxide framework is used whereas in organic phosphor a long chain organic polymer material is used as a luminescent host. Phosphors work based on excitation and emission processes. The process of light absorption and emission in phosphor materials occurs due to the formation of energy bands, that is, valence band and conduction band. Initially, the atom in the valence band absorbs some light energy and goes to the excited state; later, it makes a spontaneous transition to the ground state with the emission of light in longer wavelength regions. This emitted light is also of two kinds: fluorescence and phosphorescence [19]. If the incident light is off, then the emission also stops; this process is called fluorescence; however, in phosphorescence, materials emit light in microsecond duration after the incident light is cut off. The absorption of energy, which is utilized to excite the phosphor material, is either by the host or by intentionally doped rare-earth ions.

The rare-earth-based phosphors are widely used in various fields such as solid-state lighting, CR tubes, TV screen, display panels, biomedical fields, TL dosimetric applications, solar cells, and phototherapy lamps [8,20–24]. It plays an important role in tuning various important optical, optical, and magnetic, electrical, electronic properties. Mostly, emission arises due to rare-earth ions, when phosphor materials are excited by receiving energy from an energy source and activator ions show very weak absorption, then another type of rare-earth ion is codoped with the activator in the phosphor materials, called as sensitizers, which absorb energy and subsequently transfer the energy to activator. This process is called energy transfer [19]. By this process, the required emission colors can be achieved by choosing the appropriate activator and synthesizer.

Rare-earth ions are generally f block elements; they are also called lanthanide elements. At present, 17 rare-earth elements are founded specifically the 14 lanthanides (Ce–Lu) and scandium, yttrium, and Lanthanum. The electronic configuration of lanthanides is partially filled 4f shell. These three elements behave like rare-earth elements. Generally, rare-earth ions are found in the form of trivalent elements such as Eu^{3+}, Dy^{3+}, Ce^{3+}, and Gd^{3+} but some rare-earth ions are found in the form of divalent elements. Every element has its own physical, chemical, and luminescence properties. The role of rare-earth ions in phosphors can be explained with the help of excitation and emission type spectra. Table 9.1 shows the list of some rare-earth ions and their expected excitation, emission, and emission transitions with potential applications. Recently, different rare-earth-activated host materials such as oxides, aluminates, silicates, tungstate, vanadate, phosphate, borates, titanates are studied, and they find a potential application. Xianju Zhou and their colleagues reported Eu^{3+}-doped $LiSrVO_4$ phosphors for temperature sensing application [25]. Chaitali Mehare et al. [26] reported Eu^{2+}-activated $K_3Ca_2(SO_4)_3F$ phosphor via solid-state reaction method. Bhake et al. [27] reported Ce^{3+}-doped $Na_2ZnP_2O_7$ and $Na_4P_2O_7$ pyrophosphate phosphors. Similarly, several national/international researchers and scholars are working in rare-earth-activated phosphors, which are discussed in Section 5.

3. Synthesis method

Synthesis methods play an important role in the successful preparation of rare-earth-activated phosphors. Currently, a number of synthesis routes are available for the preparation of materials. We face many problems during material synthesis and think of many aspects such as the role of rare-earth ions, pure formation, removal of impurities, particle size, morphology behavior, low cost, behavior with the environment,

Table 9.1 List of some rare-earth ions, their oxidation state, expected excitation, emission, and emission transition with possible applications.

Rare-earth ion	Oxidation state	Expected excitation	Expected emission	Emission transition	Application
Europium	Eu^{3+}	395 nm, 466 nm	593 nm, 615 nm	$^5D_0 \rightarrow {}^7F_j$ (j = 0, 1, 2,...)	Red phosphor for LED, mercury lamp, fluorescent lamp.
	Eu^{2+}	354 nm	440 nm	$4f^65d^1 \rightarrow 4f^7$	Blue phosphor for LED, laser, mercury-vapor lamp.
Dysprosium	Dy^{3+}	348 nm, 365 nm, 385 nm	480 nm, 576 nm	$^4F_{9/2} \rightarrow {}^6H_j$, J = 15/2, 13/2	Blue phosphor for LED, laser, Mosquito's repellent lamp.
Cerium	Ce^{3+}	278 nm and 295 nm	355 nm	5d → 4f	Glass and ceramics, polishing powder
Samarium	Sm^{3+}	405 nm	567 nm, 604 nm, 608 nm	$^4G_{5/2} \rightarrow {}^6H_j$ ($j = \frac{5}{2}, \frac{7}{2}, \frac{9}{2}$)	Yellow and red phosphor
Terbium	Tb^{3+}	240, 379, 465 nm	490 nm, 545 nm	$^5D_4 \rightarrow {}^7F_j$ (j = 6, 5)	Blue and green phosphor for LED
Gadolinium	Gd^{3+}	260 nm, 280 nm	313 nm	$^6P_{7/2} \rightarrow {}^8S_{7/2}$	UV light for phototherapy lamp
Praseodymium	Pr^{3+}	448 nm	486, 531, 602, 618, and 649 nm	$^3P_0 \rightarrow {}^3H_4$, $^3P_0 \rightarrow {}^3H_5$, $^1D_2 \rightarrow {}^3H_4$, $^3P_0 \rightarrow {}^3H_6$, and $^3P_0 \rightarrow {}^3F_2$	White LEDs
Holmium	Ho^{3+}	418 nm, 454 nm, 474 nm, 486 nm	491 nm, 550 nm, 662 nm, 758 nm	($^5F_3 \rightarrow {}^5I_8$), ($^5F_4 \rightarrow {}^5S_2 \rightarrow {}^5I_8$), ($^5F_5 \rightarrow {}^5I_8$), ($^5S_2 \rightarrow {}^5I_7$)	Solid-state lighting

Neodymium	Nd^{3+}	192 nm, 265 nm, 333 nm, 390–720 nm	824 nm, 809 nm, 745 nm, 688 nm, 633 nm, 590 nm, 531 nm, 518 nm, 477 nm, 466 nm, 435 nm	$(^4I_{9/2} \rightarrow {}^4F_{5/2})$, $(^4I_{9/2} \rightarrow {}^4F_{7/2})$, $(^4I_{9/2} \rightarrow {}^4S_{3/2})$, $(^4I_{9/2} \rightarrow {}^4F_{9/2})$, $(^4I_{9/2} \rightarrow {}^2K_{15/2})$, $(^4I_{9/2} \rightarrow {}^4G_{5/2}, {}^2H_{11/2})$, $(^4I_{9/2} \rightarrow {}^2K_{13/2})$, $(^4I_{9/2} \rightarrow {}^4G_{9/2})$, $(^4I_{9/2} \rightarrow {}^2K_{11/2})$, and $(^4I_{9/2} \rightarrow {}^2P_{1/2})$	Optical temperature sensing, laser, biomedicine, solid-state lighting
Erbium	Er^{3+}	950–1000 nm	364, 377, 406, 450 and 487, 529, 550 and 656 nm	$^4I_{15/2} \rightarrow {}^4G_{9/2}$, $^4I_{15/2} \rightarrow {}^4G_{11/2}$, $^4I_{15/2} \rightarrow {}^2H_{9/2}$, $^4I_{15/2} \rightarrow {}^4F_{5/2}$, $^4I_{15/2} \rightarrow {}^4F_{7/2}$ $^4I_{15/2} \rightarrow {}^2H_{11/2}$, $^4I_{15/2} \rightarrow {}^4S_{3/2}$, and $^4F_{9/2} \rightarrow {}^4I_{15/2}$	solid-state lighting, solar cell, optical temperature sensing
Thulium	Tm^{3+}	355 nm, 467 nm	453 nm, 472 nm	$^1D_2 \rightarrow {}^3F_4$ $^1G_4 \rightarrow {}^3H_6$	Biological imaging of deep tumors, optical temperature sensing, blue-emitting phosphors

reproducibility, and possible applications. Therefore, research scholars and researchers are trying to synthesize new materials with better properties, greater functionality, and cheaper costs than the current ones. Scientists commonly use different methods to gain better control over the size and composition of particles, such as the solid-state reaction method, combustion method, sol–gel method, coprecipitation method, wet chemical method, and hydrothermal method. In this section, we are discussing some synthesis methods, which are used for the preparation of rare-earth-activated phosphor materials.

3.1 Solid-state reaction method

The solid-state reaction method has attracted much attention from researchers for the synthesis of rare-earth-activated phosphors. This method gained popularity because this is a straightforward method and it has higher productivity, lower cost, homogeneity of phase, high crystallinity, the grain size of phosphors, and luminescence properties. In this method, solid precursors, such as oxides, (oxy) fluorides, (oxy) chlorides, and (oxy) nitrides, are required [28]. This method is based on grinding, mixing, and heat treatment/temperature. The mixed powder at high temperatures facilitates the diffusion of atoms or ions in the chemical reaction. The solid-state reaction method is completed in the following steps [28]:

The synthesis of phosphors through solid-state reaction method study in the following steps:

- All required ingredients are weighted in the proper ratio (according to calculation) using an electronic balance machine and transferred in the crucibles.
- All ingredients mixed properly by using mortar and pastel and again transferred in the crucibles.
- In the next step, crucibles are put in the furnace for heat treatment. It may be a two-step process (preheating and final heating) or a single-step process (final heating at a higher temperature).
- Heating temperature depends on the materials melting points
- After heating, the furnace is cooled up to room temperature then obtained powder is again crushed (mixed) properly.
- After that the obtained white powder is our final product and it can be used for further characterization.

Currently, a number of rare-earth-activated phosphors are synthesized via the solid-state reaction method. Recently, Yatish R. Parauha and S.J. Dhoble have reported Eu^{2+}-activated blue color emitting $Ca_3Al_2O_6$ down-conversion phosphors, which were synthesized via solid-state reaction method [29]. In this work, they have used calcium carbonate precipitated ($CaCO_3$) and aluminum oxide (Al_2O_3) as starting ingredients, and europium oxide (Eu_2O_3) is used as an activator. During this work, starting ingredients and activators were crushed for around 30 min, then preheated at 400°C for 2 h, and the final heating treatment was carried out at 800°C for 24 h. The obtained powder was reduced in a reducing atmosphere in the presence of activated charcoal at 800°C for 4 h. The obtained product is characterized by X-ray diffraction, scanning electron microscope (SEM), Fourier transform infrared, and photoluminescence (PL) techniques. The PL emission spectra show a broad emission band centered at 445 nm under 334 nm excitation. The emission band observed due to allowed $4f^65d^1 \rightarrow 4f^7$ electronic transition of Eu^{2+} ions. Their features revealed that the synthesized phosphors can be useful as a blue-emitting component for white light-emitting diodes [29].

Junpeng Xue et al. reported Sm^{3+}-activated $NaSrLa(MoO_4)O_3$ phosphors, which were synthesized by a high-temperature solid-state reaction method. In this investigation, authors were used Na_2CO_3 (99.995%), $SrCO_3$ (99.90%), La_2O_3 (99.99%), MoO_3 (99.95%), and Sm_2O_3 (99.90%) chemicals as raw ingredients. These raw ingredients were homogeneously mixed for around 30 min and heated at 1200°C for 7 h. The synthesized phosphors bright reddish-orange emissions under 403 nm excitation. Junpeng Xue et al. reported synthesized phosphors were promising red-emitting materials for solid-state illumination.

3.2 Combustion method

Combustion synthesis is an attractive and popular synthesis method for the preparation of rare-earth-activated phosphors. Advanced materials such as powders, ceramics, intermetallics, composites, and functional materials can be manufactured by this method [30]. This method has gained more popularity and attention in the research world due to its advanced properties such as easy synthesis method, low cost, and energy saving. Along with these advantages, combustion synthesis has proved to be the best synthesis route for the industrial production of many inorganic materials. The combustion synthesis method consists of two steps: first, self-diffusion high-temperature synthesis; second, thermal explosion mode. Carbide, nitride, boride, sulfate, and oxynitride materials can be synthesized with the help of the combustion method [31–35]. This process involved a highly exothermic reaction between metal nitrates (oxidizers) and fuels. In this method, urea, glycine, carbohydrazide (CH), oxalyl di hydrazide (ODH), citric acid, malonic acid dihydrazide (MDH), and tetra formal trisazine (TFTA) can be used as fuel materials [36]. The chemical reaction between the reactants requires a temperature around 550°C and produces a reaction temperature of over 1200°C. In this process, a vigorous flame with a large amount of gases is ignited and the process is completed in 10 min. The detailed discussion about combustion synthesis has been reported in various papers, chapters, and books [30,37–47]. Martinez Pacheco et al. [48] reported basic steps followed during the combustion synthesis process, as shown in Fig. 9.1. They reported TiB_2-based cermets via combustion method and investigate the physical and electrical properties of the products. Several other researchers have reported various rare-earth-activated phosphors such as $MgAl_2O_4:Eu^{3+}$ [49], $Ba_3Y_4O_9:Sm^{3+}$ [50], $BaYZn_3AlO_7:Eu^{3+}$ [51], $BaMgAl_{10}O_{17}$: Eu [52], and $NaBaBO_3:Ce^{3+}$ [53].

3.3 Sol–gel method

The sol–gel method is a useful method for the synthesis of phosphors with uniform particle size distribution. Nano/micron-sized phosphors can be produced by this

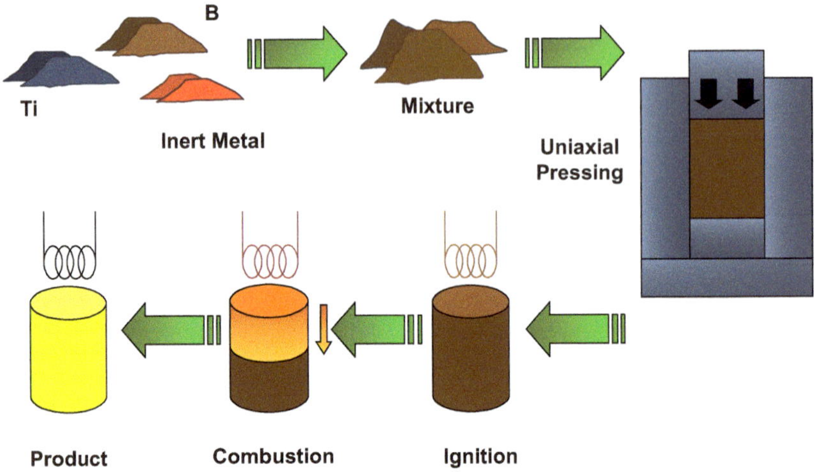

Figure 9.1 Scheme of the basic steps during the combustion synthesis process. Reproduced from Ref. [48] (open access).

method. In this method, nitride, chloride, sulfate forms, or water removable materials can be used as starting precursors. In the preparation of phosphors by this method, citric acid is used as the chelating agent, while polyethylene glycol (PEG) is best used as the surfactant. Fig. 9.2 shows a schematic diagram of the sol–gel synthesis process. The method proceeds through a series of chemical processes, including

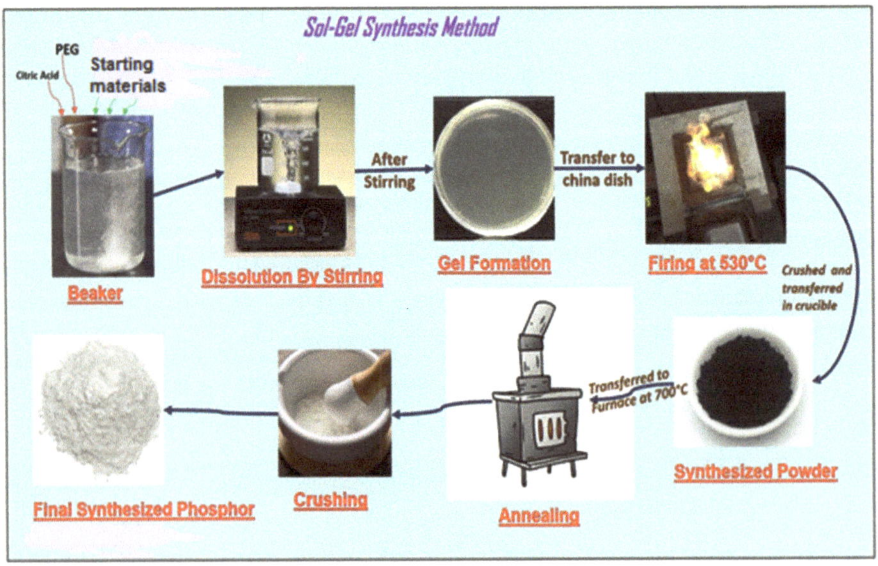

Figure 9.2 Schematic diagram of sol–gel synthesis process.

hydrolysis, gelation, drying, and thermal treatment. The process involves a sol through hydrolysis and polymerization reactions and subsequent attainment of a rigid porous gel [54–56]. The sol–gel synthesis process involves the following steps:

- Weighing all raw ingredients according to calculation
- Dissolving precursors (in the form of metal nitrates) in distilled water
- Dissolving a chelating reagent and a surfactant with a solution of precursors
- Then, the obtained solution is kept into magnetic stirrer for stirring at about 80°C
- After some time, condensation takes place to yield a polymeric network leading to the formation of sol and later on the gel.
- Then, the obtained solution is fired at 550°C
- After this process, a black powder is obtained. To remove the PEG content, it is heated to a higher temperature of 500°C.
- The nano/micron-sized white powder is obtained at the end of the process, which can be used for further characterization.

Mingjun Song et al. [57] reported sol–gel synthesized $KBaGd(MoO_4)_3:Eu^{3+}$ phosphors using KNO_3 (AR), $Ba(NO_3)_2$ (AR), Gd_2O_3 (99.999%), Eu_2O_3 (99.99%), $(NH_4)_6Mo_7O_{24}\cdot 4H_2O$ (AR), HNO_3(AR), and $NH_3\cdot H_2O$ (AR) raw ingredients. During the synthesis, the obtained final solution was dried at 80°C in a water bath to form a yellow transparent gel, then it was dried in an oven at 120°C until it turned into a black dried gel. Further, the obtained gel was annealed at different temperatures (500, 600, 700, 800, and 900°C) for 5 h in a muffle furnace, and the final products were obtained [57].

Harpreet Kaur et al. [58] reported Sm^{3+}-activated Na_2CaSiO_4 phosphors, which have been synthesized by the sol-gel technique. $Ca(NO_3)_2.4H_2O$, $NaNO_3$, tetraethyl orthosilicate, and $Sm(NO_3)_3.6H_2O$ are used as starting materials for the synthesis of Sm^{3+}-activated Na_2CaSiO_4 phosphors. The structural, morphological, and photoluminescence properties of the synthesized material have been investigated in this investigation. The synthesized phosphors show strong orange emission, confirmed by the Commission Internationale de l'Éclairage (CIE) diagram that revealed the chromaticity coordination in the orange region [58].

Similarly, several other researchers reported various rare-earth-activated phosphors, such as $NaPbLa(MoO_4)_3:Er^{3+}/Yb^{3+}$ [59], $Na_2Ca_2Si_3O_9:Eu^{3+}$ [60], $KBaY(MoO_4)_3:Dy^{3+}$ [61], $CaTiO_3:Pr^{3+}$ [62], $CaLa_2ZnO_5:Gd^{3+}$ [63], and $Ba_2SiO_4:Tb^{3+}$ [64], by using sol–gel method.

3.4 Coprecipitation method

Coprecipitation synthesis route is a very easy and effective method for the preparation of rare-earth-activated phosphors. This is a very simple synthesis approach, and it is not required for expensive instruments, special reaction conditions, or complicated processes. This method can be considered a smart technique for the development of oxide and fluoride-based phosphors because this method does not require rare earth to form oxides and fluorides. This synthesis approach is an excellent choice for researchers and students due to its amazing advantages such as high accuracy and better stoichiometric control. However, this synthesis approach has some disadvantages such as low final production yield and long consumption time. Distilled water is always required for the synthesis of materials through the coprecipitation synthesis route [65,66].

Thomas Jansen et al. [67] reported Mn^{4+}-activated K_2NbF_7 and K_2TaF_7 phosphors via coprecipitation synthesis route. The synthesis process used for the synthesis of these materials is shown in Fig. 9.3. In the first step of the synthesis, K_2MnF_6 was synthesized by using KF (Sigma Aldrich, 99%), hydrofluoric acid (Sigma Aldrich, ≥48%), and $KMnO_4$ (Sigma Aldrich, ≥99%) raw materials. Further, NbF_5 (Sigma Aldrich 99.9%) or TaF_5 (Sigma Aldrich 99.9%) was separately dissolved in the hydrofluoric acid. Then, the solutions were mixed together. After 20 min of stirring, synthesized materials were washed with cold acetone (−5°C) and dried in a vacuum oven at 40°C for 4 h, as shown in Fig. 9.3.

Premjit Singh et al. [68] reported $CdMoO_4:Dy^{3+}$ phosphors were successfully synthesized by the coprecipitation method. The chemicals $Cd(NO_3)_2$ (99%), Na_2MoO_4 (99.5%) from Thomas baker and $DyCl_3$ anhydrous (99.9%) from Alfa-Aesar were used in this investigation. The synthesized phosphors have been annealed at 300, 700, 800, 900, and 1000°C for 2 h. Further, their structural, surface morphology and luminescence properties have been investigated.

Various rare-earth-activated phosphors, such as $NaSrPO_4:Gd^{3+}$ [69], $CaAl_2O_4$: Eu^{2+}, Nd^{3+} [70], $Sr_2MgSi_2O_7$: Eu^{2+} [71], $Sr_2P_2O_7:Dy^{3+}$, Ce^{3+} [72], $CaZr_4(PO_4)_6$:

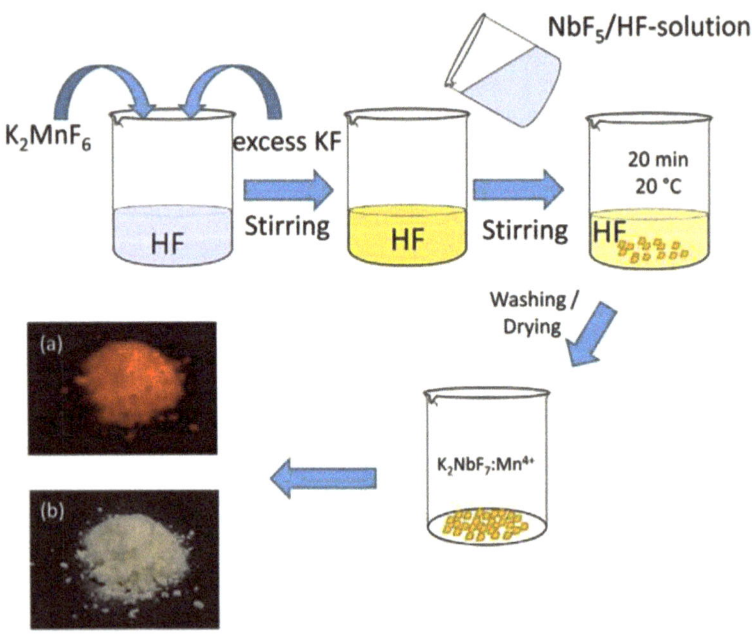

Figure 9.3 Schematic diagram of coprecipitation synthesis process. Reprinted with permission from Ref. [67], Copyright Elsevier.

Sm^{3+} [73], and $SrLaAlO_4:Er^{3+}$ [74], have been successfully synthesized by the coprecipitation synthesis route.

3.5 Hydrothermal method

High temperature and pressure are used in the process of the hydrothermal method. In this method, material dissolved in water crystallizes by high temperature and pressure. Powder, single crystal, fiber, and ceramic materials can be synthesized by this method. It is a self-purifying technology that requires low processing temperatures and is used for the detection of nanomaterials at high vapor pressures and temperatures [75]. The hydrothermal method is combined with a number of other systems, including ultrasound-assisted, electrochemical-assisted, surfactant-assisted, and microwave-assisted hydrothermal processing, to control the reaction medium and to improve physical characteristics. This method also has the potential to improve physical characteristics such as particle morphology and crystallite size. The main advantage of synthesized materials by this technique is that at low synthetic temperatures, crystalline powders are obtained without the need for further heat treatment.

Haiming Cheng et al. [76] reported hydrothermal synthesis process and red-emitting $K_2NaAlF_6:Mn^{4+}$ phosphor, which is synthesized by hydrothermal method using potassium fluoride (KF, >98%), sodium fluoride (NaF, >98%), aluminum fluoride (AlF_3, >99%), hydrofluoric acid solution (HF, 40 wt%), potassium permanganate ($KMnO_4$, >99.5%), potassium hydrogen difluoride (KHF_2, >99%) and hydrogen peroxide aqueous solution (H_2O_2, 30.0%). Fig. 9.4 shows a schematic hydrothermal synthesis process of red-emitting $K_2NaAlF_6:Mn^{4+}$ phosphor.

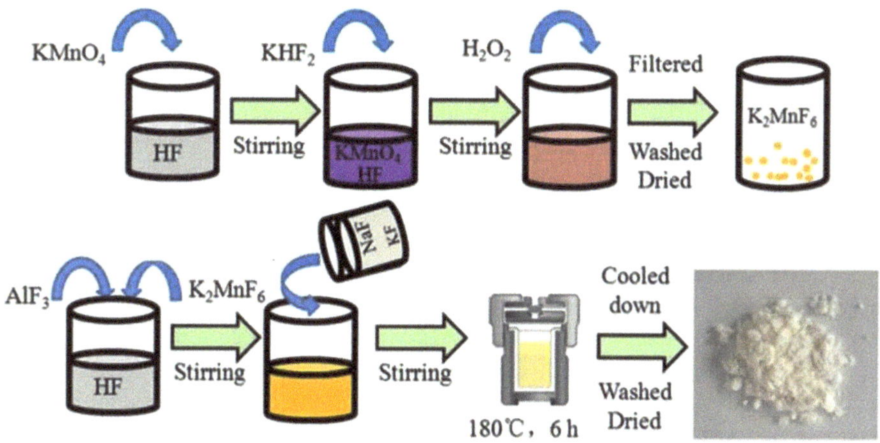

Figure 9.4 A schematic hydrothermal synthesis process of red-emitting $K_2NaAlF_6:Mn^{4+}$ phosphor.
Reproduced from Ref. [76] (open access).

Some of the rare-earth-activated phosphors prepared by the hydrothermal method are $Sr_3Al_2O_6:Sm^{3+}$ [77], $Gd_2Sn_2O_7:Eu^{3+}$ [78], Sm^{3+} and Eu^{3+} codoped $GdPO_4$ [79], $Sc_2W_3O_{12}:Eu^{3+}$ [80], $Ca_3(PO_4)_2:Dy^{3+}$ [81], $Zn4B_6O_{13}$:Tb/Eu [82], etc.

4. Phosphors for forensic applications

Rare-earth-activated phosphors with spherical morphology are important in the visualization of LFPs application. It offers superior sensitivity, low background hindrance, exceptional efficiency, low toxicity, and stress-free detection of LFPs, which was an essential requirement for forensic investigators. According to the research available in the literature, UV light excited rare-earth-activated phosphors can be performed for LFPs application. Therefore, UV light excited rare-earth-activated phosphors have a wide scope of applications in forensic science. In this section, we discuss previously reported rare-earth-activated phosphors, which have potential applicability in forensic applications.

Eluri Pavitra et al. [83] investigated Mn^{4+}-activated Ba_2LaNbO_6 (BLN) phosphors, synthesized by citric acid-assisted sol–gel route. The synthesized phosphors show excellent optical properties, showing narrow emission bands around 685 nm upon excitation at 351 and 510 nm. The High resolution scanning electron microscope (HRSEM) image of $0.25Mn^{4+}$ ions doped BLN nanophosphors revealed nanorod-like morphology. In this investigation, the practical applicability of $0.25Mn^{4+}$ ions doped BLN nanophosphors was studied by using the powder dusting technique. The latent FPs were investigated on four different surfaces like stainless steel, aluminum foil, glass, and plastic, their photographed are observed under 365 nm illumination. FPs of synthesized phosphors are captured from the same candidate for all mentioned surfaces. Fig. 9.5 A–D shows the images of latent FPs under daylight while Fig. 9.5A' and B' under 365 nm illumination. The investigation shows that the collected FPs are useful for the identification of all three required levels. Its level 1, level 2, and level 3 define the ridge orientation field and pattern configuration, fingerprint ridge, and discriminatory information (width, shape, pores, curvature, etc.), respectively. Their entire results revealed potential applicability for latent fingerprint technologies.

G.R. Revannasiddappa et al. [10] investigated Dy^{3+}-activated $Ca_2MgSi_2O_7$ (CMS) nanopowders, which were synthesized by green combustion method using Aloe vera gel as fuel. Further, their structural, spectroscopic and advanced forensic characteristics were investigated. The synthesized phosphors have akermanite structure (tetragonal shape) with space group P421m (113). The analyzed value of band gap for synthesized phosphors lies in the range of 3.52–3.68 eV. SEM and Transmission electron microscope (TEM) images of $CMS:Dy^{3+}$ (5 mol%) show irregularly shaped particles with a particle size of about 50 nm. The spectroscopic features show strong emission bands around 480 ($^4F_{9/2} \rightarrow {}^6H_{15/2}$), 578 ($^4F_{9/2} \rightarrow {}^6H_{13/2}$), and 666 nm ($^4F_{9/2} \rightarrow {}^6H_{11/2}$) are monitored under 352 nm ($^6H_{15/2} \rightarrow {}^6P_{7/2}$) excitation wavelength. In addition, the synthesized phosphors probed to visualize the LFPs on the surface of

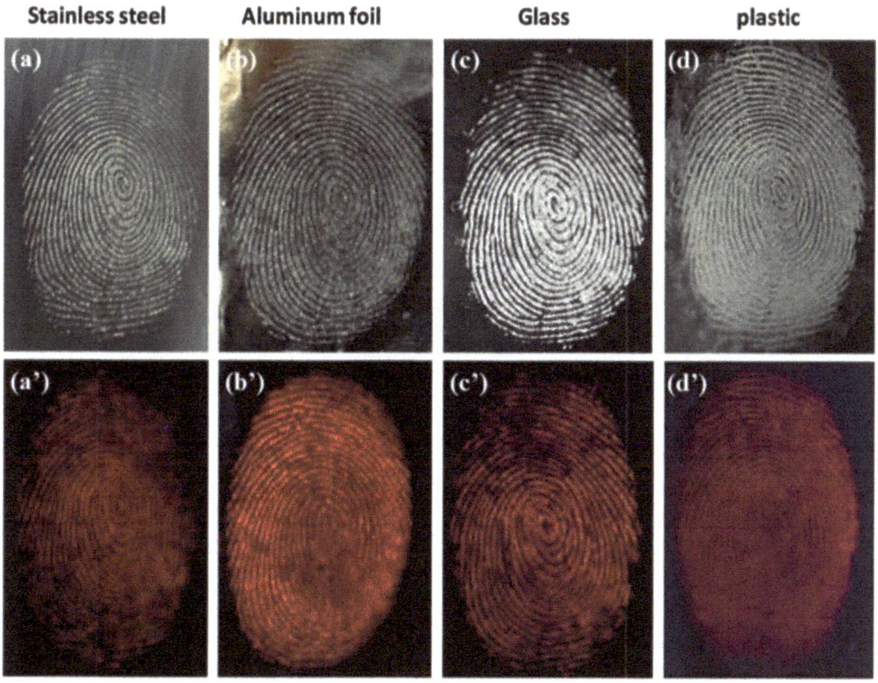

Figure 9.5 Images of latent FPs on four different surfaces like stainless steel, aluminum foil, glass, and plastic (A–D) under daylight and (A'–D') under 365 nm illumination. Reprinted with permission from Ref. [83], Copyright Elsevier.

the aluminum foil, as shown in Fig. 9.6. From the collected fingerprint images, clear ridge features can be easily observed in normal daylight without any difficulty. The investigated results of synthesized CMS:Dy^{3+} (5 mol%) phosphor revealed that the collected FPs are useful for the identification of all three ridge features.

Basavaraj et al. [84] reported rare-earth Tb- and Eu-doped cadmium silicate ($CdSiO_3$) nanophosphors, which offers new promise in LFPs and anticounterfeiting applications in surface-based science. In this investigation, nanophosphors are successfully synthesized via eco-friendly ultrasound-assisted sonochemical method. For this investigation, LFPs were deposited on various porous and nonporous surfaces, such as glass slides, aluminum foil, glass bottles, mobile screens, teacups, spatulas, currency notes, and magazine covers with various backgrounds. Fig. 9.7 is a schematic demonstration of visualization of LFPs by powder dusting method using $CdSiO_3$:Tb^{3+} (5 mol%) nanophosphors, reported by Basavaraj et al. [84]. The deposited all fingerprint samples exhibit clear ridge structures with good sensitivity and selectivity. In addition, $CdSiO_3$:Eu^{3+} (5 mol%) nanophosphors were checked for anticounterfeiting application. The authors used $CdSiO_3$:Eu^{3+} (5 mol %) NP ink for drawing different symbols with processing distilled water as shown in Fig. 9.8, and it was measured under visible light and UV light (254 nm). This investigation explores new possibilities for visualization of LFP and anticounterfeiting applications.

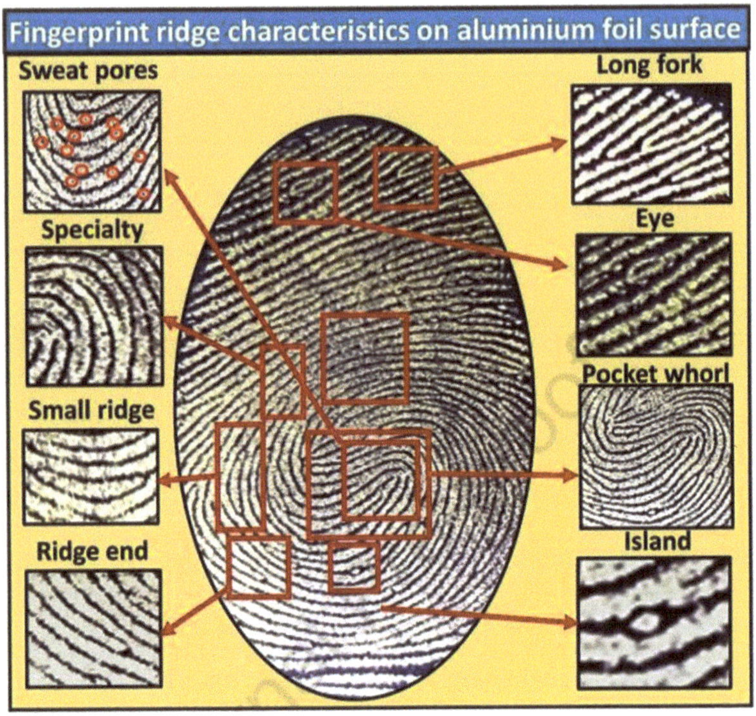

Figure 9.6 Visualization of the LFPs on aluminum foil surface by using the CMS:Dy^{3+} (5 mol%) phosphor.
Reprinted with permission from Ref. [10], Copyright Elsevier.

Navami et al. [85] investigated $CaZrO_3$:Dy^{3+} (1−11 mol%) nanophosphors, which are synthesized by a sonochemical method using bio template as a surfactant. Under 350 nm excitation, PL emission spectra exhibit three narrow bands around 481 ($^4F_{9/2} \rightarrow {}^6H_{15/2}$), 575 ($^4F_{9/2} \rightarrow {}^6H_{13/2}$), and 665 nm ($^4F_{9/2} \rightarrow {}^6H_{11/2}$). The maximum emission intensity for $CaZrO_3$:Dy^{3+} phosphors was found at 7 mol%, after that concentration quenching takes place. The distinctive features of the LFPs, such as the long thorn, ridge end, dot, hook, scar, and sweat pores, are detected. Fig. 9.9 shows collected fingerprints on different surfaces under normal light. Their distinctive characteristics suggest that the synthesized phosphors have potential suitability for advanced forensic applications.

The $BaTiO_3$:Eu^{3+} (5 mol%) nanophosphor was studied by Dhanalakshmi et al. [86] for the visualization of LFPs on multifunctional substrates and anticounterfeiting applications. In this investigation, Eu^{3+}-activated $BaTiO_3$ nonophosphors were successfully synthesized by the solution combustion method. Synthesized novel $BaTiO_3$:Eu^{3+} (5 mol%) nanophosphor exhibits emission bands around 592, 615, 654, and 702 nm, which are corresponding due to the $^5D_0 \rightarrow {}^7F_J$ (J = 0, 1, 2, and 3) transitions of Eu^{3+} ions. The visualization of LFPs using optimized $BaTiO_3$:Eu^{3+} (5 mol%) nanophosphor was studied by powder dusting method in various surfaces as porous, semiporous,

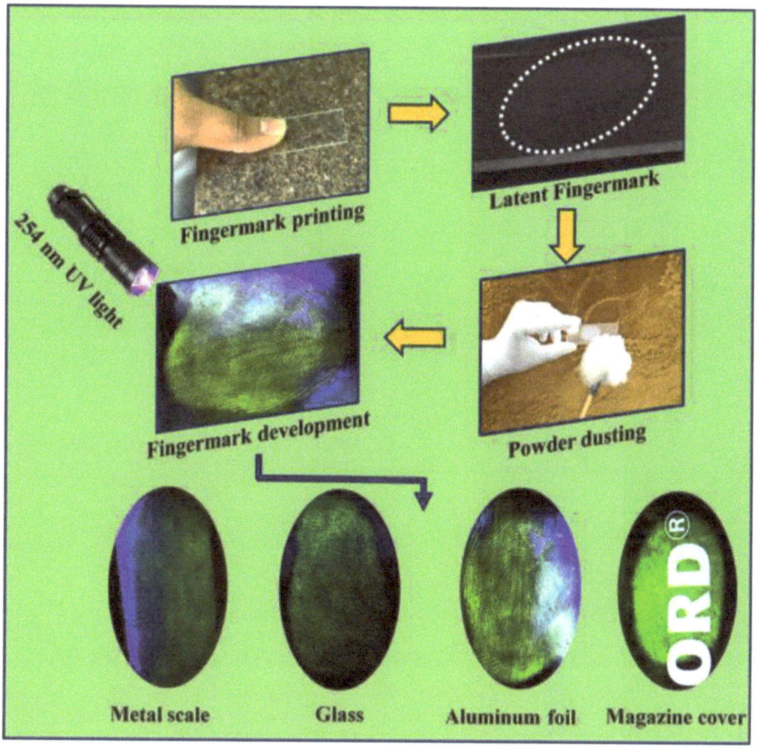

Figure 9.7 Schematic demonstration of visualization of LFPs by powder dusting method using $CdSiO_3:Tb^{3+}$ (5 mol%) nanophosphors.
Reprinted with permission from Ref. [84], Copyright Elsevier.

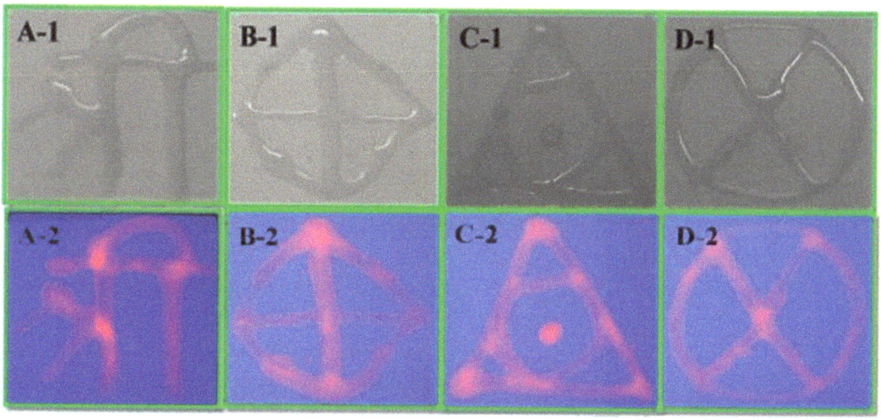

Figure 9.8 Anticounterfeiting labels painted with $CdSiO_3:Eu^{3+}$ (5 mol %) NP ink under visible light (A1–D1) and UV 254 nm light (A2–D2).
Reprinted with permission from Ref. [84], Copyright Elsevier.

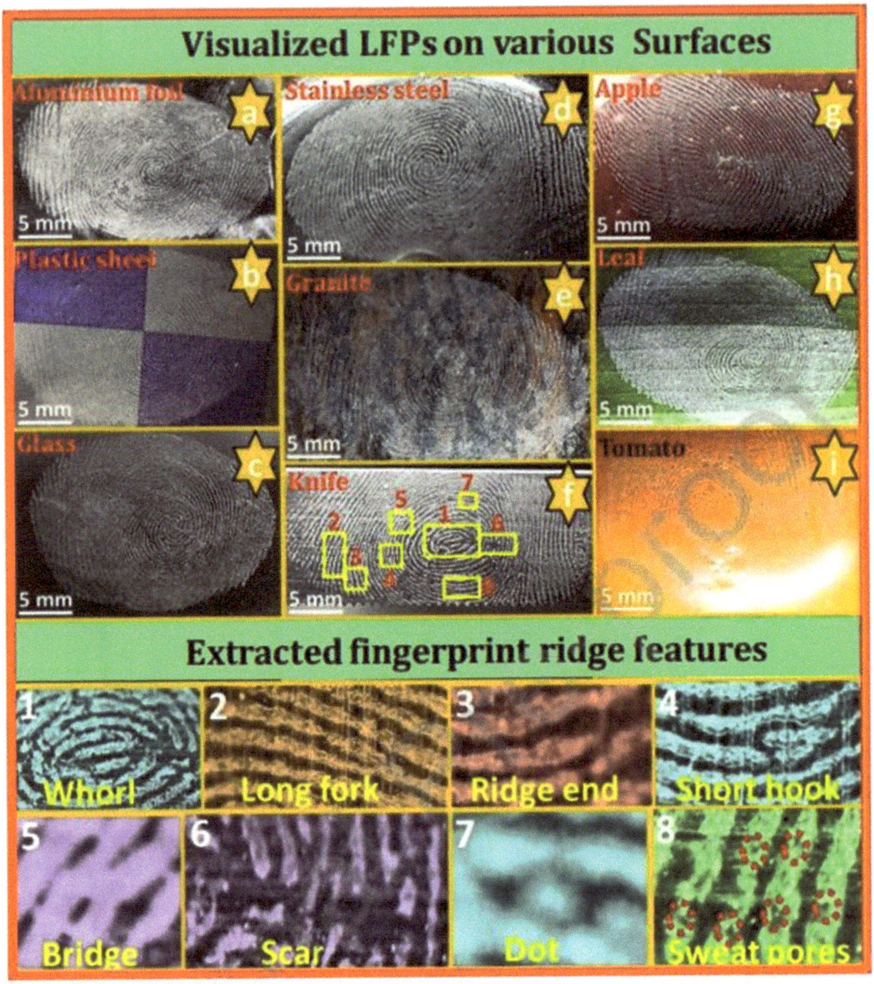

Figure 9.9 Visualized LFPs on porous (A−C), nonporous (D−F), semiporous surfaces (G−I), and extracted ridge features on knife surface (1−9) using $CaZrO_3:Dy^{3+}$ (7 mol%) nanophosphors under normal light.
Reprinted with permission from Ref. [85], Copyright Elsevier.

and nonporous surfaces. The visualized LFPs show various ridge details such as eye, ridge end, island, specialty, sweat pores, bifurcation, and scar with high sensitivity, selectively, contrast, and low background hindrance. LFPs were visualized on various porous magazine covers with different backgrounds, white paper aluminum foil, playing cards, and paper coffee cups with $BaTiO_3:Eu^{3+}$ (5 mol%) nanophosphor using the powder dusting method, as shown in Fig. 9.10. The results of this investigation have clearly demonstrated that the collected LFPs reveal unique and sharp ridge details.

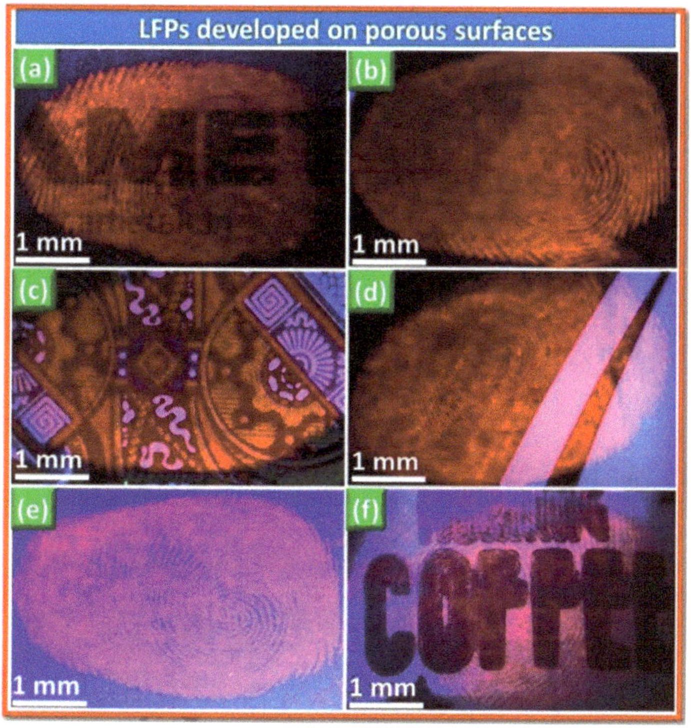

Figure 9.10 Visualized LFPs on various porous magazine covers with different backgrounds. Reprinted with permission from Ref. [86], Copyright Elsevier.

This investigation also demonstrated that the collected FPs are useful for the identification of all three required levels.

Feng Li et al. [87] reported red fluorescent $La_2(MoO_4)_3$:Eu^{3+} phosphors, which were synthesized by the hydrothermal method. The prepared phosphor exhibits a strong red emission band at 616 nm ($^5D_0 \rightarrow {}^7F_2$) under 269 nm excitation. When the Eu^{3+}-activated $La_2(MoO_4)_3$ phosphor was deposited on the glass, it was found that the ridge patterns under 254 nm excitation showed fine contrast and no interference from the substrate. Similarly, $La_2(MoO_4)_3$:Eu^{3+} phosphor was deposited on various single and multiple background color subtracts such as white ceramic tiles, aluminum alloys sheets, plastic plates, painted wood, black, red, green, and yellow-colored marbles, personal bank card, paper carton Chinese paper money, and coin. The exhaustive study and characteristics of $La_2(MoO_4)_3$:Eu^{3+} phosphor has opened a new avenue for forensic science in developing latent fingerprints.

$Gd_2Ti_2O_7$:Eu^{3+} nanophosphors synthesized by the citrate sol-gel method were investigated by Park et al. [88] for optical devices and fingerprint identification. Park et al. [88] investigated crystal structure, morphology, optical band gap, luminescent properties, Judd-Ofelt and fingerprint were analyzed. The synthesized phosphors have a cubic crystal structure with an irregular shape of particles. The synthesized

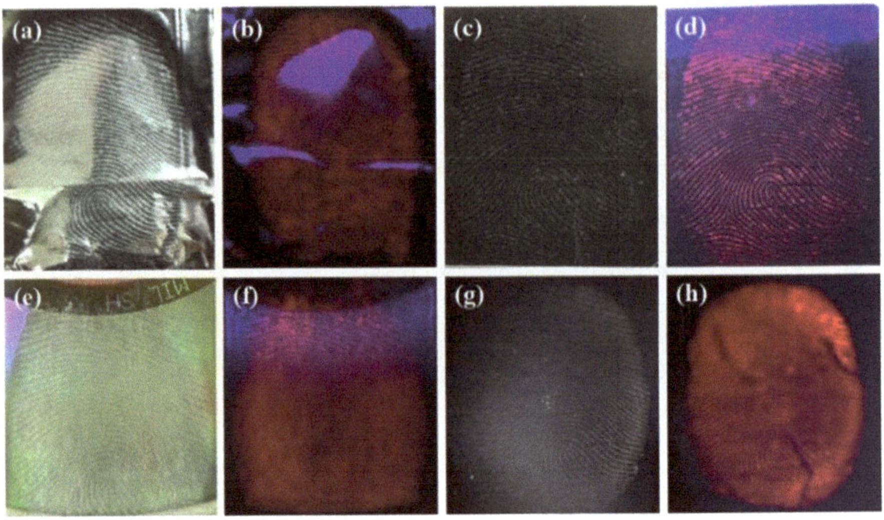

Figure 9.11 Latent finger print images with different surface such as (A, B) aluminum foil, (C, D) glass, (E, F) compact disk, and (G, H) stainless steel stained with $Gd_2Ti_2O_7:Eu^{3+}$ nanophosphors.
Reprinted with permission from Ref. [88], Copyright Elsevier.

phosphors also show excellent optical properties. The photoluminescence emission spectra show the highest emission intensity for the 0.09 mol concentration of Eu^{3+} ions. It also depicts a strong orange band (588 nm) with several other emissions under 393 nm excitation. In addition, synthesized $Gd_2Ti_2O_7:Eu^{3+}$ nanophosphors were deposited into the glass surface and monitored with UV light and without UV light. Furthermore, to confirm the practical applicability for the latent FPs with different surfaces of substrates such as aluminum foil, glass, compact disc, and stainless steel, $Gd_2Ti_2O_7:Eu^{3+}$ nanophosphors were deposited, as shown in Fig. 9.11. Fig. 9.11 shows the photographs of LFP grown by $Gd_2Ti_2O_7:0.09Eu^{3+}$ nanophosphor under a white lamp and UV lamp (365 nm). All developed fingerprint images are clearly evident for the detailed ridge with better contrast.

As discussed earlier, various rare-earth-activated phosphors have been investigated by many researchers and scholars. Some rare-earth-activated phosphors are briefly discussed earlier, and some more rare-earth-activated phosphors are listed in Table 9.2.

5. Future scope and improvements

The biggest problem with the use of rare-earth activated for dusting purposes is the cost of rare-earth metal ions. Generally, the price of host materials is relatively low compared to rare-earth ions. It is expected that it remains similar in the foreseeable future because high demand for these materials for manufacturing high-quality display devices. Most of the dusting of latent fingerprints is done in open areas and as other

Table 9.2 List of rare-earth-activated phosphors for forensic applications.

Ref. No	Compound name	Synthesis method	Particle size	Photoluminescence properties	LFPs image
[89]	$YAlO_3: Tb^{3+}$	Combustion route	27 nm	$\lambda_{exc} = 341$ nm $\lambda_{em} = 545$ nm	
[11]	$GdAlO_3: Tb^{3+}$	solution combustion method	10–100 nm	$\lambda_{exc} = 353, 378, 397$, and 491 nm $\lambda_{em} = 468, 491, 514, 543, 585$, and 620 nm	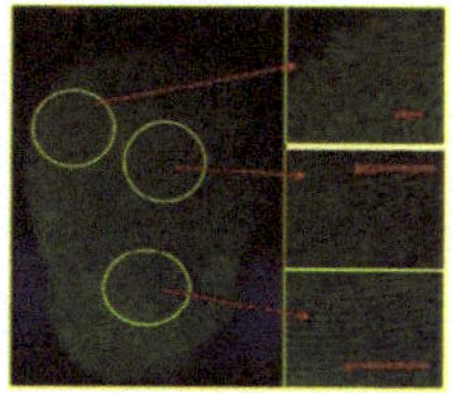
[90]	Li^+ ions doped $NaYF_4: Er^{3+}/Yb^{3+}$	Modified ball milling-assisted route	700 nm	$\lambda_{em} = 520–541$ nm and 545–565 nm (Green emission) + 650–685 nm (red emission)	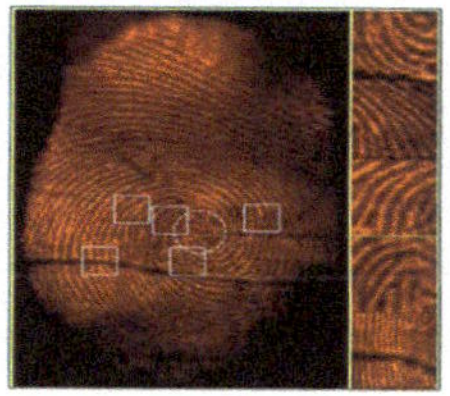
[91]	$SrMoO_4: Er^{3+}-Yb^{3+}$	Chemical coprecipitation route	200 nm	$\lambda_{em} = 523$ nm, 550 nm, 658 nm	

Continued

Table 9.2 List of rare-earth-activated phosphors for forensic applications.—cont'd

Ref. No	Compound name	Synthesis method	Particle size	Photoluminescence properties	LFPs image
[92]	$La_{10}W_{22}O_{81}$: Eu^{3+}	Hydrothermal assisted solid-state (HASS) method	400–500 nm	λ_{exc} = 361, 371, 380, 394, 415, 464, and 535 nm λ_{em} = 582, 590, 614, 649 and 698 nm	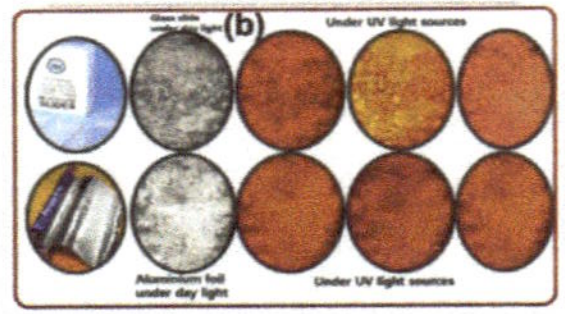
[93]	$Y_4Zr_3O_{12}$: Eu^{3+}	solvothermal reaction method	57 nm	λ_{exc} = 322, 361, 384, 393 and 416 nm λ_{em} = 575, 614, 628, 655 and 715 nm	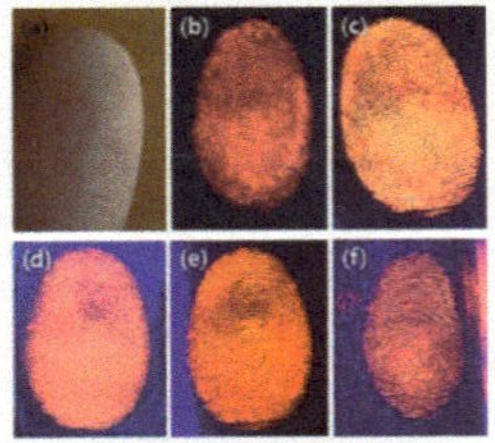
[94]	$ZnGa_2O_4$: Mn^{2+}	Microwave technique	5–7 nm	λ_{exc} = 244 and 294 nm λ_{em} = 503 nm	
[95]	$CaGdAlO_4$: Eu^{3+}	Solvothermal reaction method	submicrometers	λ_{exc} = 318, 362, 376, 382, 395, 416 and 465 nm λ_{em} = 540, 558, 590, 620, 656 and 701 nm	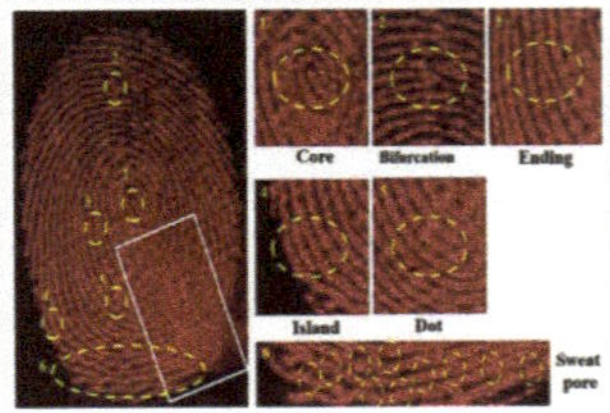

[96]	Eu^{3+}: $Y_2Ti_2O_7$/ SiO_2	sol−gel	18 nm	λ_{em} = 580 nm, 612 nm, 650 nm, 700 nm	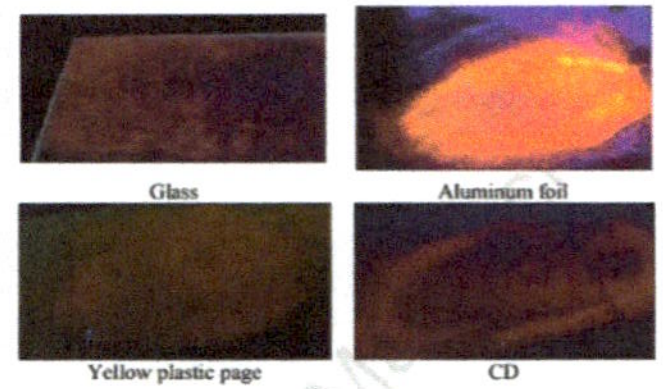
[97]	YVO_4: Bi^{3+}, Eu^{3+}	Micro-emulsion method	22 nm	λ_{em} = 539, 595, 616, 619, 652, and 699 nm	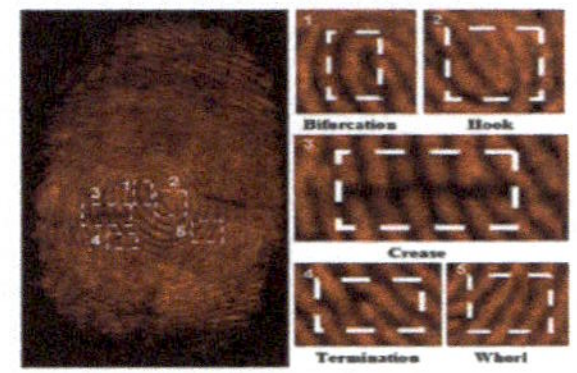
[98]	Ln^{3+}: $Y_2Zr_2O_7$/ SiO_2	sol-gel	9−40 nm	Characteristic emission band of rare earth ions	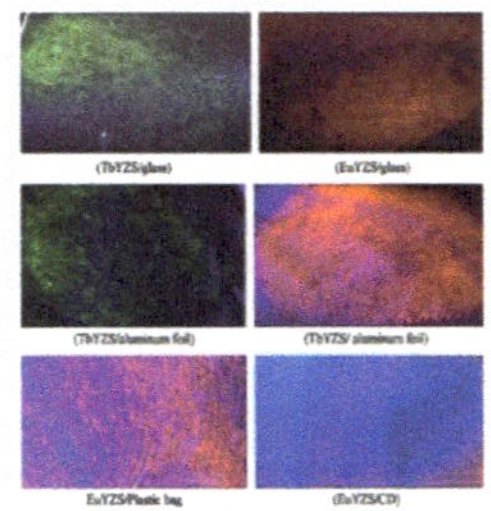

Continued

Table 9.2 List of rare-earth-activated phosphors for forensic applications.—cont'd

Ref. No	Compound name	Synthesis method	Particle size	Photoluminescence properties	LFPs image
[99]	$SrTiO_3:Pr^{3+}$	simple sonochemical route	27 nm	$\lambda_{em} = 491, 527, 607, 626, 647, 685$, and 737 nm	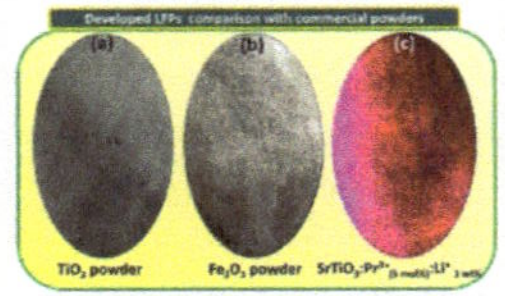
[100]	$Tm_2WO_6:Er^{3+}$	solid-state reaction method	1.2–1.5 μm	$\lambda_{exc} = 980$ nm $\lambda_{em} = 546$ nm, 657 nm	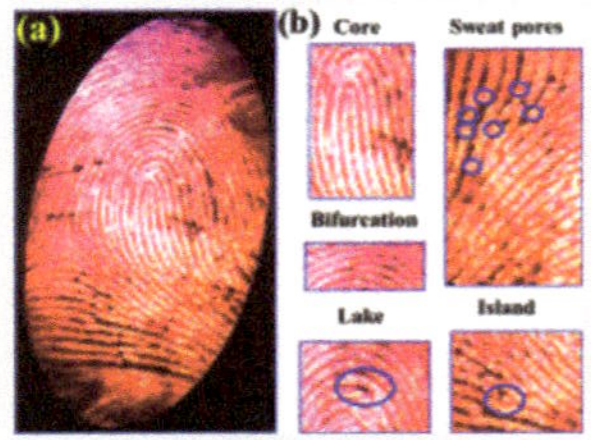
[101]	$Gd_2MoO_6:Eu^{3+}$	Coprecipitation	200 nm	Between 575 and 670 nm	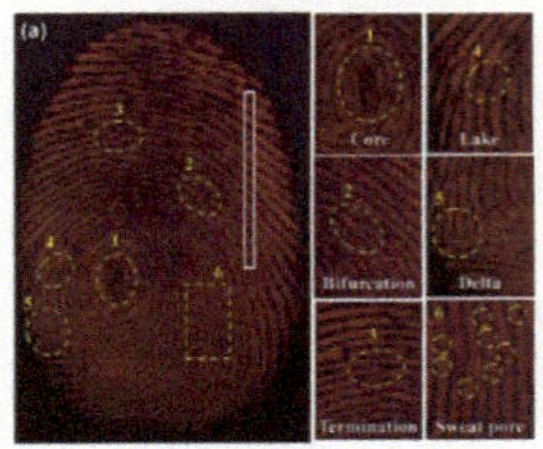

[102]	$NaYbF_4$: Ho^{3+}	Hydrothermal method	10–25 nm	λ_{em} = 540 nm, 640 nm	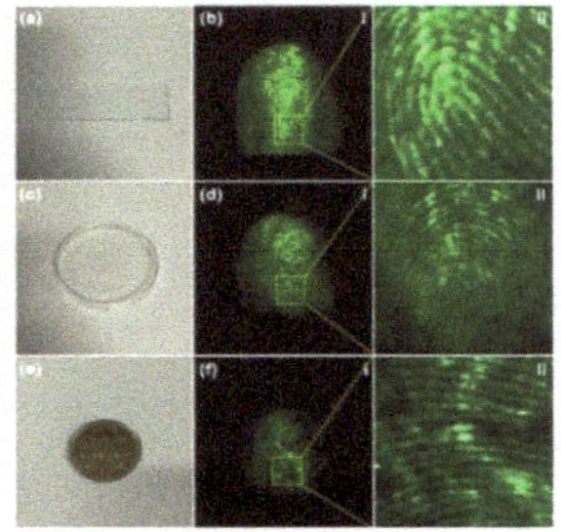
[2]	Mg_2TiO_4: Mn^{4+}	Solid-state reaction	submicrometers	λ_{exc} = 350 nm λ_{em} = 661 nm	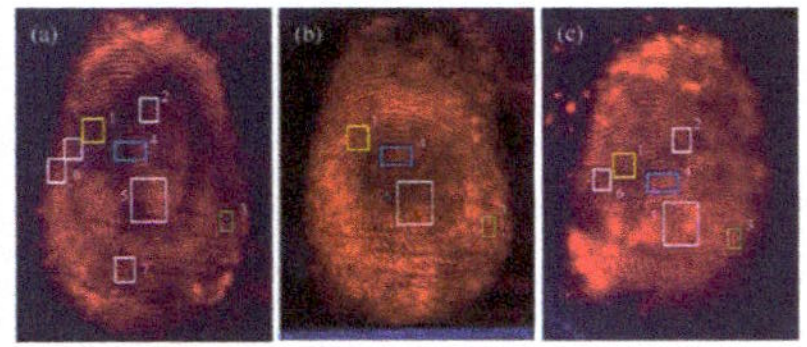
[103]	Sr_2MgMoO_6: Eu^{3+}	Solid-state reaction process	1–2 μm	λ_{exc} = 393 nm λ_{em} = 596, 616, 646, and 722 nm	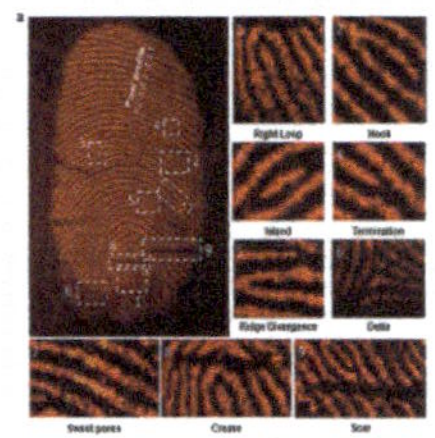

powders of comparable sizes are already in use the examiners should not be at any more risk, and indeed are already trained in the use of equivalent materials. We can explore fingerprints detection technology by using rare-earth-activated phosphors. The present study and the previous reports indicate that rare-earth-activated phosphors have great potential in the future for forensic applications, especially fingerprint detection. Studying a wide variety of surfaces, colors, and patterns will make it easier for the crime scene examiner to obtain the best evidence in the future. Simultaneously, the development of the simplest methods of phosphor manufacturing will open up new opportunities in the development of forensic applications. One of the most attractive features of rare-earth-activated phosphors is their narrow, line-like emission characteristics. From a forensic point of view, narrow emission peaks can be more effectively divided by utilizing the widest selection of photographic filters available to practitioners. Compared to other materials, rare-earth-activated phosphors provide better ridge details such as eye, ridge end, island, specialty, sweat pores, bifurcation, and scar with high sensitivity, selectively, contrast, and low background hindrance. Recent studies and previous literature revealed that rare-earth-activated phosphors can be used as the labeling agents for the visualization of LFPs and they can also play an important role in the physical identification of LFPs in the field of forensic science.

6. Conclusion

In this chapter, we have discussed rare-earth-activated phosphors in detail, and their forensic applications have been highlighted. In this chapter, we have focused on the discussion of the synthesis methods and results of phosphors reported earlier. Conventional methods of latent fingerprint detection are useful for the visualization and development of fingerprints on a wide range of substrates. However, they are nonspecific in nature, and more research is required to better understand the reaction mechanism involved in these reactions and to develop new, effective, and sensitive reagents for processing LFPs detection (aged and degraded) on difficult and unusual surfaces. Fingerprint developers use surfaces such as porous, nonporous, or semiporous to study fingerprints residues, allowing this biometric evidence to be observed and documented. No single technique is ideal for processing all kinds of surfaces, and new, sensitive, and efficient methods are being developed for enhancing LFPs on different types of surfaces. The study demonstrates that luminescent methods of visualization offer greater sensitivity than nonluminescent approaches, which means that they improve the likelihood of obtaining clear ridge details from fingerprint evidence. In this chapter, a number of rare-earth-activated phosphors are discussed, and as a result, it is concluded that rare-earth-activated phosphors have new possibilities for visualization of LFPs and anticounterfeiting applications.

References

[1] M. Ge, M. Abolhasani, C. Shi, X. Shen, Y. Zhu, X. Li, Z. Pang, Facile synthesis of a color-tunable microcrystal phosphor for anti-counterfeit applications, ACS Omega 5 (2020) 32420−32425, https://doi.org/10.1021/acsomega.0c04516.

[2] S.H. Yang, Y.C. Hung, P.C. Tseng, H.Y. Lee, Versatile deep-red $Mg_2TiO_4:Mn^{4+}$ phosphor for photoluminescence, thermometry, and latent fingerprint visualization, J. Alloys Compd. 801 (2019) 394−401, https://doi.org/10.1016/j.jallcom.2019.06.028.

[3] Y.R. Parauha, V. Chopra, S.J. Dhoble, Synthesis and luminescence properties of RE^{3+} ($RE = Eu^{3+}$, Dy^{3+}) activated $CaSr_2(PO_4)_2$ phosphors for lighting and dosimetric applications, Mater. Res. Bull. 131 (2020) 110971, https://doi.org/10.1016/j.materresbull.2020.110971.

[4] X. Huang, Q. Sun, B. Devakumar, Novel efficient deep-red-emitting $Ca_2LuTaO_6:Mn^{4+}$ double-perovskite phosphors for plant growth LEDs, J. Lumin. 222 (2020) 117177, https://doi.org/10.1016/j.jlumin.2020.117177.

[5] T. Thoř, K. Rubešová, V. Jakeš, J. Cajzl, L. Nádherný, D. Mikolášová, R. Kučerková, M. Nikl, Lanthanide-doped Lu_2O_3 phosphors and scintillators with green-to-red emission, J. Lumin. 215 (2019), https://doi.org/10.1016/j.jlumin.2019.116647.

[6] H. Yang, W. Zhao, X. Lin, Z. Liao, Z. Nie, L. Luo, W. Zhang, Z. Hu, J. Zhong, Hundreds of times of photo-stimulation with low energy light as a new reused bio-imaging phosphor from Cr^{3+}, Si^{4+}-doped $Y_3Ga_5O_{12}$, J. Lumin. 219 (2020) 116871, https://doi.org/10.1016/j.jlumin.2019.116871.

[7] A. Verma, S.K. Sharma, Down-conversion from Er^{3+}-Yb^{3+} codoped $CaMoO_4$ phosphor: a spectral conversion to improve solar cell efficiency, Ceram. Int. 43 (2017) 8879−8885, https://doi.org/10.1016/j.ceramint.2017.04.023.

[8] V.V. Shinde, R.G. Kunghatkar, S.J. Dhoble, UVB-emitting Gd^{3+}-activated M_2O_2S (where M = La, Y) for phototherapy lamp phosphors, Luminescence 30 (2015) 1257−1262, https://doi.org/10.1002/bio.2889.

[9] B.S. Rohini, H. Nagabhushana, G.P. Darshan, R.B. Basavaraj, S.C. Sharma, P. Amudha, M. Rahul, B. Daruka Prasad, Multifunctional applications of self - assembled 3D CeO_2: Cr^{3+} hierarchical structures synthesized via ultrasound assisted sonochemical route, J. Alloys Compd. 724 (2017) 897−909, https://doi.org/10.1016/j.jallcom.2017.07.054.

[10] G.R. Revannasiddappa, R.B. Basavaraj, M.S. Rudresha, G. Nagaraju, S. Kumar, N. Sasidhar, White-light emitting $Ca_2MgSi_2O_7:Dy^{3+}$ nanopowders: structural, spectroscopic investigations and advanced forensic applications, Vacuum 184 (2021) 109940, https://doi.org/10.1016/j.vacuum.2020.109940.

[11] P.K. Jisha, S.C. Prashantha, H. Nagabhushana, Luminescent properties of Tb doped gadolinium aluminate nanophosphors for display and forensic applications, J. Sci. Adv. Mater. Dev. 2 (2017) 437−444, https://doi.org/10.1016/j.jsamd.2017.10.001.

[12] W. Wang, X. Lei, Z. Ye, N. Zhao, H. Yang, The luminescent properties and latent fingerprint identification application of AlN:Ce, Tb phosphors, J. Alloys Compd. 705 (2017) 253−261, https://doi.org/10.1016/j.jallcom.2017.02.121.

[13] S.J. Park, J.Y. Park, H.K. Yang, Luminescence of a novel cyan emitting $Sr_{10}(PO_4)_6O$: Ce^{3+} phosphor for visualization of latent fingerprints and anti-counterfeiting applications, Sensor. Actuator. B Chem. 262 (2018) 542−554, https://doi.org/10.1016/j.snb.2018.02.053.

[14] D. Chávez, C.R. Garcia, J. Oliva, L.A. Diaz-Torres, A review of phosphorescent and fluorescent phosphors for fingerprint detection, Ceram. Int. 47 (2021) 10−41, https://doi.org/10.1016/j.ceramint.2020.08.259.

[15] Y. Liu, K. Ai, L. Lu, Designing lanthanide-doped nanocrystals with both up- and down-conversion luminescence for anti-counterfeiting, Nanoscale 3 (2011) 4804–4810, https://doi.org/10.1039/c1nr10752f.

[16] M. Zhao, P. Xi, X. Gu, Z. Li, M. Gao, B. Cheng, Synthesis, characterization and fluorescence properties of a novel rare earth complex for anti-counterfeiting material, J. Rare Earths 28 (2010) 75–78, https://doi.org/10.1016/S1002-0721(10)60359-6.

[17] S. Han, H.J. Bae, J. Kim, S. Shin, S.E. Choi, S.H. Lee, S. Kwon, W. Park, Lithographically encoded polymer microtaggant using high-capacity and error-correctable QR Code for anti-counterfeiting of drugs, Adv. Mater. 24 (2012) 5924–5929, https://doi.org/10.1002/adma.201201486.

[18] Z. Liu, L. Zhao, W. Chen, X. Fan, X. Yang, S. Tian, X. Yu, J. Qiu, X. Xu, Multiple anti-counterfeiting realized in $NaBaScSi_2O_7$ with a single activator of Eu^{2+}, J. Mater. Chem. C. 6 (2018) 11137–11143, https://doi.org/10.1039/c8tc04018d.

[19] K.V.R. Murthy, H.S. Virk, Luminescence phenomena: an introduction, Defect Diffusion Forum 347 (2014) 1–34, https://doi.org/10.4028/www.scientific.net/DDF.347.1.

[20] S. Verma, K. Verma, D. Kumar, B. Chaudhary, S. Som, V. Sharma, V. Kumar, H.C. Swart, Recent advances in rare earth doped alkali-alkaline earth borates for solid state lighting applications, Phys. B Condens. Matter 535 (2018) 106–113, https://doi.org/10.1016/j.physb.2017.06.073.

[21] H. Lian, Z. Hou, M. Shang, D. Geng, Y. Zhang, J. Lin, Rare earth ions doped phosphors for improving efficiencies of solar cells, Energy 57 (2013) 270–283, https://doi.org/10.1016/j.energy.2013.05.019.

[22] Y.R. Parauha, S.J. Dhoble, Thermoluminescence study and evaluation of trapping parameter of rare earth activated $Ca_3Al_2O_6$: RE (RE = Eu^{2+},Ce^{3+}) phosphors, J. Mol. Struct. 1211 (2020) 127993, https://doi.org/10.1016/j.molstruc.2020.127993.

[23] K. Soga, K. Tokuzen, K. Tsuji, T. Yamano, N. Venkatachalam, H. Hyodo, H. Kishimoto, Application of ceramic phosphors for near infrared biomedical imaging technologies, Opt. Components Mater. VII. 7598 (2010) 759807, https://doi.org/10.1117/12.841516.

[24] P. Psuja, D. Hreniak, W. Strek, Rare-earth doped nanocrystalline phosphors for field emission displays, J. Nanomater. 2007 (2007), https://doi.org/10.1155/2007/81350.

[25] X. Zhou, L. Chen, S. Jiang, G. Xiang, L. Li, X. Tang, X. Luo, Y. Pang, Eu^{3+} activated $LiSrVO_4$ phosphors: emission color tuning and potential application in temperature sensing, Dyes Pigments 151 (2018) 219–226, https://doi.org/10.1016/j.dyepig.2017.12.059.

[26] C.M. Mehare, Y.R. Parauha, N.S. Dhoble, C. Ghanty, S.J. Dhoble, Synthesis of novel Eu^{2+} activated $K_3Ca_2(SO_4)_3F$ down-conversion phosphor for near UV excited white light emitting diode, J. Mol. Struct. 1212 (2020) 127957, https://doi.org/10.1016/j.molstruc.2020.127957.

[27] A.M. Bhake, Y.R. Parauha, S.J. Dhoble, Synthesis and photoluminescence study of Ce^{3+} ion-activated $Na_2ZnP_2O_7$ and $Na_4P_2O_7$ pyrophosphate phosphors, J. Mater. Sci. Mater. Electron. 31 (2020) 548–559, https://doi.org/10.1007/s10854-019-02559-4.

[28] G.B. Nair, H.C. Swart, S.J. Dhoble, A review on the advancements in phosphor-converted light emitting diodes (pc-LEDs): phosphor synthesis, device fabrication and characterization, Prog. Mater. Sci. 109 (2020) 100622, https://doi.org/10.1016/j.pmatsci.2019.100622.

[29] Y.R. Parauha, S.J. Dhoble, Photoluminescence and electron-vibrational interaction in 5d state of Eu^{2+} ion in $Ca_3Al_2O_6$ down-conversion phosphor, Opt Laser. Technol. 142 (2021) 107191, https://doi.org/10.1016/j.optlastec.2021.107191.

[30] A. Varma, A.S. Mukasyan, Combustion synthesis of advanced materials: fundamentals and applications, Kor. J. Chem. Eng. 21 (2004) 527–536, https://doi.org/10.1007/BF02705444.

[31] Z. Cao, C. Zuo, In situ synthesis of chromium carbide nanocomposites from solution combustion synthesis precursors, J. Mol. Struct. 1175 (2019) 496–503, https://doi.org/10.1016/j.molstruc.2018.08.008.

[32] M. Qin, H. Wu, Z. Cao, D. Zhang, B. Jia, X. Qu, A novel method to synthesize vanadium nitride nanopowders by ammonia reduction from combustion precursors, J. Alloys Compd. 772 (2019) 808–813, https://doi.org/10.1016/j.jallcom.2018.09.112.

[33] H. Van Tuyen, P. Van Do, V.X. Quang, N.T. Thai An, L. Van Khoa Bao, N. Tran, Effect of B_2O_3 content on Eu^{3+} –Surrounding in the system SrO–x.B_2O_3 containing Eu^{3+} ions prepared by combustion method, Phys. B Condens. Matter 555 (2019) 36–40, https://doi.org/10.1016/j.physb.2018.11.047.

[34] G.R. Durand, N. Hakmeh, V. Dorcet, V. Demange, F. Cheviré, O. Merdrignac-Conanec, New insights in structural characterization of transparent ZnS ceramics hot-pressed from nanocrystalline powders synthesized by combustion method, J. Eur. Ceram. Soc. 39 (2019) 3094–3102, https://doi.org/10.1016/j.jeurceramsoc.2019.03.033.

[35] S. Kumar, S. Rajawat, R. Purohit, M.M. Malik, Optical studies of MWCNT doped boron carbon oxynitride (BCNO) nano composite phosphor material, Optik 176 (2019) 617–625, https://doi.org/10.1016/j.ijleo.2018.09.030.

[36] Y.R. Parauha, V. Sahu, S.J. Dhoble, Prospective of combustion method for preparation of nanomaterials: a challenge, Mater. Sci. Eng. B Solid-State Mater. Adv. Technol. 267 (2021) 115054, https://doi.org/10.1016/j.mseb.2021.115054.

[37] A. Varma, J.-P. Lebrat, Combustion synthesis of advanced materials, Chem. Eng. Sci. 47 (1992) 2179–2194.

[38] G. Liu, J. Li, K. Chen, Combustion synthesis of refractory and hard materials: a review, Int. J. Refract. Metals Hard Mater. 39 (2013) 90–102, https://doi.org/10.1016/j.ijrmhm.2012.09.002.

[39] K.C. Patil, S.T. Aruna, S. Ekambaram, Combustion synthesis, Curr. Opin. Solid State Mater. Sci. 2 (1997) 158–165, https://doi.org/10.1016/s1359-0286(97)80060-5.

[40] S.T. Aruna, A.S. Mukasyan, Combustion synthesis and nanomaterials, Curr. Opin. Solid State Mater. Sci. 12 (2008) 44–50, https://doi.org/10.1016/j.cossms.2008.12.002.

[41] M. Lackner, Combustion Synthesis: Novel Routes to Novel Materials, Bentham Science, 2010, https://doi.org/10.2174/97816080515571100101. ISBN: 978-1-60805-155-7.

[42] Y.-H. Kiang, Combustion Fundamentals and Energy Systems, Fuel Property Estimation and Combustion Process Characterization, 2018, pp. 313–362, https://doi.org/10.1016/b978-0-12-813473-3.00009-x.

[43] K.C. Patil, M.S. Hegde, T. Rattan, S.T. Aruna, Combustion Synthesis, Properties and Applications, Chemistry of Nanocrystalline Oxide Materials, 2008, https://doi.org/10.1142/6754.

[44] F. Deganello, A.K. Tyagi, Solution combustion synthesis, energy and environment: best parameters for better materials, Prog. Cryst. Growth Char. Mater. 64 (2018) 23–61, https://doi.org/10.1016/j.pcrysgrow.2018.03.001.

[45] A. Cincotti, R. Licheri, A.M. Locci, R. Orrù, G. Cao, A review on combustion synthesis of novel materials: recent experimental and modeling results, J. Chem. Technol. Biotechnol. 78 (2003) 122–127, https://doi.org/10.1002/jctb.757.

[46] R. Pampuch, L. Stobierski, Solid combustion synthesis of refractory carbides: (A review), Ceram. Int. 17 (1991) 69–77, https://doi.org/10.1016/0272-8842(91)90013-P.

[47] K.C. Patil, S.T. Aruna, T. Mimani, Combustion synthesis: an update, Curr. Opin. Solid State Mater. Sci. 6 (2002) 507–512, https://doi.org/10.1016/S1359-0286(02)00123-7.

[48] M. Martinez Pacheco, R.H.B. Bouma, L. Katgerman, Combustion synthesis of TiB_2-based cermets: modeling and experimental results, Appl. Phys. Mater. Sci. Process 90 (2008) 159–163, https://doi.org/10.1007/s00339-007-4245-9.

[49] S.M. Rafiaei, M. Shokouhimehr, Effect of fuels on nanostructure and luminescence properties of combustion synthesized $MgAl_2O_4:Eu^{3+}$ phosphors, J. Mol. Struct. 1193 (2019) 274–279, https://doi.org/10.1016/j.molstruc.2019.05.057.

[50] A. Hooda, S.P. Khatkar, A. Khatkar, R.K. Malik, M. Kumar, S. Devi, V.B. Taxak, Reddish-orange light emission via combustion synthesized $Ba_3Y_4O_9$: Sm^{3+} nanocrystalline phosphor upon near ultraviolet excitation, J. Lumin. 217 (2020) 116806, https://doi.org/10.1016/j.jlumin.2019.116806.

[51] A. Hooda, A. Khatkar, M. Kumar, R.K. Malik, S. Devi, S.P. Khatkar, V.B. Taxak, Crystal configuration and photoluminescent aspects of red-emitting combustion synthesized novel $BaYZn_3AlO_7$: Eu^{3+} nanophosphor, J. Alloys Compd. 823 (2020), https://doi.org/10.1016/j.jallcom.2020.153641.

[52] A. Verma, A. Verma, G.V. Bramhe, I.P. Sahu, Optical studies of the $Ba_{1-x}MgAl_{10}O_{17}$: Eu_x phosphors synthesis by combustion route, J. Alloys Compd. 769 (2018) 831–842, https://doi.org/10.1016/j.jallcom.2018.07.371.

[53] M. Oglakci, M. Topaksu, N. Can, Thermoluminescence glow curves of beta irradiated $NaBaBO_3:Ce^{3+}$ phosphor synthesized by combustion method, Sensors Actuators, A Phys. 315 (2020) 112299, https://doi.org/10.1016/j.sna.2020.112299.

[54] B. Wang, H. Lin, J. Xu, H. Chen, Y. Wang, $CaMg_2Al_{16}O_{27}:Mn^{4+}$-based red phosphor: a potential color converter for high-powered warm W-led, ACS Appl. Mater. Interfaces 6 (2014) 22905–22913, https://doi.org/10.1021/am507316b.

[55] N. Bajpai, A. Tiwari, S.A. Khan, R.S. Kher, N. Bramhe, S.J. Dhoble, Effects of rare earth ions (Tb, Ce, Eu, Dy) on the thermoluminescence characteristics of sol-gel derived and γ-irradiated SiO_2 nanoparticles, Luminescence 29 (2014) 669–673, https://doi.org/10.1002/bio.2604.

[56] T. Peng, L. Huajun, H. Yang, C. Yan, Synthesis of $SrAl_2O_4$:Eu, Dy phosphor nanometer powders by sol-gel processes and its optical properties, Mater. Chem. Phys. 85 (2004) 68–72, https://doi.org/10.1016/j.matchemphys.2003.12.001.

[57] M. Song, L. Wang, Y. Feng, H. Wang, X. Wang, D. Li, Preparation of a novel red $KBaGd(MoO_4)_3:Eu^{3+}$ phosphor by sol-gel method and its luminescent properties, Opt. Mater. 84 (2018) 284–291, https://doi.org/10.1016/j.optmat.2018.07.019.

[58] H. Kaur, M. Jayasimhadri, M.K. Sahu, P.K. Rao, N.S. Reddy, Synthesis of orange emitting Sm^{3+} doped sodium calcium silicate phosphor by sol-gel method for photonic device applications, Ceram. Int. 46 (2020) 26434–26439, https://doi.org/10.1016/j.ceramint.2020.04.224.

[59] C.S. Lim, A.S. Aleksandrovsky, V.V. Atuchin, M.S. Molokeev, A.S. Oreshonkov, Microwave sol-gel synthesis, microstructural and spectroscopic properties of scheelite-type ternary molybdate upconversion phosphor $NaPbLa(MoO_4)_3:Er^{3+}/Yb^{3+}$, J. Alloys Compd. 826 (2020) 152095, https://doi.org/10.1016/j.jallcom.2019.152095.

[60] Y. Zhu, C. Tong, R. Dai, C. Xu, L. Yang, Y. Li, Luminescence properties of $Na_2Ca_2Si_3O_9$: Eu^{3+} phosphors via a sol-gel method, Mater. Lett. 213 (2018) 245−248, https://doi.org/10.1016/j.matlet.2017.11.082.
[61] M. Song, Y. Liu, Y. Liu, L. Wang, N. Zhang, X. Wang, Z. Huang, C. Ji, Sol-gel synthesis and luminescent properties of a novel $KBaY(MoO_4)_3$:Dy^{3+} phosphor for white light emission, J. Lumin. 211 (2019) 218−226, https://doi.org/10.1016/j.jlumin.2019.03.052.
[62] D. Meroni, L. Porati, F. Demartin, D. Poelman, Sol-gel synthesis of $CaTiO_3$:Pr^{3+} red phosphors: tailoring the synthetic parameters for luminescent and afterglow applications, ACS Omega 2 (2017) 4972−4981, https://doi.org/10.1021/acsomega.7b00761.
[63] V. Singh, V. Natarajan, N. Singh, A.K. Srivastava, Y.W. Kwon, G. Lakshminarayana, $CaLa_2ZnO_5$:Gd^{3+} phosphor prepared by sol-gel method: photoluminescence and electron spin resonance properties, Optik 212 (2020) 164247, https://doi.org/10.1016/j.ijleo.2020.164247.
[64] A.G. Bispo- Jr., S.A.M. Lima, S. Lanfredi, F.R. Praxedes, A.M. Pires, Tunable blue-green emission and energy transfer properties in Ba_2SiO_4:Tb^{3+} obtained from sol-gel method, J. Lumin. 214 (2019) 116604, https://doi.org/10.1016/j.jlumin.2019.116604.
[65] G. Xu, X. Zhang, W. He, H. Liu, H. Li, R.I. Boughton, Preparation of highly dispersed YAG nano-sized powder by co-precipitation method, Mater. Lett. 60 (2006) 962−965, https://doi.org/10.1016/j.matlet.2005.10.052.
[66] A.B. Gaikwad, S.C. Navale, V. Samuel, A.V. Murugan, V. Ravi, A co-precipitation technique to prepare $BiNbO_4$, $MgTiO_3$ and $Mg_4Ta_2O_9$ powders, Mater. Res. Bull. 41 (2006) 347−353, https://doi.org/10.1016/j.materresbull.2005.08.010.
[67] T. Jansen, F. Baur, T. Jüstel, Red emitting K_2NbF_7:Mn^{4+} and K_2TaF_7:Mn^{4+} for warm-white LED applications, J. Lumin. 192 (2017) 644−652, https://doi.org/10.1016/j.jlumin.2017.07.061.
[68] N.P. Singh, N.R. Singh, Y.R. Devi, B. Singh Sh, T.D. Singh, N.R. Singh, N.M. Singh, Effects of annealing temperature on structural and luminescence properties of $CdMoO_4$:Dy^{3+} phosphor synthesized at room temperature by co-precipitation method, Solid State Sci. 102 (2020) 106172, https://doi.org/10.1016/j.solidstatesciences.2020.106172.
[69] V. Singh, C.B. Annapurna Devi, A.S. Rao, J.L. Rao, Narrowband ultraviolet B emitting Gd^{3+} activated $NaSrPO_4$ phosphor prepared by co-precipitation method, Optik 208 (2020) 163632, https://doi.org/10.1016/j.ijleo.2019.163632.
[70] Y. Yuan, J. Wang, J. Wang, L. Jing, Y. Zhu, X. Li, X. Song, M. Ge, Structural characterization and optical properties of long-lasting $CaAl_2O_4$:Eu^{2+},Nd^{3+} phosphors synthesized by microwave-assisted chemical co-precipitation, J. Rare Earths 35 (2017) 652−657, https://doi.org/10.1016/S1002-0721(17)60959-1.
[71] S.H. Yang, H.Y. Lee, P.C. Tseng, M.H. Lee, Photoelectric properties of $Sr_2MgSi_2O_7$:Eu^{2+} phosphors produced by co-precipitation method, J. Lumin. 231 (2021) 117787, https://doi.org/10.1016/j.jlumin.2020.117787.
[72] S.S. Liu, D.C. Zhu, J.S. Wang, C. Zhao, T. Han, The luminescence properties of $Sr_{2-1.5x-1.5y}P_2O_7$:xDy^{3+},yCe^{3+} phosphor for near-UV-based white LEDs synthesized by a chemical co-precipitation method, Luminescence 32 (2017) 1582−1592, https://doi.org/10.1002/bio.3363.
[73] V. Singh, Y.R. Parauha, S.J. Dhoble, V.K. Kummara, N. Ravi, Orange light emission from co-precipitation derived $CaZr_4(PO_4)_6$ doped with Sm^{3+} phosphor, Optik (2021) 167229, https://doi.org/10.1016/j.ijleo.2021.167229.
[74] C. Yue, Y. Pu, D. Zhu, Q. Yan, A novel green-emitting $SrLaAlO_4$:Er^{3+} phosphor synthesized by co-precipitation method for w-LEDs and optical thermometry, J. Mater. Sci. Mater. Electron. 32 (2021) 4228−4238, https://doi.org/10.1007/s10854-020-05167-9.

[75] S.M. Mohun, M. Wilkinson, The eye of the caecilian Rhinatrema bivittatum (Amphibia: Gymnophiona: Rhinatrematidae), Acta Zool 96 (2015) 147−153, https://doi.org/10.1111/azo.12061.
[76] H. Cheng, Y. Song, G. Liu, D. Li, X. Dong, J. Wang, W. Yu, Hydrothermal synthesis of narrow-band red emitting $K_2NaAlF_6:Mn^{4+}$ phosphor for warm-white LED applications, RSC Adv. 7 (2017) 45834−45842, https://doi.org/10.1039/c7ra09671b.
[77] Y. Yang, H. Pan, L. Guan, D. Wang, J. Zhao, J. Yang, Z. Yang, X. Li, Electronic structure and luminescent properties of $Sr_3Al_2O_6:Sm^{3+}$ orange phosphor prepared by hydrothermal method, J. Mater. Res. Technol. 9 (2020) 3847−3855, https://doi.org/10.1016/j.jmrt.2020.02.011.
[78] J. Liao, L. Nie, Q. Wang, S. Liu, J. Fu, H.R. Wen, Microwave hydrothermal method and photoluminescence properties of $Gd_2Sn_2O_7$: Eu^{3+} reddish orange phosphors, J. Lumin. 183 (2017) 377−382, https://doi.org/10.1016/j.jlumin.2016.11.076.
[79] J. Yang, H. Xiong, J. Dong, C. Yang, S. Gan, L. Zou, Facile hydrothermal synthesis and luminescent properties of Sm^{3+}/Eu^{3+} codoped $GdPO_4$ phosphors, J. Phys. Chem. Solid. 111 (2017) 355−363, https://doi.org/10.1016/j.jpcs.2017.08.031.
[80] X. Li, H. Xu, X. Xia, F. Xie, S. Zhong, D. Xu, One-step synthesis of $Sc_2W_3O_{12}:Eu^{3+}$ phosphors with tunable luminescence for WLED, Ceram. Int. 45 (2019) 10461−10467, https://doi.org/10.1016/j.ceramint.2019.02.107.
[81] X. Yu, D. Yuan, X. Mi, Hydrothermal synthesis and luminescent properties of $Ca_3(PO_4)_2$: Dy^{3+} white-emitting phosphors, J. Alloys Compd. 857 (2021) 157585, https://doi.org/10.1016/j.jallcom.2020.157585.
[82] W.L. Lian, P. Liang, Z.H. Liu, Controllable hydrothermal synthesis and morphology evolution of $Zn_4B_6O_{13}$:Tb/Eu phosphors with tunable luminescent properties, Adv. Powder Technol. 31 (2020) 1633−1642, https://doi.org/10.1016/j.apt.2020.02.004.
[83] E. Pavitra, G.S.R. Raju, J.Y. Park, S.K. Hussain, N.R. Chodankar, G.M. Rao, Y.K. Han, Y.S. Huh, An efficient far-red emitting $Ba_2LaNbO_6:Mn^{4+}$ nanophosphor for forensic latent fingerprint detection and horticulture lighting applications, Ceram. Int. 46 (2020) 9802−9809, https://doi.org/10.1016/j.ceramint.2019.12.253.
[84] R.B. Basavaraj, H. Nagabhushana, G.P. Darshan, B. Daruka Prasad, M. Rahul, S.C. Sharma, R. Sudaramani, K.V. Archana, Red and green emitting CTAB assisted $CdSiO_3:Tb^{3+}/Eu^{3+}$ nanopowders as fluorescent labeling agents used in forensic and display applications, Dye, Pigment 147 (2017) 364−377, https://doi.org/10.1016/j.dyepig.2017.08.011.
[85] D. Navami, G.P. Darshan, R.B. Basavaraj, S.C. Sharma, D. Kavyashree, K.N. Venkatachalaiah, H. Nagabhushana, Shape controllable ultrasound assisted fabrication of $CaZrO_3:Dy^{3+}$ hierarchical structures for display, dosimetry and advanced forensic applications, J. Photochem. Photobiol. A Chem. 389 (2020) 112248, https://doi.org/10.1016/j.jphotochem.2019.112248.
[86] M. Dhanalakshmi, R.B. Basavaraj, G.P. Darshan, S.C. Sharma, H. Nagabhushana, Pivotal role of fluxes in $BaTiO_3:Eu^{3+}$ nano probes for visualization of latent fingerprints on multifaceted substrates and anti-counterfeiting applications, Microchem. J. 145 (2019) 226−234, https://doi.org/10.1016/j.microc.2018.10.020.
[87] F. Li, S. Liu, R. Qi, H. Li, T. Cui, Effective visualization of latent fingerprints with red fluorescent $La_2(MoO_4)_3:Eu^{3+}$ microcrystals, J. Alloys Compd. 727 (2017) 919−924, https://doi.org/10.1016/j.jallcom.2017.08.182.

[88] S.J. Park, J.Y. Kim, J.H. Yim, N.Y. Kim, C.H. Lee, S.J. Yang, H.K. Yang, The effective fingerprint detection application using $Gd_2Ti_2O_7:Eu^{3+}$ nanophosphors, J. Alloys Compd. 741 (2018) 246–255, https://doi.org/10.1016/j.jallcom.2018.01.116.

[89] G.P. Darshan, H.B. Premkumar, H. Nagabhushana, S.C. Sharma, B. Daruka Prasad, S.C. Prashantha, R.B. Basavaraj, Superstructures of doped yttrium aluminates for luminescent and advanced forensic investigations, J. Alloys Compd. 686 (2016) 577–587, https://doi.org/10.1016/j.jallcom.2016.05.255.

[90] S. Som, C.Y. Yang, C.H. Lu, S. Das, Synthesis of Li^+-ion activated $NaYF_4$: Er^{3+}/Yb^{3+} phosphors via a modified solid-state process for latent fingerprint detection, Ceram. Int. (2019), https://doi.org/10.1016/j.ceramint.2018.12.035.

[91] A.K. Soni, V.K. Rai, $SrMoO_4:Er^{3+}$-Yb^{3+} upconverting phosphor for photonic and forensic applications, Solid State Sci. 58 (2016) 129–137, https://doi.org/10.1016/j.solidstatesciences.2016.06.010.

[92] K.N. Kumar, L. Vijayalakshmi, P. Hwang, A.D. Wadhwani, J. Choi, Bright red-luminescence of Eu^{3+}ion-activated $La_{10}W_{22}O_{81}$ microphosphors for noncytotoxic latent fingerprint imaging, J. Alloys Compd. 840 (2020) 155589, https://doi.org/10.1016/j.jallcom.2020.155589.

[93] J.Y. Park, H.K. Yang, Novel red-emitting $Y_4Zr_3O_{12}:Eu^{3+}$ nanophosphor for latent fingerprint technology, Dyes Pigments 141 (2017) 348–355, https://doi.org/10.1016/j.dyepig.2017.02.036.

[94] S.G. Menon, A.K. Kunti, D.E. Motaung, H.C. Swart, A new recipe for the rapid microwave synthesis of high quantum yield Mn^{2+}-doped $ZnGa_2O_4$ phosphors for potential forensic applications, New J. Chem. 43 (2019) 17446–17456, https://doi.org/10.1039/c9nj04225c.

[95] J.Y. Park, J.W. Chung, S.J. Park, H.K. Yang, Versatile fluorescent $CaGdAlO_4:Eu^{3+}$ red phosphor for latent fingerprints detection, J. Alloys Compd. 824 (2020), https://doi.org/10.1016/j.jallcom.2020.153994.

[96] M. Saif, M. Shebl, A.I. Nabeel, R. Shokry, H. Hafez, A. Mbarek, K. Damak, R. Maalej, M.S.A. Abdel-Mottaleb, Novel Non-toxic and Red Luminescent Sensor Based on Eu^{3+}: $Y_2Ti_2O_7/SiO_2$ Nano-Powder for Latent Fingerprint Detection, Elsevier B.V., 2015, https://doi.org/10.1016/j.snb.2015.05.040.

[97] J. Shao, J. Yan, X. Li, S. Li, T. Hu, Novel fluorescent label based on YVO_4: Bi^{3+}, Eu^{3+} for latent fingerprint detection, Dyes Pigments 160 (2019) 555–562, https://doi.org/10.1016/j.dyepig.2018.08.033.

[98] M. Saif, Synthesis of down conversion, high luminescent nano-phosphor materials based on new developed $Ln^{3+}:Y_2Zr_2O_7/SiO_2$ for latent fingerprint application, J. Lumin. 135 (2013) 187–195, https://doi.org/10.1016/j.jlumin.2012.10.022.

[99] S. Yeshodamma, D.V. Sunitha, R.B. Basavaraj, G.P. Darshan, B.D. Prasad, H. Nagabhushana, Monovalent ions co-doped $SrTiO_3:Pr^{3+}$ nanostructures for the visualization of latent fingerprints and can be red component for solid state devices, J. Lumin. 208 (2019) 371–387, https://doi.org/10.1016/j.jlumin.2018.12.044.

[100] R. Krishnan, G.B. Nair, S.G. Menon, L. Erasmus, H.C. Swart, Synthesis of Tm_2WO_6: Er^{3+} upconversion phosphor for high-contrast imaging of latent-fingerprints, J. Alloys Compd. 878 (2021) 160386, https://doi.org/10.1016/j.jallcom.2021.160386.

[101] J.Y. Park, J.W. Chung, H.K. Yang, Versatile fluorescent $Gd_2MoO_6:Eu^{3+}$ nanophosphor for latent fingerprints and anti-counterfeiting applications, Ceram. Int. 45 (2019) 11591–11599, https://doi.org/10.1016/j.ceramint.2019.03.030.

[102] P. Du, P. Zhang, S.H. Kang, J.S. Yu, Hydrothermal synthesis and application of Ho^{3+}-activated $NaYbF_4$ bifunctional upconverting nanoparticles for in vitro cell imaging and

latent fingerprint detection, Sensor. Actuator. B Chem. 252 (2017) 584–591, https://doi.org/10.1016/j.snb.2017.06.032.

[103] Y. Wang, Y. Ke, S. Chen, J. Luo, S. Shu, J. Gao, B. Deng, R. Yu, Luminescence investigation of red-emitting $Sr_2MgMoO_6:Eu^{3+}$ phosphor for visualization of latent fingerprint, J. Colloid Interface Sci. 583 (2021) 89–99, https://doi.org/10.1016/j.jcis.2020.09.024.

Rare-earth-activated phosphors for TL dosimetry applications

10

Karan Kumar Gupta [1], *Sudipta Som* [1], *Chung-Hsin Lu* [1] *and Sanjay J. Dhoble* [2]
[1]Department of Chemical Engineering, National Taiwan University, Taipei, Taiwan;
[2]Department of Physics, Rashtrasant Tukadoji Maharaj Nagpur University, Nagpur, Maharashtra, India

1. Introduction

In current decades, the application of rare-earth ion-doped luminescence materials has been increased in many diverse areas, for example, in lighting systems as a light-emitting diode, afterglow phosphors, scintillators, and solar cells technologies [1−3]. Rare-earth ions also play a critical role in thermoluminescence (TL). TL is the phenomenon of light generated by heating a wide band gap solid material to a temperature under that of incandescence. The energy absorption from an outward source is a central parameter to produce light. The heat is only used as a trigger. The key requirement of TL is dosimetry, as the light produced is proportional to the absorbed dose exposed to the material [4]. The application of TL is not limited to personal dosimetry, but can also be used in clinical, environmental, and high dose dosimetry [5]. Another important requirement of TL is the dating method. TL is used to determine the age of inorganic ceramic materials. TL dating also shows its usefulness in different fields such as archaeology and historical architecture, to be precise it is sediment dating. However, the application of the TL technique is not limited to dosimetry and dating, but can also be used to detect impurities in rocks and determine the effectiveness of surface catalysts [6]. TL is very sensitive to structural defects in the material, so it can be used to study these defects. The various applications of the TL technique are shown in Fig. 10.1.

The basic result of generating TL is trap charge carriers, that is, electrons and holes, which are formed by exposure to external radiation sources. The trapping of charge carriers usually occurs at the defect sites already present or intentionally created inside the material. Thus, defect sites can be categorized into two types: inherent defect sites in materials and defect sites formed by external doping impurities [4]. Negative ion vacancies are the first type, called F centers, which can trap electrons. The lattice gap is the second type, which is produced by replacing the lattice ions with higher or lower valence impurity ions. For example, replacing the divalent network with trivalent cationic impurities can induce cationic gaps to maintain charge neutrality. Cation vacancies have the highest probability of trapping holes. There are also various types of defects in the matrix, which can act as electron-trapping centers or holes. The behavior of discovering these defects is part of the research on luminescent

Rare-Earth-Activated Phosphors. https://doi.org/10.1016/B978-0-323-89856-0.00007-9

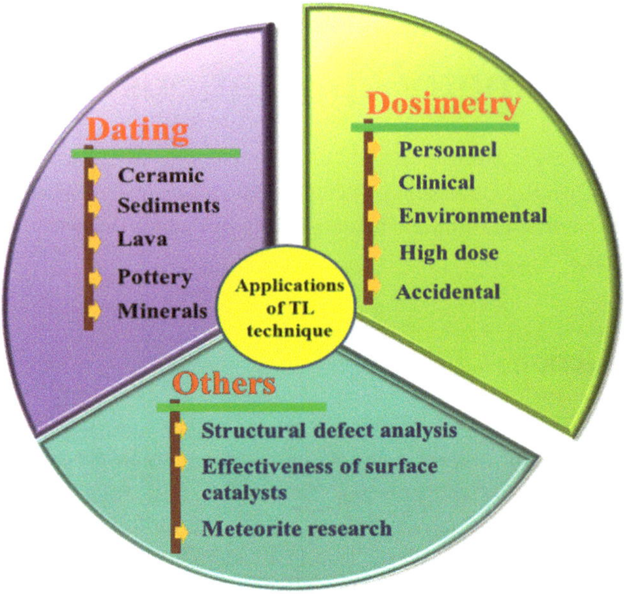

Figure 10.1 Different applications of TL technique.

materials [4,6,7]. However, there is a lack of scientific articles that specifically study the role of rare-earth ions in the thermo-luminescent material. This chapter briefly reviews the TL mechanism related to the doping of rare-earth ions and then uses a series of examples to illustrate how to change the properties of TL through the correct selection of doping impurities.

1.1 TL mechanism

A luminescence material emits light at moderate temperature after the absorption of external energy. This emitted light is called as the luminescence phenomenon, and it must not be compared with the blackbody radiation that usually happens at high temperatures [7]. The absorption of energy to luminescence materials can be carried out in different ways, and hence the word luminescence can be modified by adding a prefix to this word. An additional classification of luminescence phenomenon comes from the time delay between excitation and emission of luminescence materials [7]. For example, if the excitation source was ceased and also the emission was stopped from the material the mechanism is called as fluorescence, while if the emitted radiation or emission still continues after the removal of the excitation source the phenomenon is called as phosphorescence. All materials having this type of properties are known as phosphors. Thus, all phosphor material has a capability to change different forms of energy into light energy (UV, Vis, and NIR) [1−3]. Therefore, the research for luminescent materials will be necessary for different applications. Actually, the emitted color of light extremely depends on several factors such as intrinsic features of the material, defect content, particle size, and external factors such as temperature and external applied fields (pressure, electric, and magnetic) [4–6]. The field of applications of luminescence phenomenon in science and everyday life is tremendously huge and endlessly rising [6].

TL is the phenomenon of light emission produced by reasonable heating of an insulator or semiconductor, earlier exposed to ionizing radiation [4,6]. It usually occurs at low temperatures where blackbody light emission is insignificant. The exposure to ionizing radiation generates traps, and these traps are filled by electrons and/or holes. The heating of exposed materials enable the charge carriers to freely migrate in the lattice until they recombine or captured by other traps with a resultant photon emission. Therefore, three vital circumstances are essential for the creation of TL. The first condition is the material type which should be insulator or semiconductor. The second condition is the exposure of these materials to ionizing radiation and finally the stimulating agent that could be heat energy. Additionally, the TL phenomenon will not be observed if once the material is heated. Thus, the exposure to ionizing radiation is necessary to get the TL emission again.

Fig. 10.2 shows the detailed mechanism of the TL process. The initial interaction between the electronic structure of atoms and ionizing radiation is the ultimate creation of free charge carriers (process (a)). The free electron and/or hole travel through the crystalline structure and become trapped at the trapping centers (process (b) and (c)). Heating of TL material can stimulate electrons and holes into migration (processes (e) and (f)), which can result in recombination. Sometimes, the released charge carriers

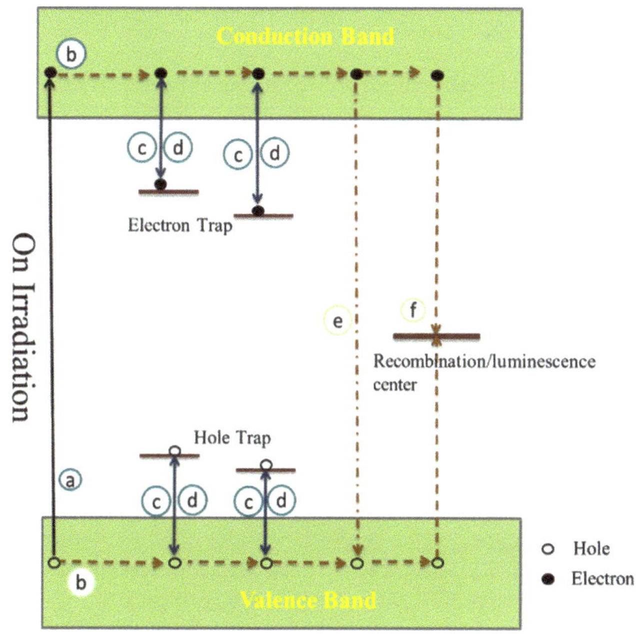

Figure 10.2 A simple band diagram showing TL mechanism.

get retrapped (process d) or recombine nonradiatively. TL phenomenon occurs if the light is produced during the recombination process. In principle, a model can be established with several traps and recombination centers, and rate equations can be developed to describe the charge traffic. Analysis of parameters such as trapping, detrapping, and recombination probabilities along with the concentration of the particular states will be essential with the involvement of additional traps and recombination centers [4]. It is very tedious work to define all these parameters by only using the shape of the TL glow curve. Three important parameters (activation energy (E), frequency factor (s), and order of kinetics (b)) are basically used to define the nature of a TL glow peak [8−11]. Several methods have been already established for estimating the trapping parameters from the shape of the TL glow curve. The advantages and disadvantages of some of these methods are shown in Table 10.1.

2. Application of TL as a dosimetry

Determining the dose of ionizing radiation via a suitable instrument is known as dosimetry. Quantitative estimation of ionizing radiation dose can establish the numerical relationship between the dose and the biological, chemical, and physical changes caused by the dose. Dosimeter systems have to satisfy the subsequent provisions, such as it should be mechanically stable, easy to carry, small, and cheap. All dosimetry should be able to estimate the radiation dose from μSv to several Sv. The dosimeter should not be dependent on the energy type, and dose-response should be independent of the position of the dosimeter [12]. Negligible fading under harsh environmental

Table 10.1 Advantages and disadvantages of trapping parameters analysis methods.

Method of analysis	Advantages	Disadvantages	References
Initial rise	Estimation of *E* without any knowledge of *s* and *b*	Inaccurate prediction of *E* values for complicated glow curves	[9]
Heating rate	*E* and *s* values can be calculated regardless of *b*	Unappropriate for multiple glow peaks	[11]
Hoogenstraaten	Similar to the heating rate method	Only applicable for first-order kinetics of TL glow curve	[10]
Peak shape	*E*, *b*, and *s* values can be calculated by utilizing the shape or geometrical properties of glow curves	Not very useful for analyzing experimental glow peaks	[8]
Curve fitting	Suitable for complex or overlapped TL glow curves	Need of correct estimation of highest peak temperature of each peak	[10]

conditions such as temperature, light, and humidity with rapid, simple, and cheap evaluation methods is the ideal properties of a dosimeter [13]. The various applications of TL dosimetry in different areas are shown below.

2.1 Personnel and clinical dosimetry

Personal dosimetry is defined in terms of the measurement of exposure from an external source of radiation of workers employed in the areas of ionizing radiation. Similarly, medical or clinical dosimetry refers to the measurements of dose to patients throughout medical treatment. The most frequently used passive personal dosimeters are TL dosimeters. The information of irradiation dose allows further radiation protection procedures in case of personnel safety or medical involvement. In personnel dosimetry, the dose range is from 10^{-5} to 10^{-1} Gy. However, in medical dosimetry, the dose range is between 10^{-5} and 1 Gy [12]. Sometimes during radiotherapy treatment, the dose rises to 10 Gy. Generally, in radiation protection clothes, a personal dosimeter is located on the left side of the chest. The dose of radiation can also be determined on other portions of the body depending on the working process. In medical dosimetry during therapeutic medical treatment, radiation doses are also determined on several portions of the body [13]. Thermoluminescence dosimeters (TLDs) are most appropriate for personal and medical dose measurements due to their small sizes.

2.2 Environmental dosimetry

Environmental monitoring of natural background radiation and manmade ionizing radiation comes under the area of environmental dosimetry. Environmental radiation depends on many factors such as the involvement of manmade ionizing radiation sources in comparison with natural radiation. Cosmic and terrestrial radiations are the main source of natural radiation dose in the environment, which is in the range of 100–200 nSv/h [14]. Cosmic radiation generally comprises high linear energy transfer (LET) and high energy ionizing radiations (neutrons). Sometimes, the inconsistency of natural radiation can surpass manmade sources. For this motive, dose measurements are carried out with the help of TLDs early at the radiological or nuclear facility boundaries to country-wide locations. The usage of passive dosimeters to conduct nationwide background monitoring can be useful as the basis for calculating the dose contribution in unexpected situations. TLDs are basically positioned out of doors for the environmental monitoring of radiation exposure from nuclear or manmade activities and natural radiation. The passive detectors (TLDs) have the advantage of determination of the integrated dose during the particular time of intervals such as a day, a month, or a year in environmental monitoring. As the dose level is very low (10^{-5} Gy), highly sensitive TL detectors are used for environmental dosimetry [12]. TLDs used in environmental dosimetry in case of a long measurement period should be highly stable toward the environmental factors such as chemical contamination, sunlight, and high moisture.

2.3 High-dose dosimetry

High-dose radiation technologies include material modification, sterilization of medical products, and preservation of agricultural products [15]. High energy gamma emitters or electron accelerators are used in the earlier technologies, which provide radiation exposure doses of the order of several thousand grays. To improve the effectiveness of radiotherapy, the absorbed dose must be correctly determined. Therefore, when radiation treatment is carried out according to the need, direction, and optimization of the process, the need for radiation dosimetry plays a vital role in radiation treatment. Nuclear power plants and liquid sludge treatment plants also require high dose measurements [15]. The intensive use of radiation in different fields has prompted researchers to develop appropriate dosimetry systems according to their fields of interest. In high-dose dosimetry, the dose ranges from 10^2 to 10^6 Gy, which can be applied to food sterilization or material processing applications [12]. Many luminescence-based detectors (TLDs) have been developed for quality control in the radiation process, and a large number of new dosimetry materials are still under study.

2.4 Retrospective dosimetry

During nuclear accidents when no dose data is available, retrospective, so-called "review" dosimetry is one of the most significant tools for accidental dosimetry. Thus, it is one of the most useful ways to estimate the amount of radiation exposure at unexpected nuclear accidents and can be termed as accidental dosimetry. The imaginable hazard of an unpredicted terror attack or nuclear accident requires suitable approaches and procedures possibly valid in an emergency. Several stuffs can be useful in the affected area as natural dosimeters. For example, in the case of the nuclear accident at Hiroshima and Nagasaki, bricks and tiles were used as accidental radiation dosimetry [16]. Recently published various retrospective dosimetry issues, including portable electronic components such as digital watches, pagers, cell phones, and telephone chip cards, consider using TL technology for dose reconstruction.

3. Rare-earth ions and their effect on TL

As we have discussed earlier that TL may arise in pure and perfect crystals due to the presence of intrinsic defects such as F center in ionic crystals (known as intrinsic luminescence) or due to the presence of extrinsic defects such as impurity induced defects (known as extrinsic luminescence). Occasionally, for scientific and application purposes, impurities and defects are purposely introduced in the desired material. Rare-earth doping can introduce several types of defects in the TL material and will be useful for the TL phenomenon. Rare-earth doping plays a very important role in the enhancement of luminescent properties of the inorganic phosphor [17]. Thus, in this present section, we will study a brief description of rare-earth ions and their role in TL properties with help of some examples.

3.1 Rare-earth ions and their applications

Rare-earth elements (REE) are those 15 elements that are located at group three and sixth period of the periodic table. REE are also called as lanthanides, and +3 oxidation state is the most energetically stable. The characteristics of electronic transitions of all trivalent REEs generally happen in the 4f layer, that is, 4f−4f transitions. Ce^{3+} is the only exception, as it holds 5d−4f transitions. As the 4f shell is protected by the optically passive outer electronic shell that acts as a shield for host lattice, the optical properties such as bandwidths and wavelengths related to 4f−4f transitions are not that much influenced by the presence of host lattice. REE-doped materials show a significant role in current optical technology [18]. The narrow spectrum of REE is ranging from the vacuum ultraviolet to far-infrared due to their partially filled f-shells that are protected by externally filled s and p shells. REE metals have found a wide range of applications, such as the production of phosphors (Y, La, Ce, Eu, Pr, Gd, Tb, Nd, Tm), optical filters (La, Ce, Pr, Nd), solid magnets (Pr, Nd, Tb, Dy), and automotive catalysts (La, Ce, Pr, Nd) [18]. However, as regards the possibility of producing white light-emitting diodes, the latter case has been widely studied. The phenomenon of photon up-conversion is understood as an anti-Stokes photoluminescence process in which the energy of two or more sequentially absorbed photons combines to produce higher energy photons, which is also confirmed by several lanthanide ions, such as Dy^{3+}, Ho^{3+}, Er^{3+}, and Tm^{3+}. Recently, REE-doped luminescent materials have received increasing attention as their potential applications as energy storage phosphors and scintillation materials for dosimetry applications. Scintillation materials generally do not involve the step of trapping free charge carriers. The most useful REE in radiation protection is shown in Fig. 10.3. The application of rare-earth ions covers many areas (Fig. 10.3), but the role and mechanism behind the incorporation of rare-earth ions into TL materials are not so clear. In the next section, we will discuss the effects of rare-earth ions on the characteristics of TL through some examples.

3.2 Rare-earth induced effect on TL phosphors

Despite successful applications, the details of how rare-earth ions are incorporated into the crystal lattice are not yet clear, and over time various increasingly complex models

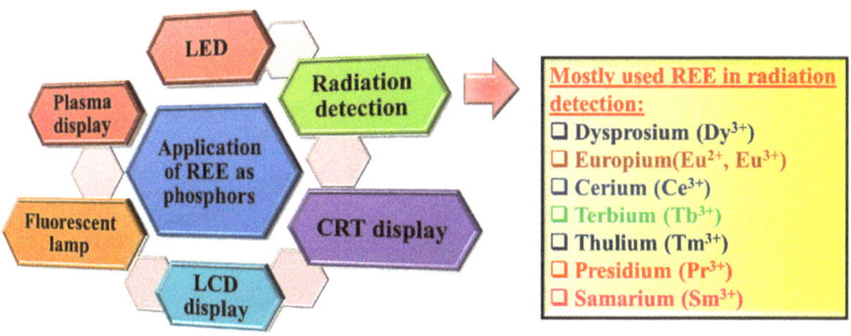

Figure 10.3 Different applications of REE and their names basically used for radiation detection.

have emerged [19]. Replacing atoms of different valences from trivalent rare-earth ions requires some charge compensation, and the shift in ion volume can also introduce stress [19]. Usually, this stress can be reduced by the combination of defect complexes, so that the luminescence response is not only sensitive to the selected rare-earth ions, but also to the concentration of dopants, their association with inherent defects, and heat treatments that allow aggregation or dissolution of the complex [19]. REEs alter the optical properties of materials as they distort the symmetry of the original position and require charge compensation. These factors are reflected in optical properties, such as wavelength, line width, lifetime, and the relative intensity of rare earth and TL transitions. Other examples of TL have pointed out that rare-earth dopants can even modify the overall structure of the lattice (i.e., not just the immediate vicinity of the dopant). Examples of changes in the overall structure of ZnO and $SrTiO_3$ are reported [19]. This sensitivity inevitably exists as the crystal supplier, crystal grain size, ion type, concentration, impurities, excitation method, and heat treatment method to improve performance. Because of this inherent complexity, we can realize that although application performance can be optimized, it is rarely possible to understand key sites in the network in detail. Phosphors always contain inherent defects, and the incorporation of impurities will generate new defects [17]. These defects play a particularly important role in the emission behavior of such phosphors. Therefore, it is very important to study different aspects of the defects existing in the phosphor. TL is a technology widely used to study the radiation dosimetry characteristics of phosphors. But recently, in the study of the role of impurities and defects in solids, an equally significant increase in this technology has been observed. This technique allows analyzing traps that exist due to defects in the phosphor and extracting information about trap depth energy and frequency factors of specific traps. As we already know, the luminescence properties of REE-doped phosphors highly depend on the position of REE energy levels in relation to the band gap of the host matrix.

An empirical model was established by Dorenbos et al. to determine the locations of 4f and 5d levels of all REE in an inorganic host [20]. Bos et al. have used a wide band gap YPO_4 material as a host lattice for different lanthanides, and experimental results were compared with the proposed model (Fig. 10.4A) [21]. The position of the energy levels of selected REE was revealed as a function of the number of electrons *n* in the 4f configuration starting from the left to the right. The lowest level of trivalent lanthanides having the electronic configuration of $4f^n$ is connected via curve 1 while the divalent lanthanides having the electronic configuration of $4f^{n+1}$ are connected via curve 2 as a function of *n*. Curves 1 and 2 display a distinctive outline that is directly connected with the strength of the binding of the 4f electrons. The ground state energy of Ce^{3+}, Pr^{3+}, and Tb^{3+} is 3.90, 2.06, and 2.24 eV. Such trivalent lanthanides will be able to trap holes from the valence band and be stable at room temperature to a certain point of temperature applied during TL experiments. On the other hand, the ground state energy position of the divalent Nd^{2+}, Sm^{2+}, Dy^{2+}, Ho^{2+}, Er^{2+} Tm^{2+}, and Yb^{2+} ions are under the lowermost of the conduction band and conforming that these trivalent lanthanides will act as electron traps. The glow peak maxima of YPO_4 doped with different lanthanides are shown in Fig. 10.4B along with the levels shown in Fig. 10.4A. The model anticipates the deepest traps for Yb^{2+} and the shallowest

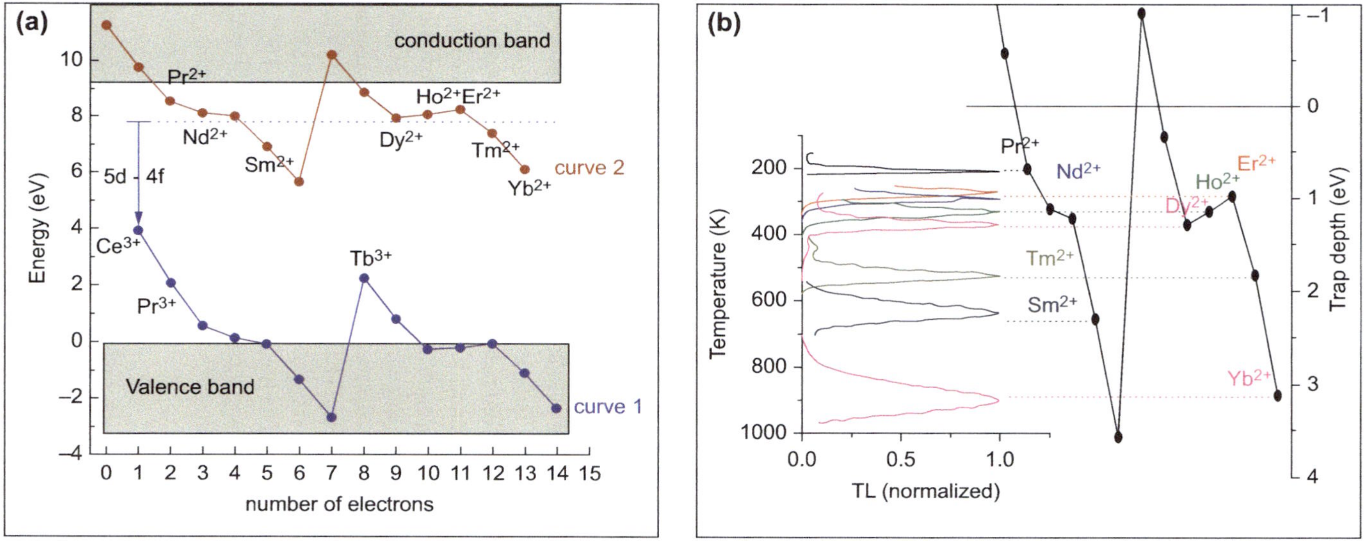

Figure 10.4 (A) Energy level scheme of the lanthanides in YPO_4, (B) Experimental glow peak maxima with energy level scheme. Reprinted with permission from Adrie J.J. Bos [21].

trap for Pr^{2+}. The TL glow peaks of the other Ln^{2+} ions similarly look as in the order anticipated by the model [21]. Therefore, with the help of this model, we can tune the TL properties of any materials according to our desired applications. The following section gives some examples of rare-earth-doped TL materials and their possible applications in dosimetry.

4. Rare-earth-activated TL dosimeters

To date, many TL dosimeters are reported and commercially utilized for the application as dosimetry. Many investigators have investigated REE-activated sulfate, fluoride, phosphate, and other oxide-based phosphors. In this section, we are going to list the name of commercially used REE-activated TL phosphors with their possible applications and drawbacks. The role of rare-earth ions toward the modification in TL properties of the phosphors is also explained in this section.

4.1 Sulfate-based materials

Sulfate-based materials are the most commonly used TLDs because of their high sensitivity toward ionizing radiation. Sulfate-based TLDs are frequently used as environmental and personal dosimeters. The TL characteristics can be tailored via the change in synthesis technique or by adding a dopant. The most common material in sulfate-based TLDs is $CaSO_4$, which is usually synthesized via the coprecipitation method. Several types of rare-earth dopants (Lu, Yb, Tm, Er, Ho, Dy, Tb, Gd, Eu, Sm, Nd, Pr, Ce, and La) has been studied by Nambi et al. for the $CaSO_4$ TLD materials [22]. Dy and Tm dopants result in maximum TL output while other REE dopants gave less TL intensity. In the $CaSO_4$ host, it is usually identified that trivalent rare-earth ions substitute Ca^{2+} ions and the ionic radii of Dy^{3+} and Tm^{3+} are nearby to the radius of Ca^{2+} ion, while other REE dopants have the largest difference. From the TL study of Nambi et al., it is clear that the doping of RE ions does not alter the basic nature of traps or glow peaks while the only change in TL peak intensity was observed [22]. A similar type of studies is also reported for the Cd, Sr, and $BaSO_4$ materials for RE (Sm, Tm, Dy, Tb, and Eu) dopants by Dixon et al. [23]. In this study, the $CdSO_4$ materials do not have sufficient emission while for other materials such as Sr and $BaSO_4$ the Tb^{3+} ions give the highest TL intensity. The high TL intensity observed for Tb^{3+} dopant and the least TL intensity for $CdSO_4$ was again explained on the basis of ionic radii and the average nearest neighbor distance. Among the all sulfate materials, $CaSO_4$:Dy and $BaSO_4$:Eu have gained the most attention due to their high TL sensitivity and dose linearity.

However, the main drawback assigned to sulfate material is its nontissue equivalency. Therefore, these TLDs are basically used for environmental and personnel dosimetry instead of clinical dosimetry. In one of our published research, we have explained the formation of different types of defect centers in Eu^{2+}-doped $LiNaSO_4$ with the help of an electron paramagnetic resonance (EPR) study [24].

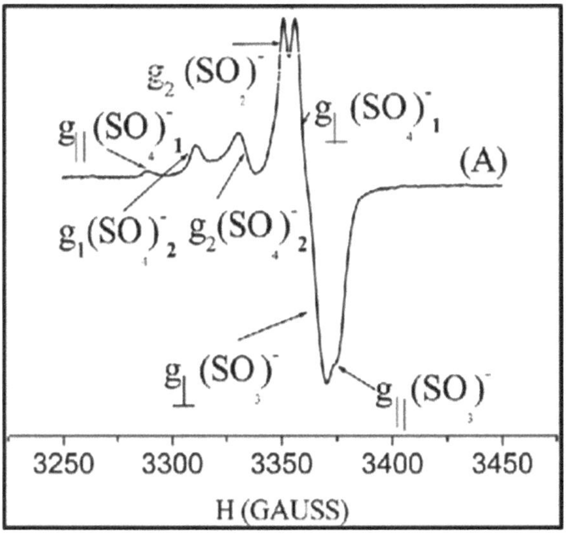

Figure 10.5 EPR spectra of $LiNaSO_4$:Eu after overnight γ-irradiation from a ^{60}Cosource, in first harmonic presentation. Reprinted with permission from K. K. Gupta [24].

Fig. 10.5 shows the EPR spectra of irradiated $LiNaSO_4$. The EPR spectra clearly represent that SO_2^-, SO_4^-, and SO_3^- defect centers are present in the irradiated $LiNaSO_4$. The presence of Eu ions creates charge compensation defects and also worked as a trapping center for holes as well as a luminescence center in $LiNaSO_4$ host. Almost a similar type of TL mechanism is observed to all sulfate-based phosphors activated with REE. The list of sulfate-based materials and their respective REE dopants with their practical application, dose linearity, and drawbacks are shown in Table 10.2.

Table 10.2 List of REE-activated sulfate-based phosphors with their major applications and drawbacks.

TL phosphor	Dose range (Gy)	Applications (dosimetry)	Drawbacks	References
$CaSO_4$:Dy	10^{-4}–3×10^3	Personnel, and high dose	Nontissue equivalency	[22]
$CaSO_4$:Tm	10^{-5}–3×10^3	Environmental and personnel	Nontissue equivalency	[23]
$LiNaSO_4$:Eu	3.4×10^{-2}–3.4	Personnel and high energy particles	High thermal fading, sensitive to light	[24,25]
$BaSO_4$:-Eu	4×10^{-3}–50	High energy particles	Nontissue equivalency, sensitive to light	[26]

4.2 Fluoride-based materials

Fluoride-based TL phosphor is considered among the topmost extensively used materials for dosimetric applications because of its tissue-equivalent effective atomic number. The TL response of that type of material is almost independent of the radiation energy. LiF is the most useful TL dosimetry, and their TL studies doped with transition metal ions were carried out in detail by many investigators [27]. However, the doping of rare-earth ions into the LiF is very less reported. In one of the studies of Seth et al., Dy^{3+} doping can be useful for the enhancement of the sensitivity of LiF for the high dose dosimetry [27]. Alkaline earth fluorides (CaF_2, SrF_2, and BaF_2) have been also widely investigated materials for TL applications [19,28]. Rare-earth doping of such fluorides has the ability to sense the very level of ionizing radiations. One of the studies of Maghrabi et al. includes samples of Ca, Sr, and Ba fluorides doped with Ce, Gd, Tb, Dy, Ho, Er, and Tm [29].

The effect of REE dopant was studied in terms of variation in glow peak position with respect to host emission. This study shows that the presence of rare-earth ions has some active role on the absolute population of the several hole traps, instead of their having a passive role in the TL process. The substitution of trivalent rare-earth ions at the site of divalent metal ions can form a surplus positive charge. Compensation of this positive ion can occur via various processes such as the creation of clusters involving impurities, interstitials, and vacancies. The creation of such defects has some uncertain effect toward the TL response. The impurities usually doped in CaF_2 are the dysprosium (Dy) and thulium (Tm) and have a positive effect toward the TL intensity. The list of several fluoride materials doped with rare-earth ions along with their applications and drawbacks are shown in Table 10.3.

4.3 Phosphate-based materials

Phosphate-based materials for TLD application have gained major attention for their suitable and efficient TL properties. These phosphors have high thermal and chemical stability, high emission intensity, and long afterglow properties. Especially, $ABPO_4$

Table 10.3 List of REE-activated fluoride-based phosphors with their major applications and drawbacks.

TL phosphor	Dose range (Gy)	Applications (dosimetry)	Drawbacks	References
CaF_2:Dy	10^{-3}–10^3	Personnel, X-rays, and photons	Nontissue equivalency, high fading	[19,28]
LiF:Mg, Dy	10–10^3	High dose	Complicated glow curve structure	[27]
MgF_2:Eu	10^{-4}–1	Personnel, neutron, and X-rays	Nontissue equivalency	[30]
YF_3:Ce, Tm	40–100	X-rays and β irradiation	Nontissue equivalency	[31]

type materials are high luminescent phosphate materials exhibiting excellent TL properties. Different material has been studied for the compound belonging to $ABPO_4$ in which A and B refer to monovalent and divalent cations, respectively. Some of the excellent materials belonging to this group are $LiCaPO_4$, $NaCaPO_4$, $LiMgPO_4$, $KMgPO_4$, and $KCaPO_4$ [32]. Among all these materials, $LiMgPO_4$ (LMP) is highly studied and shows its practical application in radiation dosimetry [33]. LMP phosphors are mainly doped with Tb, Tm, and Eu rare-earth ions [34,35]. Among all REE dopants, Tb REE along with boron has outstanding dosimetric characteristics with a least detectable dose of 20 μGy [36]. LMP phosphors show the linear response of γ-dose ranging from 1 mGy to 1 kGy. A similar phosphor has also been studied for the exposure of submicro gray dose levels via utilizing optically stimulated luminescence (OSL)-based dosimetry. LMP is one of the suitable TL phosphors for personnel and environmental dosimetry. The TL emission spectrum of LMP phosphors contained several peaks of REE dopants that confirm that the REE dopant acts as a luminescence center in the LMP host. The TL responses of differently doped LMP crystals to protons, neutrons, and α particles have also been studied [32]. Therefore, LMP can also be applicable for the dosimetry of heavy ions. These studies exposed that REE-doped LMP is an ideal material for dosimetric applications. The most apparent benefit of the LMP is its very simple glow curve structure, which had led to less difficulty in arriving at the dose linearity with radiation exposure. Other phosphate-based TL phosphors such as $LiCaPO_4$ doped with Eu, $NaCaPO_4$ doped with Dy^{3+}, and $KMgPO_4$ doped with Tb^{3+} are useful for low dose dosimetry, high dose dosimetry, and low dose β radiation dosimetry applications, respectively [32]. The list of important phosphate-based TLD materials doped with rare-earth ions is shown in Table 10.4 along with their applications and drawbacks.

Table 10.4 List of REE-activated phosphate-based phosphors with their major applications and drawbacks.

TL phosphor	Dose range (Gy)	Applications (dosimetry)	Drawbacks	References
$LiMgPO_4$:Tb, B	10^{-3}–10^{3}	Food irradiation, space, and medical	High thermal fading of TL intensity	[33,34]
$NaLi_2PO_4$:Eu	10–10^{3}	High dose and high energy radiation	High thermal fading of TL intensity	[32]
$LiCaPO_4$:Eu	7.8×10^{-3}–8.25×10^{-3}	Medical and β irradiation	Nontissue equivalency	[32]
$NaCaPO_4$:Dy	10^{-1}–10^{3}	High dose	Nontissue equivalency	[32]
$KMgPO_4$:Tb	3.9×10^{-4}–0.6	Environmental and β irradiation	Less sensitive andand nontissue equivalency	[32]

4.4 Other oxide-based materials

Oxide-based phosphors are widely studied materials, especially in TL dosimetry. However, the accomplishment of TL characteristics is highly reliant on impurities and point defects present in the host matrix. The function of impurities has been a topic of significant investigation in the past. TL induced in REE-doped Al_2O_3, $LiAlO_2$, Mg_2SiO_4, MgO, TiO_2, and BeO after irradiation with ionizing radiation has been reported by several authors [37,38]. In one of the studies of Barrows et al., several rare-earth ions (Gd, Er, Nd, Tb, Sm, and Tm) were doped into the Al_2O_3 [37]. Among these all rare-earth ions, Tm-doped aluminum oxide shows the highest TL intensity. Doping of Tb and Tm into Al_2O_3 was also studied by Andrade et al., they observed that codoping of Tb ion with Tm can produce four times more sensitive material than singly doped Al_2O_3 [39]. According to the study of Reddy et al., doping of rare-earth ions in Al_2O_3 creates additional oxygen vacancies that might act as electron donors and rare-earth ions by themselves work as hole trapping centers and luminescence centers [40]. In one of our studies, Ce^{3+}, Dy^{3+}, and Eu^{2+}-doped $LiAlO_2$ phosphors were carried out for the dosimetry of high dose measurements [15]. Eu^{2+}-doped $LiAlO_2$ samples show the highest TL intensity and high dose linearity up to 10 kGy. The possible TL mechanism behind the REE-doped $LiAlO_2$ was also explained in this paper. Through EPR study, numerous defect centers were recognized in irradiated $LiAlO_2$ phosphor (Fig. 10.6).

These defects were assigned to an aluminum/lithium-oxygen hole trapped center and F centers. The Al-OHTC/Li-OHTC center was connected with the high-temperature TL peak in undoped and Ce/Dy/Eu doped phosphors. One of the F centers is related to the low-temperature TL peak while the second F center is related to the high-temperature TL peak. In this study, we have also observed the rare-earth ion was act as hole trapping centers and luminescence centers in the given matrix. Other oxide materials are also studied with various rare-earth ions and have significant TL properties [38]. Table 10.5 shows the list of some REE-activated oxide-based TLD phosphors with their applications and drawbacks.

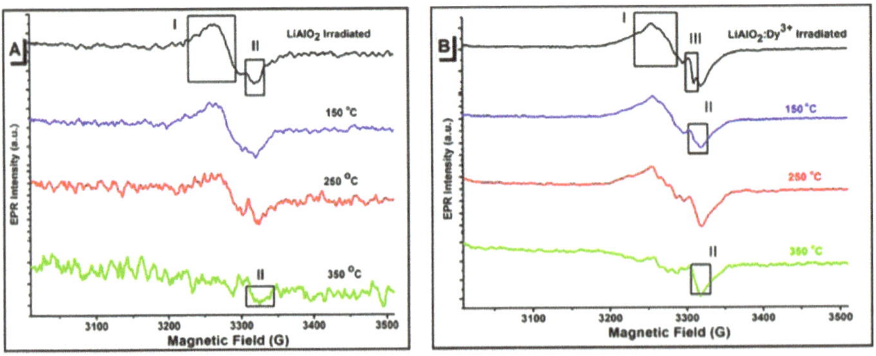

Figure 10.6 (A) EPR spectra of undoped $LiAlO_2$ and (B) Dy^{3+} doped $LiAlO_2$ at different temperatures after overnight gamma irradiation from a ^{60}Cosource.
Reprinted with permission from K. K. Gupta [15].

Table 10.5 List of REE-activated oxide-based phosphors with their major applications and drawbacks.

TL phosphor	Dose range (Gy)	Applications (dosimetry)	Drawbacks	References
Al_2O_3:Tb, Tm	0.1–5 Gy	Clinical and personnel	Nontissue equivalent and radiation energy dependency	[37,39]
Mg_2SiO_4:Tb	10^{-3}–4×10^2	High dose	Nontissue equivalent	[41]
MgB_4O_7:Dy, Na	6×10^{-8}–10^2	Personnel and medical	Sensitive to environmental factors	[42]
BeO:Na, ce, Dy	10^{-2}–50	Personnel and medical	Toxicity	[43]
$LiAlO_2$:Eu	50–10^3	High dose	Less sensitive	[15]

5. Conclusion

This chapter gives a detailed description of the role and importance of rare-earth-activated phosphors for TL dosimetry applications. From the first section, it can be easily observed that the TL technique can be utilized in different areas such as dosimetry, dating, and defect-related analysis. The TL mechanism shown through the band diagram model explained the necessity of defects required for the TL process. However, the second section includes the application of the TL technique in different dosimetry applications. The environmental and personnel dosimetry requires highly sensitive materials. In the third section, the role and importance of REE in TL materials are explained. Through this section, it can be easily observed that how the incorporation of rare-earth ions can modify the crystal lattice and improved the TL response of all REE-activated phosphors. The mostly used rare-earth ions for phosphors in TL dosimetry are also shown in this section. In the last section, some example is given based on the REE-activated sulfate, fluoride, phosphate, and other oxide-based TLD phosphors. In this section, the role of rare-earth ions in the TL mechanism is also mentioned. It can be easily observed through this section that the incorporation of REE in TL phosphors can act as both trapping centers and luminescence centers. Overall, this chapter shows the importance of rare-earth-activated phosphors in TL dosimetry applications.

References

[1] G.C. Tyrrell, Phosphors and scintillators in radiation imaging detectors, Nucl. Instrum. Methods Phys. Res. Sect. A Accel. Spectrom. Detect. Assoc. Equip. 546 (1–2) (2005) 180–187, https://doi.org/10.1016/j.nima.2005.03.103.

[2] K.K. Gupta, S. Som, C.H. Lu, Synthesis and spectroscopic characterization of strontium magnesium silicate phosphors for white afterglow applications, Optik 226 (P1) (2021) 165916, https://doi.org/10.1016/j.ijleo.2020.165916. Elsevier GmbH.
[3] C.Y. Ou, et al., Photovoltaic characteristics and computational simulation of samarium-ion doped Cu(In, Ga)Se2 thin films prepared via a non-vacuum coating process, J. Alloys Compd. (2021), https://doi.org/10.1016/j.jallcom.2021.160377. Elsevier.
[4] H. Aboud, H. Wagiran, R. Hussin, Mechanism of thermoluminescence, Int. J. Sci. Eng. Res. 3 (10) (2012) 1−9.
[5] R. Capelletti, 'Luminescence' Encyclopedia of Condensed Matter Physics, 2005.
[6] A.J.J. Bos, Thermoluminescence as a research tool to investigate luminescence mechanisms, Materials 10 (12) (2017), https://doi.org/10.3390/ma10121357.
[7] R.A. Clark, Intrinsic Dosimetry: Properties and Mechanisms of Thermoluminescence in Commercial Borosilicate Glass, 2012.
[8] G. Kitis, R. Chen, V. Pagonis, Thermoluminescence glow-peak shape methods based on mixed order Kinetics, Physica Status Solidi (A) Appl. Mater. Sci. 205 (5) (2008) 1181−1189, https://doi.org/10.1002/pssa.200723470.
[9] A.C. Coleman, E.G. Yukihara, On the validity and accuracy of the initial rise method investigated using realistically simulated thermoluminescence curves, Radiat. Meas. 117 (December 2017) (2018) 70−79, https://doi.org/10.1016/j.radmeas.2018.07.010. Elsevier Ltd.
[10] M. Bakr, M. Omer, Determination of thermoluminescence kinetic parameters of La2O3 doped with Dy^{3+} and Eu^{3+}, Materials 13 (1047) (2020) 1−23.
[11] R. Majgier, M. Biernacka, A. Chruścińska, On the application of the various heating rate method to determine the activation energy of the dominant trap from a complex TL peak, J. Lumin. 231 (November 2020) (2021), https://doi.org/10.1016/j.jlumin.2020.117776.
[12] M.G. Celik, Synthesis and Characterization of Lithium Tetraborate and Lithium Triborate Phosphors Doped with Metals and Their Thermoluminescence Propoerties, 2016. Available at: https://open.metu.edu.tr/bitstream/handle/11511/25848/index.pdf.
[13] M.R. Komor, Thermoluminescence personal and medical dosimerty, in: Radiation Safety Problems in the Caspian Region. Series IV: Earth and Environmental Sciences, 2004, p. 177. Available at: http://site.ebrary.com/id/10129526.
[14] M. Ranogajec-Komor, Thermoluminescence dosimetry application in environmental monitoring, Radia. Safety Manag. 2 (1) (2002) 2−16.
[15] K.K. Gupta, et al., A comparative investigation of Ce^{3+}/Dy^{3+} and Eu^{2+} doped $LiAlO_2$ phosphors for high dose radiation dosimetry: explanation of defect recombination mechanism using PL, TL and EPR study, J. Lumin. 188 (2017), https://doi.org/10.1016/j.jlumin.2017.03.046.
[16] D. Mesterházy, et al., Accidental and retrospective dosimetry using TL method, Radiat. Phys. Chem. 81 (9) (2012) 1525−1527, https://doi.org/10.1016/j.radphyschem.2011.11.051.
[17] Y. Patle, et al., Study of Photoluminescence,Thermoluminescence, and afterglow properties of Dy^{3+} doped $Ba_2ZnSi_2O_7$ phosphor, Optik 226 (P1) (2021) 165896, https://doi.org/10.1016/j.ijleo.2020.165896. Elsevier GmbH.
[18] W. Gieszczyk, et al., Thermoluminescence enhancement of $LiMgPO_4$ crystal host by Tb^{3+} and Tm^{3+} trivalent rare-earth ions co-doping, Materials 12 (18) (2019), https://doi.org/10.3390/ma12182861.
[19] Y. Wang, et al., Factors controlling the thermoluminescence spectra of rare earth doped calcium fluoride, J. Lumin. 184 (2017) 55−63, https://doi.org/10.1016/j.jlumin.2016.12.011. Elsevier.

[20] P. Dorenbos, A.J.J. Bos, Lanthanide level location and related thermoluminescence phenomena, Radiat. Meas. 43 (2–6) (2008) 139–145, https://doi.org/10.1016/j.radmeas.2007.10.007.
[21] A.J.J. Bos, et al., Study of TL glow curves of YPO_4 double doped with lanthanide ions, Radiat. Meas. 46 (12) (2011) 1410–1416, https://doi.org/10.1016/j.radmeas.2011.04.021. Elsevier Ltd.
[22] K.S.V. Nambi, V.N. Bapat, A.K. Ganguly, Thermoluminescence of $CaSO_4$ doped with rare earths, J. Phys. C Solid State Phys. 7 (1974) 4403–4415.
[23] R.L. Dixon, K.E. Ekstrand, Thermoluminescence of rare earth activated $CdSO_4$, $SrSO_4$ and $BaSO_4$, J. Lumin. 8 (5) (1974) 383–390, https://doi.org/10.1016/0022-2313(74)90079-9.
[24] K.K. Gupta, et al., On the study of the C^{6+} ion beam and γ-ray induced effect on structural and luminescence properties of Eu doped $LiNaSO_4$: explanation of TSL mechanism using PL, TL and EPR study, Phys. Chem. Chem. Phys. 20 (3) (2018), https://doi.org/10.1039/c7cp05835g.
[25] P.D. Sahare, S.V. Moharil, Thermoluminescence in $LiNaSO_4$, Radiat. Eff. Defect Solid 114 (1990) 167–172.
[26] J. Azorín, et al., Thermoluminescence characteristics of BaSO4:Eu, Int. J. Radiat. Appl. Instrum. Part A 42 (9) (1991) 861–863, https://doi.org/10.1016/0883-2889(91)90225-P.
[27] P. Seth, S. Aggarwal, S. Rao, Thermoluminescence study of rare earth ion (Dy^{3+}) doped LiF:Mg crystals grown by EFG technique, J. Rare Earths 30 (7) (2012) 641–646, https://doi.org/10.1016/S1002-0721(12)60105-7. The Chinese Society of Rare Earths.
[28] M.A.,E. Kolaly, EG0800321 effect of rare earth elements on the thermoluminescent properties of locally prepared calcium fluoride phosphors, in: ESENSA-08 9th International Conference of Nuclear Sciences and Applications, 2008, pp. 1–9.
[29] M. Maghrabi, P.D. Townsend, Thermoluminescence spectra of rare earth doped Ca, Sr and Ba fluorides, J. Phys. Condens. Matter 13 (25) (2001) 5817–5831, https://doi.org/10.1088/0953-8984/13/25/308.
[30] T. Rivera, Thermoluminescence in medical dosimetry, Appl. Radiat. Isot. 71 (Suppl. L) (2012) 30–34, https://doi.org/10.1016/j.apradiso.2012.04.018. Elsevier.
[31] L.G. Jacobsohn, et al., Radioluminescence and thermoluminescence of rare earth doped and co-doped YF_3, Radiat. Meas. 106 (2017) 79–83, https://doi.org/10.1016/j.radmeas.2017.05.001. Elsevier Ltd.
[32] S. Chand, R. Mehra, V. Chopra, Recent Developments in Phosphate Materials for Their Thermoluminescence Dosimeter (TLD) Applications, Luminescence, 2020, pp. 1–10, https://doi.org/10.1002/bio.3960 (August).
[33] C.B. Palan, S.K. Omanwar, A novel TL/OSL MCaPO4:Ce (M = Li,k) phosphor for radiation dosimetry, Optik 127 (18) (2016) 7137–7142, https://doi.org/10.1016/j.ijleo.2016.04.117. Elsevier GmbH.
[34] J. Guo, et al., Optically stimulated luminescence (OSL) of $LiMgPO_4$:Tm,Tb phosphor, J. Rare Earths 35 (6) (2017) 525–529, https://doi.org/10.1016/S1002-0721(17)60943-8.
[35] W. Gieszczyk, et al., Intrinsic and dopant-related luminescence of undoped and Tb plus Tm double-doped lithium magnesium phosphate ($LiMgPO_4$, LMP) crystals, Materials 13 (9) (2020) 1–14, https://doi.org/10.3390/MA13092032.
[36] S.N. Menon, et al., TL–ESR correlation studies in $LiMgPO_4$:Tb,B phosphor, Radiat. Eff. Defect Solid 173 (3–4) (2018) 210–222, https://doi.org/10.1080/10420150.2018.1424848.
[37] V.S.M. Barros, et al., Thermoluminescence study of aluminum oxide doped with therbium and thulium, Radiat. Meas. 45 (3–6) (2010) 435–437, https://doi.org/10.1016/j.radmeas.2010.01.001. Elsevier Ltd.

[38] J. Azorin Nieto, Present status and future trends in the development of thermoluminescent materials, Appl. Radiat. Isot. 117 (2016) 135–142, https://doi.org/10.1016/j.apradiso.2015.11.111. Elsevier.
[39] M.E.A. Andrade, et al., Thermoluminescence of aluminum oxide co-doped with terbium and thulium obtained via combustion synthesis, Radiat. Meas. 46 (12) (2011) 1474–1476, https://doi.org/10.1016/j.radmeas.2011.04.008. Elsevier Ltd.
[40] S.S. Reddy, K.R. Nagabhushana, F. Singh, Mechanism of thermoluminescence in high energy carbon ion irradiated Tb^{3+} doped Al_2O_3 phosphor for carbon ion beam dosimetry, Mater. Res. Express 4 (9) (2017), https://doi.org/10.1088/2053-1591/aa8928. IOP Publishing.
[41] Y. Zhao, et al., Thermoluminescence spectra of rare earth doped magnesium orthosilicate, J. Alloys Compd. 797 (2019) 1338–1347, https://doi.org/10.1016/j.jallcom.2019.05.173.
[42] C. Furetta, et al., Dosimetric characterisation of a new production of MgB_4O_7:Dy,Na thermoluminescent material, Appl. Radiat. Isot. 52 (2000) 243–250, https://doi.org/10.1021/acs.jced.6b00626.
[43] V. Altunal, et al., A systematic study on luminescence characterization of lanthanide-doped BeO ceramic dosimeters, J. Alloys Compd. 876 (2021) 160105, https://doi.org/10.1016/j.jallcom.2021.160105. Elsevier.

Upconversion luminescence behavior of rare-earth-activated phosphors: journey of infrared to visible

11

Abhilasha Jain[1] *and Sanjay J. Dhoble*[2]
[1]Department of Metallurgical and Materials Engineering, VNIT, Nagpur, Maharashtra, India; [2]Department of Physics, Rashtrasant Tukadoji Maharaj Nagpur University, Nagpur, Maharashtra, India

1. Introduction

The concept of upconversion was initially laid down by Dutch American physicist Nicholaas Bloembergen in the year 1959 [1]. Though he thrived hard, unfortunately, he could not validate the theory of photon upconversion. Later, in the year 1966, Francois Auzel successfully reported visible upconversion emission by employing an energy transfer phenomenon using rare-earth-doped sensitizers [2]. According to the literature, the phenomenon of upconversion is realized by an exclusive class of emitters whose ions with multiple energy levels consecutively absorb two or more low energy photons, undergo transitions, and eventually luminesce a high energy photon [3]. As the upconversion process is characterized by multiphoton emission and anti-Stokes emission, it is intrinsically different from conventional photoluminescence found in metal–ligand complexes and semiconductor quantum dots.

2. Principle of upconversion

Upconversion luminescence, also referred as anti-Stokes emission, is defined as a nonlinear optical phenomenon in which the sequential absorption of two or more low energy photons eventually results in high energy luminescent emission [4,5].

Anti-Stokes emission is a unique optical process in which long-wavelength excitation is converted into short-wavelength emission. According to the generalized upconversion principle, two luminescent centers in the ground state subsequently absorb energy from the excitation source (usually continuous-wave lasers or the converged sunlight) or go through the energy transfer process and reach the excited state. During the radiative transition from the attained excited state to the ground state, the upconversion process is indeed realized in the metastable intermediate excited states with the release of high energy photon emission possessing a wavelength shorter than

Rare-Earth-Activated Phosphors. https://doi.org/10.1016/B978-0-323-89856-0.00016-X

that of absorbed light [6]. As multiple real metastable energy levels are involved in the accommodation of absorbed photons, the upconversion process contributes better luminescence efficiency than the simultaneous two-photon absorption process and second-harmonic generation nonlinear process.

3. Composition of upconversion nanophosphors

3.1 Host lattice

The selection of host lattice and its interaction with dopant ions have a strong influence on the upconversion process as the host lattice regulates the distance between dopant ions, their coordination numbers, relative spatial positions, and type of anions surrounding the dopant [7]. In upconversion luminescence, the host is defined as well-known crystalline material having low phonon energies in which a reserved amount of extrinsic ions are doped to enhance the host's potential probabilities of luminescence excitation and emission [8]. For an ideal host, low phonon energy is a prerequisite to reduce nonradiative loss and upsurge radiative emission. Also, the host should maintain good resilience for sensitizer and activator ions along with comprehensibility for free migration of near-infrared (NIR) photons in lattice [9]. Efficient upconversion phosphors ever stand in need of good tuning between host lattice and dopant ions. Rare-earth or lanthanide ions act as favorable hosts for upconversion as their 4f valence electrons reveal electronic structures having energy level spacing corresponding to infrared and visible photon energies [7].

3.2 Activators

Activators are defined as "impurities" that provide levels to allow radiative transitions. These levels are near the valence band. When electrons occupy these levels, they act as traps for valence band holes. Some activators or impurities also provide levels near the conduction band. When these levels are empty, they act as electron traps [10]. In short, activators are luminescent centers in the host lattice where electron relaxes radiatively and thus prolong the emission time.

3.3 Sensitizers

Light-harvesting ions that are successfully incorporated into the host lattice to transfer their excitation energy to neighboring activator ions are called as sensitizers. Sensitizers often play a vital role when activators are unable to have significant absorption for the provided excitation energy. In such cases, sensitizers absorb the excitation energy and sequentially transfer it to an activator for characteristic luminescent emission. The addition of sensitizer in the host upconversion phosphor apparently drives out the concern of low absorption by activator ions and helps in minimizing the quenching phenomenon that fundamentally arises via cross-relaxation events [11,12].

4. Theoretical concepts of upconversion

4.1 Crystal field splitting of energy levels

Photon upconversion is hosted by optical centers with a reliable series of metastable excited states that enable a couple of steps of excitation into an upper excited state from where short-wavelength phonons are emitted [13]. Usually, in lanthanide-doped phosphors, the free ion energy levels of the $4f^N$ states are used for achieving photon energy upconversion. These free ion energy levels are demarcated by Russell–Saunders notation of $2S+1L_J$ multiples where S, L, and J are correspondingly the total spin, orbital, and angular momenta of the $4f^N$ electrons [14]. As photon upconversion is successfully realized with trivalent lanthanide ions in phosphor host materials, their energy levels in $4f^N$ configuration are calculated using an effective operator Hamiltonian employing the suitable parameters obtained by Carnall et al. [15]. To achieve multiphoton upconversion pumped by a narrow band laser in the IR spectral region, there is an urge of energy level scheme with evenly spaced intermediate states. This is facilitated by rare-earth ion Er^{3+} in $4f^{11}$ configuration that has exceptional energy level structure to expedite IR to visible multiphoton upconversion. With individual photon energy of around 1.5 μm, nearly five photon upconversion can be realized. It is well established that the free ion energy level split under the influence of crystal field interaction. As per the crystal field theory [16,17], ion lattice interactions and free ions interactions are treated in an effective Hamiltonian operator, wherein the crystal field part is defined by

$$< H_{CF} = \sum_{k,\,q,\,j} B^k q C^{(k)} q(i) \quad -1$$

where summation includes complete electrons of the open shell of the ions, $B^k q$ denotes crystal field parameters that account for the lattice contribution relying on ion–ligand distance and site symmetry, and $C^{(k)} q$ are tensor operator components acting on the free ion wave function defined in terms of special harmonics. As the off-diagonal matrix elements between the J multiplets are nonzero, the crystal field interaction provokes J mixing. Also, as the Russell– Saunders multiplets mix with the spin–orbital coupling of the same J, the crystal field splitting is not precisely considered by single $^{2s+1}L_J$ multiplets. As a consequence of the higher density of states, the upper excited states have more noteworthy L and J mixing than the lower-lying states. To reduce energy mismatch in multistep upconversion shifts, regulation of crystal field splitting by selection of suitable host material is indispensable for the development of upconversion phosphors. In nanophosphors, the excitation energy of dopant ions is controlled and tuned by varying the lattice structure, site symmetry, and particle size. Such kind of tuning is capable of enhancing the upconversion phenomenon by preventing energy mismatches. Also, it has a reflective effect on the excited state mechanism and luminescence characteristics. For instance, the charge transfer states of Eu^{3+} are tuned across the $4f^6$ states as the 5d and charge transfer states are comparatively more sensitive than the 4f energy states, to alter the luminescence properties affecting the excited state dynamics and eventually the efficiency of upconversion mechanism [18,19].

4.2 Symmetry dependence

High refractive index and low phonon energy are the fundamental prerequisites for the accomplishment of light conversion. Nevertheless, variation in crystal field symmetry leads to a difference in energy level splitting of the $^{2s+1}L_J$ multiplets and ultimately affects the upconversion transitions due to dissimilar energy mismatches. Removal of inversion symmetry is required to enable intra $4f^N$ transitions. However, as per the Judd-Ofelt theory, the crystal field-induced coupling along with unoccupied electronic configurations in higher energy states enables the parity forbidden intra 4f−4f electronic transitions [20]. But, in a strong odd crystal field component system and high energy transition regions near $4f^{N-1}$ energy states, Judd-Ofelt theory is not satisfactory. On considering $YAlO_3:Er^{3+}$ [21], the Judd-Ofelt theory failed in predicting transition intensities at C_S site symmetry as the coefficient b_{JM} relies on the energy gap between $4f^N$ and $4f^{N-1}$ 5d states, it had a larger value for the upper $4f^N$ state lying closer to $4f^{N-1}$ 5d state. Hence, the upconversion transitions in higher energy states are anticipated to have higher probabilities than those of low lying states. To develop highly efficient upconversion nanophosphors that use the energy level schemes of the rare-earth ions, breaking of symmetry in the host material is significant. Because, when the rare-earth ion is doped in a crystalline lattice, it induces lattice distortion thus breaking down the intrinsic site symmetry. Reduction in site symmetry can also be obtained by substituting a couple of ions in the host with other ions having the same oxidation state but different ionic sizes. Wang et al. [22] have shown in $NaYF_4$:Er−Yb system that $NaYF_4$ nanoparticles were successfully tuned in size, phase, upconversion color, and intensity by replacing Y^{3+} with Gd^{3+} thus enhancing the optical upconversion emission properties of the resulting system by more than two times.

4.3 Line broadening during electronic transitions

Generally, ion−phonon interactions result in broadening of the line width during electronic transitions. However, temperature-dependent phonon coupling, in weak ion−phonon interaction executes homogenous line broadening. Inhomogeneous line broadening is found due to structural defects that nonuniformly shifts the electronic energy levels [23].

4.4 Vibronic sidebands

Due to the coupling of charge-transfer states and interactions of electronic transitions including $4f^N$ spectral states result in strong vibronic sidebands. Vibronic sideband transitions are an imperative property to understand as they can profoundly improve the efficiency of upconversion by either increasing the absorption cross-section or by expanding the bandwidth which eventually compensates the energy mismatch.

Their working principle is based on Franck–Codon principle [24] and Huang–Rhys model [25], which states that the vibration frequencies of vibronic sidebands and electric dipole moment for electronic transitions are proportional. The Huang–Rhys model is applied in general to a single vibration mode with a constant frequency that is independent of electronic states. However, the intensity function for the 0–n phonon line between two states in an electronic transition separated by an energy gap E_0 and coupled to a vibrational mode having frequency v, is given by

$$f_n(\mathrm{E}) = \exp(-\mathrm{S})\frac{S^n}{n!}\frac{1}{\sqrt{2\pi n\sigma^2}}\exp\left[-\frac{E-(E_0+nhv)]^2}{2n\sigma^2}\right]$$

where in S is the Huang–Rhys parameter defining the vibronic coupling strength [26]. It also measures the lattice displacement driven by the electronic transitions. The equation described earlier characterizes the intensity and line shape of vibronic sideband transitions that harmonically progress in the multiphonon energy state of nhv where n ranges from 0 to n. The previous equation acts as a delta function for the shaping of the line of the zero phonon transition that accounts for the electronic dispersion and inhomogeneous broadening caused by lattice defects [27]. The zero phonon lines are as strong as vibronic sidebands [28].

5. Characteristics of upconversion

Energy transfer, relaxation of phonon, and lifetime of the excited state that are sensitive to particle size are key characteristic features to understand in rare-earth-doped upconversion phosphors.

5.1 Absorption and emission of photon

The theory of absorption and emission in rare-earth ions is well established [29,30]. As per the theory, the emission intensities are related to electronic transition rates between the involved energy levels. However, these transition rates depend on the initial and final states participating in the emission wherein these initial and final states are two different $4f^N$ spectral states with added components of opposite parity $4f^N$ 5d whose magnitude depends on the electric field strength of the neighboring crystal. Regarding the emission and absorption of a photon, the Judd-Ofelt approach is followed as a couple of parameters that are difficult to be estimated theoretically are replaced by experimental parameters that are easily realized by fitting emission and absorption data. It eventually enables to estimate the radiative rates for the rare-earth-doped ion–host combinations with the help of simple photoluminescence experiments. The Judd-Ofelt theory is successfully employed to obtain radiative lifetimes of excited states. Electric and magnetic dipole transitions are pronounced in absorption spectra of the rare-earth-doped upconversion host matrix. The magnetic dipole transitions are exclusively observed in the infrared spectral range. These magnetic dipole transitions that are quite weaker than electric dipole transitions are allowed between states with the same parity. The electric dipole transitions for the rare-earth ions are, in principle, forbidden and only partly allowed because of the perturbations caused by the host phosphor. Magnetic dipole transitions are slightly affected by the crystal field, and they can therefore be used as a criterion for electric dipole transitions described by the Judd-Ofelt theory [31].

5.2 Nonradiative relaxation

Due to the interaction between ion and phonon, the population of electrons in the excited electronic state decays to lower energy states. The energy difference between the two spectral states is converted into phonon energy and is termed as nonradiative relaxation. At room temperature, the excited state emission lifetime of many rare-earth-doped upconversion phosphors is influenced by nonradiative relaxation. Cross-relaxation is a type of nonradiative relaxation process of significance in upconversion luminescence. For a dipole–dipole interaction, the overall upconversion rate of an excited rare-earth activator ion is reduced after spatial averaging by the cross-relaxation rate. Upon increasing the activator concentration, cross-relaxation becomes more distinct, ultimately leading to a decrease in upconversion property after it exceeds the optimal concentration threshold [32]. This phenomenon is also known as concentration quenching [33–35]. As radiative rates of the rare-earth ions are determined by the properties of the host and ions themselves, the nonradiative rate depends on the host type, its phonons as well as its synthesis conditions. Multiphonon relaxation is significant in perfectly crystalline materials whereas in real materials there is a high density of lattice defects that significantly contribute to nonradiative energy decays. This occurs via a two-step process: first, the excited rare-earth ion transfers its energy to a nearby defect center. Later to that, the excited defect undergoes nonradiative relaxation. However, for upconversion nanophosphors that consist of small size nanoparticles where defect density comprises most of the nanoparticle volume, these defect processes may dominate nonradiative relaxation [36].

5.3 Energy transfer process

Energy transfer is an integral process of the photon upconversion process as ion–ion interaction leads to electronic excitation of ions after which they migrate in host phosphors through resonant energy transfer and photon assisted energy transfer process [37–39]. Owing to defect-induced inhomogeneous line broadening, the excitation energy varies between ions in the same as well as different electronic states. In the resonant energy transfer process, after absorption of excitation photons, the resonant energy transfer within the sensitizer–activator pair favorably transfers the excitation energy of the sensitizer through the radiative or nonradiative pathway to a nearby activator center that may either be in its ground state or in an intermediate state. Long-range dipole–dipole interactions assist nonradiative energy transfer whereas the radiative resonant energy transfer relies upon the emission and reabsorption of a photon. According to the Forster-Dexter model [40,41], the energy transfer rate between the sensitizer and activator ions, W_{SA}, is expressed using the spectral overlap between sensitizer emission and activator absorption. The resonant energy transfer rate is given by

$$W_{SA} = WS(R_0/R)^6$$

where R is defined as the distance between the sensitizer and activator ion, and R_0 stands for the Förster radius [42]. The Förster radius is the distance at which the energy transfer rate W_{SA} is equal to W_S. In the presence of activators, sensitizers relaxation occurs due to energy transfer along with a spontaneous deactivation process that includes light emission and nonradiative relaxation in the sensitizer. However, when the concentration of activators is optimized in terms of doping then the relaxation rate of sensitizers is strongly influenced by the energy transfer. Consequently, the energy transfer rate is proportional to activator concentration [32].

In the phonon-assisted nonresonant energy transfer process, the energy difference between two electronic states is converted into phonon energy wherein due to ion phonon interaction, the population of an excited spectral state decays through the nonradiative transition to lower energy states. During the upconversion process, phonon assistance is necessary to compensate energy mismatches with the aid of vibration energies of local and lattice modes for distinct upconversion transitions. Phonon-assisted energy transfer processes strongly influence the nature of luminescence centers in nanocrystals. Because of the restrictions on vibration modes in nanocrystals, luminescence lifetime and the efficiency of energy transfer are assumed to be altered when compared with bulk crystals. In nanocrystals, because of energy transfer, luminescence decay becomes nonexponential.

6. Upconversion mechanisms

Fundamentally, in the upconversion phenomenon, a system absorbs two or more photons, combines their energies, and emits a higher energy photon. Materials exhibiting the upconversion property are divided into two classes bases. The first class includes the lanthanide-doped inorganic phosphors where raw material precursor involves either a transparent crystalline powder or ceramic that has been doped with optically active ions that are incorporated into the host phosphor as substitutional defects. On the other hand, the second class consists of organic sensitizer phosphor systems. Here, we will discuss the mechanisms related to the first class of materials.

6.1 Excited-state absorption mechanism

This mechanism is based on successive absorption of two photons by an ion with the help of an intermediate excited state. As the excited state lifetime increases, the efficiency of upconversion is increased. In the excited state absorption (ESA) process, the emission intensity exhibits a nonlinear dependence on excitation intensity.

Electrons are arranged in the ground state electronic configuration inside the ion that is absorbed. As soon as the resonance conditions are met, the absorption of a single photon occurs, and the ground state electron is promoted to the first metastable excited state. This transition is known as ground state absorption (GSA). Again, when the resonance conditions are met, then due to the long-lived nature of the metastable excited state, absorption of the second photon takes place thus promoting the excited electron to a higher excited state. This transition is considered as ESA. If the energy differences between the ground state and the excited states are correspondingly spaced, then excitation may be achieved using a monochromatic excitation source as the resonance condition for both the electronic transitions are indistinguishable. Nevertheless, luminescence quantum yields from upconversion materials that employ ESA are less efficient because of relatively low dopant ion concentration and a small cross-section of absorption of the rare-earth-doped systems that are involved in emission [13,43].

6.2 Energy transfer upconversion mechanism

The energy transfer upconversion (ETU) mechanism occurs when an ion gets excited and transfers energy to a neighboring excited ion in the system. Consequently, it leads to double excitation of two resonant excited photons thus populating the highest metastable excited state, which further emits a higher energy photon. The difference between ESA and ETU mechanism is a nonradiative energy transfer pathway that exists between neighboring ions. The nonradiative energy transfer is accomplished through dipole–dipole interactions. When the energy difference between the ground state and first excitation state of the sensitizer is resonant with the first excitation to the second excitation transition of the activator, then energy transfer occurs before the sensitizer can radiatively relax to the ground state. If there is a slight energy mismatch between the transition pairs, then energy transfer is phonon-assisted to meet the resonance absorption condition. Energy transfer occurs only if there is a coulombic overlap between the two systems. Ultimately, the overlap of the electric fields that depends on the relative distance of the ion pair relatively strongly influences the efficiency of the ETU mechanism [39,40].

6.3 Energy migration upconversion

In this mechanism, four types of lanthanide ions participate namely sensitizers (type I), accumulators (type II), migrators (type III), and activators (type IV). They are spatially confined to discrete regions of the crystalline host-lattice. In the first step, the sensitizer center is excited by GSA (ground state absorption) and transfers its excitation energy to an accumulator center. The sensitizer is holding a large absorption cross-section. The energy levels of the sensitizer are closely matching the energy gaps of the accumulator's intermediate levels. The accumulator is supposed to have a long lifetime with the long-lived metastable state to accept and accumulate energy from the sensitizer. The energy migration then occurs from the high-lying excited state of the accumulator to a migrator center. As the migrator has a long lifetime, it reduces the decay rate of the excitation energy through optical transitions. Relaxation of the activator ion to its ground state results in the production of an anti-Stokes photon that subsequently emits the upconversion luminescence [44]. The ability of the optically active ions to sequentially absorb photons and thus accept energy via long-lived intermediate states allows to capture multiple photons so that they can occur over a longer timescale, thereby allowing the use of low power excitation sources.

6.4 Photon avalanche effect

The photo avalanche process is characterized by the three nonlinear events namely transmission, emission, and rise-time. The process begins with a nonresonant ground state absorption, which promotes the electron in the ground state to the second metastable excited state. Before it, relaxation of the electron to the first metastable excited state results in either the promotion of a ground-state electron to the first excited state in the same ion or in a neighboring ion through ion-pair relaxation energy transfer processes [45]. The energy of the incident excitation radiation is in resonance with the first to third electronic transition, and thus the absorption of this radiation by species with electrons residing in the first excitation level results in promotion to the third excitation state. Resonant excited state absorption of incident radiation brings the first excited state electrons back up to the third excited state. The process repeats multiple times resulting in an exponential increase in the population of electrons of the highly third excited state. Eventually, radiative relaxation from the third excited state to the ground state results in upconversion emission [46].

7. Synthesis methods

Upconversion phosphors are broadly synthesized using the methods of thermal decomposition [47–50], hydrothermal synthesis [51–53], and wet chemical-based synthesis [54]. The following methods are briefly elaborated herewith.

7.1 Thermal decomposition method

The thermal decomposition method commonly refers to high-boiling point solvents, in which the precursors are dissolved and heated. It generally dissolves organic precursors in high-boiling organic solvents with the assistance of surfactants. As the temperature rises, the precursors start decomposing to form nucleates, which later grow into target nanoparticles. The commonly used organic precursors are trifluoroacetate compounds and surfactants that are composed of polar capping groups and long hydrocarbon chains [55]. When a solution of the desired rare earth-trifluoroacetate is introduced into the hot ligand solution, reactants are rapidly released as a consequence of thermal decomposition, causing an immediate nucleation burst [49]. Reactant dissolution is assisted by the surface ligands, which are generally oleic acid that stabilizes the particles in solution by inhibiting aggregation successfully through repulsive interactions. The decomposition, nucleation, and growth processes are finely modulated by adjusting the internal parameters like temperature and concentration of the reaction systems. After a short reaction period, the thermal decomposition method results in regular shaped particle morphology with uniform size. Therefore, this route affords high-quality upconversion systems with uniform morphology and narrow size distribution. This method is used for the synthesis of $NaYF_4$-based upconversion nanophosphors [56].

7.2 Hydrothermal synthesis

Hydrothermal synthesis usually occurs in a water-based system with a lower reaction temperature (160–220°C) following an ecofriendly approach. It is an effective and comparatively convenient synthesis technique that is used to prepare inorganic materials with diverse controlled morphologies. This technique utilizes pressure and temperature that are above the critical point of the solvent. The reaction vessels used for this synthesis consist of a thick steel-walled vessel, hermetic seals, and protective inserts that are made of Teflon or titanium to prevent corrosion of the internal cavity of the vessel by solvents [57]. In this method, the heated solvents are put at very high pressure in autoclaves where the microenvironment is different than that of the conventional synthesis processes. The solubility and reactivity of the precursors get enhanced, in the autoclave thus allowing the fine crystallization of the final upconversion phosphor product usually exhibiting high crystallinity, fine homogeneity, and little defects [58]. However, the reaction time in this process is longer, typically for several hours and even days. Zhang et al. have reported an oleic-acid-assisted hydrothermal synthesis route to prepare a uniform nanostructure of $NaYF_4$-based upconversion nanophosphors with controlled morphology size and shape [59].

7.2.1 Coprecipitation method

Coprecipitation is a popularly used wet chemical technique through which nanophosphors with satisfactory phase purity can be achieved using a relatively simple synthetic procedure. Thus, this method is easily suitable for mass production. In this method, the most advantageous feature is the different solubility of salts in water. As reagents, two or more water-soluble salts are used that react with each other. The salts that are water-insoluble are formed in the liquid phase. Heat treatment is provided to improve the crystallinity of the products at temperatures ranging from 400 to 600°C. When the concentration of this product crosses the solubility product value in the reaction medium, then precipitation takes place [60]. The positive gains of the coprecipitation method include high yield, high final product purity, negligible necessity to use organic solvents, easily reproducible, and low cost [47].Nevertheless, the properties of the obtained nanoparticles, such as size, shape, and composition, are highly dependent on the reaction parameters such as temperature and nature of the basic solution. Homogeneous precipitation occurs via the separation of the nucleation and growth of the nuclei. Two steps are involved in this process. The first step includes a short break out of nucleation when the concentration of the species reaches critical supersaturation and in the second step there is a slow growth of the nuclei by diffusion of the solutes to the surface of the nanophosphor. Many research scientists have adopted the high-temperature coprecipitation method for the development of luminescent nanocrystalline upconversion phosphors of different host materials that include $NaGdF_4$ [61] and $NaTbF_4$ [62], respectively. Park et al. have prepared highly monodisperse $NaGdF_4$:Yb, Er nanoparticles using the coprecipitation method and demonstrated their dual-functionality in cell imaging [63,64].

8. Applications of upconversion phosphors doped with rare-earth ions

Upconversion nanophosphors doped with rare-earth ions find applications in an immense arena of research fields that include lighting and display, solid-state lasers, solar cells, infrared detection, and application as bio-labels for both magnetic resonance imaging and luminescence imaging. Bilir et al. [65] have reported the production of broadband that ranges from 400 to 900 nm white light following the monochromatic infrared light (803.5 and 975 nm) excitation of both nominally undoped and Nd^{3+}-doped Y_2O_3 nanopowders. The white light emission characteristics of the Y_2O_3 nanoparticles were probed in terms of the lighting standard parameters, which demonstrated that enhanced performances result under laser excitation of 803.5 nm and in the absence of Nd-doping. Correspondingly, the experimental findings revealed that, although the presence of Nd favors the generation of white light emission, undoped Y_2O_3 nanoparticles efficiently emit white light in the presence as well as the absence of rare-earth emitting centers too. This can be considered as an unprecedented experimental finding that suggests a host-related mechanism leading to white light emission. Ramasamy et al. [66] have highlighted how in an upconversion process, photons with energy lower than the bandgap of the solar cell are converted to higher energy photons that can be absorbed by the semiconductor. Shang et al. [67] have featured that light harvesting is the most crucial step to determine how much solar irradiation can be absorbed. The standard solar irradiation spectrum (AM 1.5) covers the wavelength region from UV to IR (300−2500 nm). However, most single-junction PV devices are not capable to absorb quantized light wave packets in the long-wavelength range, thus creating the performance-limiting issue of transmission loss. Upconversion materials circumvent this issue by spectral conversion, which has been applied to a wide range of solar cells. These photovoltaic devices include c-Si, amorphous Si thin-film, GaAs, and dye-sensitized solar cells. Shalav et al. [68] have revealed that $NaYF_4$:20% Er^{3+} microcrystals can be exploited as an upconverter in bifacial c-Si solar cells. Microsized $NaYF_4$:20% Er^{3+} phosphors were dispersed into an acrylic adhesive medium with matched refractive index, and then this mixture was deposited as a thick film on the rear of a bifacial c-Si solar cell. A notable increase of external quantum efficiency of 2.5% was realized under excitation at 1523 nm. Mi et al. [69] have fabricated dual-functional upconversion nanocomposites by covalently linking carboxyl-functionalized Fe_3O_4 nanoparticles and amino-functionalized silica-coated upconversion nanoparticles. As the nanocomposites hold active carboxylic and amino groups on the surface that are proved to be chemically active and useful for bio-conjugation with biomolecules, the upconversion nanoparticles in the nanocomposite emit visible light in response to the irradiation by NIR light, enabling the application of the nanocomposites in bioimaging. The nanocomposites exhibited good superpara-magnetic and excellent green upconversion photoluminescent properties that can be applied in bioimaging. Nguyen et al. [70] employed commercially available green upconversion phosphor microcrystals and functionalized them with amino groups to adsorb quantum dots. With variation in size of quantum dots, the composite materials emitted light with different colors and emission bands centering around 580, 590, 610, and 630 nm were observed, respectively, for each sample. Wang et al. [71] manifested a novel biosensor for the detection of trace amounts of avidin based on the fluorescence resonance energy transfer (FRET) process between bio-conjugated UC nanoparticles and gold nanoparticles. Both colloidal nanoparticles were functionalized with amino groups using the suitable method and then coupled it with biotin following established protocols. In a mixed solution, a different amount of avidin was added, and the PL spectra were recorded. Avidin acts as a bridge to couple the biotin-functionalized upconversion nanoparticle and gold nanoparticle to a close proximity. Ultimately, PL intensity of up conversion nano phosphor (UCNPs) was largely suppressed, and a direct correlation between the quenching efficiency and the concentration of avidin was achieved. Hence, a FRET system with upconversion-luminescent nanoparticles as energy donors and gold nanoparticles as energy acceptors were developed and applied to detect trace amounts of avidin. Dong et al. [72] have explained in their work that in bioimaging, the optimized excitation and emission wavelengths in the NIR spectral range are in between (700−1100 nm) and red region (600−700 nm) often termed as "optical window" of the biological tissues. The upconversion nanophosphors containing Er^{3+} thus exhibit two emission bands centered at 540 and 650 nm, separately. So single-band red upconversion light (650 nm) is proved to be a boon for bioimaging. Many more research on upconversion phosphors is enlisted in Table 11.1.

9. Future prospects

With rapid progress in technology, the upconverting rare-earth-doped phosphor materials have been reinvestigated for multiple applications in the present decade. As photon upconverting systems are typically composed of inorganic host materials that are doped with an optimized amount of rare-earth ions, selecting an appropriate host phosphor is extremely imperative for the development of high-performance driven upconversion phosphors with good optical properties, including high quantum efficiency and controlled emission characteristics. For expected optical properties, more efforts have to be made for enhancing growth in absorption efficiency and refined energy transfer within upconversion phosphors. Approaches to increase upconversion lifetime are also supposed to be intensely probed in the future. However, in the case of rare-earth-doped upconversion nanophosphors, it is still challenging to achieve controllable preparation of nanostructures due to constrained methodologies on

Table 11.1 Major outcomes in rare-earth-doped upconversion phosphor research that supported realization of real-time optoelectrical and biological applications.

Sr no.	Title	Authors	Research outcome	Year	References
1.	$KLaF_4$:Er an efficient upconversion phosphor	Neetu Tyagi, A. Amarnath Reddy, R. Nagarajan	The green-to-red emission ratio (GRR) of 4.9 with a decay time range of 35–70 ms (for green emission at 522 nm, under 980 nm excitation) was achieved in the $KLaF_4$:Er for the first time that proved beneficial for in vivo biolabeling and 3D flat panel displays	2010	[73]
2.	Preparation of Gd_2O_2S: Er^{3+},Yb^{3+} phosphor and its multiwavelength sensitive upconversion luminescence mechanism	Wei Fan, Xiyan Zhang, Lixin Chen, Liping Lu	Intense characteristic emissions of Er^{3+} ions were observed when excited by CW laser radiation at 980, 1064, and 1550 nm, respectively, which revealed that the obtained samples possess the outstanding advantage of application generality in the infrared detection field	2015	[74]
3.	Yb^{3+} sensitized $Na_2Y_2B_2O_7$: Er^{3+} phosphors in enhanced frequency upconversion, temperature sensing, and field emission display	Abhishek Kumar Soni, Vineet Kumar Rai, Manoj Kumar Mahata	Frequency upconversion (UC) emission study on $Na_2Y_2B_2O_7:Er^{3+}$ has been carried out upon 980 nm laser diode excitation. Optical temperature sensing study performed in the developed codoped phosphor revealed that the material can be used in temperature sensing application within 300–613 K temperature range with maximum sensitivity $7.9*10^3\ K^{-1}$ at 300 K	2017	[75]

Continued

Table 11.1 Major outcomes in rare-earth-doped upconversion phosphor research that supported realization of real-time optoelectrical and biological applications.—cont'd

Sr no.	Title	Authors	Research outcome	Year	References
4.	Near-infrared light-mediated rare-earth nanocrystals: Recent advances in improving photon conversion and alleviating the thermal effect	Linna Lyu, Haolun Cheong, Xiang Zhao ai, Wenmin Zhang, Juan Li, Huang Hao, Yang Jun Lin, Bengang Xing	Reports rational engineering of UCNs through crystal-lattice modification, energy-transfer manipulation, and design of core−shell structures as well as ingenious integration of lanthanide-based nanocrystals with other materials including plasmonic nanomaterials and organic dyes to increase the efficiency of photon upconversion for wide applications in biosensing, multimodal imaging, drug delivery, and light-controlled therapy.	2018	[76]
5.	Rare-earth-doped $Y_4Al_2O_9$ nanoparticles for stable Light-converting phosphors	Chenyang Liu, Suman Pokhrel, Christian Tessarek, Haipeng Li, Marco Schowalter, Andreas Rosenauer, Martin Eickhoff, Shuiqing Li, Lutz Madler	$Y_4Al_2O_9$ host was doped with Eu to investigate the photoluminescence properties of $Y_{4-x}Eu_xAl_2O_9$ ($x = 0.05-1.0$). The results revealed increased photoluminescence intensity with increasing Eu^{3+} concentration. Hence, $Y_{3.5}Eu_{0.5}Al_2O_9$ is proposed to develop a stable light-converting phosphor	2020	[77]
6.	Evaluation of Lanthanide-doped upconverting nanoparticles for in vitro and in vivo applications	Dana M. Samhadaneh, Gabrielle A. Mandl, Zhao Han, Maryam Mahjoob, Stephanie C. Weber, Marius Tuznik, David A. Rudko, John A. Capobianco, Ursula Stochaj	The study addressed major obstacles that hinder the therapeutic use of Ln-UCNPs in the clinic. Multiple independent lines of evidence were provided that demonstrate that the examined Ln-UCNPs are ideal platforms for in vitro and in vivo applications	2020	[78]

7.	Precision tuning of rare-earth-doped upconversion nanoparticles via droplet-based microfluidic screening	Shangkun Li, Yingchao Meng, Yujia Guo, Tian Liu, Stavros Stavrakis, Philip D. Howes, Andrew J. de Mello	A range of particles with optimized emission intensity and wide color distribution by changing the doping degree of sensitizer Yb to provide for green−orange tunability, and the ratio of Tm−Er to give green−blue tunability were synthesized. The approach was used to scan and tune the RGB emission attainable from single $NaYF_4$:Yb, Er, Tm nanocrystal composition by optimizing elemental ratios, engendering tunable true white light emission based on optimized RGB intensity ratios under single excitation for rapid UCNP synthesis, characterization, and optimization.	2021	[79]
8.	Near-infrared-triggered upconverting nanoparticles for Biomedicine applications	Manoj Kumar Mahata, Ranjit De, Kang Taek Lee	The authors reported the progress of UCNPs in biomedicine applications, such as optogenetic modulation, immunotherapy, drug-delivery, photodynamic/thermal therapy, bioimaging, and biosensing	2021	[80]
9.	Lanthanide-doped Luminescent nanophosphors via ionic liquids	Rahul Kumar Sharma, Pushpal Ghosh	Different aspects of ionic liquids (ILs) like their ability to act as templating agents, solvents, and reaction partners and sometimes their "three-in-one" use in nanomaterials synthesis are highlighted along with various photoluminescence mechanisms of Ln^{3+} ion like upconversion	2021	[81]
10.	On enhancement and control of green emission of rare-earth-codoped hydroxyapatite nanoparticles: synthesis and upconversion luminescence properties	Hoang Nhu Van, Tu Le Manh, Dung do Thi Thuy, Vuong-Hung Pham, Duy-Hung Nguyen, Diep Pham Thi Hong, Hoang Van Hung	Low-temperature hydrothermal synthesis of a series of x Mo−1%Er−10%Yb (x: mol%)-doped hydroxyapatite (HA) phosphors was studied. The UC mechanism of Mo^{6+}-doped HA:Er−Yb was determined to be three (green band) and two-photon (red band) absorption processes suggesting that the HA:Er−Yb−Mo phosphor can have potential in solar cell and biomedical applications.	2021	[82]

fabricating assemblies involving their morphology, surface properties, and the capability for chemical alterations. Correspondingly, organized toxicity investigations in terms of size, shape, and surface coating are desired so that suitable host lattice and dopant material for upconversion nanophosphors can be well-thought-out as agents in clinical preparation.

References

[1] N. Bloembergen, Phys. Rev. Lett. 2 (1959) 84−85.
[2] F. Auzel, Acad. Sci. 263 (1966) 819.
[3] X. Liu, C. Yan, J.A. Capobianco, Chem. Soc. Rev. 44 (2015) 1299−1301.
[4] J. Zhou, Q. Liu, W. Feng, Y. Sun, F. Li, Chem. Rev. 115 (2015) 395−465.
[5] C. Xu, W.W. Webb, J. Opt. Soc. Am. B 13 (1996) 481.
[6] D.R. Gamelin, H.U. Gudel, Trans. Metal Rare Earth Comp. 214 (2001) 1.
[7] M. Haase, H. Schafer, Angew. Chem. Int. Ed. 50 (2011) 5808.
[8] Q. Zhao, F. Li, C. Huang, Chem. Soc. Rev. 39 (2010) 3007.
[9] Q. Zhao, C.H. Huang, F.Y. Li, Chem. Soc. Rev. 40 (2011) 2508.
[10] D.F. Weng, X.J. Zheng, X.B. Chen, L. Li, L.P. Jin, Eur. J. Inorg. Chem. 21 (2007) 3410.
[11] L. Aboshyan- Sorgho, C. Besnard, P. Pattison, K.R. Kittilstved, A. Aebischer, J.C.G. Bunzli, A. Hauser, C. Piguet, Angew. Chem. Int. Ed. 50 (2011) 4108.
[12] H. Suzuki, Y. Nishida, S. Hoshino, Mol. Cryst. Liq. Cryst. 406 (2003) 221.
[13] F. Auzel, Chem. Rev. 104 (2004) 139−173.
[14] G.K. Liu, B. Jacquier, Springer, Berlin, 2005, pp. 1−94.
[15] W.T. Carnall, G.L. Goodman, K. Rajnak, R.S. Rana, J. Chem. Phys. 90 (1989) 3443−3457.
[16] B.G. Wybourne, Spectroscopic Properties of Rare Earths, John Wiley, New York, 1965.
[17] G. Liu, X. Chen, J.K.A. Gschneidner, J.C.G. Bunzli, V.K. Pecharsky, Elsevier, New York, vol. 37, 2007, pp. 99−169.
[18] P. Dorenbos, J. Phys. Condens. Matter 15 (2003) 8417−8434.
[19] P. Dorenbos, J. Lumin. 111 (2005) 89−104.
[20] M.F. Reid, Cambridge University Press, Cambridge, 2000, pp. 190−226.
[21] K.K. Deb, SPIE Adv. Laser Technol. & Appl. 335 (1982) 31−35.
[22] F. Wang, Y. Han, C.S. Lim, Y. Lu, J. Wang, J. Xu, H. Chen, C. Zhang, M. Hong, X. Liu, Nature 463 (2010) 1061−1065.
[23] W.M. Yen, P.M. Selzer, Laser Spectroscopy of Solids, Springer, Berlin, 1981.
[24] R. Kubo, Y. Toyozawa, Prog. Theor. Phys. 13 (1955) 160−182.
[25] K. Huang, A. Rhys, Proc. R. Soc. A 204 (1950) 406−423.
[26] M. Wagner, J. Chem. Phys. 41 (1964) 3939−3943.
[27] R.S. Meltzer, Springer, Berlin, 2005, pp. 191.
[28] X.J. Zhou, P.A. Tanner, M.D. Faucher, J. Phys. Chem. C 111 (2007) 683−687.
[29] W. Carnall, H. Cross white, Energy Level Structure and Transition Probabilities of the Trivalent Lanthanides in LaF-3, 1978.
[30] D. Mc Cumber, Phys. Rev. 136 (1964) A954.
[31] D.E. Mc Cumber, M.D. Sturge, J. Appl. Phys. 34 (1963) 1682.
[32] J. Ganem, J. Crawford, P. Schmidt, N. Jenkins, S. Bowman, Phys. Rev. B 66 (2002) 245101.

[33] O. Igoshin, A. Burshtein, J. Lumin. 92 (2000) 123–132.
[34] W. Naumann, J. Chem. Phys. 111 (1999) 2414–2422.
[35] A. Popov, V. Gladkikh, A. Burshtein, J. Phys. Chem. 107 (2003) 8177–8183.
[36] J. Zhao, Z. Lu, Y. Yin, C. McRae, J.A. Piper, J.M. Dawes, D. Jin, E.M. Goldys, Nanoscale 5 (2013) 944–952.
[37] T. Miyakawa, D.L. Dexter, Phys. Rev. B: Solid State 1 (1970) 2961–2969.
[38] W.M. Yen, in: A.A. Kaplyanskii, R.M. Macfarlane (Eds.), Spectroscopy of Solids Containing Rare Earth Ions, North Holland, Amsterdam, 1987, pp. 1–754.
[39] T. Forster, Ann. Phys. 2 (1948) 55–75.
[40] D.L. Dexter, J. Chem. Phys. 21 (1953) 836–850.
[41] T. Forster, Ann. Phys. 437 (1948) 55–75.
[42] H. Russell, F. Saunders, Astrophys. J. 61 (1925) 38.
[43] J. Chen, J.X. Zhao, Sensors 12 (2012) 2414–2435.
[44] F. Wang, R. Deng, J. Wang, Q. Wang, Y. Han, H. Zhu, X. Chen, X. Liu, Nat. Mater. 10 (2011) 968–973.
[45] F. Wang, X. Liu, Chem. Soc. Rev. 38 (2009) 976–989.
[46] J.S. Chivian, W.E. Case, D.D. Eden, Appl. Phys. Lett. 35 (1979) 124–125.
[47] G. Yi, H. Lu, S. Zhao, Y. Ge, W. Yang, D. Chen, L. Guo, Nano Lett. 4 (2004) 2191–2196.
[48] S. Heer, K. Kompe, H.U. Gvdel, M. Haase, Adv. Mater 16 (2004) 2102–2105.
[49] H. Mai, Y. Zhang, L. Sun, C.J. Yan, Phys. Chem. C 111 (2007) 13730–13739.
[50] H.X. Mai, Y.W. Zhang, R. Si, Z.G. Yan, L.D. Sun, L.P. You, C.H. Yan, J. Am. Chem. Soc. 128 (2006) 6426–6436.
[51] J.C. Boyer, F. Vetrone, L.A. Cuccia, J.A. Capobianco, J. Am. Chem. Soc. 128 (2006) 7444–7445.
[52] G. Yi, B. Sun, F. Yang, D. Chen, Y. Zhou, J. Cheng, Chem. Mater. 14 (2002) 2910–2914.
[53] J.H. Zeng, J. Su, Z.H. Li, R.X. Yan, Y.D. Li, Adv. Mater 17 (2005) 2119–2123.
[54] J.R. Taylor, M.M. Fang, S. Nie, Anal. Chem. 72 (2000) 1979–1986.
[55] Y. Wei, F. Lu, X. Zhang, D. Chen, Chem. Mater. 18 (2006) 5733–5737.
[56] H.Q. Wang, T. Nann, P. Hanninen, H. Harma (Eds.), Springer, Berlin, Heidelberg, 2011, pp. 115–132.
[57] A.X. Yin, Y.W. Zhang, L.D. Sun, C.H. Yan, Nanoscale 2 (2010) 953.
[58] M. Lin, Y. Zhao, S. Wang, M. Liu, Z. Duan, Y. Chen, F. Li, F. Xu, T. Lu, Biotechnol. Adv. 30 (2012) 1551–1561.
[59] J. Chen, C. Guo, M. Wang, L. Huang, L. Wang, C. Mi, J. Li, X. Fang, C. Mao, S. Xu, J. Mater. Chem. 21 (2011) 2632–2638.
[60] F. Zhang, J. Li, J. Shan, L. Xu, D. Zhao, Chem. Eur J. 15 (2009) 11.
[61] X. Wang, J. Zhuang, Q. Peng, Y.D. Li, Nature 437 (2005) 121.
[62] S. Gai, G. Yang, X. Li, C. Li, Y. Dai, F. He, P. Yang, Dalton Trans. 41 (2012) 11716–11724.
[63] F. Wang, Y. Zhang, X. Fan, M. Wang, J. Mater. Chem. 16 (2006) 1031–1034.
[64] Y.I. Park, J.H. Kim, K.T. Lee, K.S. Jeon, H.B. Na, J.H. Yu, H.M. Kim, N. Lee, S.H. Choi, S.I. Baik, H. Kim, S.P. Park, B.J. Park, Y.W. Kim, S.H. Lee, S.Y. Yoon, I.C. Song, W.K. Moon, Y.D. Suh, T. Hyeon, Adv. Mater. 21 (2009) 4467.
[65] G. Bilir, G. Ozen, J. Collins, M. Cesaria, B. Di Bartolo, IEEE Photonics J. 6 (2014) 8200518.
[66] P. Ramasamy, P. Manivasakan, J. Kim, RSC Adv. 4 (2014) 34873–34895.
[67] Y. Shang, S. Hao, C. Yang, G. Chen, Nanomaterials 5 (4) (2015) 1782–1809.
[68] A. Shalav, B.S. Richards, T. Trupke, Appl. Phys. Lett. 86 (2005) 013505.

[69] C.C. Mi, J.P. Zhang, H.Y. Gao, X.L. Wu, M. Wang, Y.F. Wu, Y.Q. Di, Z.R. Xu, C.B. Mao, S.K. Xu, Nanoscale 2 (2010) 1141.

[70] T.L. Nguyen, P. Spizzirri, G. Wilson, P. Mulvaney, Chem. Commun. 6 (2009) 174–176.

[71] L.Y. Wang, R.X. Yan, Z.Y. Hao, L. Wang, J.H. Zeng, J. Bao, X. Wang, Q. Peng, Y.H. Li, Angew. Chem. Int. Ed. 44 (2005) 6054.

[72] H. Dong, L.D. Sun, Y.F. Wang, J. Ke, R. Si, J.W. Xiao, G.M. Lyu, S. Shi, C.H. Yan, J. Am. Chem. Soc. 137 (2015) 6569.

[73] N. Tyagi, A. Amarnath Reddy, R. Nagarajan, Opt. Mater. 33 (2010) 42–47.

[74] W. Fan, X. Zhang, L. Chen, L. Lu, CrystEngComm 17 (2015) 1881–1889.

[75] A. Kumar Soni, V.K. Rai, M.K. Mahata, Mater. Res. Bull. 89 (2017) 116–124.

[76] L. Lyu, H. Cheong, X. Zhao Ai, W. Zhang, J. Li, H. Huang, Y.J. Lin, B. Xing, Lyu, et al., NPG Asia Mater. 10 (2018) 685–702.

[77] C. Liu, S. Pokhrel, C. Tessarek, H. Li, M. Schowalter, A. Rosenauer, M. Eickhoff, S. Li, L. Madler, ACS Appl. Nano Mater. 3 (1) (2020) 699–710.

[78] D.M. Samhadaneh, G.A. Mandl, H. Zhao, M. Mahjoob, S.C. Weber, M. Tuznik, D.A. Rudko, J.A. Capobianco, ACS Appl. Bio Mater. 3 (7) (2020) 4358–4369.

[79] S. Li, Y. Meng, Y. Guo, L. Tian, S. Stavrakis, P.D. Howes, A.J. de Mello, J. Mater. Chem. C 9 (2021) 925–933.

[80] M.K. Mahata, R. De, K. Taek Lee, Biomedicines 9 (7) (2021) 756–781.

[81] R.K. Sharma, P. Ghosh, Front. Chem. Front. 9 (2021) 715531.

[82] N.Van Hoang, T. Le Manh, D. Do Thi Thuy, V.-H. Pham, D.-H. Nguyen, D. Pham Thi Hong, H. Van Hung, New J. Chem. 45 (2021) 751–760.

Mechanoluminescence behavior of rare-earth-activated phosphors 12

Vikas Dubey[1], *Neha Dubey*[2], *Praveen Kumar Yadaw*[3], *N. Kumar Swamy*[3], *Rituraj Chandrakar*[4] *and Vikrant Tapas*[4]
[1]Department of Physics, Bhilai Institute of Technology Raipur, Naya Raipur, Chhattisgarh, India; [2]Department of Physics, Govt. V.Y.T. PG. Autonomous College Durg, Durg, Chhattisgarh, India; [3]Department of Physics, School of Science, ISBM University, Gariyaband, Chhattisgarh, India; [4]Department of Mechanical Engineering, NMDC Polytechnic College, Dantewada, Chhattisgarh, India

1. Introduction

1.1 Luminescence

Luminescence is a nonequilibrium phenomenon, whereby external energy excites the electronic state of the substance, and their excitation energy is released in the form of cold light emission. It is a type of excess emission that occurs in addition to a body's thermal emission and has a duration that is significantly longer than the period of light oscillations ($\sim 10-14$ s). Primarily, Monardes discovered it in an extract of Ligrium nephiticiem in 1565, but then it was not properly described, until 1852, when Sir G.G. Stokes published the theoretical background for the mechanism of this absorption (as an excitation) and corresponding emission.

The luminescence phenomenon is divided into three stages: (1) energy absorption, which causes electronic excitation, (2) energy storage, and (3) photon emission [1]. Fluorescence is a basic radiative transition among electronic states with the same number of multiplicity. Fluorescent emission occurs when electron transfers form a ground vibrational level of a first ground ***singlet*** state for any other of the vibrational levels of a ground electron state. Fluorescence rate constant is $\sim 10^6-10^9$ s^{-1}. Phosphorescence is a radiative, (spin) forbidden transition between electronic states with different multiplicity. Phosphorescence occurs when electron transfers from a ground vibrational level of an excited ***triplet*** state of a molecule to any other vibrational level of a ground electron state. Phosphorescence rate constant is $\sim 10^{-3}-10^2$ s^{-1}.

The light emission from fireflies, fluorescent tubes, television screens, oscilloscope screens, light-emitting diodes, self-luminous watch dials, fluorescent indication boards or roads, etc. is a few popular examples of luminescence to name. The important applications of luminescence are fluorescent tubes, television, cathode ray oscilloscope screen, detectors for ultraviolet rays, light-emitting diodes, X-ray imaging plates, fluorescent and phosphorescent pens, radiation dosimetry, thermoluminescence dating, damage sensors, fracture sensors, stress sensors, displays, lighting devices, etc.

Rare-Earth-Activated Phosphors. https://doi.org/10.1016/B978-0-323-89856-0.00011-0

In luminescence measurements, luminescence emission spectrum, excitation spectrum, luminescence polarization, luminescence power, quantum and energy yields, a rise and a fall of luminescence, and dependence of luminescence on different parameters are determined. Four basic elements for luminescence measurements are as follows: (1) a source for excitation or stimulation, (2) a light dispersing element, (3) a sample compartment, and (4) a light detector.

For example, in photoluminescence measurements, incandescent and arc lamps, as well as gas, dye, and solid-state lasers, are used as excitation sources, Prisms and gratings are used as light dispersing elements. For wavelengths greater than 320 nm, however, for shorter wavelengths, quartz or fused silica are used. For light detection, photomultiplier tubes, photodiodes, and charge-coupled devices are used.

Two models used to understand the mechanisms of luminescence are as follows: (1) configuration coordinate model and (2) energy band model. Fig. 12.1A shows the configuration of the coordinate model, wherein the coordinate represents the potential energy for the excited and ground states of the defect or luminescence center, and the abscissa represents the nearest neighbor ion separation. The curve is approximately a parabola because the atomic group behaves like a simple harmonic oscillator.

If the center absorbs a photon of energy $h\upsilon_0$ at the equilibrium position A of the ground state, then it is also raised vertically to the other excited state at B. This transition takes place vertically because according to the Franck–Condon phenomenon, the electronic transition arises in a very small time duration of about 10^{-16} s, which is several times shorter as compared to the vibrational period of nuclei, which is of the order of 10^{-13} s. Subsequently, after this absorption, the nuclei do not reside in the equilibrium position properly to the excited state; therefore, the system moves slowly to very minimum position C of the excited state curve in 10^{-10} to 10^{-11} s, and the basic energy difference among B and C is dissipated as phonon emission or

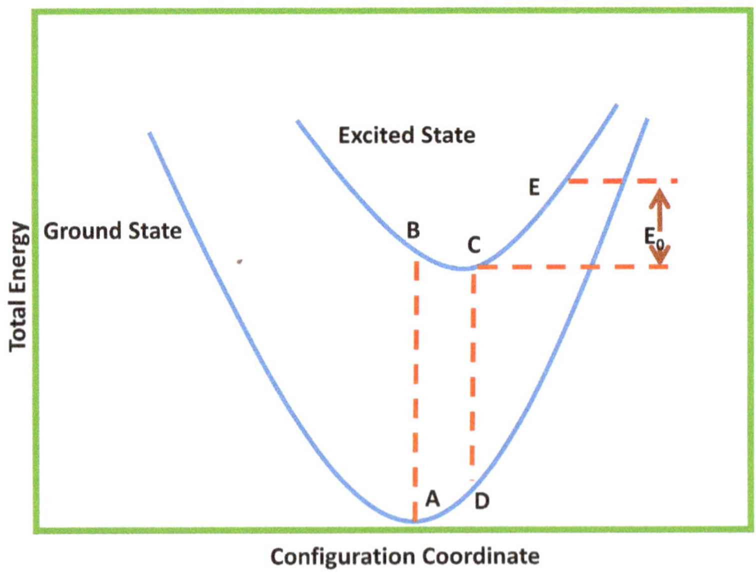

Figure 12.1a Configuration coordinate model of luminescence.

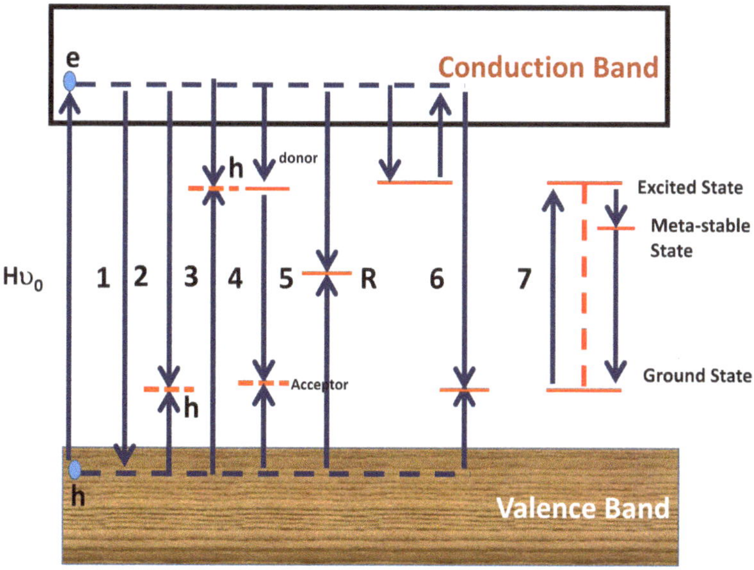

Figure 12.1b Energy band model of luminescence illustrating a schematic representation of different radiative recombination processes like (1) band-to-band recombination, (2) electron recombination with a hole stuck at the localized level above the valence band (Schon−Klasens model), (3) the recombination of the hole with the electron stuck at a localized level below the conduction band (Lambe−Klick model), (4) donor-acceptor pair recombination (Williams−Prener model), (5) electron−hole recombination center, (6) trapping, release, and recombination of the electron with a hole stuck at a localized level above the valence band, and (7) direct excitation of the impurity ion like Mn^{2+} in ZnS.

lattice vibrations. As the lifetimes of excited states are on the order of 10^{-8} s in the case of allowed transition and the period of lattice vibration is of the order of about 10^{-13} s, the excited atom can make 10^5 oscillations and transfer energy to the surrounding lattices. After reaching its novel equilibrium position on C, the center might come back to the ground state at D through the emission of quantum energy of the luminescent light, hυ. As point D is again not in the minimum energy configuration of the ground state, the system relaxes again with phonon emission to point A. As energy is transferred toward the lattice vibrations, the energy emitted is below the energy observed. Thus, there is a shift in the wavelength of emitted light as compared to that of the absorbed light and this is known as the Stokes shift.

The energy band model of luminescence might be described in the following way.

The energy states of electrons in an isolated atom are made up of distinct levels separated by forbidden energy regions. The Schrodinger equation defines the permitted states. Though, when ions or atoms are organized in an orderly way, and in very close proximity to each other to form a crystal, then their levels are mainly disturbed via mutual interaction. Consequently, the discrete states of isolated atoms are now merged into the bands of allowable energy divided by the bands of forbidden energy. The top-most band occupied with electrons is known as the valence band, and the first empty band responsible for electrical conduction is known as the conduction band. The activators such as Ag^+ and Cu^+ give rise to the activator states in the forbidden band just above the valence band and the coactivators such as Cl and Al give rise to the state in the forbidden band just below the conduction band. A schematic representation of different radiative recombination processes is illustrated in Fig. 12.1B, where steps can be self-understood.

1.1.1 Types of luminescence

Based on the modes of excitation characteristics, luminescence may be segregated into diverse types such as mechanoluminescence (ML), photoluminescence, cathodoluminescence, electroluminescence, chemiluminescence, bioluminescence, thermoluminescence, sonoluminescence, and lyoluminescence. Photoluminescence is produced by the various absorption of the photons of near-visible or visible radiation, while electroluminescence is mainly excited by the applied electric field. Similarly, the other types of luminescence are induced by the energies indicated by the prefix in luminescence (Fig. 12.2).

2. Mechanoluminescence

"ML is a type of luminescence that occurs when solids are mechanically deformed and emit cold light." ML has long been recognized as a phenomenon. Because several

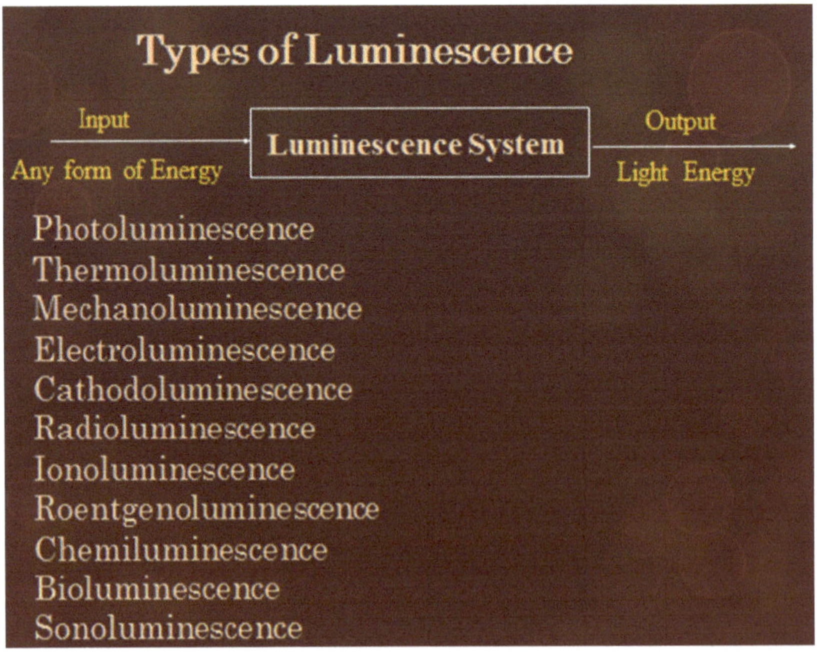

Figure 12.2 Types of luminescence.

common minerals emit very intense ML emissions, the ML was most likely discovered when humans lived in caves. The *Francis Bacon* wrote in his essay "Advancement of Learning" in 1605 that hard sugar scraped nimbly with a knife would produce a sparkling light.

ML occurs frequently in our daily lives. In the dark, we frequently witness light emissions when removing a silky garment from our bodies. This is the light that is produced when our body and silky cloth triboelectrify. This is known as triboluminescence, a kind of ML. The introduction of ML into modern adhesives and other polymers has drawn the attention of moms who use diaper tape and photographers who use rolling films.

Compressing, rubbing, stretching, crushing, grinding, cleaving, cutting, and scraping solids can all cause ML. Thermal shocks, such as those caused by rapid cooling or heating, or shock waves produced by the exposure of samples to intense laser pulses, can also stimulate it. ML can also occur as a result of phase transitions or crystal development, as well as during the separation of two different materials in contact. ML has been discovered in almost half of all inorganic solids and one-fourth to one-third of all organic solids in both crystalline and noncrystalline forms. In insulators, semiconductors, and some conductors, ML has been reported. Intense mechanoluminescent compounds include ditriphenyl phosphine oxide-manganese bromide, Eu- and Dy-doped strontium aluminate, N-acetyl anthranilic, impure saccharin, europium tetrakis (dibenzoylmethide), triethylammonium, manganese-doped zinc sulfide, and acenaphthene. The first four materials' ML is so strong that it can be seen in daylight.

3. Types of mechanoluminescence

3.1 Flow chart of types of mechanoluminescence

Basically, there are two types of ML, firstly, the *deformation ML* and secondly, the *tribo ML* (Fig. 12.3).

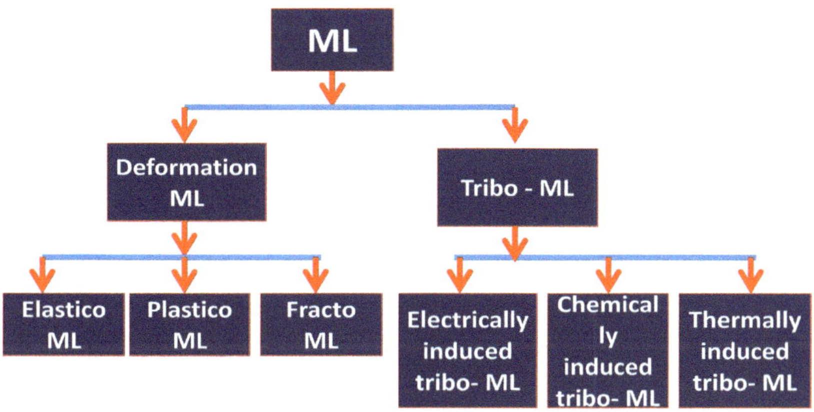

Figure 12.3 Different types of mechanoluminescence.

4. Deformation mechanoluminescence

The deformation ML is formed because of the physical process brought during the basic deformation of solids, it depends only on the materials below deformation, and it is mostly independent of the material intended for producing a deformation.

Deformation ML is further classified into three types as follows:

(i) **Elastico ML**: The luminescence formed throughout elastic deformation.
(ii) **Plastico ML**: The luminescence formed throughout plastic deformation.
(iii) **Fracto ML**: The luminescence formed throughout the fracture of solids.

5. Tribo mechanoluminescence

Tribo ML is based on contact phenomena, it is formed throughout the separation or contact of two different materials in contact and depends on the nature of both the materials in contact. The tribo ML is classified as follows:

(i) **Electrically induced tribo ML:** Electrically induced tribo ML is produced because of the triboelectric field produced through the friction between two dissimilar materials.
(ii) **Chemically induced tribo ML:** Chemically persuaded tribo ML is produced because of the chemical reaction produced through the rubbing of two dissimilar materials.
(iii) **Thermally induced tribo ML:** Thermally induced tribo ML is produced because of the heat formed through the rubbing of two different materials.

It is desirable to have a dosimeter that meets the usual requirements of precision, dose-rate independence, and postirradiation stability [2]. A dosimetric system for industrial usage must meet certain characteristics, including ease of use, speed of measurement, and low cost. When compared to the total cost of radiation processing, the cost of the dosimeter should be insignificant. Sugar has shown promising results as a dosimeter when employing the EPR approach [3–5], but the expensive cost of the equipment is a significant barrier to widespread use. Lyoluminescence [6,7] is another technology that has been employed to cut costs. ML can also help you save money. Grinding, cutting, colliding, hitting, and friction all induce ML, which is a light emission caused by mechanical stimulation [8,9]. Alkaline earth aluminate ceramics are essential host materials for luminescence applications that have been developed and researched by numerous researchers [10]. Alkaline earth aluminate (MAl2O4) is a mineral of the spinel group (MAl2O4) with the general chemical formula AB2O4, where A is a divalent atom [11]. The magnesium aluminate spinel (MgAl2O4) possesses excellent thermal, mechanical, and chemical properties, as well as high hardness and minimal electrical loss. As a result, it is currently used as a refractory for furnace walls and firebricks, and it could also be used as humidity sensors, laser materials, and a substrate in integrated circuits [12–19]. UV, β, γ, and other radiation-induced mechanoluminescent materials have been described in this chapter for personnel monitoring, environmental monitoring, radiation therapy, diagnostic radiography, and other radiation measures.

6. Mechanoluminescent materials

The first material on which ML was discovered was sugar. A wide range of organic and inorganic compounds have been found to contain ML. The phenomenon of ML has been discovered in about half of all organic and inorganic solids. ML can be found in both crystalline and noncrystalline substances. In insulators, semiconductors, and conductors, ML has been seen. The following are the six types of mechanoluminescent materials:

(i) Crystalline mechanoluminophors
(ii) Amorphous mechanoluminophors
(iii) Polymeric mechanoluminophors
(iv) Ceramic mechanoluminophors
(v) Composite mechanoluminophors
(vi) Metallic mechanoluminophors
(vii) Tribo mechanoluminophors

6.1 *Crystalline mechanoluminophors*

The noncentrosymmetric crystal, that is, the piezoelectric crystals display ML and the crystals not displaying ML are nonpiezoelectric. A few centrosymmetric crystals also exhibit ML, where the ML has a nonpiezoelectric origin [20,21]. Researchers have compiled a list of 451 mechanoluminescent crystals, of which 154 are inorganic compounds, 16 are minerals, 93 are aliphatic organic compounds, and 188 are aromatic organic compounds [22]. Mr. Chandra with his coworkers has described the crystal structure, the spectroscopy, and intensity, ML correlation of 234 organic crystals and 162 inorganic crystals [23–32].

6.2 *Amorphous mechanoluminophors*

Silicate, borate, quartz, and supplementary glasses display the phenomenon of ML [33–37]. ML has been observed throughout the fracture of the tissues of the human body [38]. Many researchers have described that the bones discharge light during their fracture [21].

6.3 *Polymeric mechanoluminophors*

The polymers like polyvinyl chloride, polyethylene, polypropylene polyvinylidene fluoride, and polystyrene display ML throughout their fracture. The ML of N-isopropylcarbazole and its derivatives has been reported [39]. Researchers reported ML during the plastic deformation of a few certain varieties of rubbers [40–42]. Tokhmetov and Vettergren have described ML during the breaking of bonds in polymethylmethacrylate [37]. Researchers reported ML during the stretching of polytetrafluoroethylene and nylon–6 [43].

6.4 Ceramic mechanoluminophors

ML has been reported in ceramics of Al_2O_3, $SiCZrO_2$, lead zirconate titanate (PZT), and Si_3N_4 [43,44]. Rare-earth-doped $SrAl_2O_4$ ceramics exhibit intense ML [45–48].

6.5 Composite mechanoluminophors

ML has been reported by many workers in Kevlar-filled epoxy fiber-reinforced epoxy boron epoxy composite, glass–epoxy composite, glass-filled polycarbonate, glass-filled polybutadine, and angle-ply glass fiber composite [49,50].

6.6 Metallic mechanoluminophors

It has been reported in the recent past that certain metals such as copper exhibit ML during their deformation and fracture.

6.7 Tribo mechanoluminophors

On the basis of contact phenomena, two materials are required to produce tribo ML, in which the luminescence is produced during the separation or contact of the materials. The renowned examples of the tribo-mechanoluminophors are as follows: diamond impregnated saw blade and glass, blood and steel rod, air and mercury, human body and silky cloth, luminescent layer and mercury, tape and substrate, steel (or sapphire) rod, and alkali halide crystal [36,44,51–55]. To date, the ML of noncolored and colored alkali halide crystals, piezoelectric crystals, and the II–VI compounds have been briefly studied in certain aspects of luminescence [56].

6.7.1 Materials for mechanoluminescence dosimetry

Different authors have reported various materials for the application of ML, such as alkali halides, oxides, nitrates, borates, aluminates, phosphates, silicates, and functional materials for ML dosimeters. Some of the studies are tabulated in Table 12.1.

6.7.2 Synthesis of mechanoluminescence dosimetry phosphors

Various synthesis methods, for example, solid-state reaction, combustion, basic sol–gel synthesis, and chemical route are proposed by various authors for the preparation of ML dosimetry phosphors.

The Few materials synthesis methods are discussed below:

(i) ***Solid-state method:*** *The conventional* solid-state synthesis method also popularly known as the Ceramic method is a traditional synthesis technique on high temperature often used for the production of as-prepared materials at a larger scale. This synthesis route is relatively inexpensive, has a short reaction time, and does not require complex apparatus. Moreover, the large yield of desired materials can be prepared in a relatively simple manner. The products are generally microcrystalline with a pure phase. The oxides and carbonates are used as initial precursors and are mixed in mortar and pestle or via ball milling technique. The

Table 12.1 The comparison of diverse types of materials for mechanoluminescence dosimeter.

Ref. No.	Authors and year	Study of materials	Application	Remarks
[57]	Bhujbal and Dhoble (2012)	$KBr:Ce^{3+}$ phosphor for radiation dosimetry	Accidental radiation Dosimetry	Gamma irradiated phosphor
[58]	Bhujbal and Dhoble (2011)	The γ-ray irradiated, Dy^{3+}, activated potassium-chloride phosphor	Basic accidental radiation dosimetry	Thermo. Lylo and high-dose measurements using the mechanoluminescence technique
[59]	Bhujbal et al. (2012)	Dy^{3+} activated $(K_{0.5}Na_{0.5})Cl$ phosphor	Accidental radiation Dosimetry	Up to ~5 kGy exposure dose and least fading in peak
[60]	Bhujbal and Dhoble (2012)	γ-rays irradiated $NaBr:Ce^{3+}$ phosphor	Ionizing radiation Dosimetry	ML dosimetry for high dose ionising radiation measurement
[61]	Rai et al. (2011)	$BaAl_2O_4$:Eu phosphor	ML dosimetry	Phosphor for radiation dosimetry
[62]	Satapathy et al. (2015)	$ZnAl_2O_4$:Eu novel phosphor	Dosimetric use due to its linear dose response	Synthesized by solution combustion technique
[63]	Satapathy et al. (2015)	Tb^{3+} doped $CaAl_2O_4$	ML and TL dosimetry	Detrapping and trapping of charge carriers in the host material can be calculated
[64]	Tigga et al. (2016)	$SrMgAl_{10}O_{17}:Eu^{2+}$ phosphor	ML dosimetry	Fracto-ML study
[65]	Tiwari et al. (2016)	$Ca_2Al_2SiO_7:Ce^{3+}$ phosphors	Phosphor for radiation Dosimetry	ML and TL dosimetry
[66]	Tiwari et al. (2016)	$Ca_2Al_2SiO_7:Ce^{3+}$ phosphors	ML dosimetry	PL and mechanoluminescence study

Continued

Table 12.1 The comparison of diverse types of materials for mechanoluminescence dosimeter.—cont'd

Ref. No.	Authors and year	Study of materials	Application	Remarks
[67]	Saluja et al. (2016)	CdS- and CdSe-doped phosphors	PL and mechanoluminescence study	Mn, Au, Ag, and Cu doped phosphor
[68]	Tiwari et al. (2015)	Ce^{3+}-doped $SrZrO_3$ phosphors	Thermoluminescence and mechanoluminescence studies	Combustion synthesis technique
[69]	Kaur et al. (2013)	Photoluminescence and mechanoluminescence spectra of (Zn, Cd)S mixed phosphor	Basic (Zn, Cd)S-mixed phosphors shifted to longer wavelengths with growing CdS content	—
[70]	Shrivastava et al. (2015)	Mechanoluminescence of $Ba_2MgSi_2O_7$ doped with Eu^{2+} and Dy^{3+}	Analysis with different impact velocity	—
[71]	Tiwari et al. (2015)	Impulsive distortion in zirconium: Ti phosphor	ML study of ZrO_2:Ti	Various analyses of mechanoluminescence peak
[72]	Tiwari et al. (2014)	Exact model for EML for II—VI semiconductors	Theoretical analysis	Elastic deformation of II—VI crystals
[73]	Tiwari et al. (2015)	Fracto ML	II—VI semiconductors	Fracto mechanoluminescence model
[74]	Kaur et al. (2010)	Mechanoluminescence spectra of gold-doped (Zn, Cd)S-mixed phosphors	PL and mechanoluminescence comparison	—
[75]	Tiwari et al. (2016)	Europium doped $CaZrO_3$ phosphor	ML study in various Eu ions	Total mechanoluminescence intensity calculation
[76]	Kaur et al. (2011)	Au-doped (Zn, Cd)S-mixed phosphors	Outcome of temperature on the ML	Temperature dependence study on ML

higher synthesis temperature may be lowered by using fluxing agents such as NH_4Cl, NaOH, KOH, and boric acid. These well-grounded materials are then mainly subjected to a higher temperature in furnaces for their calcination at various temperatures for the synthesis of phosphors.

(ii) ***Hydro(solvo)thermal method:*** Hydrothermal synthesis is a relatively lower temperature synthesis route for the fabrication of nanomaterials with varying shapes and sizes. In this method, the condensation and hydrolysis of reactants take place in a stainless steel autoclave with a solvent. If the solvent is any other than water medium, then this method is known as the solvothermal method. The autoclave is placed in a furnace or an oven at varying temperatures for different holding durations. To vary the morphology, pH is varied and different types of solvents, surfactants, and templates are used.

(iii) ***Combustion method:*** Combustion method is a facile lower temperature synthesis route, which results in highly uniform, crystalline, and pure end products. This synthesis route is quite economical and gives a high yield of nanoparticles. Various reducing agents and oxidizing also known as the fuels are used during the synthesis. Metal nitrates are used as oxidizing agents and urea glycine, citric acid, and hydrazine-based compounds are the most commonly used reducing agents. These fuels provide conditions for the very high exothermic reaction that took place. The stoichiometric amounts of fuel and oxidizing agents as calculated based on propellant chemistry. When this valency of the reducing agents balances the valency of the oxidizing agents and their ratio is 1, then the reaction is considered to be stoichiometrically balanced. The precursors and fuels are then added to an aqueous solution and are heated at room temperature (if needed). This aqueous solution is stayed in a preheated furnace, whereas the liberation of gases such as N_2 and CO_2 takes place with a flame. The temperature of the flame is sufficiently high that it outcomes in the formation of the desired products.

Fig. 12.4 shows the flow chart of characterization and preparation of phosphor by the combustion synthesis method.

(iv) ***The sol–gel method:*** The sol–gel synthesis is basically a wet chemical route known for its yield of uniform size, high purity, and crystalline nanoparticles. This method uses a chemical solution referred to as sol to produce an integrated network called gel. This method uses a chelating or complexing agent to form gel via carboxyl and hydroxyl groups. Usually, citric acid is used here as a chelating agent, thus this method is similarly called the citrate–gel method. To maintain the pH and upsurge the polymerization, ethylene glycol is used here as the medium. Entirely the precursors are commonly mixed with the chelating agent in an aqueous solution and are subjected to heating at 80–100°C (i.e., depending on the vaporization) and this results in gel formation. This gel is dried and subjected to various temperatures to obtain the desired end product

6.7.3 Characterization of mechanoluminescence dosimetry phosphors

The synthesized phosphor by various synthesis techniques such as solid-state reaction method and combustion synthesis method is characterized by ML pattern and ML spectra of the phosphor. Some of the ML patterns and spectra are reported here for the comparison of various phosphors.

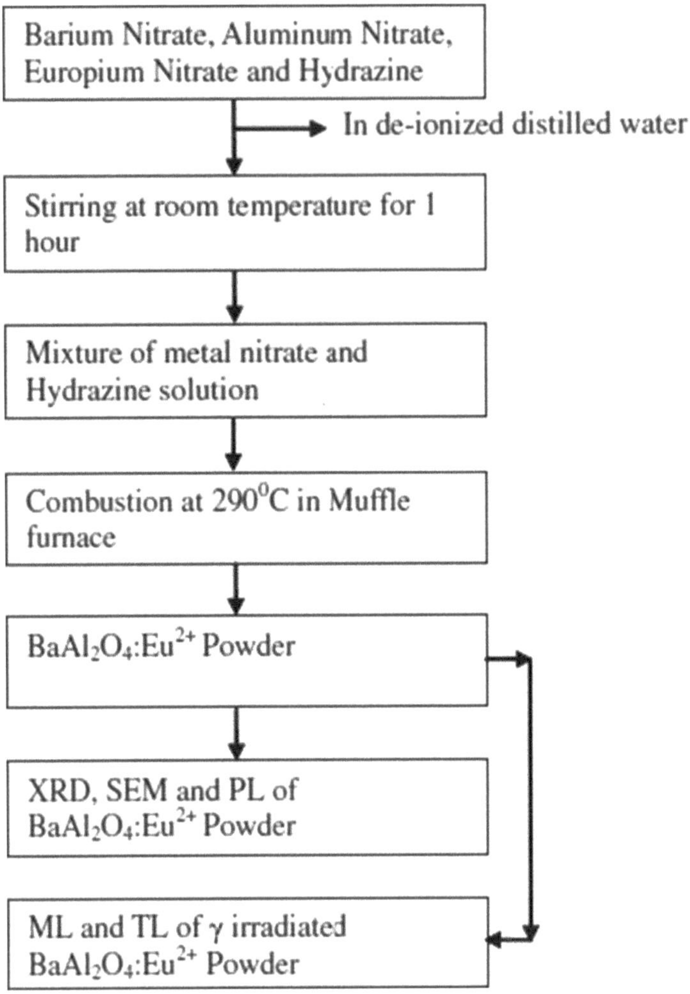

Figure 12.4 The flow chart for the characterization and preparation of the $BaAl_2O_4$:Eu phosphor.
Reuse with permission @copyright Elsevier [61].

The γ-irradiated $BaAl_2O_4:Eu^{2+}$ (0.25% mol percent) phosphor's ML glow curve (Fig. 12.5). In the ML intensity versus time graph, there are two peaks that may be seen. The strength of both peaks is also seen to rise as the g dose is increased [61].

In Fig. 12.6, (a) ML intensity versus time curve for various molar concentrations of europium metal in phosphor material, (b) mechanoluminescence intensity versus Eu concentration at its maximum, (c) the intensity of mechanoluminescence versus time curve for the dropped mass (400 g) at various impact heights, and (d) ML intensity at its maximum versus impact height are shown. The link between ML intensity and load impact height is linear, and an increase in impact height causes a proportional rise in ML intensity [77].

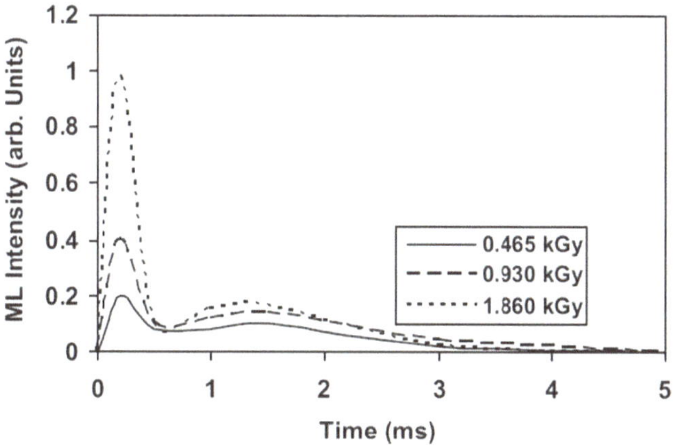

Figure 12.5 The dependence of mechanoluminescence intensity of $BaAlO_4$:Eu (~0.25 mol%) phosphor on the gamma dose.

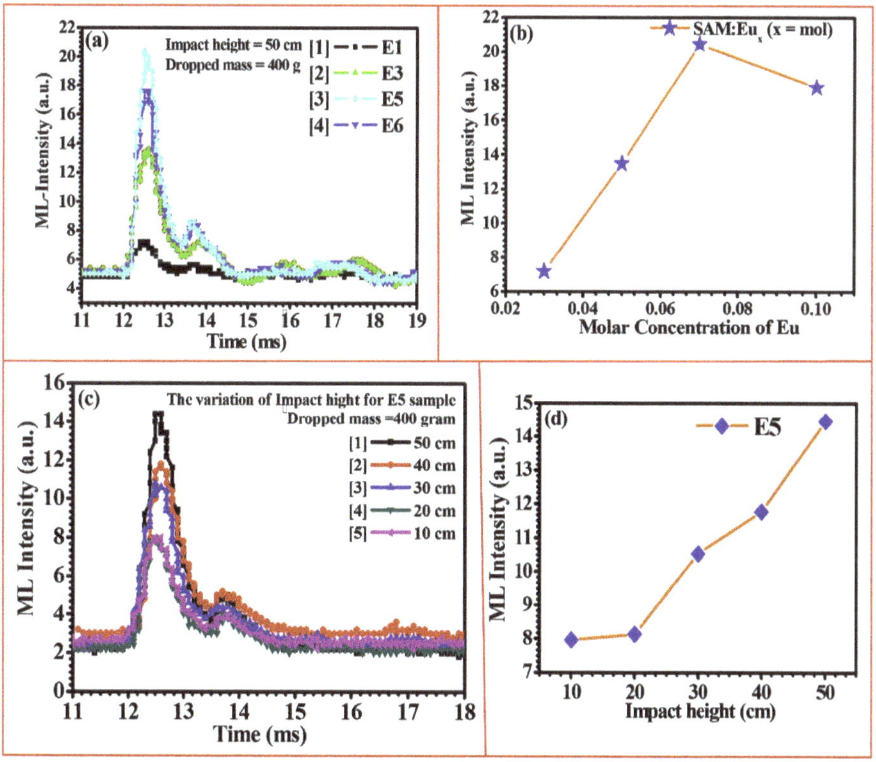

Figure 12.6 (A) Mechanoluminescence intensity versus time curve for various molar concentrations of europium metal in phosphor material. (B) Mechanoluminescence intensity versus Eu concentration at its maximum. (C) The intensity of mechanoluminescence versus time curve for the dropped mass (400 g) at various impact heights. (D) Mechanoluminescence intensity at its maximum versus impact height [77].

7. Mechanoluminescence study of $CaZrO_3:Eu^{3+}$ phosphor

ML is stimulated impulsively, and ML emission is largely due to the deformation of powder particles, according to Tiwari et al. [65]. In such instances, the transient ML intensity versus time curve only has one peak. The features of $CaZrO_3:Eu^{3+}$ phosphor ML caused by the impact of a moving piston of 500 g on the phosphors were measured (Fig. 12.7). In the ML intensity versus time graph, a single peak can be seen. The existence of a single peak suggests that ML emission involves a charge transfer process. The luminescence intensity is determined by the piston's impact velocity and dopant concentrations. Varied dopant concentrations at constant impact velocity and different impact velocities at constant dopant concentration were used in the experiment [75].

Fig. 12.7 shows the temporal relationship of mechanoluminescence intensity of $CaZrO_3:Eu^{3+}$ phosphor for various concentrations at constant impact velocity. The intensity of mechanoluminescence increases with time, reaches a peak, and then decreases with time. Fig. 12.8 shows that the mechanoluminescence intensity is maximum at 1.5 mol percent dopant Eu^{3+} concentration, then falls as the dopant concentration increases.

For a given dopant concentration (1.5 mol percent) and varying impact velocities, Fig. 12.9 depicts the time dependency of the ML intensity of $CaZrO_3:Eu^{3+}$ phosphor. It can be seen that the intensity of ML increases with time, reaches a peak, and then drops with time. With increasing piston impact velocity, the ML intensity I_m, which corresponds to the peak of the ML intensity versus time curve, increases (Fig. 12.10).

With increasing impact velocity, the value of time tm corresponding to the peak of ML intensity versus time curve at varying impact velocity of $CaZrO_3:Eu^{3+}$ phosphor decreases (Fig. 12.11). The plot between t_m and ln(1000/Vo), which is a straight line with a positive slope, is shown in Fig. 12.12.

For the $CaZrO_3:Eu^{3+}$ phosphor, Fig. 12.13 depicts the influence of total ML intensity IT (i.e., the area below the ML intensity vs. time curve) on impact velocity. The

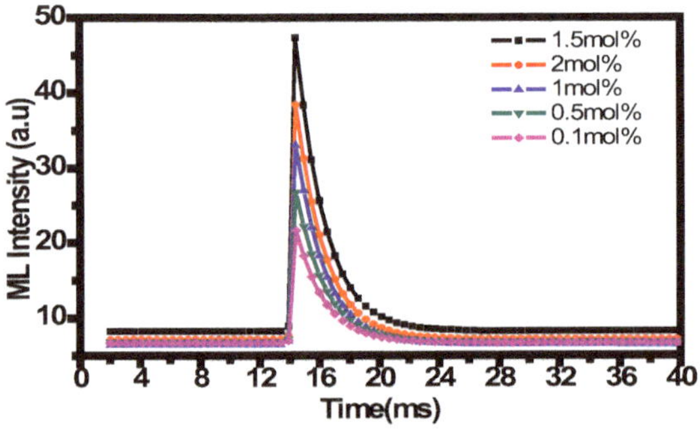

Figure 12.7 The time dependence of mechanoluminescence intensity of $CaZrO_3:Eu^{3+}$ phosphor for diverse concentrations of dopant (0.1, 0.5, 1, 1.5.2 mol%). Reuse with permission copyright Springer nature [75].

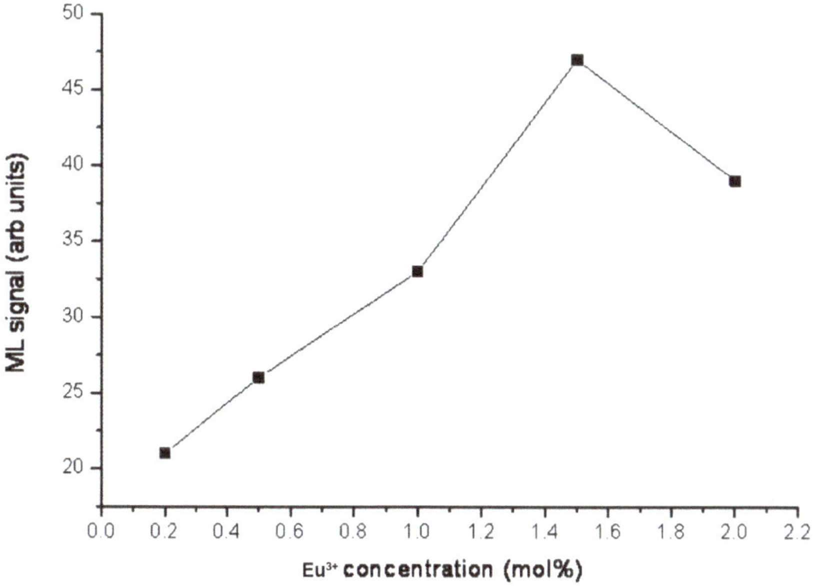

Figure 12.8 The variation in the mechanoluminescence intensity with concentration of dopant (Eu^{3+}).
Reuse with permission copyright@springer nature [75].

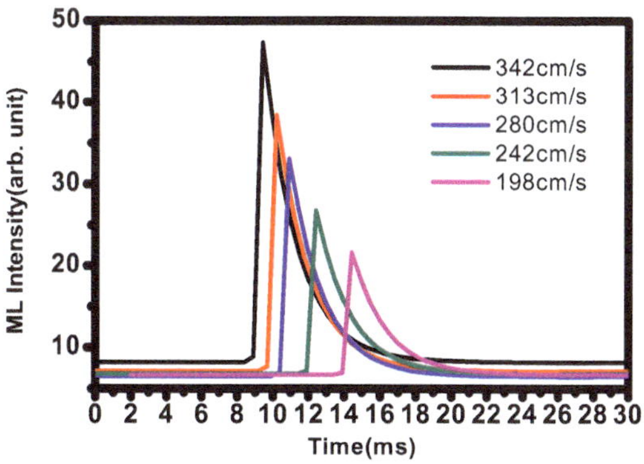

Figure 12.9 Time dependence of mechanoluminescence intensity of $CaZrO_3$:Eu^{3+}(~1.5 mol %) phosphor for different impact velocities.
Reuse with permission copyright@springer nature [75].

figure shows that total ML intensity IT grows with increasing impact velocity at first, but subsequently tends to reach a saturation value for higher impact velocity values (Figs. 12.14–12.16).

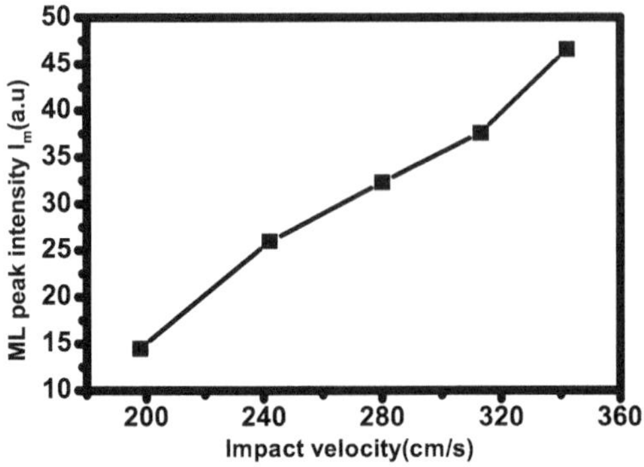

Figure 12.10 The dependence of mechanoluminescence peak intensity of $CaZrO_3$:Eu^{3+} phosphor on diverse impact velocities.
Reuse with permission copyright@springer nature [75].

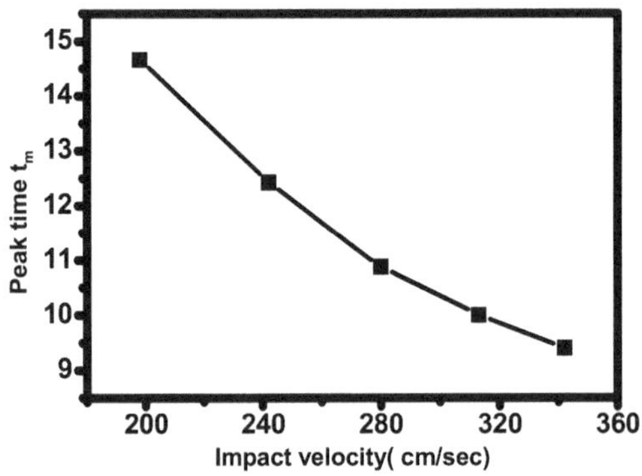

Figure 12.11 The dependence of t_m on impact velocity for $CaZrO_3$:Eu^{3+} phosphor.
Reuse with permission copyright@springer nature [75].

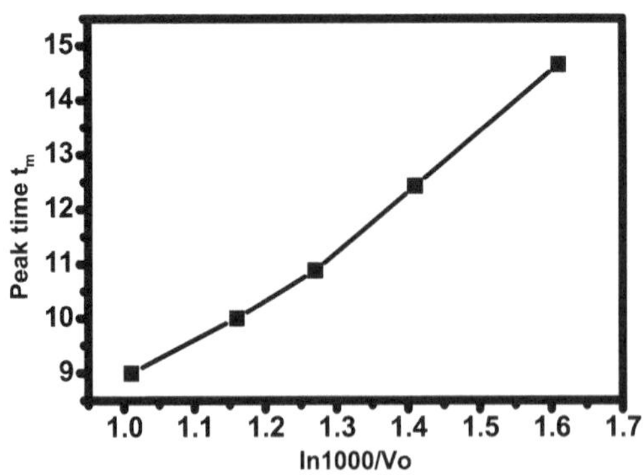

Figure 12.12 The Plot of t_m versus ln1000/Vo for $CaZrO_3$:Eu^{3+} phosphor.
Reuse with permission copyright@springer nature [75].

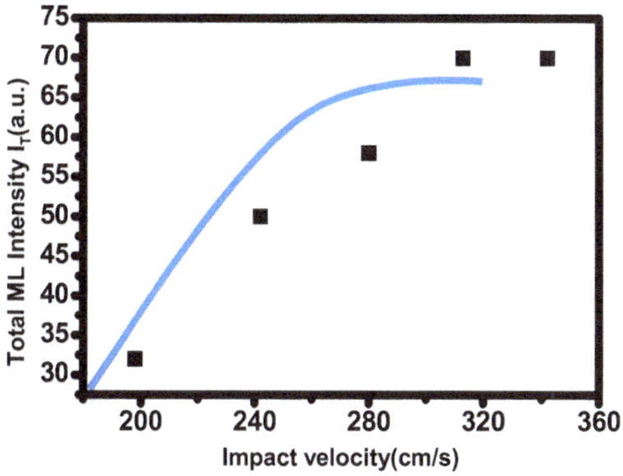

Figure 12.13 The dependence of entire mechanoluminescence intensity I_T on the impact velocity for $CaZrO_3:Eu^{3+}$ phosphor.
Reuse with permission copyright@springer nature [75].

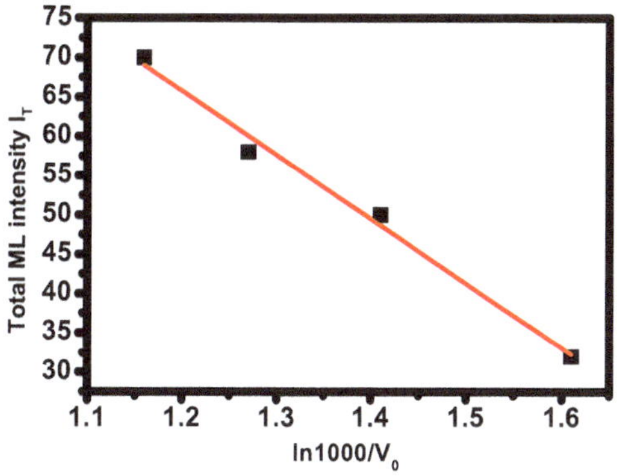

Figure 12.14 Plot of I_T versus ln1000/Vo for $CaZrO_3:Eu^{3+}$ phosphor.
Reuse with permission copyright@springer nature [75].

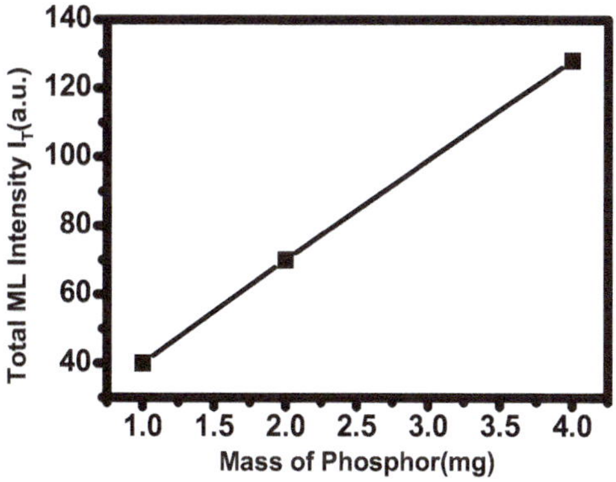

Figure 12.15 The dependence of entire mechanoluminescence intensity on the mass of phosphor($CaZrO_3:Eu^{3+}$).
Reuse with permission copyright@springer nature [75].

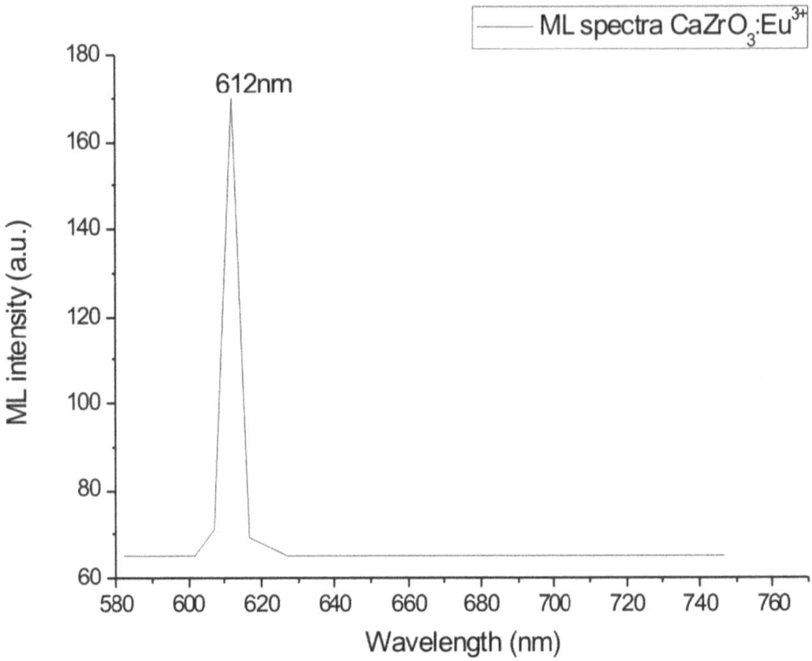

Figure 12.16 Mechanoluminescence spectrum of the $CaZrO_3:Eu^{3+}$ (~1.5mol%) phosphor. Reuse with permission copyright@springer nature [75].

8. Mechanoluminescence study of $SrZrO_3:Ce^{3+}$ phosphor

The properties of $SrZrO_3:Ce^{3+}$ phosphor ML caused by the impact of a moving piston weighing 500 g were measured (Figs. 12.17–12.22). In the ML intensity versus time graph, a single peak can be seen. The existence of a single peak suggests that ML emission involves a charge transfer process. The luminescence intensity is determined by the piston's impact velocity and dopant concentrations. Different dopant concentrations at constant impact velocity and different impact velocities at constant dopant concentration were used in the experiment [68].

Fig. 12.17 shows the temporal relationship of ML intensity of $SrZrO_3:Ce^{3+}$ phosphor for various concentrations at constant impact velocity. The intensity of ML increases over time, reaches a peak, and then decreases. Fig. 12.18 shows that mechanoluminescence intensity is highest at 0.2 mol percent dopant Eu^{3+} concentration, then falls as dopant concentration increases.

For a given dopant concentration (0.2 mol%) and varied impact velocities, Fig. 12.19 depicts the time dependency of the ML intensity of $SrZrO_3:Ce^{3+}$ phosphor. It can be seen that the intensity of ML increases with time, reaches a peak, and then drops with time. With increasing piston impact velocity, the ML intensity I_m, which corresponds to the peak of the ML intensity versus time curve, increases (Fig. 12.20).

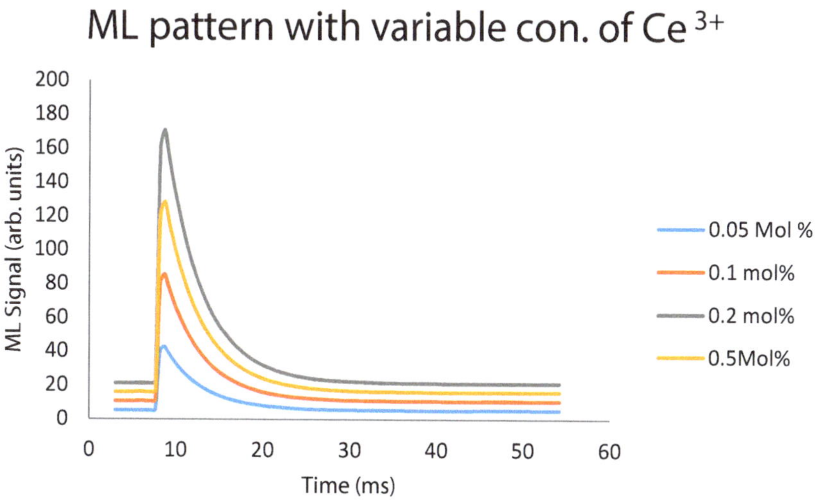

Figure 12.17 Time dependence of mechanoluminescence intensity of $SrZrO_3:Ce^{3+}$ phosphor for variable dopant concentration (0.05, 0.1, 0.2, 0.5 mol%) copyright@elsevier [68].

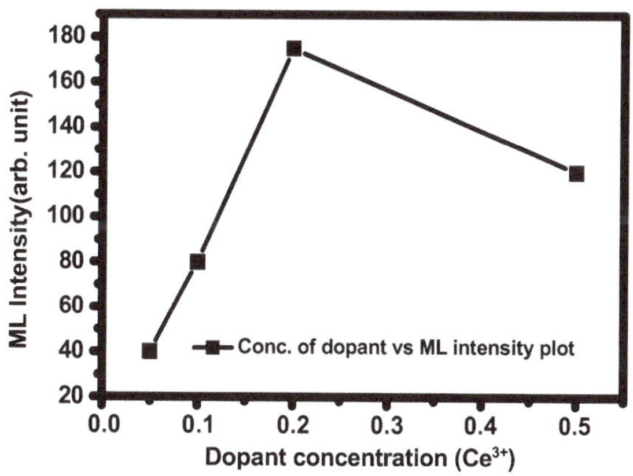

Figure 12.18 Peak mechanoluminescence intensity versus dopant (Ce^{3+}) concentration plot copyright@elsevier [68].

For $SrZrO_3:Ce^{3+}$ phosphor, Fig. 12.21 depicts the influence of total Ml intensity IT (i.e., the area below the ML intensity vs. time curve) on impact velocity.

The figure shows that total ML intensity IT grows with increasing impact velocity at first, but subsequently tends to reach a saturation value for higher impact velocity values.

Fig. 12.22 demonstrates the relationship between total ML intensity IT and phosphor mass. With increasing phosphor mass, the total mechanoluminescence intensity grows linearly.

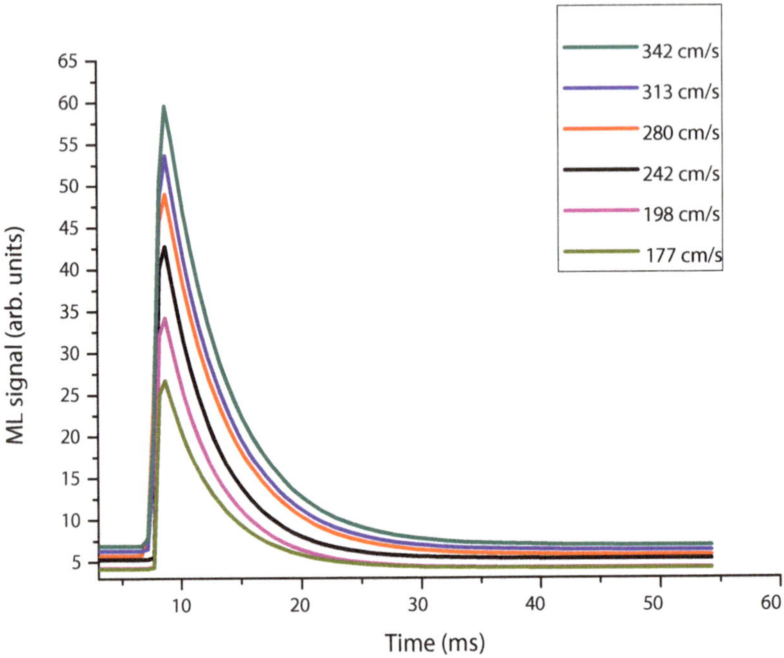

Figure 12.19 Time dependence of mechanoluminescence intensity of $SrZrO_3$:Ce^{3+} (0.2 mol%) phosphor for different impact velocities copyright@elsevier [68].

Fig. 12.23 shows ML spectra for $SrZrO_3$:Ce^{3+} phosphor, and a sharp single peak is detected at about 390 nm in ML spectra.

The influence of temperature on the ML of Au-doped phosphors of (Zn, Cd)S was studied by Kaur et al., who discovered that the ML disappears beyond a certain temperature [76]. The relationship between the temperature dependence of ML intensity and the relationship between the temperature dependence of ML intensity and the relationship between the temperature, $I = I_T^0 (1-T/T_C)^n$.

With a rising temperature of the phosphors, the peak corresponding to ML intensity versus time curve of (Zn, Cd)S: Au, Cl phosphors decreases and shifts toward lower values of time, Fig 12.23. The plot of the log I against time "t" curve for (Zn, Cd)S: Au, Cl phosphors is shown in Fig. 12.24. With increasing temperature, the slope of the log I against "t" curve remains unchanged. The slope of the log I versus "t" curve increases as the temperature of the phosphors rises for the decaying component of ML. The influence of temperature on total ML intensity IT is shown in Fig 12.25. It is noticed that IT drops with temperature and eventually vanishes above a certain temperature TC, which is substantially lower than the phosphors' melting point. The plot of log IT versus log(1−T/TC) in Fig. 12.26 is a straight line with positive slope. It was discovered that as the temperature rises, the intensity of ML decreases.

$$I_T = I_T^0 (1 - T/T_C)^n \qquad (12.1)$$

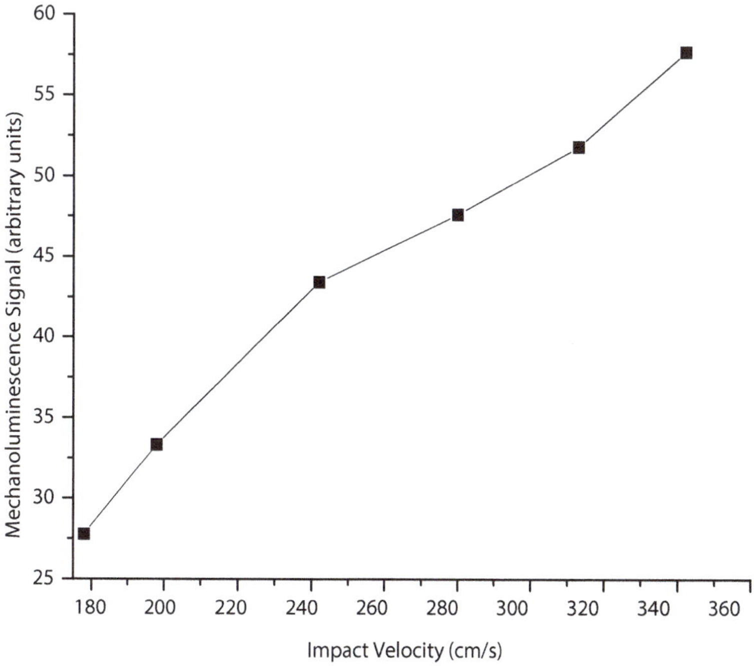

Figure 12.20 Dependence of peak mechanoluminescence intensity on different impact velocities for $SrZrO_3:Ce^{3+}$ (0.2 mol%) phosphor copyright@elsevier [68].

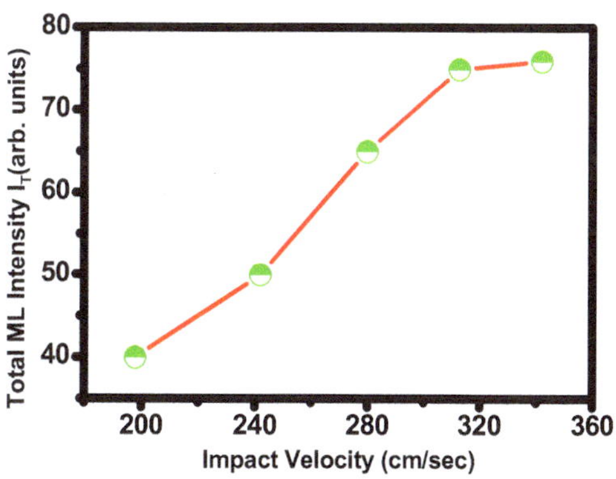

Figure 12.21 Dependence of total mechanoluminescence intensity I_T on the different impact velocities for $SrZrO_3:Ce^{3+}$ phosphor copyright@elsevier [68].

where I_T^0 is a constant, T_C is the temperature at which ML vanishes, and n is the slope of the log I_T against $\log(1-T/T_C)$ plot, which ranges from 0.9 to 1.10 [76].

Near-infrared ML crystals [78] were evaluated by Xiong et al. ML is a fascinating physical phenomenon in which a substance (typically a solid) generates light in

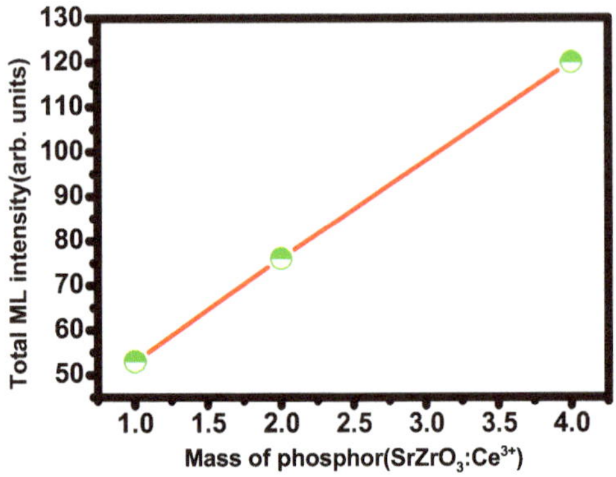

Figure 12.22 The dependence of entire mechanoluminescence intensity on the mass of phosphor ($SrZrO_3:Ce^{3+}$) copyright@elsevier [68].

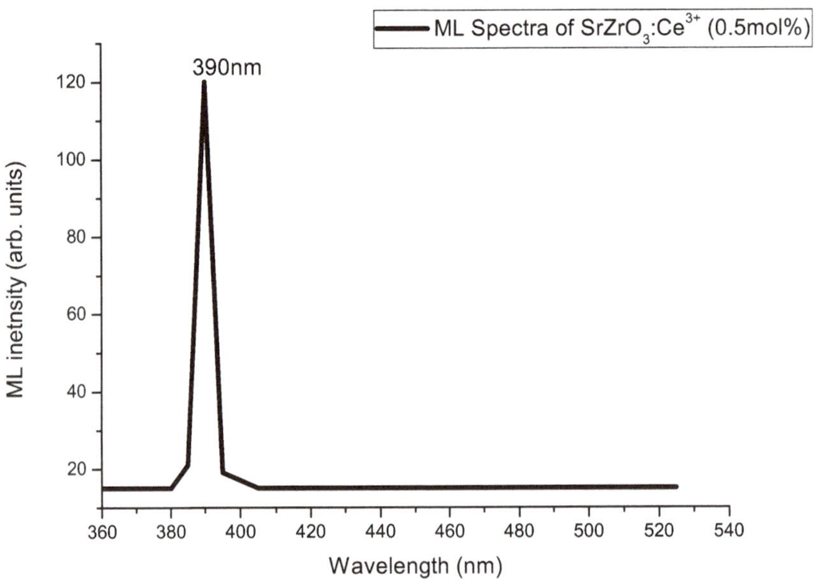

Figure 12.23 Mechanoluminescence spectrum of $SrZrO_3:Ce^{3+}$ (0.2 mol%) phosphor copyright@elsevier [68].

response to mechanical stimuli like friction, impact, compression, and bending [79,80]. Both inorganic and organic substances include it [78,81]. Fig. 12.27 depicts recent developments in ML materials.

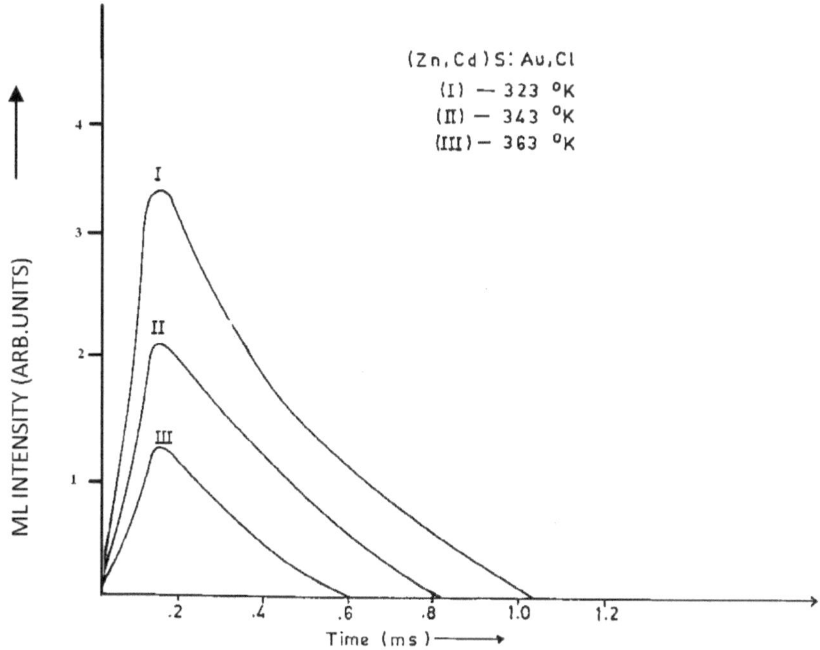

Figure 12.24 The time dependence of the mechanoluminescent intensity of (Zn, Cd)S: Au, Cl phosphors for diverse temperatures.

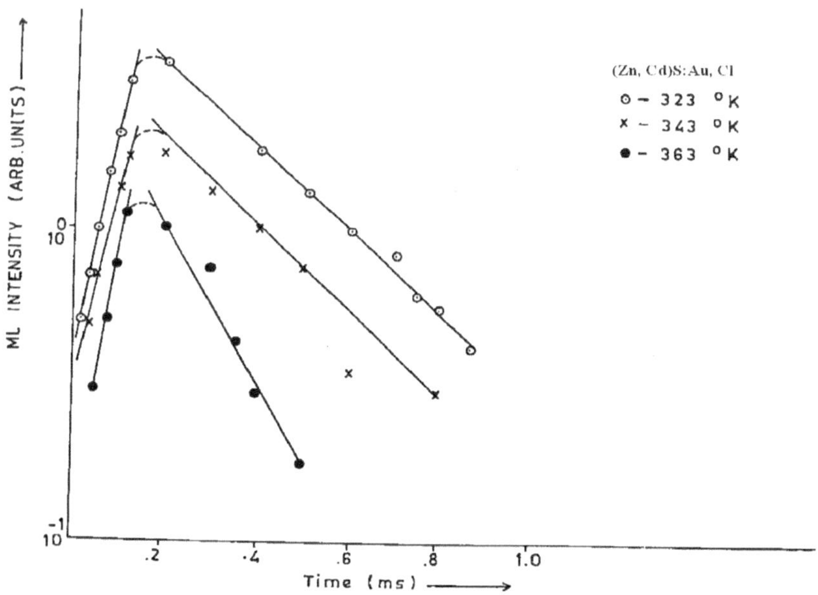

Figure 12.25 The plot of LOG I versus t, for (Zn, Cd)S: Au, Cl phosphors at diverse temperatures.

Reuse with permission copyright@elesevier [76].

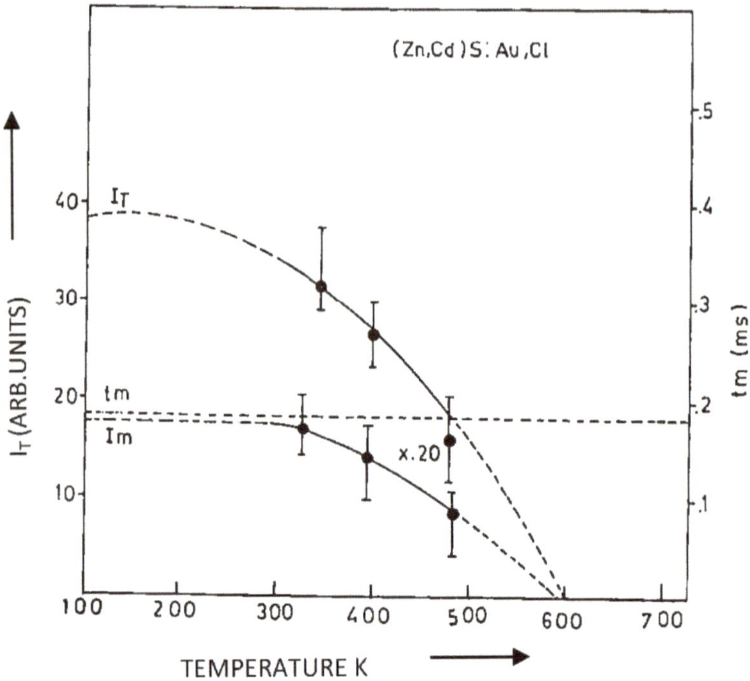

Figure 12.26 Effect of temperature on total mechanoluminescence intensity I_T and I_m of (Zn, Cd)S: Au, Cl phosphors.
Reuse with permission copyright@elesevier [76].

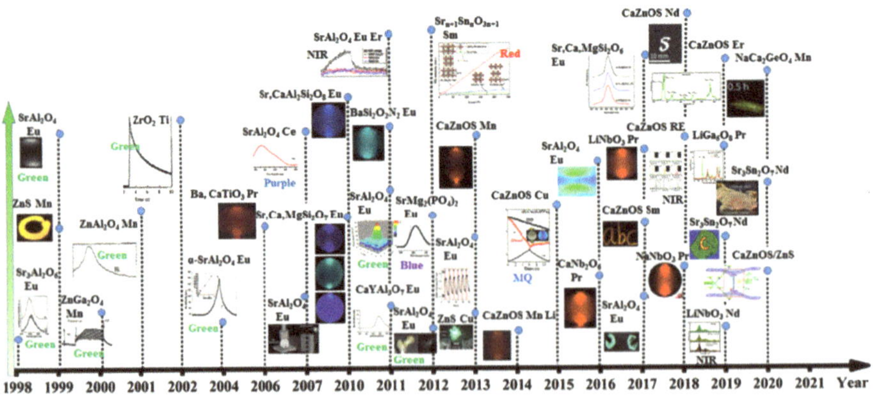

Figure 12.27 Recent advances in the mechanoluminescence material's development (year: 1998–2021).
Reuse with permission copyright@cellpress [78].

8.1 Devices for mechanoluminescence excitation

8.1.1 Devices for ML measurement devices used for ML measurement

For the ML measurements, two basic types of devices are usually needed: one for the spectral measurements and the other for deforming the samples.

8.1.1.1 Deformation techniques

Diverse deformation techniques designed for the systematic measurement of ML are as follows: stretching technique, loading technique, bending technique, pressure step technique, or compression technique.

8.1.1.2 Spectral measurement

ML spectra were first measured using a spectrograph and a photographic film, illustrates a photographic approach. The image intensifier is used to magnify the ML emission, and the image intensifier's output phosphor screen is imaged onto photographic film, with the resulting picture developed and densitometer to produce the intensity versus wavelength information. This technique's data analysis takes a long time. An array of diodes positioned at the phosphor screen turned the optical image at the phosphor screen into an electrical signal, removing the requirement for photographic imaging and densitometry. The ML spectra were recorded using a dual detector spectrometer by several researchers. Spectra are obtained one point at a time using this technique, which compares the light at each wavelength to the overall light output.

9. Characteristics of mechanoluminescence

9.1 Temporal characteristics

When crystals are got impulsively deformed, the ML intensity initially increases with time, reaches a peak, and then gradually drops. Two peaks arise in the ML intensity versus time curve of X or γ—irradiated crystals like alkali halides and phosphorescent crystals. The decay time of ML intensity after the first peak in impulsive deformation is related to the relaxation time of the piston velocity used to deform the crystal, but the decay time after the second peak is related to the lifetime of electrons in shallow traps of phosphorescent crystals or the lifetime of electrons moving along the dislocations axis in X or γ—irradiated alkali halide crystals.

9.2 Strain-rate characteristics

The ML intensity in colored II—VI semiconductors and alkali halide crystals rises linearly with the strain rates. The rate of appearance of ML pulses increases with increasing strain rate in a tiny fracto ML.

9.3 Deformation characteristics

When a solid is deformed at a constant strain rate and exhibits elastico or plastico ML, the intensity of the ML grows with time until it reaches saturation. When the deformation is stopped, the intensity of the ML drops exponentially or according to a power-law decline. When a solid that exhibits fracto ML is deformed, however, the ML arises only during the steps in the stress-strain curve, that is, only when cracks in the solid move.

9.4 Spectral characteristics

ML spectra imitate other types of luminescence spectra such as photoluminescence (PL), electroluminescence spectra, cathodoluminescence, and so on, or gas discharge spectra, or they combine both solid-state ML and gas discharge spectra.

9.5 Thermal characteristics

In general, the intensity of ML diminishes as the temperature of crystals rises. The ML intensity of X or γ-irradiated materials increases with temperature at first, reaches a peak, and then falls with subsequent temperature increases.

9.6 Pressure characteristics

The ML intensity increases with increasing pressure applied to the crystals.

9.7 Dopant characteristics

The ML intensity increases with dopant concentration in calcium fluoride, colored alkali halides, II-VI Semiconductors, and other crystals at first, then reaches an optimum value for a specific dopant concentration, and subsequently drops with further dopant concentration increases.

9.8 Newly created surface characteristics

The intensity of ML rises directly with the area of freshly formed surfaces in fracto ML.

9.9 Cleavage characteristics

The intensity of ML is determined by the crystal cleavage orientation.

9.10 Size characteristics

The ML intensity rises with the size of crystals.

9.11 Irradiation characteristics

The irradiation dose provided to the crystals improves the ML intensity of thermoluminescent materials.

9.12 Preparation characteristics

The intensity of phosphors' ML is determined by their processing circumstances.

10. Mechanisms of mechanoluminescence

10.1 Mechanisms of elastico ML

10.1.1 The dislocation mechanical interaction-induced elastico ML

The bending of dislocation segments between the pinning points occurs during the elastic deformation of a solid. As a result of the interaction between the bending dislocation segments and the color centers, electrons are transferred to the dislocation bands. As a result of the recombination of dislocation electrons with holes, light is produced [82]. Mechanical interaction−induced elastico mechanoluminescence (EML) may be responsible for the EML of X- or γ-irradiated alkali halide crystals [82,83].

10.1.2 The dislocation electrostatic interaction-induced EML

The interaction between the charged dislocations and filled traps may cause the transfer of trapped carriers to the conduction or valence band. Subsequently, the electron−hole recombination gives rise to luminescence [84].

10.1.3 The thermal excitation-induced EML

In some specific type of materials, large strain produced during elastic deformation may cause fluctuational displacements along the configuration coordinate such that there occurs crossing of the potential curves of ground state and excited state. The transfer of electrons from the ground state to the excited state may take place due to thermal energy and this may give rise to luminescence.

10.2 Mechanisms of plastico ML

10.2.1 Dislocation mechanical interaction-induced plastico ML

Electrons are transferred from defect centers to the dislocations through the mechanical interaction between defect centers and dislocations. Then the subsequent movement of dislocations transfers electrons to the recombination centers and thereby luminescence is produced. The ML produced during plastic deformation of X- or γ-irradiated alkali halide crystals is an example of dislocation mechanical interaction induced plastico ML [82,85].

10.2.2 Dislocation electrostatic interaction-induced plastico ML

When electrons in the filled traps interact electrostatically with the charged dislocations, then electrons tunnel from the trap to the conduction band. Later on, electrons recombine with hole at luminescence centers gives rise to luminescence. The stationary ML produced during plastic deformation of impurity−doped II−VI compounds is an example of dislocation electrostatic interaction induced plastico ML [84].

10.2.3 Charged dislocation electrification-induced plastico ML

Charged dislocations transfer to the surface of some crystals during plastic deformation. The surface of crystals becomes charged as a result of this. During the surface disintegration of the sample, the intense electric fields that arise are capable of causing bursts of luminescence. The charged dislocation electrification−induced plastico ML has been related to the pulsed ML observed during plastic deformation of II−VI compounds and LiF crystals [86,87].

10.2.4 Thermal-excitation-induced plastico ML

In the places where the crystal's slip bands cross, high local tensile stresses might occur. There's a chance that thermally induced electronic transitions could happen in high-stress areas. When the potential curves of the ground and excited states cross each other due to fluctuational displacements along the configuration co-ordinate, these transitions can occur. The thermal excitation−induced plastico ML has been shown to cause the ML produced during plastic deformation of nonirradiated LiF, NaCl, and MgO crystals [88,89]. The radiative decay of dislocation excitations may be responsible for the ML in this situation.

10.3 Mechanisms of fracto ML

10.3.1 Piezoelectrification-induced fracto ML

When a piezoelectric crystal is cleaved, fracture surfaces are formed, one of which is positively charged while the other is negatively charged. As a result, a powerful electric field on the order of 10^6-10^7 V cm^{-1} is generated [27,90]. The electric breakdown of gases and solids can be caused by such an electric field. Gaseous discharge, recombination luminescence, cathodoluminescence, and electroluminescence can all be produced depending on the conditions. All piezoelectric crystals, in general, have ML. Piezoelectrification−induced fracto ML is created during the fracture of sucrose, tartaric acid, uranyl nitrate hexahydrate, phenanthrene, and other materials.

10.3.2 Defective phase piezoelectrification-induced fracto ML

Some local portions of centrosymmetric crystals may be in the noncentrosymmetric phase due to a physical or chemical fault. As a result of the piezoelectrification, the passage of cracks across these locations may cause ML. Defective phase piezoelectrification-induced fracto ML is seen in arsenic trioxide and impure saccharin, for example. It should be emphasized that pure saccharin crystals do not exhibit ML [90,91].

10.3.3 Charged dislocation electrification–induced fracto ML

According to the mechanism proposed by Molotskii [92], a plastic zone with a large concentration of dislocations exists close to the tip of a fracture in any crystal, including brittle ionic crystals. Not all dislocation sources near the crack's tip succeed in evolving dislocations during fast fracture. As a result, the density of dislocations reduces as the fracture propagation velocity increases. The density of dislocation in LiF surpasses $10^8 cm^{-2}$ even at fracture propagation velocities near to 5×10^{-4} cm/s [56]. The thin part of the crystal buckles to a larger amount during asymmetric cleavage. As a result, the thin section experiences more shear stresses than the thick component, and these stresses cause the movable dislocations to move toward the thicker part. The emergence of symmetric opposite charges on the cleft region of the crystals is caused by the transport of charged dislocations across the cleavage plane.

The surface charge density produced by the asymmetric movement of dislocations in some crystals, such as LiF and NaCl, produces an electric field strong enough to trigger the dielectric breakdown of intervening gases and solids. As a result of the charged dislocation electrification produced during crystal fracture, ML may occur [56].

10.3.4 Baro–diffusion electrification–induced fracto ML

The baro–diffusion of charge carriers near the vertex of a moving crack may cause the charging of freshly formed surfaces in some nonpiezoelectric materials [93]. Vacancies and interstitial atoms serve as charge carriers. The surface charge density generated by this technique may generate an electric field strong enough to trigger dielectric breakdown of the gases and solids in between. ML can be detected due to the radiative decay of excitons and the excitation of alkali atoms existing between the charged planes, analogous to the charged dislocation electrification hypothesis. It should be noted that the ML produced during fracture of noncolored alkali halide crystals could be due to both charged dislocation electrification and baro–diffusion electrification, and it is currently unknown whether the charged dislocation electrification or baro–diffusion electrification processes are dominant.

10.3.5 Chemically-induced fracto ML

During crystal fracture, a chemical reaction may occur between the freed atoms or ions, and the subsequent release of energy may result in light. During the fractures, molecules from the surrounding gases may be absorbed or adsorbed on the newly formed surfaces, and the release of energy in this process may result in light. During the chemisorption of chlorine on sodium, Kasemo and Walden discovered the spontaneous emission of photons and electrons [94]. Kasemo has also detected light on aluminum and magnesium surfaces during oxygen chemisorption [94].

10.3.6 Thermally stimulated fracto ML

When a crystal fractures, the labor done is converted to heat, and the crystals are warmed locally. This causes the electronic system to be excited, resulting in light emission.

10.3.7 Incandescent fracto ML

In plastically strained portions of a solid, at the tip of the fracture, a high temperature is produced [95]. It was previously reported that a temperature on the order of 104 K could be maintained within a spot with a diameter of 10^{-6} m by solving thermal conduction equations [51,52]. Because these temperatures are higher than the samples' melting points, it is possible that the time it takes for a crack to grow to a length equal to the diameter of the local hot area is less than the time it takes for heat to be radiated or carried away through the crystal's body.

In vacuum, flashes of high-temperature luminosity are frequently seen when alkali halide crystals fracture [51,52]. The spectrum of the flash produced by cleavage of LiF and NaF in vacuum corresponds to the spectrum of the emission from a perfect black body at 10,000 K, but the spectrum of the emission produced by cleavage of NaCl corresponds to the spectrum of the emission from a perfect black body at 6000 K.

10.3.8 Stress perturbation-induced fracto ML

When fractures propagate through particular crystals, a high pressure is created toward the tips of mobile cracks, which may cause a new ground state to form due to thermal electron transfer when the pressure is suddenly released. The electron can either relax to the ground state or stay in the excited state long enough to relax to 1-atmosphere ground state, resulting in luminescence [96].

10.3.9 Thermal excitation-induced fracto ML

Thermally activated electronic transitions can happen in areas with a lot of strain or a lot of stress. These transitions can occur when the potential curves of the ground and excited states cross each other due to fluctuational displacement along the configuration coordinate, and the following transition of excited electrons to the ground state causes light. This action is the inverse of the excited state quenching caused by heat radiation. Many researchers have demonstrated ML during semiconductor cleavage [97]. It was discovered that the breakage of 1015 bonds produces approximately 108–109 photons. The thermal excitation mechanism causes fracto ML in Ge, SiGaP, and other semiconductors, where the interatomic distance between some of the atoms in broken bands may be quite high, allowing the thermal population of electrons from the ground state to the excited state. The electronic transition has been shown to occur in highly disordered regions of crystals subjected to mechanical stress [98].

10.3.10 Mechanically excited fracto ML

The mechanical energy supplied to real mechanoluminophors should be directly transferred to the constituent atoms via changed lattice vibrations, for example, and excite the atoms to luminescence. This mechanism may exist, but it has yet to be discovered.

10.3.11 Deep trap–induced fracto ML

When a solid fractures, new surfaces are produced, resulting in deep traps. Luminescence may result from the recombination of charge carriers with deep traps. When certain metals, such as copper, are fractured, dislocations develop on the newly generated surfaces, forming holes. The liberated electrons eventually recombine with holes and create light [56].

Longchambon devised a new setup in which the crystalline sample powder is violently thrown onto a quartz plate by a pressurized gas flow to avoid excessive sample consumption. Large exhaust tubes are used to remove both gas and dust from the crushed sample (Fig. 12.28).

To our knowledge, there is no commercially available integrated mechanoluminescent spectrometer that combines mechanical excitation and luminescence

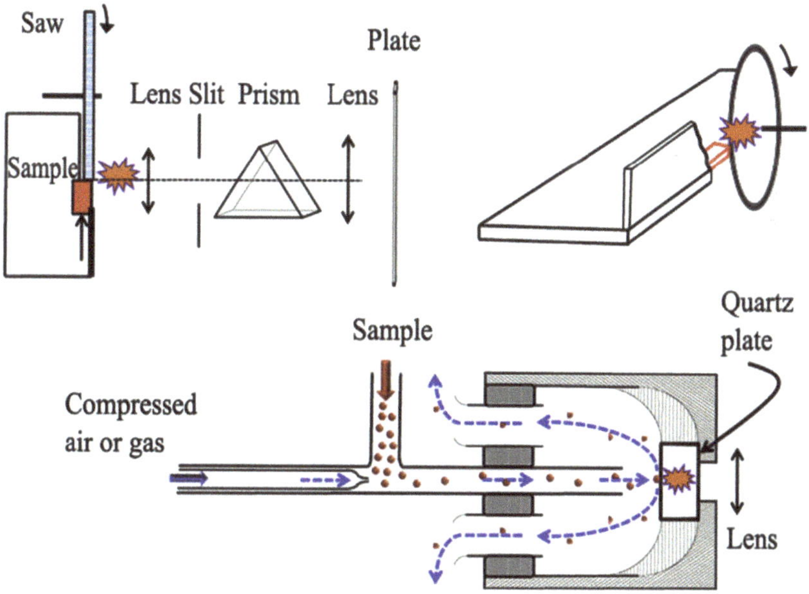

Figure 12.28 Longchambon's designs for measuring mechanoluminescence are depicted schematically. Top: abrasion of a sample with a circular metal saw. Bottom: microcrystals are crushed by gas flow against a quartz window.
Redrawn after Ref. [78].

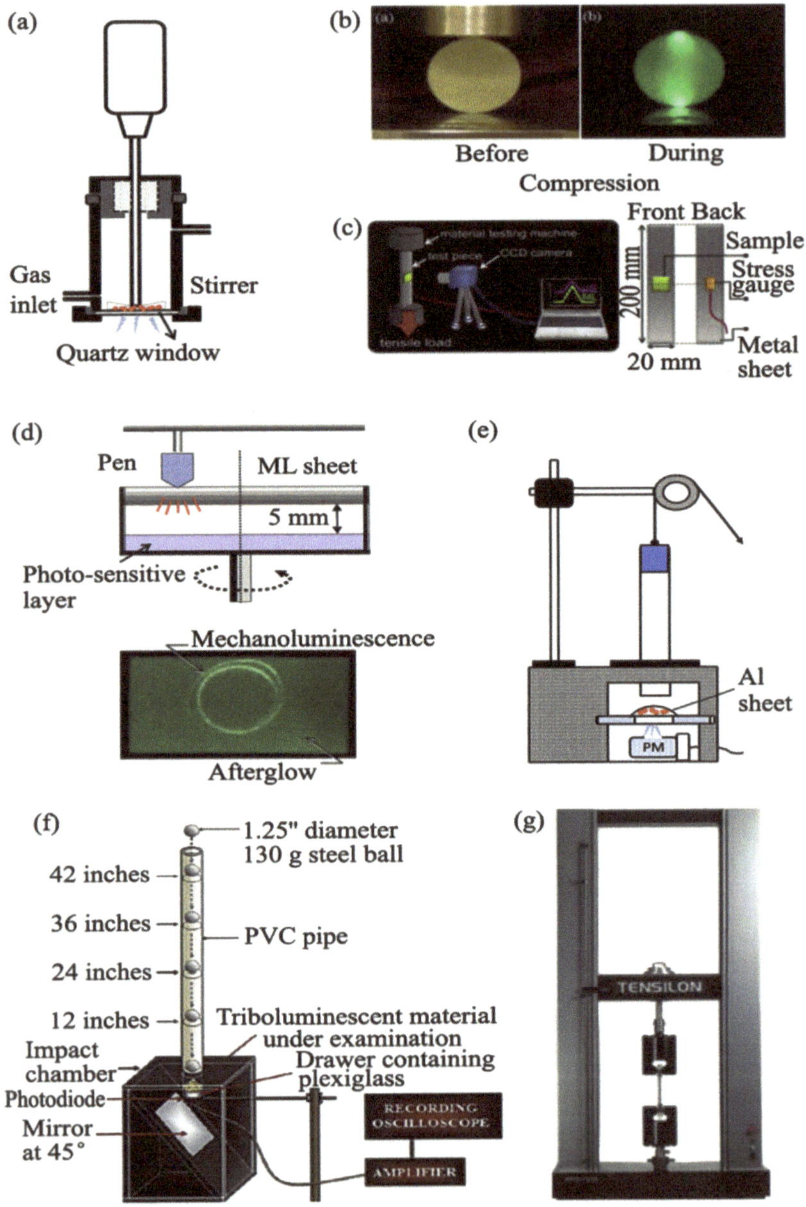

Figure 12.29 (A) Cylindrical cell with a four-blade PTFE stirrer and gas inlet; (B) Epoxy disk (25 mm diameter, 15 mm thickness) doped with a-$SrAl_2O_4$:Eu^{2+} before and during compression in a vise; (C) Set-up for visualization of stress distribution in a metal sheet upon compression or tensile load using an elasticoluminescent sensor and a high-speed camera; (D) Rotating device for scratching a sample, the right panel shows both ML and afterglow emission under 365-nm excitation; (e) Weight drop tower; (f) Ball drop tower for studying deformation ML and triboluminescence; (G) Commercial universal testing machine [106,107].

(A) Redrawn from Ref. [101]; (B) Reproduced with permission from Ref. [102], © 2006, Trans Tech Publications; (C) Reproduced with permission from Ref. [103], © 2014 World Scientific; (D) Adapted, with permission, from Ref. [104], © 2009 The Japan Society of Applied Physics; (E) Adapted from Ref. [105]; (F) Reproduced with permission from Ref. [106], © 2012, Elsevier, B.V., Amsterdam.

measurement, for a variety of reasons, one of which is the variety of ways the sample can be excited. As a result, investigators typically build their own set-up by connecting an optical measurement device to a mechanical excitation set-up. In terms of the latter, mechanical stress can be delivered in a variety of methods (Fig. 12.29) [99–107].

11. Conclusion

The study of ML materials and their application in ML dosimeters has led to the conclusion that ML can be used in a variety of applications. ML is projected to find practical applications in bioimaging to help investigate the complicated biomechanical behaviors of the human body or matching artificial implants as an unique nondestructive, in situ, real-time, and noncontactmechano-opto-responsive approach. Gamma- and UV-induced materials, such as alkali halides, zirconium-based materials, and titanium-based materials, can also be used in ML dosimeters. Some authors have reported EML behavior in alkali halides. In radiation dosimetry, the EML intensity of X- or γ-irradiated alkali halide crystals can be employed. When the crosshead of the testing machine deforming an X- or γ-irradiated crystal is stopped, the EML intensity of the X- or γ-irradiated crystals increases linearly with the strain of the crystals, and when the crosshead of the testing machine deforming an X- or γ-irradiated crystal is stopped, the EML intensity decreases with time. New applications, such as artificial intelligence, the internet of things, and biotechnology, have fueled the advancement of stress sensing technology, according to some authors. Stretchability, remoteness, stress dispersion, a multimodal nature, and biocompatibility are crucial performance properties of stress sensors for these new applications. Because of its distant and scattered response to mechanical stimuli, as well as its significant potential for stretchability, biocompatibility, and self-powering, ML-based stress sensing has gotten a lot of interest.

References

1. S. Shinoya, Yen, Phosphor Hand-Book, in: V.M. . (Ed.), CRL Press, Boca Roton, 1999.
2. T. Nakajima, T. Otsuki, Dosimetry for radiation emergencies: radiation-induced free radicals in sugar of various countries and the effect of pulverizing on the ESR signal. Int. J. Radiat. Appl. Instrum. Part A Appl. Radiat. Isotopes, 41 (4) (1990), 359–365.
3. T. Nakajima, ESR of sugar as a personnel monitor for radiation emergencies, Appl. Radiat. Isot. 46 (8) (1995) 819–825.

4. A. Tchen, C.L. Greenstock, A. Trivedi, The use of sugar pellets in ESR dosimetry, Radiat. Prot. Dosim. 46 (2) (1993) 119–121.
5. N.A. Atari, K.V. Ettinger, Lyoluminescence of irradiated saccharides, Radiat. Eff. 20 (1–2) (1973) 135–139.
6. T. Nickel, E. Pitt, A. Scharmann, Lyoluminescence dosimetry with sugar after accidental gamma ray exposures, Radiat. Prot. Dosim. 35 (3) (1991) 173–177.
7. A.J. Walton, Triboluminescence, Adv. Phys. 26 (6) (1977) 887–948.
8. G. Alzetta, I. Chudáček, R. Scarmozzino, Excitation of triboluminescence by deformation of single crystals, Phys. Status Solidi A 1 (4) (1970) 775–785.
9. Y. Enomoto, H. Hashimoto, Emission of charged particles from indentation fracture of rocks, Nature 346 (6285) (1990) 641–643.
10. T. Matsuzawa, Y. Aoki, N. Takeuchi, Y. Murayama, A new long phosphorescent phosphor with high brightness, SrAl2 O 4: Eu2+, Dy3+, J. Electrochem. Soc. 143 (8) (1996), 2670.
11. R. Dekkers, C.F. Woensdregt, Crystal structural control on surface topology and crystal morphology of normal spinel (MgAl2O4), J. Cryst. Growth 236 (1–3) (2002) 441–454.
12. C. Baudin, R. Martínez, P. Pena, High-temperature mechanical behavior of stoichiometric magnesium spinel, J. Am. Ceram. Soc. 78 (7) (1995) 1857–1862.
13. M. Sindel, N.A. Travitzky, N. Claussen, Influence of magnesium-aluminum spinel on the directed oxidation of molten aluminum alloys, J. Am. Ceram. Soc. 73 (9) (1990) 2615–2618.
14. J.T. Bailey, Sintered spinel ceramics, Am. Ceram. Soc. Bull. 47 (1968) 1025–1029.
15. C.C. Wang, Growth and characterization of spinel single crystals for substrate use in integrated electronics, J. Appl. Phys. 40 (9) (1969) 3433–3444.
16. K. Petermann, R. Clausen, E. Heumann, M. Ledig, Time resolved excited state absorption of Mn2+, Opt. Commun. 70 (6) (1989) 483–486.
17. R. Clausen, K.L.A.U.S. Petermann, Mn/sup 2+/as a potential solid-state laser ion, IEEE J. Quantum Electron. 24 (6) (1988) 1114–1117.
18. C. Wyon, J.J. Aubert, F. Auzel, Czochralski growth and optical properties of magnesium-aluminium spinel doped with nickel, J. Cryst. Growth 79 (1–3) (1986) 710–713.
19. A. Jouini, A. Yoshikawa, A. Brenier, T. Fukuda, G. Boulon, Optical properties of transition metal ion-doped MgAl2O4 spinel for laser application, Phys. Status Solidi C 4 (3) (2007) 1380–1383.
20. B.P. Chandra, Phys. Status Solidi 64 (1981) 395.
21. B.P. Chandra, in: D.R. Vij (Ed.), Luminscence of Solids, Plenum Press, New York, 1998, p. 361.
22. G. Wolff, G. Gross, L.A. Stranskii, Z. Electrochem. 56 (1952) 420 (In German).
23. B.P. Chandra, M. Elyas, J. Phys. C Solid State Phys. 12 (1979), 6951698.
24. B.P. Chandra, K.K. Shrivastava, Pramana J. Phys. 13 (1979) 611.
25. B.P. Chandra, M. Elyas, K.K. Shrivastava, D. Verma, Solid State Commun. 36 (1980) 931.
26. B.P. Chandra, R.D. Verma, Kristal. U. Tech. 15 (1980) 1197.
27. B.P. Chandra, Nucl. Tracks 10 (1985) 225.
28. B.P. Chandra, J.I. Zink, Phys. Rev. B21 (1980) 816.
29. B.P. Chandra, J.I. Zink, J. Chem. Phys. 73 (1980) 5933.
30. B.P. Chandra, J.I. Zink, Inorg. Chem. 19 (1980) 3098.
31. B.P. Chandra, J.I. Zink, J. Phys. Chem. Solids 82 (1982) 4138.
32. B.P. Chandra, R.D. Verma, A.R. Jaiswal, B. Majumdar, T.R. Chadrakar, S.V. Deshpande, Phys. Status Solidi A 72 (1982) 361.
33. L. Sodomka, Phys. Status Solidi A 1 (1971) K65–K66.
34. A.M. Avilov, V.D. Volovik, A.P. Evtvkh, S.I. Ivanav, Sov. Phys. Technol. 27 (1982) 1271.

35. J.I. Zink, W. Beese, J.W. Schindler, A.J. Smile, Appl. Phys. Lett. 40 (1982) 110.
36. G.N. Chapman, A.J. Walton, J. Phys. C. Solid State Phys. 16 (1983), 5543.
37. A.T. Tokhmetov, V.I. Vettegren, Sov. Phys. Sol. State 32 (1990) 17–19.
38. U.E. Krauya, I.V. Knets, V.B. Laizan, Polym. Phys. (USA) 201 (1978) 157–159.
39. R. Nowack, A. Krajewek, M. Samoc, Chem. Phys. Lett. 94 (1982) 270.
40. N. Kitamura, O. Saraswani, H.B. Kim, S. Tzuke, Chem. Phys. Lett. 125 (1986) 360.
41. I. Grabec, Polym. Phys. Lett. 12 (1974) 573.
42. V.M. Parfeev, M.Y. Tutan, U.E. Krauya, A.A. Scrukovskis, Mekhanika Kompozitnykh Materialov 1 (1988) 125.
43. I. Nakayama, Szuk noboiaka and hashimoto, J. Phys. D Appl. Phys. (1992), 25303.
44. J.T. Dickinson, L.C. Jensen, J. Polym. Sci. Poly Phys. Ed. 23 (1986) 873.
45. M. Akiyama, C.N. Xu, K. Nonaka, T. Watanabe, Appl. Phys. Lett. 73 (1998) 3046.
46. C.N. Xu, H. Yamada, X. Wang, IEEE Ultrasonic Symposium, 2004, p. 914.
47. Y. Jia, M. Yei, W. Jia, Opt. Mater. 28 (2006) 974.
48. J.S. Kim, Y.N. Kwon, N. Shin, K.S. Sohn, Appl. Phys. Lett. 90 (2007) 241.
49. J.T. Dickinson, L.C. Jensen, M.K. Park, Appl. Phys. Lett. 41 (1982) 443.
50. A.S. Crasto, R. Corey, J.T. Dickinson, R.V. Subramaniam, Y. Eckstein, Compos. Sci. Technol. 30 (1987).
51. K. Meyer, D. Orbikat, M. Rossberg, Kristall U. Tech. 5 (1970) 5.
52. K. Meyer, D. Orbikat, M. Rossberg, Kristall U. Tech. 5 (1970) 181.
53. C.P. Keszthely, A.J. Bard, Electrochem. Soc. 120 (1973) 1726.
54. G.N. Chapman, A.J. Walton, J. Appl. Phys. 54 (1983) 5961.
55. V.E. Orel, Y.A.Z. Popov, E.K. Gogaiskii, I.V. Leshchinskii, D.M. Khazanovish, Meditsinskaya Technika 4 (1989) 34.
56. M.I. Molotskii, Sov. Rev. Chem. 13 (1989) 1.
57. P.M. Bhujbal, S.J. Dhoble, Correlation between mechanoluminescence and lyoluminescence in gamma-irradiated KBr:Ce^{3+} phosphor for radiation dosimetry, Radiat. Meas. 47 (2012) 512–517.
58. P.M. Bhujbal, S.J. Dhoble, Lyoluminescence, thermoluminescence and mechanoluminescence studies in γ-ray irradiated Dy3+ activated potassium chloride phosphor for accidental radiation dosimetry, Luminescence 27 (5) (2012) 357–361.
59. P.M. Bhujbal, S.J. Dhoble, G. Singh, R.S. Kher, Mechanoluminescence characterizations in c-irradiated and Dy3+ activated (K0.5Na0.5)Cl phosphor for accidental radiation dosimetry, Nucl. Instrum. Methods Phys. Res. B 289 (2012) 28–33.
60. P.M. Bhujbal, S.J. Dhoble, Mechanoluminescence characterization of c-rays irradiated NaBr:Ce3+ phosphor for ionising radiation dosimetry, Indian J. Phys. 86 (5) (May 2012) 383–386.
61. R.K. Rai, A.K. Upadhyay, R.S. Kher, S.J. Dhoble, M. Mehta, BaAl2O4:Eu- Phosphor for mechanoluminescence dosimetry, Radiat. Meas. 46 (2011) 1393–1396.
62. K.K. Satapathy, G.C. Mishra, F. Khan, ZnAl2O4:Eu novel phosphor: SEM and mechanoluminescence characterization synthesized by solution combustion technique, Luminescence 30 (2015) 564–567.
63. K.K. Satapathy, G.C. Mishra, R.S. Kher, S.J. Dhoble, Mechanoluminescence and thermoluminescence characterization of Tb^{3+} doped CaAl2O4: a theoretical and experimental study, RSC Adv. 5 (2015), 79391.
64. S. Tigga, N. Brahme, D.P. Bisen, Effect of gamma irradiation on thermoluminescence and fractomechanoluminescence properties of $SrMgAl_{10}O_{17}$:Eu^{2+} phosphor, Opt. Mater. 53 (2016) 109–115.

65. G. Tiwari, N. Brahme, R. Sharma, D.P. Bisen, Sanjay Kumar Sao, Ugendra Kumar Kurrey, Enhanced long-persistence of Ca2Al2SiO7:Ce3+ phosphors for mechanoluminescence and thermoluminescence dosimetry, J. Mater. Sci. Mater. Electron. 27 (2016) 6399−6407.
66. G. Tiwari, N. Brahme, R. Sharma, D.P. Bisen, S. Kumar Sao, I. Prasad Sahu, Ca2Al2SiO7: Ce3+ phosphors for mechanoluminescence dosimetry, Luminescence 31 (8) (2016) 1479−1487.
67. J.K. Saluja, Y. Parganiha, N. Tiwari, V. Dubey, R. Tiwari, A. Prabhath, Mechano and photoluminescence spectra of cadmium sulphide and cadmium selenide doped phosphors, Optik 127 (19) (2016) 7958−7966.
68. N. Tiwari, R.K. Kuraria, S.R. Kuraria, R. kumar Tamrakar, Mechanoluminescence, photoluminescence and thermoluminescence studies of SrZrO3: Ce phosphor, J. Radiat. Res. Appl. Sci. 8 (1) (2015) 68−76.
69. J. Kaur, V. Dubey, N.S. Suryanarayana, Comparative study of Mechano-luminescence and PL spectra of different impurity-doped (Zn, Cd) S mixed phosphors, Res. Chem. Intermed. 39 (9) (2013) 4337−4349.
70. R. Shrivastava, J. Kaur, B.P. Chandra, Mechanoluminescence of Ba2MgSi2O7 doped with Eu2+ and Dy3+ phosphor by impulsive deformation, Luminescence 30 (8) (2015) 1207−1211.
71. N. Tiwari, R.K. Kuraria, S.R. Kuraria, Mechanoluminescence induced by impulsive deformation in zirconium: Ti phosphor, Radiat. Eff. Defects Solids 170 (7−8) (2015) 679−689.
72. R. Tiwari, V. Dubey, B.P. Chandra, Exact model for the elastico mechanoluminescence of II-VI phosphors, Mater. Phys. Mech. 19 (1) (2014) 25−38.
73. R. Tiwari, V. Dubey, M. Ramrakhiani, B.P. Chandra, Fracto-mechanoluminescence induced by impulsive deformation of II−VI semiconductors, Luminescence 30 (6) (2015) 883−890.
74. J. Kaur, N.S. Suryanarayana, V. Dubey, ML spectra of gold doped (Zn, Cd) S mixed phosphors, Chalcogenide Lett. 7 (5) (2010) 329−334.
75. N. Tiwari, V. Dubey, R.K. Kuraria, Mechanoluminescence study of europium doped CaZrO3 phosphor, J. Fluoresc. 26 (4) (2016) 1309−1315.
76. J. Kaur, N.S. Suryanarayana, V. Dubey, Effect of temperature on the Mechano-luminescence of Au doped (Zn, Cd) S mixed phosphors, Chin. Chem. Lett. 22 (6) (2011) 709−712.
77. A. Verma, A. Verma, Synthesis, characterization, mechano-luminescence, thermoluminescence, and antibacterial properties of $SrMgAl_{10}O_{17}$:Eu phosphor, J. Alloys Compd. 802 (2019) 394−408.
78. P. Xiong, M. Peng, Z. Yang, Near-infrared mechanoluminescence crystals: a review, Iscience 24 (1) (2021) 101944.
79. C. Wang, D. Peng, C. Pan, Mechanoluminescence materials for advanced artificial skin, Science Bulletin 65 (14) (2020) 1147−1149.
80. J.C. Zhang, X. Wang, G. Marriott, C.N. Xu, Trap-controlled mechanoluminescent materials, Progress in Materials Science 103 (2019) 678−742.
81. A. Feng, P.F. Smet, A review of mechanoluminescence in inorganic solids: compounds, mechanisms, models and applications, Materials 11 (4) (2018) 484.
82. Y.A. Ossipyan, S.Z. Shmurak, Defects in insulating crystal, in: V.M. Turchkevich, K.K. Schrarts (Eds.), Proc. Int. Conf. Riga Zinotne Publishing House, Riga, Springer Verlag, Berlin, 1981, pp. 135−160.
83. B.P. Chandra, J. Lumines. 128 (2008) 1217.
84. S.I. Bredikhin, S.Z. Shmurak, JETP Lett. 21 (1975) 156.

85. B.P. Chandra, S. Tiwary, M. Ramrakhiani, M.H. Ansari, Cryst. Res. Technol. 26 (1991) 767.
86. S.I. Bredikhin, S.Z. Shmurak, Sov. Phys. JETP 46 (1979) 768.
87. A.V. Poletaev, S.Z. Shmurak, Sov. Phys. Solid State 26 (1984), 2147.
88. V.A. Zakrevskii, A.V. Shuldiner, Sov. Tech. Phys. Lett. 10 (1984) 56.
89. V.A. Zakrevskii, A.V. Shuldiner, Sov. Phys. Solid State 27 (10) (1985), 1826.
90. A.J. Walton, Adv. Phys. 26 (1977) 887.
91. G.E. Hardy, G.I. Zink, Inorg. Chem. 15 (1976), 3061.
92. M.I. Molotskii, Fiz. Tverd. Tela (Leningrad) 18 (1976), 1765.
93. D.V. Alekseev, Sov. Phys. Sol. State 34 (1992) 1085.
94. B. Kasemo, L. Walden, Solid State Commun. 15 (1974) 571.
95. H.P. Rumpf, Chemie Ing. Techn. 32 (1960) 129.
96. G.E. Hardy, W.C. Kaska, B.P. Chandra, J.I. Zink, J. Am. Chem. Soc. 103 (1981) 1074.
97. D.G. Li, N.S. Mcalpine, D. Haneman, Surf. Soc. 303 (1994) 171.
98. Y.P. Butyagin, V.I. Beresteitskaya, V.I. Kalvanev, Kinet. Katal. 24 (1983) 441.
99. H. Longchambon, Recherches experimentales sur les phenomenes de triboluminescence et de cristalloluminescence, Bull. Soc. Fr. Mineral 48 (1925) 130.
100. I. Sage, G. Bourhill, Triboluminescent materials for structural damage monitoring, J. Mater. Chem. 11 (2001) 231.
101. G.L. Sharipov, A.A. Tukhbatullin, A.R. Bagautdinova, Quenching of electronically excited N2 molecules and Tb3þ/Eu3þ ions by polyatomic sulfur-containing gases upon triboluminescence of inorganic lanthanide salts, Luminescence 32 (2017) 824.
102. C.-N. Xu, C. Li, Y. Imai, H. Yamada, Y. Adachi, K. Nishikubo, Development of elastico-luminescent nanoparticles and their applications, Adv. Sci. Technol. 45 (2006) 939.
103. Y. Fujio, C.N. Xu, M. Nishibori, Y. Teraoka, K. Kamitani, N. Terasaki, N. Ueno, Development of highly sensitive mechanoluminescent sensor aiming at small strain measurement, J. Adv. Dielectr. 4 (2014). Art. Nr. 1450016.
104. N. Terasaki, C.-N. Xu, Mechanoluminescence recording device integrated with photosensitive material and europium-doped SrAl2O4, Jpn. J. Appl. Phys. 48 (2009). Art. Nr. 04C150.
105. R.K. Rai, A.K. Upadhyay, R.S. Kher, S.J. Dhoble, Mechanoluminescence, thermoluminescence and photoluminescence studies on Al2O3:Tb phosphors, J. Lumin. 132 (2012) 210.
106. R.S. Fontenot, W.A. Hollerman, M.D. Aggarwal, K.N. Bhat, S.M. Goedeke, A versatile low-cost laboratory apparatus for testing triboluminescent materials, Measurement 45 (2012) 431.
107. J.-C.G. Bünzli, K.-L. Wong, Lanthanide mechanoluminescence, J. Rare Earths 36 (2018) 1–41.

Rare-earth-activated phosphors for energy-efficient solar cell

13

Abhijeet R. Kadam and Sanjay J. Dhoble
Department of Physics, Rashtrasant Tukadoji Maharaj Nagpur University, Nagpur, Maharashtra, India

1. Introduction

In view of the current scenario of development and growth, solar energy is the most important renewable energy source and so are wind energy and hydropower. With the aim of approximately 110 TWh of electricity production per year, 100 GW of solar photovoltaic (PV) power was installed in 2012 all over the world. Germany ranked first in the world ranking of PV power installation with 32,411 MV photovoltaic installed. Italy, China, the United States, and Japan hold the ranks second, third, fourth, and fifth, respectively, after Germany in PV installation [1].

It was found that Africa and some South Asian countries have a higher potential to receive solar radiation of nearly about 2000–2500 kWh/m^2. It is to be noted that solar power is available in the intrinsic form in many countries, which can be made very efficient, but it is not used properly and is basically wasted [2]. The use of nonrenewable energy sources such as fossil fuels is leading to damage to the ecosystem; however, consumption of solar energy does not make any harmful effect on the ecosystem, and it is safe for human beings as well. In addition, this world is now carried toward solar systems because fossil fuels are running out up to 2300 due to an increase in energy demand. The whole world is also leading toward the recovery of energy from fossil fuels to carry out diverse activities because of intensifying sovereignty of the universal population on fossil fuels [3].

Notwithstanding of abundant compensation of solar cells, there are still some issues with the installation of PV cells for using the solar cell as an important source of energy in the global context. The first one is the irregular nature of the ecological situation. To resolve this issue, first of all, there is a need to identify the places where more solar radiation falls throughout the year. Identification of radiated parts in the country is important to fulfill the energy demand of people using renewable energy sources. Identification of hotspots in several parts of the country was done by NASA for SSE datasets for renewable energy potential assessment using high-resolution satellite data [4]. It gives some photographs of India's monthly basis global normal irradiance that give information about regions receiving favorable normal irradiance for electricity generation from solar cells.

Rare-Earth-Activated Phosphors. https://doi.org/10.1016/B978-0-323-89856-0.00015-8

1.1 Downconversion

As described earlier in this article, downconversion is used in converting high-energy photons into low-energy photons (Fig. 13.1). Downconversion can be proficient through the use of host lattice or ions that can lead to quantum efficiency greater than unity [6−8]. In host lattices, the effect of downconversion is associated with the impact of the ionization process [8]. At this point, a single high-energy photon produces multiple electron−hole pairs accredited to an inter-band Auger process beneath suitable circumstances that can be useful to produce electrical energy [9,10]. It needs the absorption of photons with energy as twice or more times the host material bandgap. The other downconversion system considers a single ion as an optically active center with three energy levels. When a high-energy photon is absorbed, a transition to the maximum energy level takes place.

Cooperative energy transfer is a downconversion process of key significance in solar cells performance [11−14]. In this procedure, a luminescent center divides its excited state energy into two by concurrently transferring its energy to two of the proximity acceptor centers acquiescent to low-energy photons for each high-energy photon absorbed [15−17]. Monte Carlo is the most used method in recent times for cooperative energy transfer [18,19]. Ordinarily, Monte Carlo technique is utilized to put the dopants arbitrarily at permitted lattice positions and afterward generate diverse configurations to simulate averaged luminescence decay curves [20]. This method was used in different downconversion phosphors for better understanding of the transfer rate between nearest neighbors in systems as $CsGd(MoO_4)_2:Yb^{3+},Er^{3+}/Ho^{3+}$ [21], $NaYF_4:Tb^{3+},Yb^{3+}$ [22], and Ce^{3+}-Yb^{3+}-codoped YAG phosphors [23], among others. To explain the impact of downconversion in a conservative solar cell, Trupke

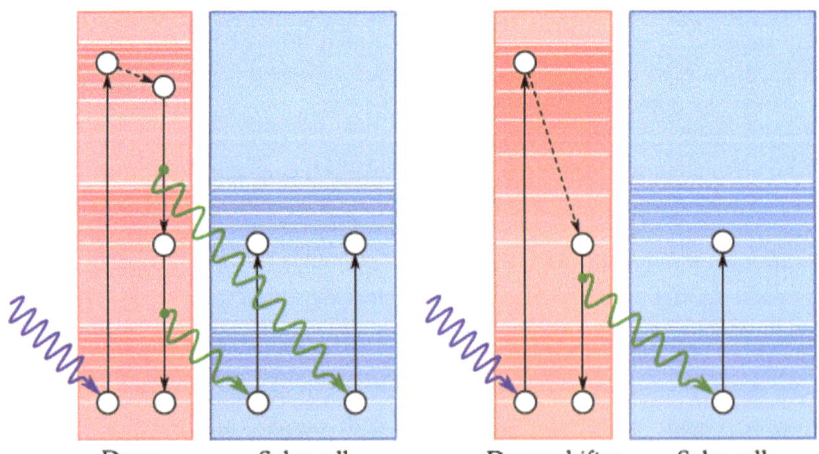

Figure 13.1 Energy diagram that shows schematically different behaviors among a downconversion and a downshifting material used to harvest the efficiency in a solar cell. Reproduced with the permission from Ref. [5] © Elsevier publications.

et al. [24] added a layer that demonstrated the downconversion effect and used a complete energy balance subsequent to the Shockley and Queisser formalism to compute its efficiency. A more sensible model, which comprises revolutionizing in radiation fluxes when light passes through interfaces of different refractive indices, can be found in Badescu et al. and others' work [25–27].

Although there are such incredible advantages for PV cells, there are some disadvantages as well. Ecological conditions and stumpy energy acquiesce in PV systems that increase area-related costs are the two major problems in PV technology use [5]. A country like India where 58% of the area receives global insolation annually [4] with an average solar energy receiving rate of 5 $kWh/m^2/day$ will easily overcome the first disadvantage. Solar cells, modules, utility meters, inverters, and performance monitoring are the main units in the PV system [5].

Theoretical solar efficiency was reported according to the previous research by the hypothesis of a single semiconductor absorber under standard 1.5 solar spectra and external quantum efficiency being 1, the photo-generated electron would be the outcome by one absorbed photon [28]. Practically convectional solar cells give lower efficiency as compared to the theoretical one, and this may be due to some problems such as reflection loss, contact voltage, and recombination [29,30]. To resolve these low-efficiency problems in solar cells and enhance better energy supply, the whole world is making different plans and strategies such as the concentration of sunlight, multijunction, multiple exciton production, and spectral modification through upconversion or downconversion mechanisms [14,31–33]. Usual solar cells are manufactured to utilize the visible spectrum, which contains an extensive portion of the solar energy spectrum. But if UV or/and IR part of the spectrum is used, then the efficiency of solar cells could be enhanced even more. Some materials have the capability to give emission in the visible range with more than one spectrum with infrared emission by absorbing UV photons; this is known as the downconversion process [34–36].

1.2 Upconversion

Upconversion (UC) is an interaction where the emission of one high-energy photon happens by the consecutive absorption of two low-energy photons. By and large, UC materials are doped with optically active sensitizer and activator particles. In this system, the absorption of subbandgap photons prompts the generation of electron–hole pairs. The electron–hole pairs at that point recombine by means of the emission of photons with energies at the band hole. There are diverse energy transfer instruments engaged with the UC interaction named as ground-state absorption or excited-state absorption (GSA/ESA), energy transfer upconversion (ETU), photon torrential slide (PA), and cooperative upconversion (CUC). Among these, the fundamental UC mechanism is the GSA/ESA, where the electron is raised to the principal energized state from the beginning by retaining a photon and afterward to a second higher energy level by the absorption of the second photon. The ESA interaction happens in single RE particles and, by and large, noticed uniquely at low doping centralization of the activators ($<1\%$), however not at high activator doping because of focus concentration quenching by means of nonradiative relaxation. The following mechanism is ETU that includes two kinds of particles. One is a sensitizer, and the other is an activator particle. In this mechanism, first, the sensitizer particle energizes and transfers energy to the activator particle and the activator drops back to the ground state that brings about UC emission. The standards for proficient energy transfer from the sensitizer to activator ought to have resounding energy levels with the nearer distance between them. Henceforth, ETU is considered as more proficient than GSA/ESA. The PA interaction was first seen in Pr^{3+} particle-based UC have in the year 1979. The PA cycle happens just at high siphon power thickness and is more perplexing contrasted with both ESA and ETU measures. At adequate high pump densities, at first, the middle degree of particles gets populated by GSA measure, trailed by full ESA or ETU from another energized particle that populates the UC emanating level. Afterward, proficient cross-relaxation happens between the energized and ground-state particles that outcomes in expanding the number of inhabitants in the repository level and the UC-producing level, causes a "torrential slide" impact of creating more energized particles and making PA measure more effective. The CUC cycle includes two sensitizers and one activator particle, which is comparative but less powerful than the ETU measure. Here, in this cycle, the agreeable energy transfer from the two adjoining sensitizers to the single activator particle helps in populating the UC transmitting level of the activator. The solitary contrast between CUC and ETU is that the activator particle in the CUC measure does not have adequate extensive middle-of-the-road energy levels full to the sensitizers that lower the effectiveness (Fig. 13.2).

1.3 Synthesis methods

Different types of synthesis methods have been employed for the synthesis of phosphors for the enhancement of solar cell efficiency. The pointwise explanation of each method has been given below.

- **Melt quenching technique**

  Melt quenching technique is one of the oldest techniques of methodology for producing phosphors that also contains phosphors for solar cell efficiency enhancement. During this technique, starting materials and minerals were mixed meticulously and melted in this solution by keeping it in the furnace. This melt solution is then poured on the solar cell; then, it is pressed instantly with a brass plate to form a thin transparent glass. This obtained film-formed coating of the solar cell shows color depending on the starting material and therefore the dopant we tend to use throughout the preparation process. The rate of quenching is another factor that affects film formation; the quicker the rate of melt quenching, the better the film formation. There are many examples within which the melt quenching technique is employed for the material preparation [38].

- **Wet chemical method**

  Wet chemical method is a chemical-based method that requires materials in the form of nitrates usually because they are easily dissolvable in water. Wet chemical method is also a method for material synthesis, but under this name "Wet Chemical" few more methods such as sol–gel method, coprecipitation method, and combustion method are added. The

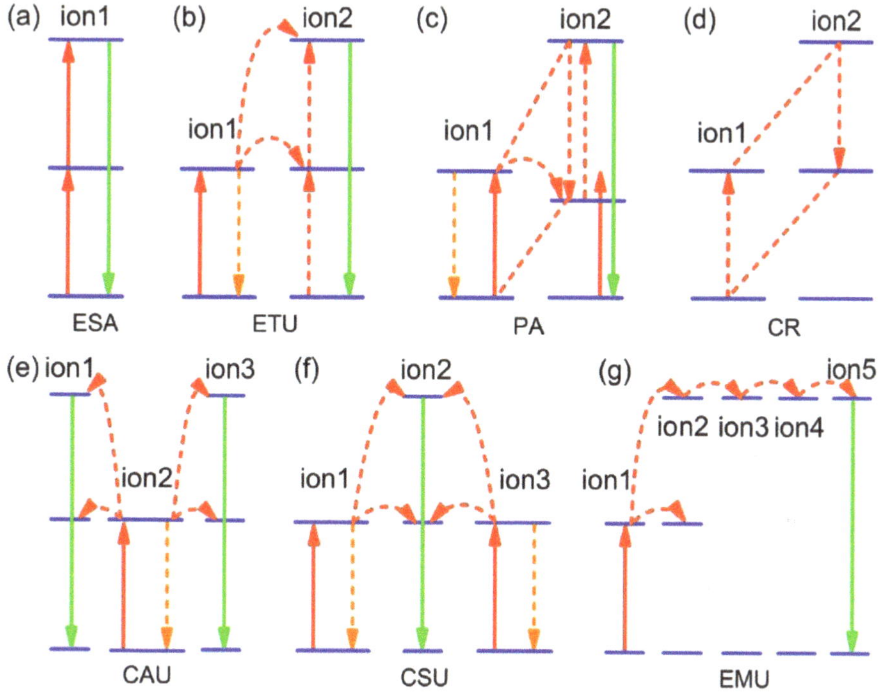

Figure 13.2 Schematic views of typical upconversion processes lanthanide-doped UCNPs: (A) excited state absorption (ESA); (B) energy transfer upconversion (ETU); (C) photon avalanche (PA); (D) cross-relaxation (CR); (E) cooperative activation upconversion (CAU); (F) cooperative sensitization upconversion (CSU); (G) energy migration-mediated upconversion (EMU). The *red solid lines*, *red dotted lines*, *orange dotted lines*, and *green lines* represent photon absorption, energy transfer, spontaneous emission, and upconversion emission, respectively.
Reproduced with the permission from Ref. [37] © Elsevier Publications.

combustion method contains more submethods; therefore, we describe this method separately at the end of this section. So, here, we explain the sol–gel method, the main wet chemical method separately.

❖ **Sol–gel method**

In this method, all water dissolvable materials were used for the synthesis of phosphor. All starting materials were dissolved in the beaker while keeping it on constant stirring at 80°C. After dissolving all the reagents, citric acid and polyethylene glycol (PEG) were added to the solution dropwise with an interval of 30 minutes. Here, citric acid is used as a chilling agent, and PEG is used as a cross-linking agent in the synthesis process. The ratio of citric acid and PEG is to be maintained at 1:1 in the synthesis process. After mixing all the reagents, this solution is kept on constant stirring with 80°C temperature of the hotplate. After some time, a gel-type solution is formed. This solution is then kept in the oven for drying at 100°C. After drying, a black floppy sol was obtained. This black powder is due to the chilling agent. This black powder is then annealed at above 500°C. After annealing, a white fine powder is obtained; this is further used for coating phosphor on the solar cell to enhance the efficiency of the cell. Few interesting phosphors prepared by this technique to enhance solar cell efficiency are reported.

❖ **Combustion synthesis**

Combustion synthesis is one of the main types of wet chemical-based synthesis and is further divided into two categories, that is, simple combustion synthesis method and solution combustion synthesis method. In the combustion synthesis method, the type of the synthesis technique also differs by the fuel we are using for the synthesis. Different fuels such as urea, citric acid, glycine, ethanol, and Diformyl hidrazine (DFH) are used for the preparation material by both solution and simple combustion methods [39–43]. The method for preparation of sample by both methods is explained in a subsequent part of this chapter.

➢ **Simple combustion method**

In the simple combustion method, all the starting materials were weighed in stoichiometric ratio and fuel which is used can be used in a particular ratio and dissolve all the materials in a mortar pistel using distilled water. After dissolving all the samples in distilled water, transfer this solution to a china dish and fire this sample in an open furnace at 500–550°C. An enormous flame was analyzed in the open furnace while the sample was burning. Once this sample was burned completely, the flame was damped for 5 minutes. Then, the sample was taken out from the oven and cooled at room temperature. After cooling, the sample can be crushed for 5 minutes and kept in the furnace for annealing for 24 hours for removing impurities. The sample is then cooled down at room temperature after 24 hours, and this sample is used for further studies.

➢ **Solution combustion method**

Solution combustion technique is another type of combustion technique using fuel. Similar to the simple combustion technique, the same types of fuels were used in this method as well. In this technique, all the starting materials were dissolved in a beaker by keeping it on constant stirring at constant temperature on magnetic stirrer. Once the starting materials and fuel get dissolved, transfer this solution into the china dish and repeat the same procedure as described in the simple combustion method.

It is to be noted that when citric acid is used as fuel, the sample turns black after firing it in the open furnace in both methods. Once it is kept for annealing, it turns into white powder and is useful for studies.

- **Main wet chemical method**

In this method, we do not need any type of fuel for synthesis as we used before in the sol–gel technique. Here, we require starting materials that can dissolve in a minimum amount of water. Take all starting materials in stoichiometric ratio and dissolve them separately in a beaker in minimum amount of water with constant stirring on a hotplate keeping it at 80°C. After dissolving it completely, add them in one beaker one by one with constant stirring and keep it on a hotplate. After this, a white precipitate was found; this precipitate was kept for drying in a hot air oven over the night at 90°C. After that, collect the dried sample and anneal it for 2 hours for removing impurity. By cooling it at room temperature collect the final product for further characterizations.

- **Coprecipitation method**

This is another wet chemical-based synthesis method that also required water dissolvable salts for synthesis. In this process, all the salts are dissolved into a double-distilled water solution, and rare earths are also dissolved into water solution in the presence of ethanol. Ethanol is generally used as a surfactant to avoid agglomeration, control the size of the particles, and maintain uniformity in the particle size. The coprecipitation process in the absence of ethanol causes particles to undergo aggregation thereby resulting in the nonuniformity and increase in particle size [44]. All solutions are mixed in the presence of ethanol. After mixing, the precipitate is then collected by centrifuging it. After getting the final precipitate, wash it several times with distilled water to remove impurity. This washed precipitate is then dried in an oven at 100°C for 24 hours. This dried sample is then calcined for 2 hours and a final product is obtained after cooling it at room temperature.

- **Recrystallization method**

  This is another wet chemical method in which we require starting materials of the same nature as we used in the previous methods. The sample is prepared by dissolving the appropriate constituents in a common solvent followed by recrystallization, which can be achieved by boiling off the solvent. The difficulty in maintaining chemical homogeneity is serious, as in homogeneities have a harmful consequence on the properties of the phosphors. Moreover, there is a problem of agglomeration. So, to triumph over this problem, particle size is reduced initially by grinding, dry milling, or wet milling [45].

- **Self heat-generating synthesis and similar methods**

  In this method, lower reaction temperatures are attained by obtaining the reactants in fine form. An ingenious way of lowering the operating temperature is to use the heat produced in the exothermic chemical reaction itself for synthesis. The foremost benefit of this technique is that heat is not supplied externally. This eradicates the use of refractory muffles, insulators, and crucible materials. In this method, the reactants are systematically mixed and pressed to form a bar. One end of the bar is heated to a high temperature using some sort of flame. Once the exothermic reaction starts, it proliferates across the length of the bar.

2. Role of DC/UC in solar cells

Fig. 13.3 displays the limitation of energy conversion in solar cells is inattentiveness to a complete solar spectrum. The dispersion of spectral response of sunlight at AM 1.5G contains photons with wavelengths from ultraviolet to infrared though the present solar cells use just a moderately little part of the solar photons. As a matter of fact, the solar

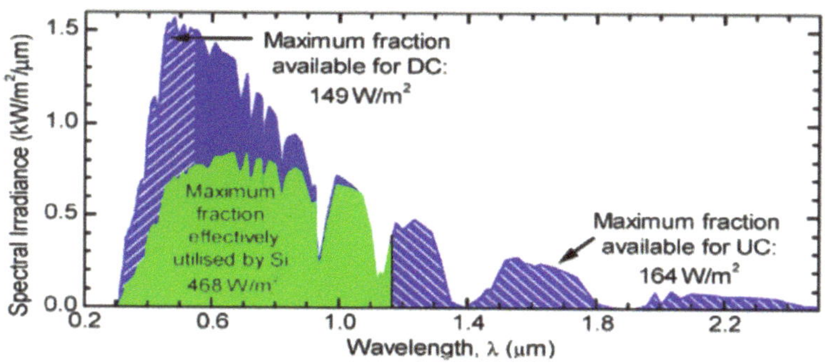

Figure 13.3 Regions of the AM1.5G spectrum that can be used in a solar cell through DC and UC (*blue with lines*). The fraction absorbed by a thick solar silicon device is shown in green. Reproduced with the permission from Ref. [5] © Elsevier publications.

cell materials could react to a narrow range of solar photons with energy coordinating with equivalent to the bandgap of the material. In UC, at least two photons retention toward emission of light of shorter wavelengths could permit toward utilization of photons with energy lower than the bandgap. The photons with energy higher than the bandgap are assimilated, and the overabundance energy is not adequately used and is unconfined using heat. UC materials conquer this dispute, which could transform 900−1600 nm light, inside the reaction scope of solar cells, and this UC converter is covered at the rear of the solar cell. Although, in the downconversion (DC) process during quantum cutting, the emission of two photons, after absorption of one high-energy photon, could be exploited for energy conversion in solar cells. Habitually, the DC layer is being positioned at the front surface of the solar cell.

Furthermore, the DC/UC film serves as an overcrowding layer that provides the payback of lessening the charge reconsolidation, enabling quicker charge transfer, lower electron transport resistance, and withholding the dark current in dye-sensitized solar cells and perovskite solar cell devices. The device behavior has been enhanced corresponding to the clean energy conversion by light-collecting capability through converting the UV and NIR radiations to visible emission. These DC/UC layers not just assume a significant part as spectral modifiers yet, in addition, limits the thermalization losses happening inside the cells by the scattering impact offered by them. This scattering approach will upgrade the connection of incident photons with a light reaper and additionally broadens the photons' travel way by keeping the light spread.

3. Working principle of DC/UC-based solar cells

3.1 Working principle of upconversion solar cells

UC nanostructures play an important role in formative luminescence efficiency restrains dopants and host matrices. Dopants give a luminescence center, while host matrices provide a proposal for energy transfer amid the dopants and expel them into the most advantageous situation. Concerning the UC with convoluted representations, energy transfer governs the UC luminescence processes.

Usually, four mechanisms are normally implicated in the UC process viz ESA, energy ETU, cooperative energy transfer (CET), and the photon avalanche effect. All these four mechanisms will be discussed one by one in detail.

- **Excited-State Absorption:**

  Among all the four mechanisms that describe the working principle for UC, ESA is the first mechanism. ESA is the mechanism in which two photons are successively absorbed by an electron affiliated with the multiple energy levels. Thus, it is preferred to possess energy levels of luminescence centers amid a sufficient amount of electrons allowing it may ease the absorption of a further photon. A UC material with large absorption unvarying absorbs the utmost number of photons and enhances the possibility of electronic transitions. Accessibility of numerous suitable photons per unit volume assists in stimulating electrons to higher states consistently absorbing a constant velocity of uphill transitions. As the temperature enhances, the higher energy levels turn into less occupied diminishing the probability of photon intensification. All these necessities are hard to accomplish and can obstruct development. This is as well significant stating that the chances of incidence of this procedure are superior for materials with low doping.

- **Energy Transfer Upconversion:**
  Energy transfer upconversion is a further technique of accomplishing photon upconversion. It is a method that includes the excitation of a particle to a level by basic assimilation of an excited photon. The necessary extra energy is moved from another active particle going through NR deexcitation. The particle is by and large energized through a functioning solid-state laser. The two particles are named as sensitizer and activator, individually. ETU, which is viewed as more productive than ESA, includes two principal thoughts: upconversion and energy transfer. The intriguing ETU is essentially a dipole–dipole energy transfer that happens because of a grouping of particles inside the host medium and is in some cases called as concentration quenching. In this way, the Ln^{3+} particles having various energized states are of more prominent benefits for ETU.
- **Cooperative Energy Transfer:**
  Cooperative energy transfer is an upconversion mechanism wherein both sensitizer and activator give a crucial performance. It is like ETU aside from that there exists no real intermediate energy level in the activator. In this component, two sensitizers are agreeably initiated to an excited state to satisfy a simultaneous energy transfer to an adjoining activator. Aside from fundamental UC emission from the two activators, low energy helpful UC outflows are likewise likely from the sensitizers. CET component is pertinent predominantly to inorganic materials, polymers, and glasses.
- **Photon Avalanche Effect:**
  There is a variance between the energy of the excited photon and the bandgap between the ground and the intermediate state on providing the photon to the system; the luminescent centers pursue ESA and inhabit any higher excited state. Then, there intervenes a resonant cross-relaxation amid this excited ion and adjacent ground-state ion by which the intermediate states of both ions are engaged. This procedure prolongs and after some time the intermediate states are exponentially occupied above the excitation entrance. In this method, the number of avalanche transitions goes on increasing until the consumption of so excited ions is less than the ground-state ions. To the extent that nanomaterials are involved, this mechanism does not appear very proficient.

3.2 Working principle of downconversion solar cells

Over and above losses from subbandgap photons, it is predictable that an had further dimensions of the solar spectral energy lost as of photons occupying a lot of energy. In silicon solar cells, a photon with energy exceeding the bandgap will generate an electron–hole pair, but erstwhile, in the conduction band, the charge carrier overindulgence energy will frequently be lost as degenerate heat. Cells also have the finest external quantum efficiency that fluctuates with wavelength consequently in theory receiving as often of the spectrum as promising at that energy would be an outstanding approach to develop efficiency. Even though opposing to UC, lower bandgap cells or methods with a poor UV response benefited the most from a proficient DC layer. Owing again to their different energy structure, rare-earth materials have extensive potential for DC devices. A large number of the capable instruments for DC seem undifferentiated from UC yet in turn around. One of the creators depicts five cycles as regular. Right off the bat, an instrument comparable to ground-state absorption and excited-state absorption in converse, a UV photon is caught up in a solitary particle, advancing an electron from the ground state to an optional excited state before rotting successively, emitting two NIR photons. There additionally exist three potential courses for DC by means of resounding energy transfer between two unmistakable particles. At long last, a converse couseable energy transfer happens when a particle retains a photon and transfers the energy to two extra particles elevating them to an excited state. These states at that point all the while rot emitting two low-energy photons. Notwithstanding these likenesses, it has been shown not normal for UC, DC will in general be a direct interaction free of light intensity.

4. Rare-earth-activated DC/UC phosphors for dye sensitized solar cell (DSSC)/perovskite solar cell (PSC) applications

Elevating light-gathering efficiency in a more limited and longer wavelength (UV, NIR) locale is an evergreen issue in the field of photovoltaic innovation. The utilization of RE-doped DC or UC phosphors is an elective method to expand the phantom reaction and light dissipation. As of now, UC materials-based examination is reached out from economically accessible solar cells to further developed solar cells like DSSCs and PSCs. Effective UC photoluminescence works within the age of more electron-opening sets, and more extensive absorption is advantageous to use the bigger part of the solar range in photovoltaics. The DC/UC phosphors as dissipating layers adjust the photon ways and expand the voyaging distance of occurrence light in the photo-terminals, along these lines upgrading the cosharpening for example expanding the photons absorption by the sensitizers. Numerous reports suggested that the mix of the UC materials into the TiO_2 photoanodes helps in advancing the light gathering capacity and lessens the electron-opening recombination. Choice of reasonable host materials for DC/UC phosphors is a significant errand to accomplish effective outflow. Despite the fact that when similar activator particles are utilized, the emanation wavelength and UC intensity are affected by having arrangements. Up to now, various materials including fluorides, chlorides, sulfides, and oxides have been considered as the UC have. Among these host materials, fluorides have shown an advantage over the others because of their low phonon energies. In the previous few years, the majority of the specialists gave concentrated consideration on $NaYF_4$ have and had been end up being quite possibly the best host cross-sections for lanthanide particles to upgrade their UC glow. The benefit of low phonon energy is to diminish the instrument of multiphonon relaxation. Despite the fact that the fluorides have the value of low phonon energy, they endure with feeble chemical bonds and warm steadiness and are not reasonable for down to earth applications because of their contaminated nature.

4.1 Rare-earth-activated DC/UC for DSSC applications

In the year 2012, Zhu et al. [46] investigated Y_3Al_5O:Ce phosphors as a scattering layer and UV change material for DSSCs and accomplished a PCE of 7.91%, higher than the gadget without scattering layer (6.97%). Under 456 nm excitation Y_3AlO_{12}: Ce phosphor shows a wide pinnacle situated at 526 nm, which is in acceptable concurrence with the assimilation scope of the N-719 color. This prompts the improvement of incident photon to current efficiency (IPCE) esteem in the reach 400–450 nm for the DSC gadget. Zalas et al. adjusted the permeable TiO_2 electrodes with every single RE component (Y, La, Ce, Nd, Tb, Dy, Sm, Er, Nd, Ho, Yb and Lu) and contemplated their effect on the gadget performance. The best photovoltaic execution was noticed for tests altered with Ce and Yb particles with a PCE of 8.88% and 8.80% separately when contrasted with the unmodified cell. Here, the RE layer goes about as a scattering layer that forestalls electron back-move and improves the energized electron transportation cycle and color adsorption capacity. In 2013, He et al. arranged working electrodes with a long-determined phosphor, $SrAl_2O_4:Eu^{2+}$, Dy^{3+} on the highest point of the TiO_2 nanoparticle layer, and the gadget accomplished a higher productivity of 6.85% as contrasted and exposed gadget (5.04%). These materials have a great optical match with generally utilized color in DSSCs and, as a light-scattering layer, improve the light collecting capacity coming about in improved J_{sc}. $SrAl_2O_4:Eu^{2+}$, Dy^{3+} phosphor decreases I_3^- focus in the electrolyte that builds the open-circuit voltage [47].

In another investigation, $LaPO_4$:Ce, Tb and (Mg, Zn) $Al_{11}O_{19}$:Eu green phosphors were brought into TiO_2 electrode by Hong et al., and $LaPO_4$:Ce, Tb phosphor-based DSC displayed an increment in J_{sc} and efficiency of 30% when contrasted with immaculate TiO_2-based devices [48]. The particles size of green phosphor is as demonstrated in field emission scanning electron microscope (FE-SEM) pictures that offer the adsorption of higher level of dye molecules and furthermore uphold further entrance of the I^-/I_3^- redox couple into the TiO_2 photoelectrode. These phosphors do not just have expansive optical absorption compared with (Mg, Zn) $Al_{11}O_{19}$:Eu phosphors yet, in addition, offer decreased inside resistances that show an incredible effect on the current density of the DSSC device.

In this manner, group of J. Wu arranged europium-doped yttrium fluoride (Eu^{3+}: YF_3) and applied it into the photoanode of DSSC revealing an efficiency of 7.74% [49]. $Eu^{3+}:YF_3$ as p-type dopant collects more UV light contrasted with that of the DSSC without $YF_3:Eu^{3+}$. Shen et al. [50] blended Eu^{3+}-doped $NaGdF_4$ NCs with polymethyl methacrylate as DC layer and covered on the posterior of TiO_2 electrode recording a PCE of 9.34% compared with DSSC without DC layer.

Later in the year 2016, Puntambekar and Chakrapani [51] announced enormous area solar cells with 63% expansion in the J_{sc} alongside a 54% increment in PCE by applying Eu- and Dy-doped $SrAl_2O_4$ long-tenacious phosphor as DC layer in the DSSC. In another investigation, Vinod Kumar et al. detailed Eu-doped TiO_2 downshifting layered-based DSSC that shows higher Current Density (J_{SC}) because of the ascent in the quantity of photons in the noticeable area changed over from UV district because of the doping of Eu ions [52]. His examination likewise uncovered that the deformity-related emanation has diminished with expanded doping centralization of Eu particles compared with imperfection-related discharge noticed for undoped TiO_2.

Hereafter, the DC cycle modified the UV properties of the solar cell under the solar spectrum. By and large, the regularly utilized colors have less absorption in the IR region of the spectrum that restricts the J_{sc} produced by DSSC. Likewise, when presented to UV light, there is debasement in the DSSC execution. To conquer these issues, Yue et al. adjusted the working photoanode tin oxide (SnO_2) reduced layer by europium particle-doped ytterbium fluoride (Eu^{3+}:YbF_3) and created long haul stable DSSCs without precedent for the year 2017. A short time later, the Duan group manufactured a DSSC gadget that can remain to all sorts of climate conditions with the help of long persistent $SrAl_2O_4$:Eu^{2+}, Dy^{3+}, and $Sr_4Al_{14}O_{25}$:Eu^{2+}, Dy^{3+} DC phosphors and announced an efficiency of 8.86% and 26% under sun enlightenment and in obscurity conditions [53]. Recently, in 2018, Llonas showed improved light reaping capacity with the fuse of Eu^{3+}-doped Y_2WO_6/TiO_2 layer as the working electrode in DSC devices. These DSSC devices with DC layer displayed better photovoltaic execution of efficiency 3.9% compared with uncovered TiO_2 (3.1%) [54].

4.2 Rare-earth-activated DC/UC for PSC applications

The act of consolidating DC phosphors in DSSC was subsequently stretched out to PSCs as well. In this unique situation, Chander et al. [55] presented a YVO_4:Eu^{3+} nanophosphor DC layer in TiO_2-based PSC to alleviate UV degradation and detailed an efficiency of 7.93% compared with control PSC because of way length upgrade of long-frequency light due to scattering. The DS phosphor-blended PSC showed improved security over half of its underlying efficiency under enlightenment contrasted with the PSC without the phosphor layer (35%). Later, Hou et al. used Eu^{3+}-doped $ZnGa_2O_4$ DC material that energizes perovskite ($CH_3NH_3PbI_3$) safeguard to create more photograph produced electron-opening sets, which in turn increment the PCE of PSCs. The phone created with 8 mg/mL of $ZnGa_2O_4$:Eu^{3+} has shown better photovoltaic execution with more noteworthy strength contrasted with the phone without DC layer [56]. In another investigation, Chen et al. chipped away at Eu^{3+}-doped CeO_2 nanophosphor DC layer-based PSCs and announced a PCE of 10.8% with great dependability compared with the reference exposed TiO_2 cell [57]. The contemplated nanoparticles of size 70 nm and octahedral shape give amazing light dissipating and catching capacity. Au@Y_2O_3:Eu^{3+} nanoparticle monolayer for PSC was planned by Kim and coworkers who consolidate the impacts of intensification of the changed over wave by the limited surface plasmon reverberation and frequency conversion. Here, the Au nanoparticles were covered on the outside of Y_2O_3:Eu^{3+} phosphor monolayer, to improve the efficiency to 16.1% from 15.2% (uncovered PSCs) and long haul steadiness of PSCs. Dai and coworkers showed effective PSC by covering Eu-4,7-diphenyl-1,10-phenanthroline (Eu-complex) as DC layer on the nonleading side of FTO glass and detailed 11.8% improvement in J_{sc} and 15.3% in efficiency. The proposed Eu-complex aids in improving the gadget execution just as diminishing UV degradation for PSCs [58].

5. Future perspective of rare-earth-activated materials

Rare-earth-activated up/down conversion materials can be orchestrated effectively at low temperatures, and they discovered different applications such as transparent conducting oxide (TCO), tandem solar cells, photocatalysis, water parting, and drug conveyance frameworks. The optical properties of the TCOs can be finely tuned by doping luminescent rare-earth components into the wide bandgap semiconductors that outcomes in improving their electrical property. Coming to tandem solar cells, the use of these lanthanide-doped materials as dissipating layers decreases the deformity communities transporters, that is, thermalization misfortunes that in turn upgrade the gadget efficiency. When these rare-earth-doped luminescent materials are combined with monetarily accessible wide bandgap semiconductors such as TiO_2, it helps in the quick debasement of inorganic and organic poisons and water parting under visible light irradiation. Notwithstanding the previously mentioned applications, as of late they discovered the potential for in vivo bioimaging, photodynamic treatment, and photothermal treatment just as profound 3D bioimaging because of frequency tuning. When the photograph responsive mixtures and polymeric materials are combined with UCNPs, they go about as an appropriate candidate for drug conveyance frameworks. With the use of UC/DC materials in solar photovoltaics, the greatest force conversion efficiency that can be accomplished might be slightly higher than 30%.

6. Concluding remark

This chapter gives an idea about different RE-doped phosphor materials and their promising applications in DSC and PSC devices detailed before. The presentation of these phosphors materials ends up being the positive possibility to acknowledge profoundly efficient solar cells. DC materials have the ability to change over UV light to visible radiation while the UC materials retain NIR light and afterward convert to visible light. Consequently, wide photoresponse to UV and NIR region was accomplished by means of DC and UC measure and incites electron-opening pair age in DSC and PSC devices. The light gathering is additionally advanced by consolidating the ghostly modifier with light-dissipating impact or surface plasmon resonances. The converters were applied on the outside of the TiO_2 film, which fills in as a boundary for charge recombination in solar cells. The strength, soundness, and lifetime of the devices can be improved by lessening direct exposure to UV or NIR light. Besides, RE-doped-nanostructured TiO_2 offers better electron transport and less recombination misfortunes compared with the exposed devices. In view of these outcomes, clearly the fuse of RE-doped phosphor materials is a plausible choice to acknowledge more efficient solar cells.

Acknowledgments

One of the authors SJD is thankful to the Department of Science and Technology (DST), India (Nano Mission) (Project Ref. No. DST/NM/NS/2018/38(G), dated January 16, 2019) for financial assistance and R. T. M. Nagpur University, Nagpur for constant encouragement.

References

[1] Where is Solar Power Used the Most? - Energy Informative, n.d. https://energyinformative.org/where-is-solar-power-used-the-most/. (Accessed 14 April 2019).

[2] F.D.J. Nieuwenhout, A. Van Dijk, P.E. Lasschuit, G. Van Roekel, V.A.P. Van Dijk, D. Hirsch, H. Arriaza, M. Hankins, B.D. Sharma, H. Wade, Experience with solar home systems in developing countries: a review, Prog. Photovoltaics Res. Appl. 9 (2001) 455–474, https://doi.org/10.1002/pip.392.

[3] N. Kannan, D. Vakeesan, Solar energy for future world: - a review, Renew. Sustain. Energy Rev. 62 (2016) 1092–1105, https://doi.org/10.1016/j.rser.2016.05.022.

[4] T.V. Ramachandra, R. Jain, G. Krishnadas, Hotspots of solar potential in India, Renew. Sustain. Energy Rev. 15 (2011) 3178–3186, https://doi.org/10.1016/j.rser.2011.04.007.

[5] M.B. de la Mora, O. Amelines-Sarria, B.M. Monroy, C.D. Hernández-Pérez, J.E. Lugo, Materials for downconversion in solar cells: perspectives and challenges, Sol. Energy Mater. Sol. Cells 165 (2017) 59–71, https://doi.org/10.1016/j.solmat.2017.02.016.

[6] M.J. Crane, D.M. Kroupa, D.R. Gamelin, Detailed-balance analysis of Yb^{3+}: $CsPb(Cl_{1-x}Br_x)_3$ quantum-cutting layers for high-efficiency photovoltaics under real-world conditions, Energy Environ. Sci. 12 (2019) 2486–2495, https://doi.org/10.1039/c9ee01493d.

[7] W. Chang, L. Li, M. Dou, Y. Yan, S. Jiang, Y. Pan, M. Cui, Z. Wu, X. Zhou, Dual-mode downconversion luminescence with broad near-ultraviolet and blue light excitation in Tm^{3+}/Yb^{3+} codoped oxy-fluoride glasses for c-Si solar cells, Mater. Res. Bull. 112 (2019) 109–114, https://doi.org/10.1016/j.materresbull.2018.12.011.

[8] C.S. Erickson, M.J. Crane, T.J. Milstein, D.R. Gamelin, Photoluminescence saturation in quantum-cutting Yb^{3+} -doped $CsPb(Cl_{1-x}Br_x)_3$ perovskite nanocrystals: implications for solar downconversion, J. Phys. Chem. C 123 (2019) 12474–12484, https://doi.org/10.1021/acs.jpcc.9b01296.

[9] G. Alymov, V. Vyurkov, V. Ryzhii, A. Satou, D. Svintsov, Auger recombination in Dirac materials: a tangle of many-body effects, Phys. Rev. B 97 (2018) 1–13, https://doi.org/10.1103/PhysRevB.97.205411.

[10] A. Solodovnyk, C. Kick, A. Osvet, H.-J. Egelhaaf, E. Stern, M. Batentschuk, K. Forberich, C.J. Brabec, Computational optimization and solution-processing of thick and efficient luminescent down-shifting layers for photovoltaics, Phys. Simul. Photonic Eng. Photovolt. Dev. V. 9743 (2016) 97430G, https://doi.org/10.1117/12.2211737.

[11] Y. Wang, H. Zhang, S. Qu, C. Su, Downconversion and upconversion emissions of $GdPO_4:Yb^{3+}/Tb^{3+}$ and its potential applications in solar cells, J. Alloys Compd. 677 (2016) 266–270, https://doi.org/10.1016/j.jallcom.2016.03.242.

[12] A.K. Rufino Souza, A.P. Langaro, J.R. Silva, F.B. Costa, K. Yukimitu, J.C. Silos Moraes, L. Antonio de Oliveira Nunes, L. Humberto da Cunha Andrade, S.M. Lima, On the efficient $Te^{4+} \rightarrow Yb^{3+}$ cooperative energy transfer mechanism in tellurite glasses: a potential

material for luminescent solar concentrators, J. Alloys Compd. (2019) 1119−1126, https://doi.org/10.1016/j.jallcom.2018.12.038.
[13] F. Zhao, Y. Liang, J.B. Lee, S.J. Hwang, Applications of rare earth Tb^{3+}-Yb^{3+} co-doped down-conversion materials for solar cells, Mater. Sci. Eng. B 248 (2019) 114404, https://doi.org/10.1016/j.mseb.2019.114404.
[14] K. Li, R. Van Deun, Enhancing the energy transfer from Mn^{4+} to Yb^{3+} via a Nd^{3+} bridge role in $Ca_3La_2W_2O_{12}$:Mn^{4+},Nd^{3+},Yb^{3+} phosphors for spectral conversion of c-Si solar cells, Dye Pigment 162 (2019) 990−997, https://doi.org/10.1016/j.dyepig.2018.11.030.
[15] C.L. Cortes, Z. Jacob, Fundamental figures of merit for engineering Förster resonance energy transfer, Opt. Express 26 (2018) 19371, https://doi.org/10.1364/oe.26.019371.
[16] J. Bohlen, A. Cuartero-González, E. Pibiri, D. Ruhlandt, A.I. Fernández-Domínguez, P. Tinnefeld, G.P. Acuna, Plasmon-assisted Förster resonance energy transfer at the single-molecule level in the moderate quenching regime, Nanoscale 11 (2019) 7674−7681, https://doi.org/10.1039/c9nr01204d.
[17] R. Marin, L. Labrador-Paéz, A. Skripka, P. Haro-González, A. Benayas, P. Canton, D. Jaque, F. Vetrone, Upconverting nanoparticle to quantum dot Förster resonance energy transfer: increasing the efficiency through donor design, ACS Photonics 5 (2018) 2261−2270, https://doi.org/10.1021/acsphotonics.8b00112.
[18] J. Hu, Y. Zhao, B. Chen, H. Xia, Y. Zhang, H. Ye, Efficient near-infrared downconversion and energy transfer mechanism in Pr^{3+}/Yb^{3+} co-doped $Na_5Y_9F_{32}$ single crystals, J. Lumin. 205 (2019) 500−507, https://doi.org/10.1016/j.jlumin.2018.10.002.
[19] S. Mahama, D.K.P Asiedu, K. Lee, Simultaneous wireless information and power transfer for cooperative relay networks with battery 5, IEEE Access, 2017, pp. 13171−13178.
[20] D. Wu, X. Dong, W. Xiao, Z. Hao, J. Zhang, Efficient visible-to-NIR spectral conversion for polycrystalline Si solar cells and revisiting the energy transfer mechanism from Ce^{3+} to Yb^{3+} in $Lu_3Al_5O_{12}$ host, Inorg. Chem. 58 (2019) 234−242, https://doi.org/10.1021/acs.inorgchem.8b02304.
[21] K. Li, R. Van Deun, Mutual energy transfer luminescent properties in novel $CsGd(MoO_4)_2$ Yb^{3+}, Er^{3+}/Ho^{3+} phosphors for solid-state lighting and solar cells, Phys. Chem. Chem. Phys. 21 (2019) 4746−4754, https://doi.org/10.1039/c8cp06538a.
[22] L. Lin, Z. Yu, Z. Wang, Z. Feng, F. Huang, L. Huang, Q. Dai, F. Zhang, Z. Zheng, Plasmon-enhanced upconversion and quantum-cutting of water-soluble Au@SiO_2/$NaYF_4$:Tb^{3+},Yb^{3+}@$NaYF_4$ nanocomposites, J. Lumin. 214 (2019) 1−7, https://doi.org/10.1016/j.jlumin.2019.116598.
[23] S.K. Karunakaran, C. Lou, G.M. Arumugam, C. Huihui, D. Pribat, Efficiency improvement of Si solar cells by down-shifting Ce^{3+}-doped and down-conversion Ce^{3+}-Yb^{3+} co-doped YAG phosphors, Sol. Energy 188 (2019) 45−50, https://doi.org/10.1016/j.solener.2019.05.076.
[24] T. Trupke, M.A. Green, P. Würfel, Improving solar cell efficiencies by down-conversion of high-energy photons, J. Appl. Phys. 92 (2002) 1668−1674, https://doi.org/10.1063/1.1492021.
[25] V. Badescu, A. De Vos, Influence of some design parameters on the efficiency of solar cells with down-conversion and down shifting of high-energy photons, J. Appl. Phys. 102 (2007), https://doi.org/10.1063/1.2785026.
[26] K. Znajdek, M. Sibiński, Z. Lisik, A. Apostoluk, Y. Zhu, B. Masenelli, P. Sedzicki, Zinc oxide nanoparticles for improvement of thin film photovoltaic structures' efficiency through down shifting conversion, Opto-Electron. Rev. 25 (2017) 99−102, https://doi.org/10.1016/j.opelre.2017.05.005.

[27] B. Fetliński, M. Malinowski, Potential energy yield increase of a solar spectra down-converter equipped photovoltaic device in real operational conditions, Sol. Energy 165 (2018) 148−158, https://doi.org/10.1016/j.solener.2018.02.071.

[28] W. Shockley, H.J. Queisser, Detailed balance limit of efficiency of p-n junction solar cells, J. Appl. Phys. 32 (1961) 510−519, https://doi.org/10.1063/1.1736034.

[29] A. Manjceevan, J. Bandara, Systematic stacking of PbS/CdS/CdSe multi-layered quantum dots for the enhancement of solar cell efficiency by harvesting wide solar spectrum, Electrochim. Acta 271 (2018) 567−575, https://doi.org/10.1016/j.electacta.2018.03.193.

[30] P.S. David, P. Panigrahi, G.S. Nagarajan, Enhanced near IR Downconversion Luminescence in Eu^{3+}-Yb^{3+} co-doped V activated ZnO host: an effort towards efficiency enhancement in Si-Solar cells, Mater. Lett. (2019), https://doi.org/10.1016/j.matlet.2019.03.046.

[31] N.U. Rahman, W.U. Khan, W. Li, S. Khan, J. Khan, S. Zheng, T. Su, J. Zhao, M.P. Aldred, Z. Chi, Simultaneous enhancement in performance and UV-light stability of organic-inorganic perovskite solar cells using a samarium-based down conversion material, J. Mater. Chem. A. 7 (2019) 322−329, https://doi.org/10.1039/c8ta09362h.

[32] S.J. Mofokeng, V. Kumar, R.E. Kroon, S.-H. Cho, O.M. Ntwaeaborwa, Enhanced red emission of Eu^{3+} in ZnO-TiO_2:Dy^{3+}, Eu^{3+} nanocomposites by UV downconversion process, J. Vac. Sci. Technol. B 37 (2019) 022901, https://doi.org/10.1116/1.5081953.

[33] X. Luo, J.Y. Ahn, S.H. Kim, Aerosol synthesis and luminescent properties of $CaAl_2O_4$: Eu^{2+}, Nd^{3+} down-conversion phosphor particles for enhanced light harvesting of dye-sensitized solar cells, Sol. Energy 178 (2019) 173−180, https://doi.org/10.1016/j.solener.2018.12.029.

[34] J.C.G. Bünzli, A.S. Chauvin, Lanthanides in Solar Energy Conversion, 2014, https://doi.org/10.1016/B978-0-444-62711-7.00261-9.

[35] R.T. Wegh, H. Donker, K.D. Oskam, A. Meijerink, Visible quantum cutting in $LiGdF_4$: Eu^{3+} through downconversion, Science 283 (1999) 663−666, https://doi.org/10.1126/science.283.5402.663.

[36] S.P. Ghawade, K.A. Deshmukh, D.R. Peshwe, S.J. Dhoble, A.D. Deshmukh, Efficient red downshifting in layered structure: a broad spectral converter for enhancing photo-response of solar cell, J. Inorg. Organomet. Polym. Mater. 27 (2017) 173−180, https://doi.org/10.1007/s10904-017-0684-8.

[37] Q.C. Sun, Y.C. Ding, D.M. Sagar, P. Nagpal, Photon upconversion towards applications in energy conversion and bioimaging, Prog. Surf. Sci. 92 (2017) 281−316, https://doi.org/10.1016/j.progsurf.2017.09.003.

[38] F.U. Xinjie, S. Lixin, L.I. Jiacheng, Radiation induced color centers in cerium-doped and cerium-free multicomponent silicate glasses, J. Rare Earths 32 (2014) 1037−1042, https://doi.org/10.1016/S1002-0721(14)60180-0.

[39] A.R. Kadam, S.J. Dhoble, Synthesis and luminescence study of Eu^{3+}-doped $SrYAl_3O_7$ phosphor, Luminescence 34 (2019) 846−853, https://doi.org/10.1002/bio.3681.

[40] J. Malleshappa, H. Nagabhushana, B.D. Prasad, S.C. Sharma, Y.S. Vidya, K.S. Anantharaju, Structural, photoluminescence and thermoluminescence properties of CeO2 nanoparticles, Accepduscrt, Opt. Int. J. Light Electron Opt. (2015), https://doi.org/10.1016/j.ijleo.2015.10.114.

[41] V. Ramón, O. Barrón, F. María, E. Ochoa, C. Cruz, R. Bernal, Thermoluminescence of novel MgO − CeO_2 obtained by a glycine-based solution combustion method, Appl. Radiat. Isot. (2016) 1−5, https://doi.org/10.1016/j.apradiso.2016.02.002.

[42] G. Sahu, A.S. Gour, R.K. Chandrakar, Thermoluminescence and optical properties of Dy^{3+} doped MgO nanoparticles, Prepar. Solut. Combust. Synth. Method 7 (2017) 115−127.
[43] N.J. Shivaramu, B.N. Lakshminarasappa, K.R. Nagabhushana, F. Singh, Abstract, SAA, 2015, https://doi.org/10.1016/j.saa.2015.09.019.
[44] M. Singh, P.D. Sahare, Nuclear instruments and methods in physics research B redox reactions in Cu-activated nanocrystalline LiF TLD phosphor, Nucl. Instrum. Methods Phys. Res. B. 289 (2012) 59−67, https://doi.org/10.1016/j.nimb.2012.08.003.
[45] S. Bahl, S.P. Lochab, P. Kumar, $CaSO_4$: DY, Mn: a new and highly sensitive thermoluminescence phosphor for versatile dosimetry, Radiat. Phys. Chem. 119 (2016) 136−141, https://doi.org/10.1016/j.radphyschem.2015.10.004.
[46] G. Zhu, X. Wang, H. Li, L. Pan, H. Sun, X. Liu, T. Lv, Z. Sun, $Y_3Al_5O_{12}$:Ce phosphors as a scattering layer for high-efficiency dye sensitized solar cells, Chem. Commun. 48 (2012) 958−960, https://doi.org/10.1039/c1cc16089c.
[47] W. He, T.S. Atabaev, H.K. Kim, Y.H. Hwang, Enhanced sunlight harvesting of dye-sensitized solar cells assisted with long persistent phosphor materials, J. Phys. Chem. C 117 (2013) 17894−17900, https://doi.org/10.1021/jp307954n.
[48] S. Park, M. Vosguerichian, Z. Bao, A review of fabrication and applications of carbon nanotube film-based flexible electronics, Nanoscale 5 (2013) 1727−1752, https://doi.org/10.1039/c3nr33560g.
[49] J. Wu, J. Wang, J. Lin, Y. Xiao, G. Yue, M. Huang, Z. Lan, Y. Huang, L. Fan, S. Yin, T. Sato, Dual functions of YF_3:Eu^{3+} for improving photovoltaic performance of dye-sensitized solar cells, Sci. Rep. 3 (2013) 1−5, https://doi.org/10.1038/srep02058.
[50] J. Shen, Z. Li, R. Cheng, Q. Luo, Y. Luo, Y. Chen, X. Chen, Z. Sun, S. Huang, Eu^{3+} -Doped $NaGdF_4$ Nanocrystal Down-Converting Layer for E Ffi Cient Dye-Sensitized Solar Cells, 2014.
[51] A. Puntambekar, V. Chakrapani, Excitation energy transfer from long-persistent phosphors for enhancing power conversion of dye-sensitized solar cells, Phys. Rev. B 93 (2016) 1−7, https://doi.org/10.1103/PhysRevB.93.245301.
[52] V. Kumar, S.K. Swami, A. Kumar, O.M. Ntwaeaborwa, V. Dutta, H.C. Swart, Eu^{3+} doped down shifting TiO_2 layer for efficient dye-sensitized solar cells, J. Colloid Interface Sci. 484 (2016) 24−32, https://doi.org/10.1016/j.jcis.2016.08.060.
[53] J. Duan, J. Wang, Q. Tang, B. He, W. Wang, Long persistence phosphor assisted all-weather solar cells. Electricity generation beyond sunny days, Chem. Commun. 53 (2017) 3209−3212, https://doi.org/10.1039/c7cc00537g.
[54] J. Llanos, I. Brito, D. Espinoza, R. Sekar, P. Manidurai, A down-shifting Eu^{3+}-doped Y_2WO_6/TiO_2 photoelectrode for improved light harvesting in dye-sensitized solar cells, R. Soc. Open Sci. 5 (2018) 1−9, https://doi.org/10.1098/rsos.171054.
[55] N. Chander, A.F. Khan, P.S. Chandrasekhar, E. Thouti, S.K. Swami, V. Dutta, V.K. Komarala, Reduced ultraviolet light induced degradation and enhanced light harvesting using YVO_4:Eu^{3+} down-shifting nano-phosphor layer in organometal halide perovskite solar cells, Appl. Phys. Lett. 105 (2014), https://doi.org/10.1063/1.4891181.
[56] X. Hou, T. Xuan, H. Sun, X. Chen, H. Li, L. Pan, High-performance perovskite solar cells by incorporating a $ZnGa_2O_4$:Eu^{3+} nanophosphor in the mesoporous TiO_2 layer, Sol. Energy Mater. Sol. Cells 149 (2016) 121−127, https://doi.org/10.1016/j.solmat.2016.01.021.

[57] W. Chen, Q. Luo, C. Zhang, J. Shi, X. Deng, L. Yue, Z. Wang, X. Chen, S. Huang, Effects of down-conversion $CeO_2:Eu^{3+}$ nanophosphors in perovskite solar cells, J. Mater. Sci. Mater. Electron. 28 (2017) 11346–11357, https://doi.org/10.1007/s10854-017-6928-0.

[58] L. Jiang, W. Chen, J. Zheng, L. Zhu, L. Mo, Z. Li, L. Hu, T. Hayat, A. Alsaedi, C. Zhang, S. Dai, Enhancing the photovoltaic performance of perovskite solar cells with a down-conversion Eu-complex, ACS Appl. Mater. Interfaces 9 (2017) 26958–26964, https://doi.org/10.1021/acsami.7b10101.

Rare-earth-activated phosphors for drug delivery

14

Megha Jain and Palwinder Singh
Department of Physics, Chungbuk National University, Cheongju, Republic of Korea

1. Introduction

Multifunctional material has become a buzzword in the scientific industry due to its unique and tunable properties leading to its potential applications in different fields. These materials have been increasingly employed in biomedical applications, due to their ability to combine two or more functions, such as different types of imaging or imaging with drug delivery, targeting, or various therapies [1–4]. Recently, the development of rare-earth-activated multifunctional materials combining distinct magnetic and fluorescent nanoparticles as well as mesoporous/hollowed structures in a single nano-object has become the center of attraction, and these are promising candidates for biomedical applications, such as diagnostic analysis, magnetic resonance imaging (MRI), fluorescent labeling, and drug delivery carriers [5,6]. Mesoporous structures, specifically hollow spherical particles with mesoporous shells, combine the characteristics of both macroporous and mesoporous structures in one single unit, in which a hollow cavity can store the cargos and shell provides controlled release pathways for encapsulated drugs [7]. Several advantages offered by such materials such as combination of magnetic and fluorescent properties enable them as dual-imaging probes providing non-invasive and high spatial resolution MRI for real-time monitoring in addition to bifunctionality of the high sensitivity fluorescence imaging [8,9]. Thus, using the multifunctionality of a material enables the advancement of platforms for simultaneous imaging diagnosis and drug delivery. Among such materials, luminescent nanomaterials are worth mentioning due to their versatile applications such as optoelectronics, sensors, field-effect transistors, optical storage devices, medical and in vitro and in vivo biological labels, solar cells, and light-emitting diodes [10]. Semiconductor quantum dots (QDs) and organic dyes have been employed as biological labels using luminescence property; however, with limitations. QDs such as Ag_2S, CdS, PbS, and Ag_2Se have a low resistance to chemical degradation, are toxic, and show fluorescence intermittence, while dyes exhibit rapid photobleaching [11]. So, rare-earth-activated nanophosphors offer promising alternative fluorescent materials for biomedical applications owing to their splendid properties such as low toxicity, low production cost, high chemical stability, and tunable optical properties achievable through variation of rare-earth dopants and the host matrix [6]. Luminescent properties of rare-earth ions originate from their electronic configuration [Xe]$4f^n$, $n = 0-14$. These 4f electronic levels are highly shielded by $5s^2 5p^6$ orbitals and are least affected by their chemical environment, and therefore exhibit high photostability, high

Rare-Earth-Activated Phosphors. https://doi.org/10.1016/B978-0-323-89856-0.00018-3

luminescence quantum yield, and sharp emission bands, providing more selectivity for bioassays [12]. Upconversion luminescence is a mechanism through which high energy photons are obtained from low energy photons assisted by a host matrix and rare-earth elements. Rare-earth dopant can serve as either an activator that is a light center or a sensitizer that absorbs energy and transfers it to the activator. The role of the host matrix is to provide an appropriate crystal field for the luminescence center. Fig. 14.1 depicts the energy levels of rare-earth ions and pathways to realize upconversion emissions. Various energy transfer mechanisms involved in the upconversion mechanism are depicted by Fig. 14.2 [13,14]. Upconversion rare-earth-activated fluorescence materials are beneficial over downconversion counterparts due to their merits such as the following:

* NIR excitation is in optical transparency window (700−1100 nm) giving a high signal-to-noise ratio and detection sensitivity.
* Fluorescence from biological tissues or drugs on infrared (IR) excitation is extremely low while interfering biomolecules in question absorb in the ultraviolet (UV), not in the IR region. Thus, autofluorescence noise is reduced.
* Detection process is not affected by the external environment, as upconversion happens only inside the host matrix.

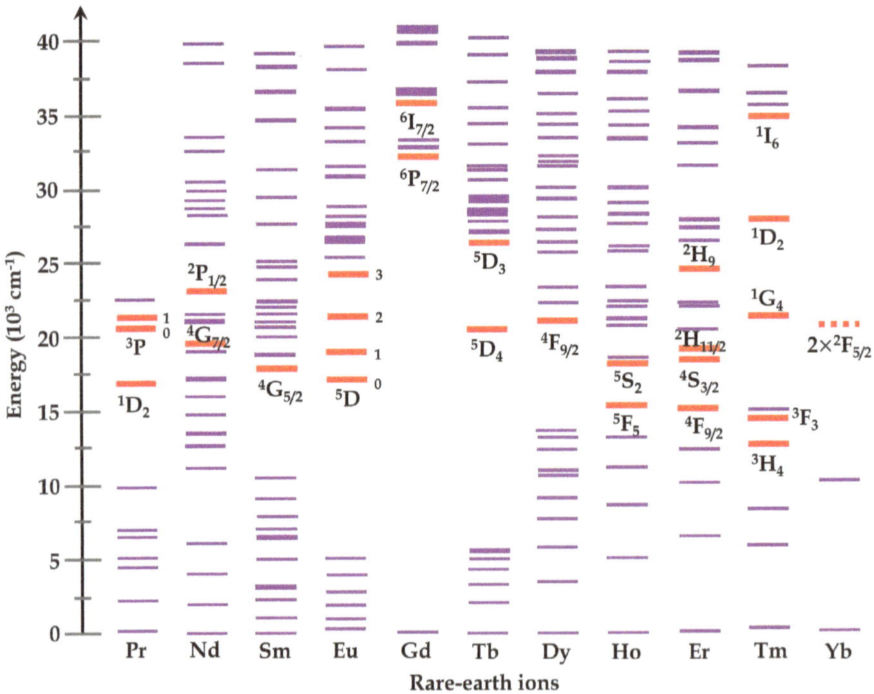

Figure 14.1 Energy-level diagram of rare-earth ions. Typical upconversion emissive excited states are highlighted in pink color.

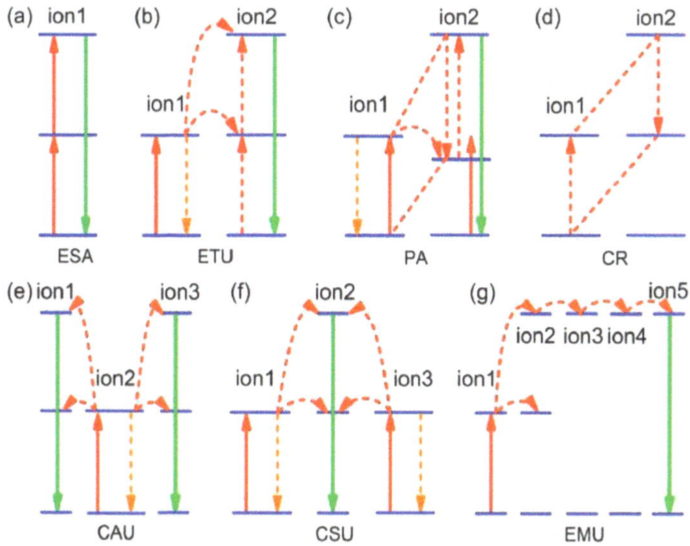

Figure 14.2 Schematic view of typical upconversion processes in rare-earth-activated phosphors: (A) excited state absorption (ESA), (B) energy transfer upconversion (ETU), (C) photon avalanche (PA), (D) cross-relaxation (CR), (E) cooperative activation upconversion (CAU), (F) cooperative sensitization upconversion (CSU), (G) energy migration-mediated upconversion (EMU). The red solid lines, red dotted lines, orange dotted lines, and green lines represent photon absorption, energy transfer, spontaneous emission, and upconversion emission, respectively.
From Ref. [14].

* Superior chemical stability and remarkable photostability are free of on–off blinking and measurable photobleaching under prolonged single-particle excitation. Manifold emission color can be tuned by changing the host lattices and activator concentration under single NIR light excitation.
* NIR excitation can be performed using a portable, energy-efficient, and cheap IR laser.
* NIR radiation is less harmful to cells, minimizes autofluorescence from biological tissues, and has high penetration depth in tissues.

Field of drug delivery is quite broad as it embraces the development of compatible and efficient drug delivery carrier materials. Efficient drug delivery systems can deliver precise quantities of therapeutic drugs to the targeted cells or tissues in a controlled release way to enhance the efficiency of delivered drugs and reduce toxicity simultaneously [15]. This field is interdisciplinary with the objective of proper administration of bioactive agents to achieve a desired clinical response in the body, enhancing drug therapy. Enhanced product safety, improved patient convenience, and compliance resulting from controlling the drug delivery are achieved by designing and developing technologies that modify the temporal and spatial drug release profile [16]. Growth in the area of drug delivery is very rapid representing an ever-evolving field for biomedical materials science, bringing both commercial and therapeutic values to healthcare products. Subsequent sections of this chapter include a discussion about the drug delivery mechanism and research findings concerned with the employment of several rare-earth-activated phosphors for efficient drug delivery and cell therapy applications.

2. Drug targeting and delivery

Drug delivery is a rapidly expanding field, surpassing the borders between pharmaceutical sciences, biosciences, and engineering prompted by the evolution of computational methods. It covers a broad range of techniques used to get therapeutic agents into bioorganism. The strategies used to control the drug release from pharmaceutical systems are based on biological, physicochemical, and mathematical principles [17]. There are several methods by which the drug release can be administered and controlled depending upon the particular application. Conventionally, drug delivery systems/formulations integrated with drugs exist either in bulk, micro, or nanoforms such as tablets, sprays, suspensions, powders, and capsules are administrated to the organism in invasive (surgical injections) or noninvasive (oral, mucus, nasal, transdermal, etc.) routes [18]. When the active agent can be administered at its preferred site of action and kept away from sites associated with toxicity, the therapy becomes safer, efficient, and more acceptable improving patient compliance. Main attention is attracted by sustained or controlled drug delivery systems in the modern medical fields as they offer higher drug delivery efficiency and reduced toxicity in comparison to conventional drug administration in dosage forms. Drug targeting is the ability of an active agent/carrier to accumulate in the target site selectively and quantitatively, independent of the site and methods of administration [19]. The local concentration of the drug at the site(s) should be high, while its concentration in other nontarget organs and tissues should be below certain minimal levels to prevent negative side effects. Thus, drug targeting is a very important mechanism of drug delivery control. It employs a strategy in which platform and nanoscale devices are developed for the selective delivery of small drug molecules and therapeutic genes to the cells of interest. Such strategies may differ but all utilize nanometer-sized particles or nanocarriers to deliver therapeutic payloads to the targeted cells [20]. This is because drug formulations in bulk form offer some disadvantages such as poor solubility, reduced half-life, poor cell interactions, low bioavailability, large size, difficulty to administer, and lack of selectivity. The concept of therapeutic delivery relies on the following key mechanistic elements, each playing a significant role for efficient delivery [15].

* Specific cellular binding.
* Intracellular uptake of drug-carrying nanomaterials by targeted cells.
* Controlled release of carried drug molecules in an active form.

Nanocarriers can be in any shape being unique in chemical and physical aspects such as synthetic methods, surface functionality and modification, core-shell structure, size, and shape. It is extremely important to understand their physicochemical properties and interactions with biological systems to avoid undesired effects due to their intrinsic toxicity or immunogenicity. Nanocarriers serve as skeletons in the systems and are capable of loading and releasing drugs, keeping intact framework in blood circulation, and possessing good biocompatibility and pharmacological properties [20]. Nanocarriers are designed in such a way so that they enter/accumulate in the disease cells. Latest drug delivery systems are designed to be sensitive to stimuli, stimuli being classified into endogenous stimuli and exogenous stimuli. Endogenous stimuli correspond to the microenvironment surrounding the defected cells originating from the genetic mutations. Mutation leads to misfunctioning of the cells. Exogenous stimuli refer to disease irrelevant physical stimuli such as light, heat, magnetic field, or ultrasound [21]. Fig. 14.3 schematically illustrates the endogenous and exogenous stimuli used for designing drug delivery systems. Nanocarriers containing exogenous stimuli sensitive materials can respond under controlled applied stimuli such as in photodynamic responsive systems, photothermal responsive systems, and magneto responsive systems. The next step is the drug release by the carrier, which is realized in a smart way on targeted sites. Such ways include chemical bond cleavage, structure change, or physiochemical property change. Inorganic materials contain modifiable chemical groups, which are linked with therapeutic agents through stimuli-sensitive bonds [18]. Chemical bond cleavage leads to the release of these agents on

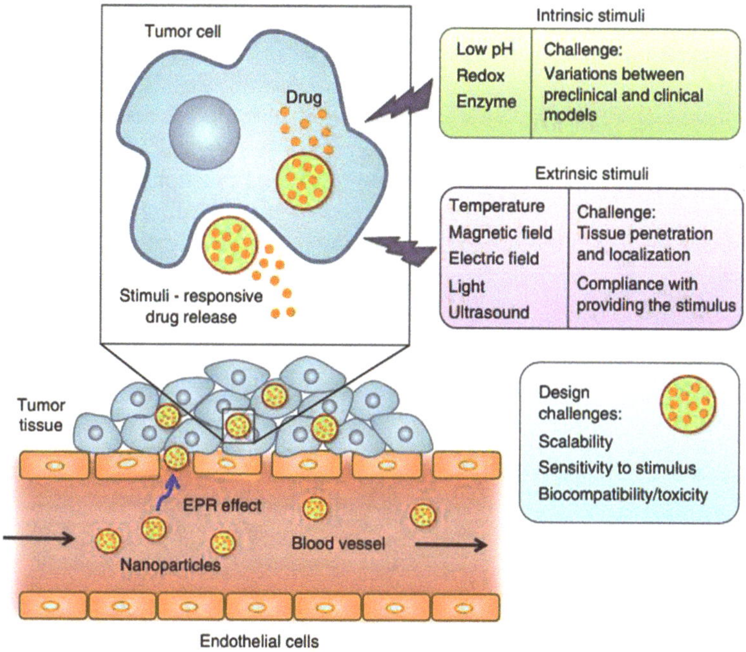

Figure 14.3 Schematic illustration of endogenous and exogenous stimuli commonly used in designing smart drug delivery systems.
From Ref. [21].

encountering the stimuli. Some popular chemical bonds sensitive to stimuli are pH-responsive bonds, photo-sensitive bonds, or redox stimuli bonds. Nanocarriers are structurally designed in a way to load therapeutic agents and can change the structure through framework disintegration, size or morphology change, gatekeeping strategy by modifying the pore sensitivity to stimuli to unload the agents. Physiochemical property change involves the dissolution of nanocarrier, surface charge reversal under stimuli [21]. Rare-earth-activated phosphors have been found to be quite useful in representing stimuli-responsive drug release leading to targeted drug delivery offering an additional advantage of in situ imaging/monitoring of the whole process.

3. Phosphors for drug delivery

Rare-earth elements comprise 15 lanthanide series elements from lanthanum to lutetium in addition to yttrium and scandium. Rare-earth ions exhibit peculiar luminescence properties accompanied by intra-4f or 4f to 5d transitions. Specifically, upconversion luminescence is one of the most intriguing features of rare-earth luminescence attracting attention for its applicability in various fields. Mesoporous phosphors activated by rare-earth ions also have potential applications in the fields of drug delivery, disease diagnosis, and therapy [5,12]. Additionally, rare-earth-activated persistent luminescence is also explored for in vivo optical imaging of target area and delivery of drug precisely [6,11,12]. However, the only limitation is in situ excitation of persistent phosphors, which can be overcome by combining rare-earth dopants suitable for upconversion/antistokes process allowing the in situ excitation by less harmful NIR radiation [13,22]. Fig. 14.4 depicts the recent approaches being utilized to construct rare-earth activated drug delivery systems [6] and Fig. 14.5 illustrates some modifications including core–shell structure and functionalization proposed in upconversion nanoparticles for drug delivery systems [23]. In this section, some of the phosphors activated by rare-earth ions explored for drug delivery application are discussed.

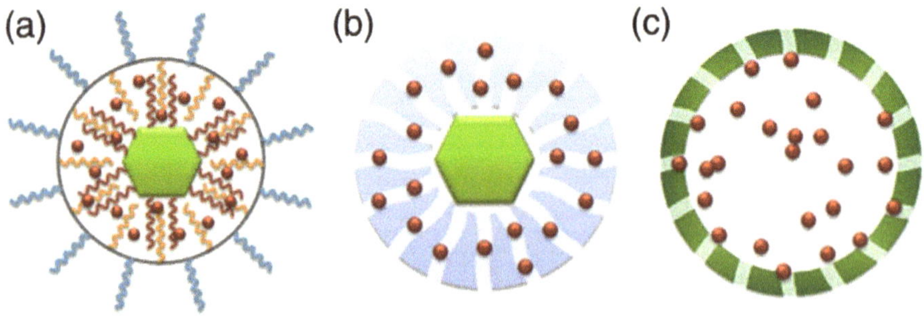

Figure 14.4 Schematic representation of current approaches to construct rare-earth-based drug delivery systems: (A) hydrophobic pockets, (B) mesoporous shell structure, and (C) hollow mesoporous-coated spheres.
From Ref. [6].

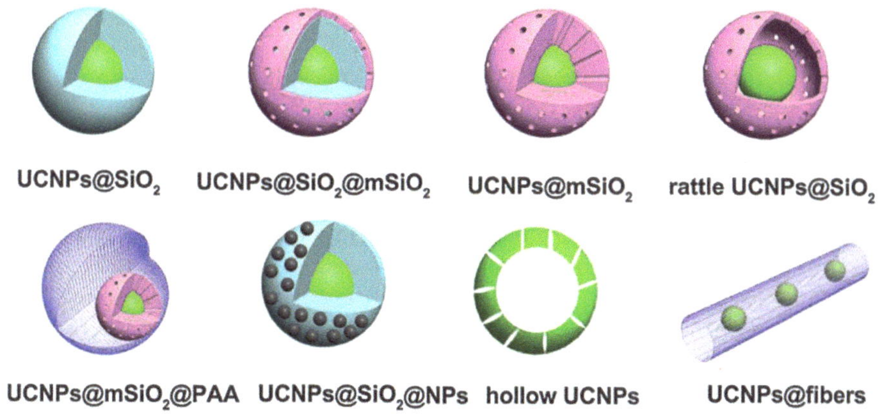

Figure 14.5 Schematic illustration of upconverted nanoparticles (UCNPs)-based drug delivery carriers ($mSiO_2$: mesoporous silica, PAA: polyacrylic acid, NPs: functional nanoparticles such as Fe_3O_4, $NaYF_4$, and Au).
From Ref. [23].

3.1 Rare-earth-activated silica

An efficient drug delivery system should be able to transport therapeutic drugs to the desired destination with little loss and release the same in a controlled manner. However, hydrolysis of most biodegradable polymer-based drug delivery systems induces erosion of carrier and release immediately after the dispersion of such systems in water, reducing the drug-releasing efficiency [24]. But this drawback has been overcome in recent decades by employing ordered mesoporous silica materials. These materials have gained attention owing to their stable mesoporous structure, uniform and tunable pore size, modifiable surface features, biocompatibility, and nontoxic nature [25]. However, pure mesoporous silica suffers some limitations in the study of targeted drug delivery mechanisms, tracking the efficiency of drug release, drug kinetics marker, etc. With further advancement in the field, surface-functionalized mesoporous silica materials as hosts for several drug molecules have gained intensive attention as these provide a platform for simultaneous targeted drug delivery, quick diagnosis, and efficient therapy [7]. Luminescent materials such as organic dyes, QDs, and rare-earth phosphors or magnetic materials have been successfully utilized to functionalize the mesoporous silica in the form of core-shell, embedded, and rattle-type structures [25].

In a research report, mesoporous silica particles were functionalized with luminescent $YVO_4:Eu^{3+}$ layers to investigate the ibuprofen (IBU) drug delivery efficiency of the system [26]. Drug loading and release rate largely depended upon specific surface area and pore volume of the functionalized silica. No leaching of Eu^{3+}-activated YVO_4 layers was observed indicating good stability of the carrier. Red color emission under UV excitation enhanced with the cumulative release of IBU, making the tracking and monitoring of drug release easier. In a similar study, mesoporous silica nanoparticles were functionalized using red luminescent $GdVO_4:Eu^{3+}$ nanophosphor [27]. The functionalized carrier was found to be exhibiting pH-responsive drug release behavior for doxorubicin (DOX) that is increased with decreased pH, with good biocompatibility and efficient drug delivery to ovarian cancer cells. Functionalization helped in brightening the T_1-weighted images and enhanced the r_1 relaxivity of water protons, suggesting their use as T_1 contrast agents for MRI due to the presence of Gd^{3+} ions. The luminescence intensity of Eu^{3+} ions was highly quenched on loading the DOX due to the creation of an environment with high phonon frequency, which was recovered as soon as the drug was released. Correlation of luminescence intensity with cumulative drug release and efficient drug delivery predicted the functionalized silica to be a multifunctional drug carrier possessing luminescence tagging, MRI, and pH-controlled release property. Another interesting study involved the investigation of a drug delivery system in a novel bifunctional hollow sphere formed out of coating luminescent YBO_3:Eu^{3+} nanoparticles onto uniformly hollow mesoporous silica spheres derived by etching from spherical Fe_3O_4 templates [28]. Prepared functionalized carrier exhibited mesoporous shell structure, hollow space inside, high surface area, well-dispersed nanospheres of controlled size, tunable specific surface, and pore volume. The composite showed characteristic red emission lines of Eu^{3+} ($^5D_0$–$^7F_{1-4}$) upon UV excitation, even after loading the anticancer drug DOX on the surface and cavity. It further exhibited a sustained drug release property, further releasing in a texture and pH-sensitive patterns. Here, as well, the downconversion emission intensity increased with the release of the drug, enabling the easy monitoring of drug release.

A different strategy was utilized by Chen et al. [29], as shown by the schematic in Fig. 14.6, in which efficiency of multifunctional envelope type mesoporous silica

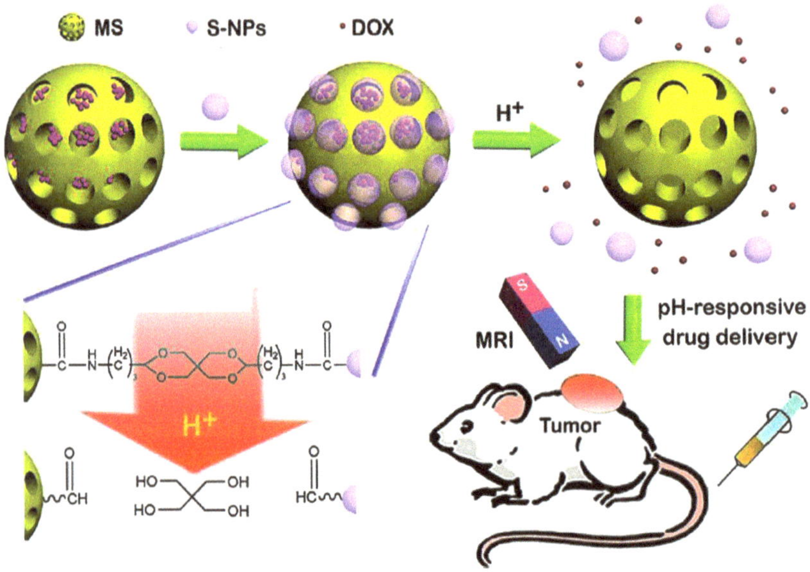

Figure 14.6 Schematic illustration of the mesoporous silica nanoparticle system for pH-responsive drug delivery and magnetic resonance imaging. MS— mesoporous silica, S-NPs—small nanoparticles composed of $NaGdF_4$:Yb^{3+}/Tm^{3+}, and DOX—doxorubicin drug. From Ref. [29].

nanoparticle system possessing pH-sensitivity, nontoxicity, and biological specificity was demonstrated for drug delivery and MRI. Herein, functionalized ultra-small rare-earth-doped nanoparticles, that is, $NaGdF_4$:Yb/Tm@$NaGdF_4$, with ultra-thin TaO_x layer were grafted onto pores of mesoporous silica with the acid-labile acetals on its surface. Acetals were used as protecting group for the carbonyl group, which on hydrolyzation at lower pH enabled the removal of nanoparticle layers. Anticancer drug DOX was loaded in the pores and was released with burst under acidic environment due to hydrolyzation reactions of acetals. This nanogated strategy was found to be highly efficient for cancer therapy for both in vitro and in vivo. Further, this nanocomposite was observed to be metabolized without any harm, and it degraded into nontoxic products in a few days. Nanocomposite was further modified with polyethylene glycol (PEG) to achieve longer circulation in blood vessels. Passive tumor targeting was permitted by enhanced permeation and retention effect of nanosized composite, revealing its potential for cancer therapy, that is, chemotherapy.

3.2 Rare-earth-activated $NaYF_4$

Sodium yttrium fluoride ($NaYF_4$) is one of the most important upconversion participants in drug delivery systems [30]. Among the cubic (α) and hexagonal (β) crystalline phases, β-phase $NaYF_4$ serves as an efficient host matrix for upconversion. Several rare-earth dopants in $NaYF_4$ host lattice have been explored from time to time to tune the upconversion emission properties. $NaYF_4$ doped with Yb^{3+} and Er^{3+} is quite popular among phosphors for infrared to visible upconversion. In an attempt to produce first-ever upconversion nanoparticles in serum-based dispersion being stable under physiological conditions in contrast to silica, polyacrylic acid, or polyethylene imine coatings, Budijono et al. [31] prepared the colloidal β-phase $NaYF_4$:Yb^{3+}, Er^{3+} nanoparticles through single-step cothermolysis using oleic acid/trioctyl phosphine ligand. Prepared upconversion nanoparticles were stable in buffers and serum media through flash nanoprecipitation using PEG coatings. Irradiation by NIR ($\lambda \sim 980$ nm) light produced two strong emission bands at 540 nm (green) and 660 nm (red). The stability of PEG-coated nanoparticles under different physiological conditions allows such formulations to be utilized in deep-penetrating bioimaging, NIR photodynamic therapy, and drug delivery to solid cancer tumor cells. However, the main challenges were the particle size which does not allow easy permeation through the cell, and on the other hand, decreasing size leads to reduced emission. Additionally, large polymer groups utilized to stabilize the nanoparticles can be replaced with smaller hydrophobes helping in increasing the permeability but decreasing the stability. Surface cross-linking can be explored as an alternative in such systems.

In yet another study by Dai et al. [32], DOX-conjugated $NaYF_4$:Yb^{3+}/Tm^{3+} nanoparticles were inspected for performance in anticancer drug delivery systems. Donor uniform spherical-shaped $NaYF_4$:Yb^{3+}/Tm^{3+} nanoparticles demonstrated the ability to release DOX through hydrazone bond cleavage in mildly acidic environments, helpful to reduce the side-effect of chemotherapeutics. Spectral overlap between donor emission at 452 nm ($^1D_2 \rightarrow {}^3F_4$) and 477 nm ($^1G_4 \rightarrow {}^3H_6$) with broad absorbance band of acceptor DOX at ~480 nm enabled efficient energy transfer among them. Luminescence resonance energy transfer (LRET) mechanism enabled the upconversion luminescence in $NaYF_4:Yb^{3+}/Tm^{3+}$, whose quenching and recovery were utilized as optical probes to monitor the conjunction and release of DOX. It was observed that DOX-conjugated $NaYF_4:Yb^{3+}/Tm^{3+}$ nanoparticles exhibited a cytotoxic effect on SKOV3 ovarian cancer cells. Thus, pH-triggered drug release and upconverting-based LRET property were combined for an efficient drug delivery system.

Ordered mesoporous materials with sustained-release properties are considered ideal candidates for drug delivery; however, these candidates in bulk form offer some disadvantages as well in targeted drug delivery and drug kinetics markers. This drawback can be overcome by designing multifunctional core–shell structured composites with luminescent and mesoporous properties. Mesoporous structure helps in loading and releasing the drug molecules, while the luminescence intensity can be tracked to evaluate the efficiency of drug release. A similar designing strategy has been utilized by Kang et al. [33], in which mesoporous SiO_2 layers of varying thickness were coated over $NaYF_4:Yb^{3+}/Er^{3+}$ upconversion nanoparticles to synthesize core–shell structured upconversion luminescent and mesoporous $NaYF_4:Yb^{3+}/Er^{3+}$@$nSiO_2$@$mSiO_2$ nanospheres. IBU was utilized as a model drug, whose adsorbed amount on mesoporous channels increased with increased thickness of SiO_2 coating. IBU drug delivery tracking revealed a two-step process composed of initial diffusion-controlled release followed by a slower release rate. On excitation by 980 nm NIR laser, nanostructures emitted green ($^2H_{11/2}$ and $^4S_{3/2} \rightarrow {}^4I_{15/2}$) and red ($^4F_{9/2} \rightarrow {}^4I_{15/2}$) fluorescence of Er^{3+} even after loading of IBU, the intensity of which increased with an increase in the cumulative released amount of IBU. This paved the path for the identification, tracking, and monitoring of drug release. A similar study was reported by Tian et al. [34], in which a drug delivery system for colon targeting was designed using biocompatible, upconversion, and pH-responsive core–shell-structured $NaYF_4:Yb^{3+}/Er^{3+}$@SiO_2@PMAA nanocomposite. 5-aminosalicylic acid was employed as a therapeutic drug whose release from nanocomposites was monitored under different pH conditions. It was observed that a little amount of drug was released in the simulated gastric fluid of pH ~1.2, while a large amount of drug was released in the simulated colonic fluid of pH ~7.4, being monitored by change of upconversion emission intensity. The efficiency of drug delivery depended upon the state of the PMAA network and interaction between carrier and drug, and emission intensity increased with the release of the drug.

On similar lines in another study [35], uniform-sized α-$NaYF_4:Yb^{3+}$, Er^{3+} nanospheres with mesoporous shells and hollow interior structures were prepared for drug delivery. Polyethylenimine ligands were a major contributor to the formation of hollow structured nanospheres, and folic acid (FA), a model drug for targeting cancer cells, was conjugated on the surface of nanospheres through free amine groups. Bright green emission ($^2H_{11/2} \rightarrow {}^4I_{15/2}$, $^4S_{3/2} \rightarrow {}^4I_{15/2}$, $^4H_{9/2} \rightarrow {}^4I_{15/2}$ electronic transitions of Er^{3+} ions) without background noise was obtained on 980 nm IR laser excitation, enabling the tracking of drug delivery. The conjugated system exhibited greater cytotoxicity than DOX-conjugated α-$NaYF_4$ owing to specific cell uptake by HeLa cells via FA receptor-mediated endocytosis. In another study [36], following the core−shell-structured strategy, mesoporous silica coated with $NaYF_4$:$Yb_{0.4}$/$Tm_{0.02}$@$NaGdF_4$:$Yb_{0.1}$@$NaNdF_4$:$Yb_{0.1}$ upconversion nanoparticles was investigated for 808-nm NIR light-responsive in vivo and in vitro cancer treatment. The surface of nanocomposite was functionalized with β-cyclodextrin rings bridged by light cleavable platinum (IV) prodrug and adamantane, blocking the anticancer DOX drug inside the pores. Upon excitation by NIR light, emitted UV light (346 nm, $^1I_6 \rightarrow {}^3F_4$ and 363 nm, $^1D_2 \rightarrow {}^3H_6$) activated the platinum (IV) prodrug to achieve higher toxic platinum (II) complexes, unblocking the pores and releasing DOX. The main advantage of this designed structure is that both DOX, as well as platinum (II) complexes, can kill cancer (HeLa) cells. Thus, enabling a dual-drug delivery system based on photoactivated and photo-controlled cancer therapy is depicted in Fig. 14.7.

Gai et al. [37] modified the core−shell architecture for the fabrication of multifunctional nanocomposites using silica-coated Fe_3O_4 spheres as core and ordered mesoporous silica as shell, functionalized by deposition of $NaYF_4$:Yb^{3+}, Er^{3+}/Tm^{3+} upconversion phosphors. Modification of Fe_3O_4 core was done to prevent the fluorescence from being quenched by Fe_3O_4. Inert silica coating being a good candidate for protective matrix due to reliable chemical stability, biocompatibility, and reactivity

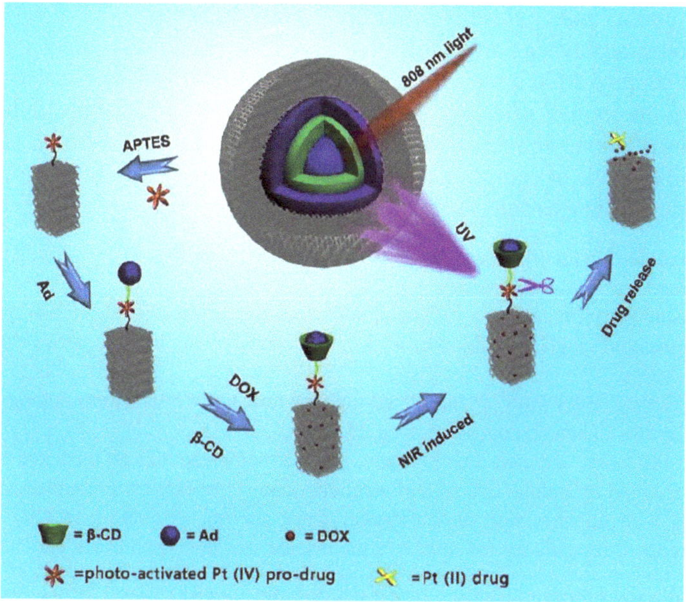

Figure 14.7 Schematic illustrating preparation, functionalization protocol, and controlled release process of UCNP@mSiO2-DPP-DOX@CD. UCNP: $NaYF_4$:$Yb_{0.4}$/$Tm_{0.02}$@$NaGdF_4$:$Yb_{0.1}$@$NaNdF_4$:$Yb_{0.1}$, DPP: trans, trans, trans-[$Pt(N_3)_2(NH_3)$ (py) $(CCH_2CH_2COOH)_2$] (Pt(IV) prodrug), *DOX*, doxorubicin; *CD*, cyclodectrin.
From Ref. [36].

with different coupling agents was employed for the purpose. This strategy led to the formation of Fe_3O_4@$nSiO_2$@$mSiO_2$@$NaYF_4$:Yb^{3+}, Er^{3+}/Tm^{3+} nanocomposites possessing mesoporous, upconversion luminescent, and magnetic properties. The high saturation magnetization (38.0 emu/g) exhibited by these nanocomposites enabled the drug targeting under an external magnetic field and unique upconversion green emission for Yb^{3+}/Er^{3+} and blue for Yb^{3+}/Tm^{3+} under 980 nm laser excitation even after loading the IBU drug. Tests indicated the drug release was in a controlled manner and emission intensity enhanced with the release of IBU, thus enabling the tracking and monitoring of drug release.

3.3 Rare-earth-activated metal-organic frameworks

Another important emerging class in drug delivery applications is hybrid organic–inorganic materials known as metal–organic frameworks (MOFs). MOFs being composed of metal-containing nodes and organic ligands have gained considerable attention as delivery vectors owing to the large surface area, ordered and tunable pore pattern, tunable structure, good biocompatibility, and functional versatility [38]. Rare-earth-containing MOFs have recently become the subject of attraction throughout the biotechnology field. These can be formed from either rare-earth ions or serve as host structures in which rare-earth ions can be incorporated supported by the highly porous and oriented structures of MOFs. MOFs can support several drugs and are biodegradably supporting them for biosensing and drug delivery applications. Rare-earth-coupled MOFs have been successfully utilized as nanocarriers encapsulating drug molecules offering increased permeability of biofilms influence the in vivo distribution, controlled drug release rate, improved stability, and broad bioavailability [39]. In comparison to other drug carriers, rare-earth-coupled MOFs offer unique advantages such as the following.

* Easy to simulate and design them based on drug characteristics by simply varying the composition or/and architecture geometry.
* Good biocompatibility and easy to functionalize through the attached ligands or metal nodes.
* Affinity to capture high loading doses of drug molecules owing to large pore volume-to-surface area ratios.
* MOF chemistry paves ways to control and tune the host-guest interactions among MOFs and drug molecules to enable controlled drug release.
* Drug quantification, targeted drug delivery, and chemotherapeutics all mediated by imaging can be exploited by unique optical and magnetic features of rare-earth coupled MOFs functionalized with upconversion luminescence. Thus, serving as dual-mode drug carriers as well as contrast imaging agents improves the disease diagnosis and treatment channel.

Liu et al. [40] prepared a series of MOFs containing Gd^{3+}/Tm^{3+} ions, enabling the composite to show upconversion luminescence and superior positive MRI properties. Gd^{3+}/Tm^{3+}-MOF as multifunctional drug delivery system was inspected in delivering the DOX drug. It was found that drug loading and pH responsiveness were highly improved on modifying the Gd^{3+}/Tm^{3+}-MOFs with uniform mesoporous silica shells and FA. Upon 980 nm laser excitation, imaging results exhibited blue and red emission, inferring the potential of composite for drug delivery as well as upconversion-MRI dual-mode imaging. Considering the prospects of rare-earth-activated upconversion materials, Wang et al. [41] developed novel multifunctional water-soluble $NaLnF_4$@MOF-Ln (Ln = Y^{3+}, Tm^{3+}, Yb^{3+}, and Eu^{3+}) nanocomposites with dual-mode luminescence. Dual mode here refers to the combination of upconversion blue luminescence from $NaYF_4:Tm^{3+}/Yb^{3+}$ with the downconversion red luminescence of MOF-Y:Eu^{3+}. Composites offered an advantage over quenching and bleaching of organic molecules in QDs, through fluorescence imaging ability as in vivo bioprobe. Nontoxic nanocomposites were labeled with HeLa cells successfully, carrying the DOX drug simultaneously for controlled and efficient drug release in a simulated cancer cell environment. The pH sensitivity of drug loading and releasing was also observed under different pH environments. Kundu et al. [42] prepared a Gd^{3+}-based porous MOF in which a three-dimensional structure interpenetrated with a one-dimensional nano ranged open channel filled with hydrogen-bonded water assemblies. Synthesized nanocomposites exhibited high thermal stability, porosity, good water stability, organic solvent, and stability with retention of crystallinity under mild acidic−basic conditions. In vitro and in vivo studies of pH-responsive drug DOX loading and release capability enabled the MOFs to be applicable in cancer cell therapy.

3.4 Other rare-earth-activated systems

Another class of materials involves rare-earth-based inorganic crystalline materials, among which Gd_2O_3 owing to low phonon energy, proton relaxation, combination of optical and magnetic properties, and scintillation properties emerged as an excellent candidate for fluorescence imaging, MRI, multifunctional bioprobes, etc. [43]. Nanoparticles of Gd_2O_3 have been approved as effective T_1 MRI contrast agents as Gd^{3+} owns a large number of unpaired electrons. Gd_2O_3-based nanoparticles are easier to fabricate as compared to multistep synthetic procedures and synthetic conditions involved in designing multicomponent composites. Performance degradation of individual components and inhomogeneity of morphology and properties in multicomponent composites pose limitations on their applications. Another advantage of Gd_2O_3 is it being a good host matrix to be doped by other rare-earth ions for upconversion and downconversion, proving it promising for in vitro/in vivo optical imaging. Di et al. [44] synthesized single-phased spherical mesoporous $Gd_2O_3:Eu^{3+}$ nanoparticles, proving them simultaneously applicable in drug delivery and cell imaging. The mesoporous structure enabled the high DOX drug loading efficiency and a sustained release of the same. Intrinsic luminescence helped in imaging the cancer cell detection, drug release processes indicating the potential of simple compound Gd_2O_3 as an effective carrier for cancer cell treatment. In another study, $Gd_2O_3:Eu^{3+}$ functionalized with FA was efficiently utilized for detection and treatment of breast cancer, the second leading cause of cancer death [45]. Functionalized nanoparticles exhibited strong red emission at 613 nm when excited at 254 nm. In vitro studies indicated that the compound poses a negligible cytotoxic effect on normal 293T and T-47D breast cancer cells. Cellular uptake analysis inferred quite high internalization of FA-conjugated rare-earth nanoparticles into T-47D cells as compared to MDA-MB-231 breast cancer cells. In vivo confocal and CT imaging studies showed that functionalized nanoparticles accumulated more efficiently in T-47D tumor xenograft in comparison to MDA-MB-231 tumor, revealing it to be a promising material for breast cancer detection. Similarly, Gd_2O_3:Yb^{3+}/Er^{3+} hollow spheres were synthesized through a simple synthesis method for investigating its drug delivery efficiency and optical imaging property [46]. These were observed to be exhibiting bright multicolored upconversion luminescence that was tunable from green to red by varying the codopants Yb^{3+}/Er^{3+} ratio under 980 nm NIR laser excitation. T_1-weighted images were brightened by Gd_2O_3 nanospheres, and r_1 relaxivity of water protons was also enhanced, suggesting Gd_2O_3 to be a good T_1 contrast agent for MRI. Hollow spheres were tested for drug storage and delivery efficiency using IBU as a model drug. Cooperation among internal and external surface adsorption enabled high drug loading amount, while mesoporous holes on ruptured hollow spheres facilitated the IBU transport toward the internal surface. The system revealed a burst release of the drug (60% in 0.5 h) followed by a slow release (80% in 2 h) and complete release after 10 h. Burst release was assigned to rapid leaching of IBU from the outer surface, while slower one was from the internal surface of hollow spheres.

In addition to the previously discussed versatile materials designed for drug delivery systems, some less popular materials activated by rare-earth ions have also been explored. One among them is the class of solid inorganic fluorides, apart from $NaXF_4$ (X = Gd, Lu, Y, etc.), possessing potential candidature in the medical field due to low phonon frequencies, high resistivity, and anionic conductivity [47]. To tackle the problem of free diffusion governed burst drug release, Dai et al. [48] designed a controlled drug release system based on CaF_2:Ce^{3+}/Tb^{3+}-polyacrylic acid (PAA) composite microspheres by filling the pH-responsive PAA inside the hollow CaF_2 microspheres through the UV photopolymerization route. Designed composites exhibited strong green luminescence for Tb^{3+} ions under UV excitation, which changed upon drug loading and releasing thereby helping in monitoring the drug delivery mechanism. DOX was employed as a model drug to evaluate the loading and release behavior of composite microspheres. PAA-modified CaF_2:Ce^{3+}/Tb^{3+} microspheres effectively prevented the sudden burst drug release by potentially selecting the disease site with the acid environment such as tumor or stomach pH, decreasing the dose-dependent biotoxicity and adverse effects of DOX. MTT assay for L929 fibroblast cells viability indicated good biocompatibility of carrier. In vitro cell cytotoxicity test on ovarian cancer cells verified the DOX loaded composite nanospheres exhibiting comparable cytotoxicity compared with the free drug at the same concentration. In a similar study, CaF_2:Yb^{3+}/Er^{3+}/Mn^{2+}-poly(2-Aminoethyl methacrylate) (PAMA) hybrid hollow microspheres were fabricated based on multimodal medical imaging combined with cancer therapy within a single system [49]. Upon 980 nm NIR laser excitation, orange upconverted fluorescence was obtained enhanced by Mn^{2+} addition, which was utilized for imaging modality, while the presence of Mn^{2+} and Yb^{3+} also offered the enhanced T_1-weighted MRI and CT imaging. Amine group containing PAMA insertion inside the hollow spheres helped in loading the Pt(IV) prodrug conveniently through bonding with the polymer network constructing a nanoscale anticancer drug carrier. Prepared nanocomposite carrier exhibited effective inhibition for HeLa cells through MIT assay; however, Pt(IV) prodrug and microspheres showed little cytotoxicity to HeLa cells attributed to the fact that anticancer ability is recovered only once Pt(IV) prodrug is reduced to Pt(II) drug in the cellular environment. In vivo antitumor tests demonstrated that the composite microsphere carrier loaded with prodrug was superior to the control groups for tumor suppression and reducing the side effects to normal tissue. Thus, this study proved $CaF_2:Yb^{3+}/Er^{3+}/Mn^{2+}$-PAMA-Pt(IV) a potential candidate for a multifunctional platform providing drug delivery, upconversion cell imaging, T_1-weighted MRI, and CT imaging for simultaneous trimodal bioimaging and therapeutic applications.

Anionic-type layered double hydroxides have also shown great capabilities for delivery of therapeutic drugs and bioactive materials in vivo as well as in vitro, offering high efficiency and drug loading density, protection of loaded molecules from undesired degradation. Moreover, such layered hydroxides are biocompatible with widely used nanoparticles such as silica, carbon nanotubes, and iron oxide. In a first-ever such example, Gu et al. [50] investigated thermostability, luminescence, cytotoxic effect, and controlled drug release behavior of layered europium hydroxide loaded with naproxen (NPX) drug, that is, NPX-LEuH. Different deprotonation degrees lead to NPX-LEuH composites with diverse structures, the thermal stability of organic species enhanced after intercalation. For NPX-LEuH in the 1:0.5 ratio, Eu^{3+} luminescence was markedly increased, while slight luminescence increase was observed for other different deprotonations (NPX-LEuH ~ 1:0.7, 1:1, 1:1.2), ascribed to a more efficient energy transfer from NPX to Eu^{3+} due to well matching of triplet energy level between NPX and Eu^{3+} in 1:0.5 case. Cytotoxic investigations through sulforhodamine B colorimetric assay revealed that LEuH has low cytotoxic effects on most of the cells. A detailed study on drug release behavior carried out in $Na_2HPO_4-NaH_2PO_4$ buffer solution (pH ~ 6.86 and 37°C) inferred the controlled complete release of NPX in 200 min, the mechanism being the ion-exchange reaction between NPX and HP/H_2P in bulk solution, which was fast in beginning and slowed down gradually.

Nanodiamonds (NDs) are yet another versatile class of multifunctional materials whose precise engineering can provide platforms enabling simultaneous diagnostic imaging and targeted drug delivery and hence, improving the therapeutic efficacy, avoiding the adverse side effects [51]. NDs possess chemical stability, good biocompatibility, functionalization versatility, large specific surface area, and unique surface electrostatics. Qin et al. [52] fabricated multifunctional NDs-based nanoparticles possessing brilliant fluorescence and paramagnetic properties through covalently functionalizing the NDs with rare-earth (Eu^{3+}, Gd^{3+}) 2-thenoyltrifluoroacetone complexes. Such composite NDs exhibited intense red emission with lifetime (~0.68 ms) much longer than that of autofluorescence from bio-tissues, enhancing the T_1-weighted MRI due to coordinated rare-earth ions. Rare-earth-activated NDs demonstrated good drug DOX storage capability and a pH-responsive drug release behavior. MIT assay revealed the low toxicity in a particular concentration range and effective chemotherapy toward gastric cancer cells. Thus, dual-mode imaging and drug delivery capability of rare-earth-activated NDs make them promising nanomaterials for multifunctional applications. Some of the versatile multifunctional rare-earth activated drug carriers are listed in Table 14.1.

4. Outlook and challenges

This chapter has summarized the rare-earth-activated phosphors applicable in drug delivery systems. A lot of research has been carried out on solid inorganic phosphors owing to their stability, pore tunability, and ease to control the structure. However, the practical applicability of these phosphors is still lagging behind in reality due to some disadvantages or limitations of inorganic phosphor-based carriers. The main disadvantage of rare-earth-doped phosphors is the relatively low intensity of their luminescence, which is attributed to the nature of electronic transitions of rare-earth ions. Since, most of them involve a redistribution of electrons within the 4f subshells and electron dipole selection rules forbid such transitions, which is the most common reason for weak luminescence although rules being relaxed by several mechanisms. Other factors contributing to reduced luminescence intensity are quenching processes, among which surface quenching and concentration quenching are the most contributing ones in nanocarriers. Surface defects, presence of various ligands, solvent molecules on surface passivate the surface or give rise to nonradiative de-excitation processes via multiphononic processes leading to a decrease in rare-earth luminescence. Considering the sum of contributions from rare-earth dopants from surface and interior of particles, a decrease in nanocarrier's size leads to an increase of surface dopant concentration and hence, a decrease in luminescence intensity. Interaction of rare-earth dopants with host matrix (possessing high phonon energies favoring nonradiative relaxations) itself can also become the cause of reduced luminescence intensity. Increased concentration of active luminescence cations is expected to increase luminescence intensity. However, beyond a particular concentration of dopants reverse effect happens. This may be attributed to the energy transfer between the same type of rare-earth ions increasing the nonradiative relaxation processes, influenced by host material, rare-earth dopant, and its distribution in the host. Apart from luminescence intensity, rare-earth-activated phosphors also offer other disadvantages such as limited drug loading efficiency, high toxicity, poor biocompatibility, uncontrollable drug release, difficult excretion, and toxic degradation products. Additionally, although large-scale research is ongoing worldwide on single or multifunctional phosphor-based drug carriers with special chemical and physical properties enabling high drug loading efficiency and controlled release rates, the limitation is large-scale synthesis putting a practical bar on their real-time application. Regarding some drug carriers, their reported biological effects are conflicting sometimes, which may be due to different methods of preparation and approaches to treat the disease cells. Furthermore, toxicology on the nanoscale cannot be ignored as most of the research is carried out in an in vitro environment only. Multifunctional systems although seem exciting, combining imaging, targeting and drug delivery within one system,

Table 14.1 Summarized information of rare-earth-activated drug delivery carriers.

Drug carrier	Model drug	Release experiment	Characteristic application	Target	References
$NaYF_4:Er^{3+}$ functionalized with TPGS[a]	DOX	In vivo	Imaging and chemotherapy	MCF-7 breast cancer cells	[53]
$NaLuF_4:Yb^{3+}$, Tm^{3+} modified with citrate acid	–	In vivo	CT imaging, MRI, drug delivery	Lymphatic vessel	[54]
α-$NaYF_4$:20%Yb^{3+}, 2%Er^{3+} nanospheres	DOX	In vitro	Targeted anticancer drug delivery and cell imaging	HeLa cells	[55]
$NaLuGdF_4:Yb^{3+}$, Er^{3+}, Cr^{3+} modified with Al–NH-PEG-NH polymers	DOX	In vitro	Chemotherapy	KB cells	[56]
$NaYF_4:Yb^{3+}/Er^{3+}$@$NaYF_4:Yb^{3+}/Nd^{3+}$@ $NaYF_4$ modified using SiO_2–Au	Merocyanine540	In vivo	PDT, PTT[b]	Oral cancer cells	[57]
$NaYF_4:Yb^{3+}$, Er^{3+} modified with PEG[c]	Mitoxantrone	In vivo	Chemotherapy and PDT	Hepatocellular carcinoma	[58]
$NaYF_4$:20%Yb^{3+}, 2%Er^{3+} modified with lipids	MC540	In vitro	PDT	Breast cancer cells	[59]
$NaYF_4:Yb^{3+}$, Er^{3+} modified with DEVD peptide	DOX	In vivo	Chemotherapy	4T1 cells	[60]
$NaYF_4:Yb^{3+}$, Er^{3+} functionalized with Au/PEG	–	In vitro	Chemotherapy	Bladder cancer cells	[61]
$NaYF_4:Yb^{3+}$, Er^{3+}, Nd^{3+} core–shell structures modified with PEG	Polydopamine	In vivo	PTT	H22 cells	[62]
$NaYF_4$: Yb^{3+}, Er^{3+} core with polymersome lipid shell	ZnPc	In vitro	PDT	HeLa cells	[63]
$NaYF_4:Yb^{3+}$, Er^{3+}@ $NaYF_4:Yb^{3+}$@ $NaYF_4$: Yb^{3+}, Nd^{3+} core–shell–shell nanostructures	IR806	In vitro	PTT	MDA-MB-231 cells	[64]
$Gd_2O_3:Er^{3+}$@nSiO_2@mSiO_2 mesoporous nanocomposite	IBU	In vitro	Controlled drug delivery	L929 fibroblast cells	[65]

Continued

Table 14.1 Summarized information of rare-earth-activated drug delivery carriers.—cont'd

Drug carrier	Model drug	Release experiment	Characteristic application	Target	References
α-$NaYF_4$:80%Yb^{3+}, 2%Er^{3+}@CaF_2 core-shell nanoparticles loaded with PEG and imidazole	–	In vitro and in vivo	PDT	A549 tumor cells	[66]
Mesoporous silica nanoparticles functionalized with YVO_4:Eu^{3+} layers	IBU	In vitro	Controlled slow drug delivery	Biological drug distribution	[26]
SiO_2@YBO_3:Eu^{3+} hollow spheres with Fe_3O_4 core templates	DOX	In vivo	pH-responsive and controlled drug delivery	L929 fibroblast cells	[28]
Mesoporous silica envelope functionalized with $NaGdF_4$:Yb^{3+}/Tm^{3+}@$NaGdF_4$@TaO_x layer	DOX	In vitro and in vivo	Chemotherapy	HeLa and 4T1 cells	[29]
β-$NaYF_4$:Yb^{3+}/Er^{3+} core with Fe-based MOF as shell functionalized by AS1411 aptamer	DOX	In vitro	Chemotherapy and diagnostics	MCF-7 and 293 (embryonic kidney) cells	[67]
$NaGdF_4$:Yb^{3+}, Tm^{3+}@$NaGdF_4$:Yb^{3+}, Nd^{3+}@Fe-MOF	DOX	In vivo and in vitro	PTT, PDT, chemotherapy	L929 fibroblast and HeLa cells	[68]
$NaYF_4$:Yb^{3+}, Er^{3+}@$NaYF_4$:Nd^{3+} as core and MOF as shell modified by NH_2-PEG-COOH and folic acid	DOX	In vitro	pH-responsive chemotherapy, bioimaging	HeLa cells	[69]
Mesoporous silica-coated $NaYF_4$:Yb^{3+}, Er^{3+} nanospheres	ZnPc, MC540	In vivo and in vitro	Targeted PDT	B16–F0 melanoma cells	[70]

yet special attention is needed to be paid toward the interactions between drug and imaging probe, control of drug release, strategies for drug release and accurate imaging of whole mechanism.

The large gap between high scale research reports and clinical application can be reduced only by addressing the previously mentioned limitations. To deal with the reduced luminescence intensity from rare-earth phosphors, new strategies should be developed. For instance, downconversion and upconversion exhibiting materials can be grouped into two groups: one dealing with the enhancement of absorption through energy or charge transfer and the other concerned with the minimization of quenching effects, that is, core–shell architecture type. Core–shell structures can efficiently reduce/eliminate the quenching effects by physically separating the surface defects and surface-associated ligands from rare-earth emitters. Coating with silica or silica-based materials is another remedy providing functionalization as well as reducing the quenching effects. Coatings based on plasmonic materials such as gold or silver can also help in enhancing the rare-earth luminescence, as plasmonic electric fields influence the local electric fields changing the crystal field around the sensitizer and activator rare-earth ions. Additionally, there is a need to develop efficient, large-scale, and cost-effective production methods to meet the requirements. Detailed structure-function studies, development of standardized methods for evaluating the biological impacts of carriers before clinical trials, in vivo in addition to in vitro investigations are required. Investigations on rare-earth-based drug delivery systems are still in a nascent stage and a large scope of research is available in this exciting field to produce additional breakthrough discoveries for multifunctional drug delivery systems [53].

References

[1] F. Alexandra, Multifunctional nanodevice reverses drug resistance, Nat. Rev. Drug Discov. 14 (5) (2015), https://doi.org/10.1038/nrd4617, 309–309.

[2] L. Geon-Hui, M. Hanul, K. Hyemin, L.G. Hwang, K. Woosung, Y. Seunghyup, M. David, Y.S. Hyun, B. Zhenan, H.S. Kwang, Multifunctional materials for implantable and wearable photonic healthcare devices, Nat. Rev. Mater. 5 (2) (2020) 149–165, https://doi.org/10.1038/s41578-019-0167-3.

[3] W. Liping, J. Grace, B.D. Kumar, S. Vrinda, S. Jay, K. Sami, P. Nirav, Y. Qingqing, L. Joon, J. Woraphong, W. Shanshan, P. H. B., G. Gennadi, L. Ratneshwar, Multifunctional stimuli responsive polymer-gated iron and gold-embedded silica nano golf balls: nanoshuttles for targeted on-demand theranostics, Bone Res. 5 (1) (2017) 1–14, https://doi.org/10.1038/boneres.2017.51.

[4] T.V. P., Multifunctional, stimuli-sensitive nanoparticulate systems for drug delivery, Nat. Rev. Drug Discov. 13 (11) (2014) 813–827, https://doi.org/10.1038/nrd4333.

[5] B.K. Gupta, S. Singh, P. Kumar, Y. Lee, G. Kedawat, T.N. Narayanan, S.A. Vithayathil, L. Ge, X. Zhan, S. Gupta, A.A. Martí, R. Vajtai, P.M. Ajayan, B.A. Kaipparettu, Bifunctional luminomagnetic rare-earth nanorods for high-contrast bioimaging nanoprobes, Sci. Rep. 6 (1) (2016) 1–12, https://doi.org/10.1038/srep32401.

[6] S. Jie, Z. Liang, H. Gang, Lanthanide-doped upconverting luminescent nanoparticle platforms for optical imaging-guided drug delivery and therapy, Adv. Drug Deliv. Rev. 65 (5) (2013) 744−755, https://doi.org/10.1016/j.addr.2012.05.007.

[7] H. Chih-Chia, H. Wileen, Y. Chen-Sheng, Shell-by-shell synthesis of multi-shelled mesoporous silica nanospheres for optical imaging and drug delivery, Biomaterials 32 (2) (2011) 556−564, https://doi.org/10.1016/j.biomaterials.2010.08.114.

[8] X. Can, Q. Xiaogang, Cerium oxide nanoparticle: a remarkably versatile rare earth nanomaterial for biological applications, NPG Asia Mater. 6 (3) (2014), https://doi.org/10.1038/am.2013.88 e90−e90.

[9] D.J. Naczynski, M.C. Tan, R.E. Riman, P.V. Moghe, Rare earth nanoprobes for functional biomolecular imaging and theranostics, J. Mater. Chem. B 2 (20) (2014) 2958−2973, https://doi.org/10.1039/c4tb00094c.

[10] L. Yang, G. Mindaugas, Q. Jianrong, Long persistent phosphors—from fundamentals to applications, Chem. Soc. Rev. 45 (8) (2016) 2090−2136, https://doi.org/10.1039/C5CS00582E.

[11] F. Jiangli, W. Suzhen, S. Wen, G. Shigang, K. Yao, D. Jianjun, P. Xiaojun, Anticancer drug delivery systems based on inorganic nanocarriers with fluorescent tracers, AIChE J. 64 (3) (2018) 835−859, https://doi.org/10.1002/aic.15976.

[12] E. Alberto, B.A. I., C.-C. Carolina, N.N. O., Z.M. V., L. Mariano, G.-M. Daniel, O. Manuel, J. P. W., Rare earth based nanostructured materials: synthesis, functionalization, properties and bioimaging and biosensing applications, Nanophotonics 6 (5) (2017) 881−921, https://doi.org/10.1515/nanoph-2017-0007.

[13] J. Feifei, L. Guoling, Y. Bo, Y. Bing, S. Youqing, C. Hailin, Investigation of rare earth upconversion fluorescent nanoparticles in biomedical field, Nanotechnol. Rev. 8 (1) (2019) 1−17, https://doi.org/10.1515/ntrev-2019-0001.

[14] S. Qi-C., D.Y. C., S.D. M., N. Prashant, Photon upconversion towards applications in energy conversion and bioimaging, Prog. Surf. Sci. 92 (4) (2017) 281−316, https://doi.org/10.1016/j.progsurf.2017.09.003.

[15] W.P. T., C.S. Ki, Mechanisms of drug release in nanotherapeutic delivery systems, Chem. Rev. 115 (9) (2015) 3388−3432, https://doi.org/10.1021/cr5004634.

[16] R.V. Kalaydina, K. Bajwa, B. Qorri, A. Decarlo, M.R. Szewczuk, Recent advances in "smart" delivery systems for extended drug release in cancer therapy, Int. J. Nanomed. 13 (2018) 4727−4745, https://doi.org/10.2147/IJN.S168053.

[17] A.D. Yohanes, L.L. Yeng, W. Chi-Hwa, Mathematical modeling and simulation of drug release from microspheres: implications to drug delivery systems, Adv. Drug Deliv. Rev. 58 (12−13) (2006) 1274−1325, https://doi.org/10.1016/j.addr.2006.09.007.

[18] T.M. Allen, P.R. Cullis, Drug delivery systems: entering the Mainstream, Science 303 (5665) (2004) 1818−1822, https://doi.org/10.1126/science.1095833.

[19] M. Rubiana, S. Luciano, Drug delivery systems: past, present, and future, Curr. Drug Targets 5 (5) (2004) 449−455, https://doi.org/10.2174/1389450043345407.

[20] W.A. Z., N. Katarzyna, M.K. H., C. Halina, Nanoparticles as drug delivery systems, Pharmacol. Rep. 64 (5) (2012) 1020−1037, https://doi.org/10.1016/s1734-1140(12)70901-5.

[21] R. Daniel, J. Nitin, T. Wei, M. K. J., P. Dan, Progress and challenges towards targeted delivery of cancer therapeutics, Nat. Commun. 9 (1) (2018) 1−12, https://doi.org/10.1038/s41467-018-03705-y.

[22] M. Guan, H. Dong, J. Ge, D. Chen, L. Sun, S. Li, C. Wang, C. Yan, P. Wang, C. Shu, Multifunctional upconversion-nanoparticles-trismethylpyridylporphyrin-fullerene nanocomposite: a near-infrared light-triggered theranostic platform for imaging-guided photodynamic therapy, NPG Asia Mater. 7 (7) (2015) e205, https://doi.org/10.1038/am.2015.82.
[23] D. Yang, P. Ma, Z. Hou, Z. Cheng, C. Li, J. Lin, Current advances in lanthanide ion (Ln3+)-based upconversion nanomaterials for drug delivery, Chem. Soc. Rev. 44 (6) (2015) 1416−1448, https://doi.org/10.1039/c4cs00155a.
[24] M.M. T., S. Yosi, K. Hiroto, R. Neal, S. Maurizio, H.D. A., Targeted drug delivery strategies for precision medicines, Nat. Rev. Mater. (2021) 1−20, https://doi.org/10.1038/s41578-020-00269-6.
[25] Y. Piaoping, G. Shili, L. Jun, Functionalized mesoporous silica materials for controlled drug delivery, Chem. Soc. Rev. 41 (9) (2012) 3679−3698, https://doi.org/10.1039/c2cs15308d.
[26] Y. Piaoping, Q. Zewei, L. Lanlan, H. Shanshan, L. Jun, Luminescence functionalization of mesoporous silica with different morphologies and applications as drug delivery systems, Biomaterials 29 (6) (2008) 692−702, https://doi.org/10.1016/j.biomaterials.2007.10.019.
[27] S. Huang, Z. Cheng, P. Ma, X. Kang, Y. Dai, J. Lin, Luminescent GdVO4: Eu3+ functionalized mesoporous silica nanoparticles for magnetic resonance imaging and drug delivery, Dalton Trans. 42 (18) (2013) 6523−6530, https://doi.org/10.1039/c3dt33114h.
[28] Y. Guixin, G. Shili, Q. Fengyu, Y. Piaoping, SiO2@YBO3:Eu 3+ hollow mesoporous spheres for drug delivery vehicle, ACS Appl. Mater. Interfaces 5 (12) (2013) 5788−5796, https://doi.org/10.1021/am401349z.
[29] C. Yan, A. Kelong, L. Jianhua, S. Guoying, Y. Qi, L. Lehui, Multifunctional envelope-type mesoporous silica nanoparticles for pH-responsive drug delivery and magnetic resonance imaging, Biomaterials 60 (2015) 111−120, https://doi.org/10.1016/j.biomaterials.2015.05.003.
[30] Y. Sheng, L.D. Liao, N. Thakor, M.C. Tan, Rare-earth doped particles as dual-modality contrast agent for minimally-invasive luminescence and dual-wavelength photoacoustic imaging, Sci. Rep. 4 (1) (2014) 1−8, https://doi.org/10.1038/srep06562.
[31] B.S. J., S. Jingning, Y. Nan, M. Yutaka, H. Thomas, H. A. R., J. Yiguang, K. P. R., Synthesis of stable block-copolymer-protected NaYF4:Yb3+, Er3+ up-converting phosphor nanoparticles, Chem. Mater. 22 (2) (2010) 311−318, https://doi.org/10.1021/cm902478a.
[32] D. Yunlu, Y. Dongmei, M. Ping'an, K. Xiaojiao, Z. Xiao, L. Chunxia, H. Zhiyao, C. Ziyong, L. Jun, Doxorubicin conjugated NaYF4: Yb3+/Tm3+ nanoparticles for therapy and sensing of drug delivery by luminescence resonance energy transfer, Biomaterials 33 (33) (2012) 8704−8713, https://doi.org/10.1016/j.biomaterials.2012.08.029.
[33] K. Xiaojiao, C. Ziyong, L. Chunxia, Y. Dongmei, S. Mengmeng, M. Ping'an, L. Guogang, L. Nian, L. Jun, Core−shell structured up-conversion luminescent and mesoporous NaYF4:Yb3+/Er3+@ nSiO2@mSiO2 nanospheres as carriers for drug delivery, J. Phys. Chem. C 115 (32) (2011) 15801−15811, https://doi.org/10.1021/jp203039t.
[34] T. Boshi, L. Shaohua, L. Wei, J. Lin, L. Qingfeng, S. Yurong, L. Chunyang, W. Zhenling, D. Yaping, Construction of pH-responsive and up-conversion luminescent NaYF4:Yb3+/Er3+@SiO2@PMAA nanocomposite for colon targeted drug delivery, Sci. Rep. 6 (1) (2016) 1−11, https://doi.org/10.1038/srep21335.

[35] C. Li, J. Lin, Hollow structured upconversion luminescent NaYF4: Yb3þ, Er3þ nanospheres for cell imaging and targeted anti-cancer drug delivery, Biomaterials 34 (5) (2013) 1601−1612, https://doi.org/10.1016/j.biomaterials.2012.11.004.
[36] D. Yunlu, B. Huiting, D. Xiaoran, L. Chunxia, H. Fei, M. Ping'an, Y. Piaoping, L. Jun, 808 nm near-infrared light controlled dual-drug release and cancer therapy in vivo by upconversion mesoporous silica nanostructures, J. Mater. Chem. B 5 (11) (2017) 2086−2095, https://doi.org/10.1039/C7TB00224F.
[37] G. Shili, Y. Piaoping, L. Chunxia, W. Wenxin, D. Yunlu, N. Na, L. Jun, Synthesis of magnetic, up-conversion luminescent, and mesoporous core-shell-structured nanocomposites as drug carriers, Adv. Funct. Mater. 20 (7) (2010) 1166−1172, https://doi.org/10.1002/adfm.200902274.
[38] H. Liangcan, L. Yuan, L. Joseph, F. Wenpei, L. Qunying, Z. Chao, H. Pintong, C. Xiaoyuan, Recent progress in nanoscale metal-organic frameworks for drug release and cancer therapy, Nanomedicine 14 (10) (2019) 1343−1365, https://doi.org/10.2217/nnm-2018-0347.
[39] C. Jian, L. Xuejiao, T. Hongqi, Metal-organic framework (MOF)-Based drug delivery, Curr. Med. Chem. 27 (35) (2020) 5949−5969, https://doi.org/10.2174/0929867326666190618152518.
[40] Y. Liu, C. Zhang, C. Xu, C. Lin, K. Sun, J. Wang, X. Chen, L. Li, A.K. Whittaker, H.B. Xu, Controlled synthesis of up-conversion luminescent Gd/Tm-MOFs for pH-responsive drug delivery and UCL/MRI dual-modal imaging, Dalton Trans. 47 (32) (2018) 11253−11263, https://doi.org/10.1039/c8dt02436g.
[41] D. Wang, C. Zhao, G. Gao, L. Xu, G. Wang, P. Zhu, Multifunctional NaLnF4@ MOF-Ln nanocomposites with dual-mode luminescence for drug delivery and cell imaging, Nanomaterials 9 (9) (2019) 1274, https://doi.org/10.3390/nano9091274.
[42] K. Tanay, M. Shouvik, P. Prasun, G. Arunava, D.D. David, B. Rahul, Mechanical downsizing of a Gadolinium(III)-based metal-organic framework for anticancer drug delivery, Chem. Eur J. 20 (33) (2014) 10514−10518, https://doi.org/10.1002/chem.201402244.
[43] S. Arindam, M.S. Chandra, D. Kashmiri, D. Pritam, D.P. Sujatha, Surface-engineered multifunctional Eu:Gd2O3 Nanoplates for targeted and pH-responsive drug delivery and imaging applications, ACS Appl. Mater. Interfaces 9 (4) (2017) 4126−4141, https://doi.org/10.1021/acsami.6b12804.
[44] D. Weihua, R. Xinguang, Z. Haifeng, S. Naoto, S. Yoshio, Q. Weiping, Single-phased luminescent mesoporous nanoparticles for simultaneous cell imaging and anticancer drug delivery, Biomaterials 32 (29) (2011) 7226−7233, https://doi.org/10.1016/j.biomaterials.2011.06.019.
[45] A. Jain, P.G.J. Fournier, V. Mendoza-Lavaniegos, P. Sengar, F.M. Guerra-Olvera, E. Iñiguez, T.G. Kretzschmar, G.A. Hirata, P. Juárez, Functionalized rare earth-doped nanoparticles for breast cancer nanodiagnostic using fluorescence and CT imaging, J. Nanobiotechnol. 16 (1) (2018) 1−18, https://doi.org/10.1186/s12951-018-0359-9.
[46] T. Gan, G. Zhanjun, L. Xiaoxiao, Z. Liangjun, Y. Wenyan, Y. Liang, J. Shan, R. Wenlu, X. Gengmei, L. Shoujian, Z. Yuliang, Facile fabrication of rare-earth-doped Gd2O3 hollow spheres with upconversion luminescence, magnetic resonance, and drug delivery properties, J. Phys. Chem. C 115 (48) (2011) 23790−23796, https://doi.org/10.1021/jp209055t.
[47] C. Yu, G. Yu, C. Hangrong, Z. Deping, L. Yaping, Z. Yuanyi, L. Faqi, J. Xiufeng, W. Xia, C. Feng, H. Qianjun, Z. Linlin, S. Jianlin, Engineering inorganic nanoemulsions/nanoliposomes by fluoride-silica chemistry for efficient delivery/Co-delivery of hydrophobic agents, Adv. Funct. Mater. 22 (8) (2012) 1586−1597, https://doi.org/10.1002/adfm.201102052.

[48] D. Yunlu, Z. Cuimiao, C. Ziyong, M. Ping'an, L. Chunxia, K. Xiaojiao, Y. Dongmei, L. Jun, pH-responsive drug delivery system based on luminescent CaF2: Ce3+/Tb3+-poly(acrylic acid) hybrid microspheres, Biomaterials 33 (8) (2012) 2583–2592, https://doi.org/10.1016/j.biomaterials.2011.12.014.

[49] D. Xiaoran, D. Yunlu, L. Jianhua, Z. Ying, M. Ping'an, C. Ziyong, C. Yinyin, D. Kerong, L. Xuejiao, H. Zhiyao, L. Chunxia, L. Jun, Multifunctional hollow CaF2:Yb3+/Er3+/Mn2+-poly(2-Aminoethyl methacrylate) microspheres for Pt(IV) pro-drug delivery and tri-modal imaging, Biomaterials 50 (2015) 154–163, https://doi.org/10.1016/j.biomaterials.2015.01.040.

[50] G. Qingyang, C. Wen, D. Fei, J. Ruijun, Fabrication of a nano-drug delivery system based on layered rare-earth hydroxides integrating drug-loading and fluorescence properties, Dalton Trans. 45 (30) (2016) 12137–12143, https://doi.org/10.1039/C6DT01875K.

[51] B.I. Kharisov, O.V. Kharissova, L. Chávez-Guerrero, Synthesis techniques, properties, and applications of nanodiamonds, Synth. React. Inorg. Metal Org. Chem. 40 (2) (2010) 84–101, https://doi.org/10.3109/10799890903555665.

[52] Q. Shi-Rong, Z. Qi, C. Zhen-Guo, Z. Dong-Xue, Z. Kui-Kui, S. Li-Xia, F. Hui-Jie, W. Yao-He, S. Chong-Xin, Rare earth-functionalized nanodiamonds for dual-modal imaging and drug delivery, Diam. Relat. Mater. 91 (2019) 173–182, https://doi.org/10.1016/j.diamond.2018.11.015.

[53] T. Gan, Z. Xiaopeng, Z. Xiao, Y. Wenyan, Y. Jie, W. Dongliang, Z. Zhiping, Y. Xiangliang, G. Zhanjun, Z. Yuliang, TPGS-stabilized NaYbF4: Er upconversion nanoparticles for dual-modal fluorescent/CT imaging and anticancer drug delivery to overcome multi-drug resistance, Biomaterials 40 (2015) 107–116, https://doi.org/10.1016/j.biomaterials.2014.11.022.

[54] Y. Sun, J. Peng, W. Feng, F. Li, Upconversion nanophosphors Naluf4: Yb,Tm for lymphatic imaging in vivo by real-time upconversion luminescence imaging under ambient light and high-resolution x-ray CT, Theranostics 3 (5) (2013) 346–353, https://doi.org/10.7150/thno.5137.

[55] Y. Dongmei, K. Xiaojiao, M. Ping'an, D. Yunlu, H. Zhiyao, C. Ziyong, L. Chunxia, L. Jun, Hollow structured upconversion luminescent NaYF4: Yb3+, Er3+ nanospheres for cell imaging and targeted anti-cancer drug delivery, Biomaterials 34 (5) (2013) 1601–1612, https://doi.org/10.1016/j.biomaterials.2012.11.004.

[56] T.S. M., S. Mirkomil, H.B. The, G. Zayakhuu, B.-S. Dana, L. Yong-Ill, Naturally modified nonionic alginate functionalized upconversion nanoparticles for the highly efficient targeted pH-responsive drug delivery and enhancement of NIR-imaging, J. Ind. Eng. Chem. 57 (2018) 424–435, https://doi.org/10.1016/j.jiec.2017.08.051.

[57] C. Ming-Hsien, C. Shao-Pou, C. Chieh-Wei, C. Yung-Chieh, L.R. Jie, T.D. Ping, H. Michael, C. Ren-Jei, C. Xueyuan, L. Ru-Shi, Single 808 nm laser treatment comprising photothermal and photodynamic therapies by using gold Nanorods hybrid upconversion particles, J. Phys. Chem. C 122 (4) (2018) 2402–2412, https://doi.org/10.1021/acs.jpcc.7b10976.

[58] H. Yong, A. Yanli, J. Gang, W. Xihui, H. Chen, D. Yinan, T. Qiusha, Theranostic micelles based on upconversion nanoparticles for dual-modality imaging and photodynamic therapy in hepatocellular carcinoma, Nanoscale 10 (14) (2018) 6511–6523, https://doi.org/10.1039/C7NR09717D.

[59] W. Hanjie, D. Chunhong, Z. Peiqi, W. Sheng, L. Zhongyun, C. Jin, Lipid coated upconverting nanoparticles as NIR remote controlled transducer for simultaneous photodynamic therapy and cell imaging, Int. J. Pharm. 46 (1–2) (2014) 307–313, https://doi.org/10.1016/j.ijpharm.2014.03.029.

[60] Z. Na, W. Baoyan, H. Xianglong, X. Da, NIR-triggered high-efficient photodynamic and chemo-cascade therapy using caspase-3 responsive functionalized upconversion nanoparticles, Biomaterials 141 (2017) 40–49, https://doi.org/10.1016/j.biomaterials.2017.06.031.
[61] C.S. K., S. Lih-Jen, M. Chenchen, W.C. D., F.T. W., P. Wounjhang, Multifunctional nanoclusters of NaYF4: Yb3+, Er3+ upconversion nanoparticle and gold nanorod for simultaneous imaging and targeted chemotherapy of bladder cancer, Mater. Sci. Eng. C 97 (2019) 784–792, https://doi.org/10.1016/j.msec.2018.12.113.
[62] D. Xing, L. Jianhua, L. Dapeng, L. Junqi, W. Fan, L. Leijiao, W. Yinghui, S. Shuyan, Z. Hongjie, Multifunctional core/satellite polydopamine@Nd3+-sensitized upconversion nanocomposite: a single 808 nm near-infrared light-triggered theranostic platform for in vivo imaging-guided photothermal therapy, Nano Res. 10 (10) (2017) 3434–3446, https://doi.org/10.1007/s12274-017-1555-x.
[63] H. Beibei, Z. Bin, Y. Weitao, D. Chunhong, W. Hanjie, C. Jin, Construction of near infrared light triggered nanodumbbell for cancer photodynamic therapy, J. Colloid Interface Sci. 494 (2017) 363–372, https://doi.org/10.1016/j.jcis.2017.01.053.
[64] S.L. Lin, Z.R. Chen, C.A. Chang, Nd3+ sensitized core-shell-shell nanocomposites loaded with ir806 dye for photothermal therapy and up-conversion luminescence imaging by a single wavelength nir light irradiation, Nanotheranostics 2 (3) (2018) 243–257, https://doi.org/10.7150/ntno.25901.
[65] X. Zhenhe, L. Chunxia, M. Ping'an, H. Zhiyao, Y. Dongmei, K. Xiaojiao, L. Jun, Facile synthesis of an up-conversion luminescent and mesoporous Gd2O3: Er3+@nSiO2@mSiO2 nanocomposite as a drug carrier, Nanoscale 3 (2) (2011) 661–667, https://doi.org/10.1039/C0NR00695E.
[66] L. Fangyuan, D. Yang, L. Jianan, S. Heng, W. Jin, L. Ruiqing, K. Dokyoon, H. Taeghwan, L. Daishun, Responsive assembly of upconversion nanoparticles for pH-activated and near-infrared-triggered photodynamic therapy of deep tumors, Adv. Mater. 30 (35) (2018) 1802808, https://doi.org/10.1002/adma.201802808.
[67] D. Kerong, H. Zhiyao, L. Xuejiao, L. Chunxia, Z. Yuanxin, D. Xiaoran, C. Ziyong, L. Jun, Aptamer-mediated up-conversion core/MOF shell nanocomposites for targeted drug delivery and cell imaging, Sci. Rep. 5 (1) (2015) 1–7, https://doi.org/10.1038/srep07851.
[68] Y. Dan, X. Jiating, Y. Guixin, Z. Yuan, J. Hongjiao, B. Huiting, G. Shili, H. Fei, Y. Piaoping, Metal-organic frameworks join hands to create an anti-cancer nanoplatform based on 808 nm light driving up-conversion nanoparticles, Chem. Eng. J. 344 (2018) 363–374, https://doi.org/10.1016/j.cej.2018.03.101.
[69] C. Hai-Lin, J. Fei-Fei, W. Song, Y. Ming-Tao, S. You-Qing, Y. Bing, Core–shell upconversion Nanoparticle@Metal–organic framework nanoprobes for targeting and drug delivery, Integr. Ferroelectr. 20 (1) (2020) 66–78, https://doi.org/10.1080/10584587.2020.1728627.
[70] N.M. Idris, M.K. Gnanasammandhan, J. Zhang, P.C. Ho, R. Mahendran, Y. Zhang, In vivo photodynamic therapy using upconversion nanoparticles as remote-controlled nanotransducers, Nat. Med. 18 (10) (2012) 1580–1585.

Persistent luminescent behavior of rare-earth-activated phosphors

15

D.S. Kshatri and Shubhra Mishra
Department of Applied Physics, Shri Shankaracharya Institute of Professional Management and Technology, Raipur, Chhattisgarh, India

1. Introduction

Luminescence or ("weak glow") "powerless shine" Wiedernann [1] is the radiation produced by a material inordinate in contrast with thermal radiation. It results from the progress of an energized electronic state to a ground state or another lower energy state, the span of which is quite longer than the time of light vibrations (10^{-10} s). Luminescence is the outflow of light by a substance not coming about because of heat; it is subsequently a type of cold body radiation. It very well may be brought about by chemical responses, electrical energy, subatomic movements, or stress on the crystal. This markedly differentiates luminescence from incandescence, which is light emanated by a substance because of heating. There can be two kinds of light sources, specifically, the hot light sources comprising the sun, sweltering iron, burning pieces of coal, tungsten fiber, and so forth and the cold light sources consisting of firefly, oscilloscope screen, television screen, light emanating diodes, fluorescent tubes, self-radiant watch dial, etc. We can differentiate between the hot and cold sources of light by touching the sources or recording the spectra of light. While the light coming from hot sources has a persistent range, the light from cold sources has a band spectrum or line spectrum, or a blend of both. In fact, luminescence is the phenomenon of the cold emanation of light, and the phenomenon of hot emission of light is known as incandescence. Alternately stated, any kind of emission other than incandescent is termed as luminescence. Deductively, luminescence is the nonbalance phenomenon of abundance outflow well beyond the thermal emission of a body, in which emission has a span significantly surpassing the time of light oscillations. Although surplus emission well beyond the thermal background happens in reflected and refracted light, Rayleigh and Raman scattering, and Cherenkov and laser radiation, they are not the phenomenon of luminescence as the outflow spans in these phenomena are of the order of the period of light oscillations.

Although the force of incandescent emission increments with temperature, usually, the intensity of luminescence diminishes with the increasing temperature of the materials. Luminescence might happen in translucent and nonglass-like solids just as in fluids and gases. Luminescent solids are typically denoted as phosphors. The process of luminescence comprises three steps: (a) energy concentration causing electronic excitation, (b) stockpiling of assimilated energy, and (c) discharge of photons.

Rare-Earth-Activated Phosphors. https://doi.org/10.1016/B978-0-323-89856-0.00010-9

Furthermore, depending upon the storage time, for example, the time delay between the excitation and emission, luminescence can be categorized into two classifications. In the event that the storage time is more limited than 10^{-8} s, the process is named as fluorescence; else, it is called phosphorescence. The interval 10^{-8} s is picked as it gives the lifetime of an atom in the energized state for which back to the ground state is complemented by dipole radiation.

Since ancient times, luminescence is known as a frail cold glow from decaying woods, certain bugs, mushrooms, fishes, microorganisms, sea algae, minerals, and solar lights. It was in the middle of the 19th century, methodical investigation of luminescence started, after the work of E. Becquerel on the decay of luminescence and of G. Stokes up on the nature of luminescence spectra. In 1852, Stoke's law was framed that stated "the wavelength of luminescence is greater than the wavelength of the exciting radiation." Thenceforth, persistent investigation on various features of luminescence has been conducted around the world.

2. Experimental

2.1 Reactant materials and bicombustion synthesis technique

For the preparation of nanocrystalline $Sr_xAl_2O_4{:}Eu^{2+}{}_{0.01}$, $Dy^{3+}{}_y$ ($x = 0.985$ and $y = 0.005$, $x = 0.98$ and $y = 0.01$, $x = 0.97$ and $y = 0.02$, $x = 0.96$ and $y = 0.03$) phosphors, stoichiometric amounts of $Sr(NO_3)_2$ (1.4319 g, Merck, 99.9999%), $Al(NO_3)_3 \cdot 9H_2O$ (1.0196 g, Merck, 99.9999%) and $CO(NH_2)_2$ (2.5611 g) are used as raw materials with a proportionate amount of distilled water. In addition, Eu_2O_3 (0.0351 g, Merck, 99.9999%) and Dy_2O_3 (0.0745 g, Merck, 99.9999%) are taken as coactivators and dissolved in concentrated HNO_3 before being transferred to a crucible. A little quantity of H_3BO_3 (AR grade) is used as a flux and the $CO(NH_2)_2$ (2.5611 g, AR grade) as a fuel. These reagents are added to 100 mL of distilled water, and then water is mechanically stirred at 100°C to evaporate superfluous water until the volume of the solution decreases to 50 mL. After that, the solution is transferred to the crucible and placed in a furnace at a temperature of $\sim 600 \pm 5$°C. Within 5 min, the reaction begins to give yellow flames. The mixture froths and swells into foam-like material and breaks out with a flame and glows into incandescence. This continues for a few seconds, and as it is finished, the crucible is removed from the furnace and kept open to allow cooling. After cooling, we get a fluffy material shape [2,3]. The resultant phosphor is a mono-combustion product ($SrAl_2O_4$:Eu, Dy), and the process is termed as combustion synthesis technique (CST) [2]. As-synthesized $SrAl_2O_4$:Eu, Dy mono-combustion product is then again mixed with urea $CO(NH_2)_2$ and transferred to a crucible that is further kept in the furnace at $\sim 600 \pm 5$°C. Thereafter, the process is repeated to obtain the final product named bicombustion. Using B-CST, four compositions ($Sr_{0.985}Al_2O_4{:}Eu^{2+}{}_{0.01}$, $Dy^{3+}{}_{0.005}$, $Sr_{0.98}Al_2O_4{:}Eu^{2+}{}_{0.01}$, $Dy^{3+}{}_{0.01}$, $Sr_{0.97}Al_2O_4{:}Eu^{2+}{}_{0.01}$, $Dy^{3+}{}_{0.02}$, $Sr_{0.96}Al_2O_4{:}Eu^{2+}{}_{0.01}$, $Dy^{3+}{}_{0.03}$) of nanophosphors are synthesized to show the behavior of persistent luminescence.

2.2 Characterization techniques

The diffractometer used in this study is a Bruker make D8 Advance X-ray diffractometer with Si trip high-speed detector (Lynx EYE). The source of the X-ray radiation is Cu K_α with a wavelength of 1.54 Å, operated at 40 kV and 40 mA with a scan step of 0.05 degrees in 2θ and a scan speed of 0.7 degrees/s. High-resolution transmission electron microscopy (HRTEM-JEM 2100) operated at 200 kV is used to determine the crystallite size precisely and accurately. To conduct HRTEM, B-CST nanophosphors dissolved in acetone are gently brought into a copper grid coated with carbon and are allowed to dry in the air overnight. The morphology of prepared samples was studied using German make ZEISS-EVO 60 field emission scanning electron microscope (FESEM). The excitation and emission spectra are measured at RT using the Hitachi F-2500 fluorescence spectrophotometer. The afterglow measurements are made with an indigenous experimental set-up comprising an RCA-931 photomultiplier tube and a digital nanoammeter (model DNM-121).

3. Results and discussions

3.1 X-ray diffraction

The X-ray diffractograms explore what crystalline phases are in the materials, the quantity of each crystalline phase in the materials and the presence of an amorphous substance in the materials. The peak broadening in XRD highlights noticeable features like the formation of smaller crystallite size, microstrain, stacking faults, an inhomogeneous composition in a solid solution or alloy, etc. The XRDs of B-CST $SrAl_2O_4$:Eu, Dy products are shown in Fig. 15.1. It is observed from both the diffractograms that peaks position of all main planes (020), (−211), (220), (211), (031), (040), (231), (−303), and (−432) corresponding to 2θ values at 23 degrees, 33 degrees, 34 degrees, 35 degrees, 41 degrees, 47 degrees, 51 degrees, 72 degrees, and 75 degrees, respectively, is exactly similar and determines that the monoclinic structure (International Center for Diffraction Data, PDF No-00-009-0039) of $SrAl_2O_4$:Eu, Dy nanophosphors remains conserved. It is also found that full width at half maxima (FWHM) in the XRD pattern is comparatively wide which may be due to narrow size distribution or some microstrain or stacking faults. But in this case, the peak broadening is due to narrow size distribution only, which is supported by HRTEM results, being analyzed in the next section.

The data for planes (020), (−211), (220), (211), (031), (040), (−303), and (−432) are tabulated in Table 15.1(A−D). The crystalline sizes (*D*) of various samples are estimated using Scherrer's formula [4]:

$$D = \frac{K\lambda}{\beta\cos\theta} \tag{15.1}$$

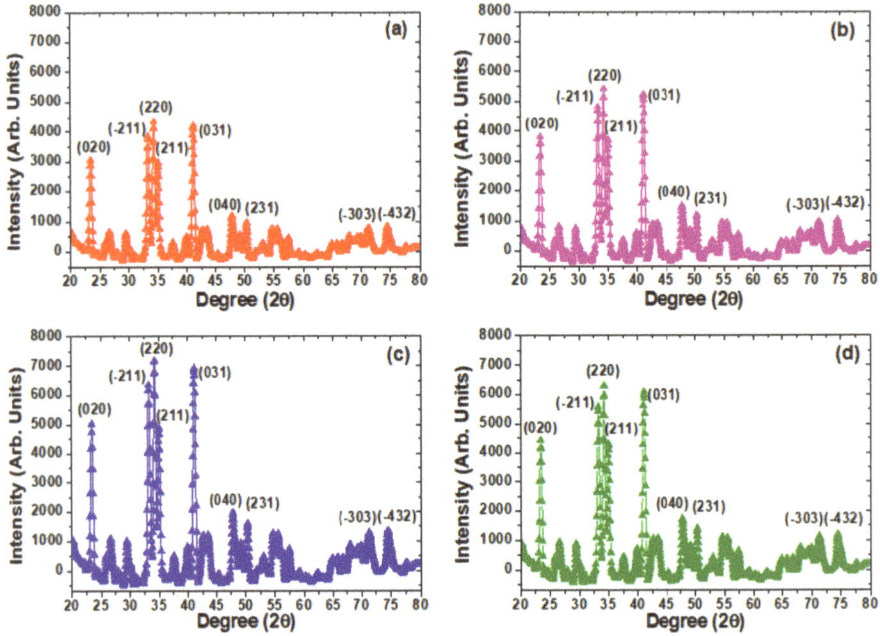

Figure 15.1 X-ray diffractograms of nanophosphors (A) $Sr_{0.985}Al_2O_4$:$Eu^{2+}_{0.01}$, $Dy^{3+}_{0.005}$, (B) $Sr_{0.98}Al_2O_4$:$Eu^{2+}_{0.01}$, $Dy^{3+}_{0.01}$, (C) $Sr_{0.97}Al_2O_4$:$Eu^{2+}_{0.01}$, $Dy^{3+}_{0.02}$, (D) $Sr_{0.96}Al_2O_4$: $Eu^{2+}_{0.01}$, $Dy^{3+}_{0.03}$.

where K is a constant having different values for different grain shapes, λ the wavelength of X-rays, β the FWHM, and θ is the Bragg's angle. The average crystallite sizes for different $Sr_xAl_2O_4$:$Eu^{2+}_{0.01}$, Dy^{3+}_y ($x = 0.985$ and $y = 0.005$, $x = 0.98$ and $y = 0.01$, $x = 0.97$ and $y = 0.02$, $x = 0.96$ and $y = 0.03$) nanophosphors are calculated to be as 22.6681 nm, 23.9556 nm, 23.9565 nm, and 22.9926 nm, respectively, which support nanocrystalline nature of prepared phosphors.

3.2 Field emission scanning electron microscope

The FESEM studies were conducted to examine the surface morphology and the crystalline size of the synthesized phosphors. The FESEM micrographs of $Sr_xAl_2O_4$: $Eu^{2+}_{0.01}$, Dy^{3+}_y ($x = 0.985$ and $y = 0.005$, $x = 0.98$ and $y = 0.01$, $x = 0.97$ and $y = 0.02$, $x = 0.96$ and $y = 0.03$) nanophosphors are shown in Fig. 15.2A–D. As noticed, all the samples have comparatively unpredictable morphology with angularity and corners with differing grain sizes. It is additionally seen that the crystalline sizes are almost the same regardless of the arrangement of the sample. Thus, it is concluded that the doping process does not bring amount marked variations to the morphology and size of the nanostructures. The findings of Ayvacikli et al. [5] had already established this. They detailed that the phosphor powders were sporadic particles with various shapes and sizes. In the current case additionally, the surfaces of the foams

Table 15.1 (A), (B), (C), and (D) shows values of various parameters calculated from the XRD data of different $SrAl_2O_4$:Eu, Dy phosphors.

(A) $Sr_{0.985}Al_2O_4$: $Eu_{0.01}$, $Dy_{0.005}$

2θ (degree)	FWHM (β) (radian)	Lattice spacing (*d*) (Å)	*hkl*	Crystallite size (*D*) (nm)
23.3899	0.0076	4.41284	(020)	21.2351
33.2444	0.0044	3.52210	(−211)	32.1628
34.2549	0.0082	3.12692	(220)	27.5423
35.0108	0.0076	3.03732	(211)	23.5934
41.0638	0.0074	2.97373	(031)	21.6742
47.0638	0.0099	2.55036	(040)	14.0645
51.0124	0.0089	2.49234	(231)	29.1730
72.1003	0.0092	2.39460	(−303)	16.2324
75.0032	0.0079	2.28112	(−432)	18.3354

(B) $Sr_{0.98}Al_2O_4$:$Eu_{0.01}$, $Dy_{0.01}$

2θ (degree)	FWHM (β) (radian)	Lattice spacing (*d*) (Å)	*hkl*	Crystallite size (*D*) (nm)
23.3899	0.0074	4.41284	(020)	22.0238
33.2444	0.0045	3.35465	(−211)	31.1523
34.2549	0.0081	3.84102	(220)	27.8632
35.0108	0.0078	3.21155	(211)	21.2319
41.0638	0.0072	2.51152	(031)	23.2846
47.0638	0.0096	2.01205	(040)	17.5871
51.0124	0.0086	2.56833	(231)	34.4006
72.1003	0.0090	2.87515	(−303)	17.9121
75.0032	0.0079	2.47811	(−432)	17.6815

(C) $Sr_{0.97}Al_2O_4$:$Eu_{0.01}$, $Dy_{0.02}$

2θ (degree)	FWHM (β) (radian)	Lattice spacing (*d*) (Å)	*hkl*	Crystallite size (*D*) (nm)
23.3899	0.0077	4.54785	(020)	20.1454
33.2444	0.0045	3.11454	(−211)	31.0154
34.2549	0.0082	3.28825	(220)	27.2683
35.0108	0.0078	3.57215	(211)	21.0016
41.0638	0.0074	2.31246	(031)	21.0155
47.0638	0.0096	2.02465	(040)	17.4687
51.0124	0.0087	2.57820	(231)	34.2187
72.1003	0.0089	2.78215	(−303)	18.2487
75.0032	0.0072	2.78787	(−432)	25.2187

Table 15.1 (A), (B), (C), and (D) shows values of various parameters calculated from the XRD data of different $SrAl_2O_4$:Eu, Dy phosphors.—cont'd

(D) $Sr_{0.96}Al_2O_4$: $Eu_{0.01}$, $Dy_{0.03}$				
2θ (degree)	**FWHM (β) (radian)**	**Lattice spacing (*d*) (Å)**	***hkl***	**Crystallite size (*D*) (nm)**
23.3899	0.0079	4.15487	(020)	19.2154
33.2444	0.0044	3.38785	(−211)	31.5765
34.2549	0.0084	3.35988	(220)	27.0015
35.0108	0.0077	3.05687	(211)	20.5846
41.0638	0.0071	2.54898	(031)	24.2488
47.0638	0.0099	2.84879	(040)	15.8731
51.0124	0.0087	2.04586	(231)	34.4846
72.1003	0.0090	2.85865	(−303)	18.0158
75.0032	0.0068	2.12356	(−432)	27.5782

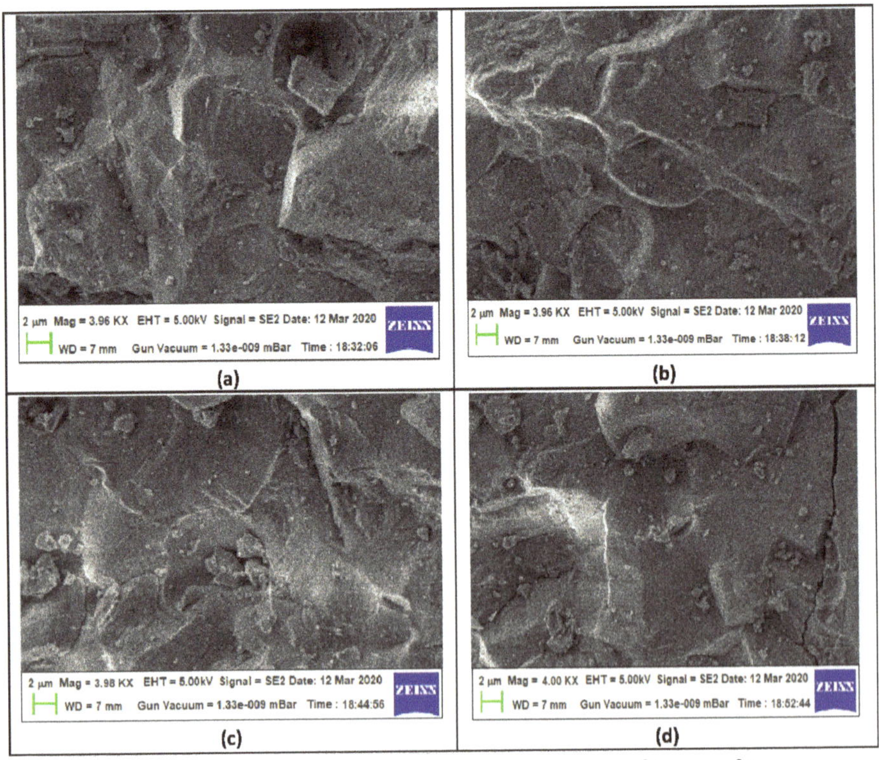

Figure 15.2 FESEM images of nanophosphors (A) $Sr_{0.985}Al_2O_4$:$Eu^{2+}{}_{0.01}$, $Dy^{3+}{}_{0.005}$, (B) $Sr_{0.98}Al_2O_4$:$Eu^{2+}{}_{0.01}$, $Dy^{3+}{}_{0.01}$, (C) $Sr_{0.97}Al_2O_4$:$Eu^{2+}{}_{0.01}$, $Dy^{3+}{}_{0.02}$, (D) $Sr_{0.96}Al_2O_4$: $Eu^{2+}{}_{0.01}$, $Dy^{3+}{}_{0.03}$.

show a great deal of breaks, voids, and pores shaped by the gases evading during the ignition response. Indeed, the enormous measure of escaping gases disperses heat and subsequently keeps the material from sintering and in this manner gives conditions to the formation of the nanocrystalline stage. Other than this, it is additionally stated that nanorod crystallites have a bigger surface region than blossom-like crystallites [6]. The enormous surface area has the genuine disadvantage that the luminescent intensity is disabled because of the introduction of countless imperfections into the phosphor crystal. The deformities have genuine ramifications on glow materials as they give nonradiative recombination courses for electrons and holes. The expanding level of Dy in the original sample brings about the grains with poor limits, which is upheld by optical studies where luminescence intensity drops after a specific Dy concentration.

3.3 High-resolution transmission electron microscopy

High-resolution transmission electron microscopy (HRTEM) images are undoubtedly the most convenient means to find out about the particle size [7]. The particle size and crystal structure of the synthesized $Sr_xAl_2O_4{:}Eu^{2+}{}_{0.01}$, $Dy^{3+}{}_y$ ($x = 0.985$ and $y = 0.005$, $x = 0.98$ and $y = 0.01$, $x = 0.97$ and $y = 0.02$, $x = 0.96$ and $y = 0.03$) nanophosphor powders are determined with HRTEM at 200 kV using JEM-2100 HRTEM. Fig. 15.3A–D portrays HRTEM analysis of nanophosphors. The crystallites of as-synthesized nanophosphors are round in shape having a lopsided surface and are profoundly agglomerated nonetheless the sizes of crystallites attained are surprisingly unique. Fig. 15.3 clearly depicts that the nanophosphors synthesized by B-CST have a size range of 2–30 nm with a normal size of 10–20 nm. This observation upholds significantly diminished crystallite sizes.

3.4 Photoluminescence characteristics

Fig. 15.4A–D indicates the emission spectra of $Sr_xAl_2O_4{:}Eu^{2+}{}_{0.01}$, $Dy^{3+}{}_y$ ($x = 0.985$ and $y = 0.005$, $x = 0.98$ and $y = 0.01$, $x = 0.97$ and $y = 0.02$, $x = 0.96$ and $y = 0.03$) nanoparticles at a thrilling frequency of 365 nm. These spectra display a wideband emanation joined by the peak at 528 nm, which is attributed to the regular $4f^6$ $5d^1$–$4f^7$ changes of Eu^{2+}. No special emissions of Dy^{3+} and Eu^{3+} in the spectra suggest that Eu^{3+} ions have been changed to Eu^{2+} completely [8]. On the other hand, the position of the emission peak in the phosphorescence curve depicts insignificant change, irrespective of variation in the amount of Dy^{3+} ions doping. It is seen here that Dy^{3+} actuates the formation of hole trap levels and prolongs the afterglow. Hence, in $SrAl_2O_4{:}Eu^{2+}$, Dy^{3+} nanophosphors with higher concentrations of Dy^{3+}, production of increasingly more hole-trap levels takes place with greater PL intensities [9]. Presumably, this corresponds to $4f^6$ $5d^1$ to $4f^7$ change of Eu^{2+} ions. As the energized $4f^6$ $5d^1$ configurations of Eu^{2+} particles are amazingly sensitive to the change in the lattice environment, the 5d electron might couple strongly with the lattice [10]. Henceforth, the blended conditions of 4f and 5d configurations will be split by the crystal, which might prompt the blue shift of its emission peak [11,12]. This might be credited to the progressions of crystal lattice around Eu^{2+} and quantum size impact. The doping

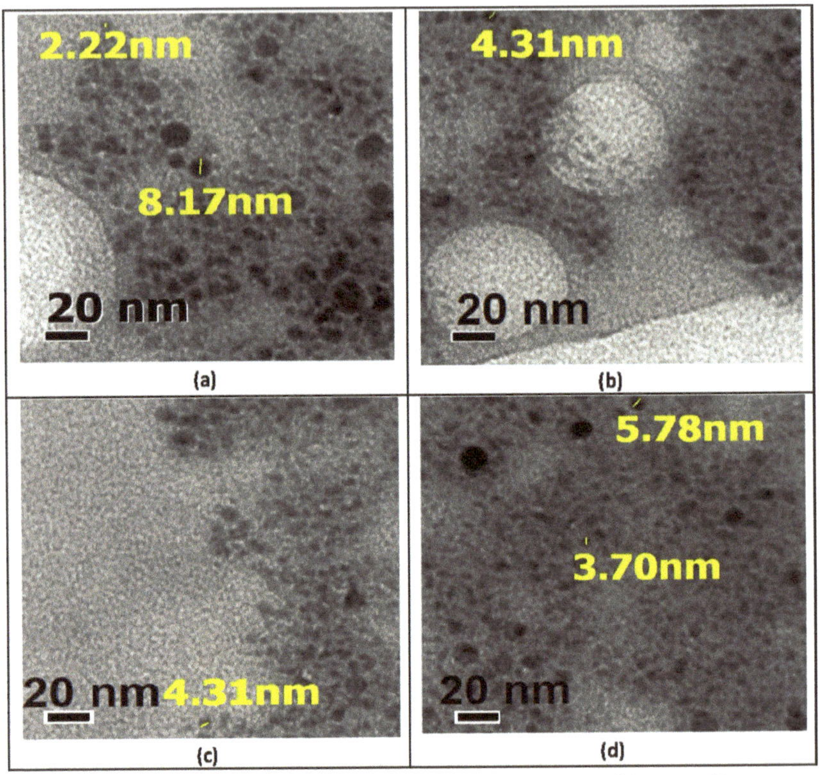

Figure 15.3 HRTEM images of nanophosphors (A) $Sr_{0.985}Al_2O_4:Eu^{2+}{}_{0.01}$, $Dy^{3+}{}_{0.005}$, (B) $Sr_{0.98}Al_2O_4:Eu^{2+}{}_{0.01}$, $Dy^{3+}{}_{0.01}$, (C) $Sr_{0.97}Al_2O_4:Eu^{2+}{}_{0.01}$, $Dy^{3+}{}_{0.02}$, (D) $Sr_{0.96}Al_2O_4:Eu^{2+}{}_{0.01}$, $Dy^{3+}{}_{0.03}$.

of Dy^{3+} ions further produces intense trapping energy levels in the crystal matrix, which trap and store electrons. It upsets the process of relaxation changes of energized electrons of Eu^{2+} particles from invigorated state getting back to ground state [13]. At the point when the molar centralization of Eu^{2+} and Dy^{3+} ions arrive at a specific value, the best synergistic impact somewhere in the range of Eu^{2+} and Dy^{3+} ions is accomplished. In the meantime, both the thickness and depth of trapping and the amounts of electrons stockpiled are fitting and the best emission power is acquired relating to this value. As displayed in Fig. 15.4A, when the molar concentration of Eu^{2+} and Dy^{3+} ions are 0.01 and 0.02, respectively, the emission power of rare earth strontium aluminate is discovered to be the maximum. On the other hand, for any higher molar concentration ($y > 0.02$ in $Sr_xAl_2O_4$: $Eu_{0.01}$ Dy_y), the phenomenon of concentration extinguishing happens, which brings about a decline in discharge power [14]. The excitation spectrum of $Sr_{0.97}Al_2O_4:Eu_{0.01}$, $Dy_{0.02}$ phosphor (highly intense) recorded at RT is displayed in Fig. 15.4B. The excitation spectrum shows a wide band (250–450) with maxima focused at 364 nm, prompting the green luminescence of the materials. The bandwidths of the noticed peaks are very enormous, which are because

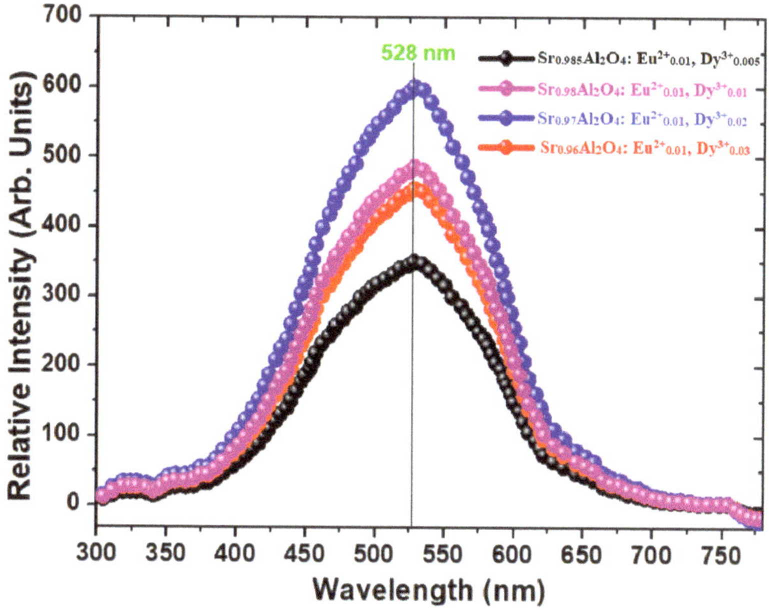

(a)

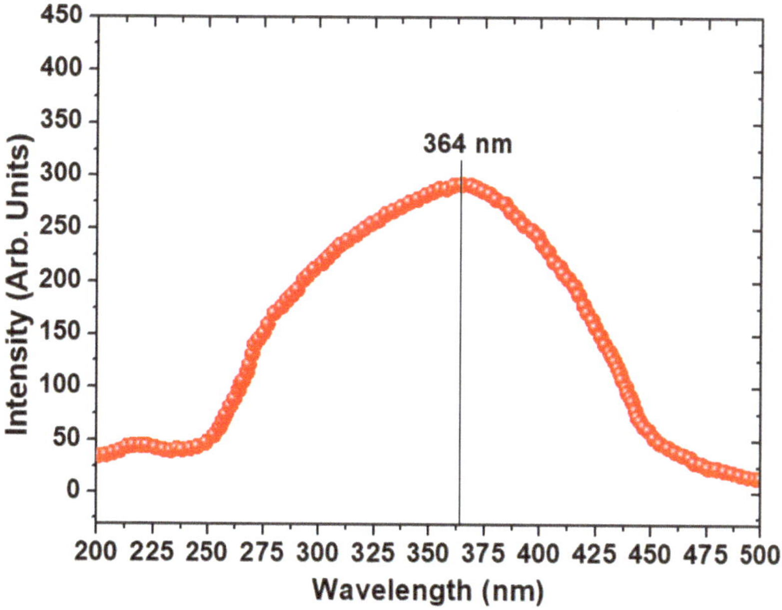

(b)

Figure 15.4 Photoluminescence spectra of different nanophosphors: (A) emission (B) excitation.

of changes between the $4f^6$ $5d^1$ and $4f^7$ electron arrangements of Eu^{2+} [15]. The excitation range shows that as-prepared $SrAl_2O_4:Eu^{2+}$, Dy^{3+} phosphor can be invigorated by a wide range of light including noticeable light proposing its fittingness for applications in various fields.

3.5 Afterglow characteristics

Fig. 15.5A–D shows the luminescent decay curves of $SrAl_2O_4:Eu^{2+}$ phosphors with increasing Dy concentration and illuminated by UV light ($\lambda = 365$ nm) for 5 min at RT. Hypothetically, the phosphorescence phenomenon is seen only when fitting trap energy level exists. If the trap energy level is thin, electrons in the trap are energized effectively and returned to the invigorated state, bringing about short luminosity time. On the other hand, if the trap energy level is more profound, higher energy is required when energized electrons in the trap return to the invigorated state, so electrons are put away only in trap energy level and the afterglow phenomenon is not noticed. It is clear from the figure, that all the afterglow decay curves are made out of three systems: the fast, intermediate, and the slow-decaying process. The fast-decaying process is because of the shore endurance time of electrons in Eu^{2+} state while the slow decaying process is due to the profound trap energy focal point of Dy^{3+} [3]. The Eu^{2+} and Dy^{3+} ions in aluminates phosphors are the luminescent centers and the traps, respectively. The long afterglow property generally results from the trap energy level produced by doping of Eu^{2+} and Dy^{3+} ions in the crystals. In effect, $SrAl_2O_4:Eu^{2+}$, Dy^{3+} phosphor is monoclinic construction, which brings about the formation of a suitable trap for creating phosphorescence [16]. All the $Sr_xAl_2O_4:Eu^{2+}{}_{0.01}$, $Dy^{3+}{}_y$ ($x = 0.985$ and $y = 0.005$, $x = 0.98$ and $y = 0.01$, $x = 0.97$ and $y = 0.02$, $x = 0.96$ and $y = 0.03$) nanophosphors show a quick decay followed by durable brightness. The decay curves relating to various phosphor samples show the underlying luminescent intensity and phosphorescence decay rate to be not the same as one another, which are mostly impacted by the Dy^{3+} concentration and host structure.

Decay times are determined by a curve fitting procedure (Fig. 15.5B) in view of the accompanying third-order exponential decay equation [17]:

$$I = A_1 \exp\left(\frac{-t}{\tau_1}\right) + A_2 \exp\left(\frac{-t}{\tau_2}\right) + A_3 \exp\left(\frac{-t}{\tau_3}\right) \quad (15.2)$$

where I is afterglow intensity; A_1, A_2, and A_3 are constants; t is time, and τ_1, τ_2, and τ_3 are decay times for the fast, intermediate, and slow exponential components separately. The information summed up in Table 15.2 shows that these phosphors display diverse decay times. The decay times determined for $Sr_xAl_2O_4:Eu^{2+}{}_{0.01}$, $Dy^{3+}{}_y$ ($x = 0.985$ and $y = 0.005$, $x = 0.98$ and $y = 0.01$, $x = 0.97$ and $y = 0.02$, $x = 0.96$ and $y = 0.03$) nanophosphors include three exponential constituents (Fig. 15.5B) with various decay times and the maximum value reaching up to 2425.19 s.

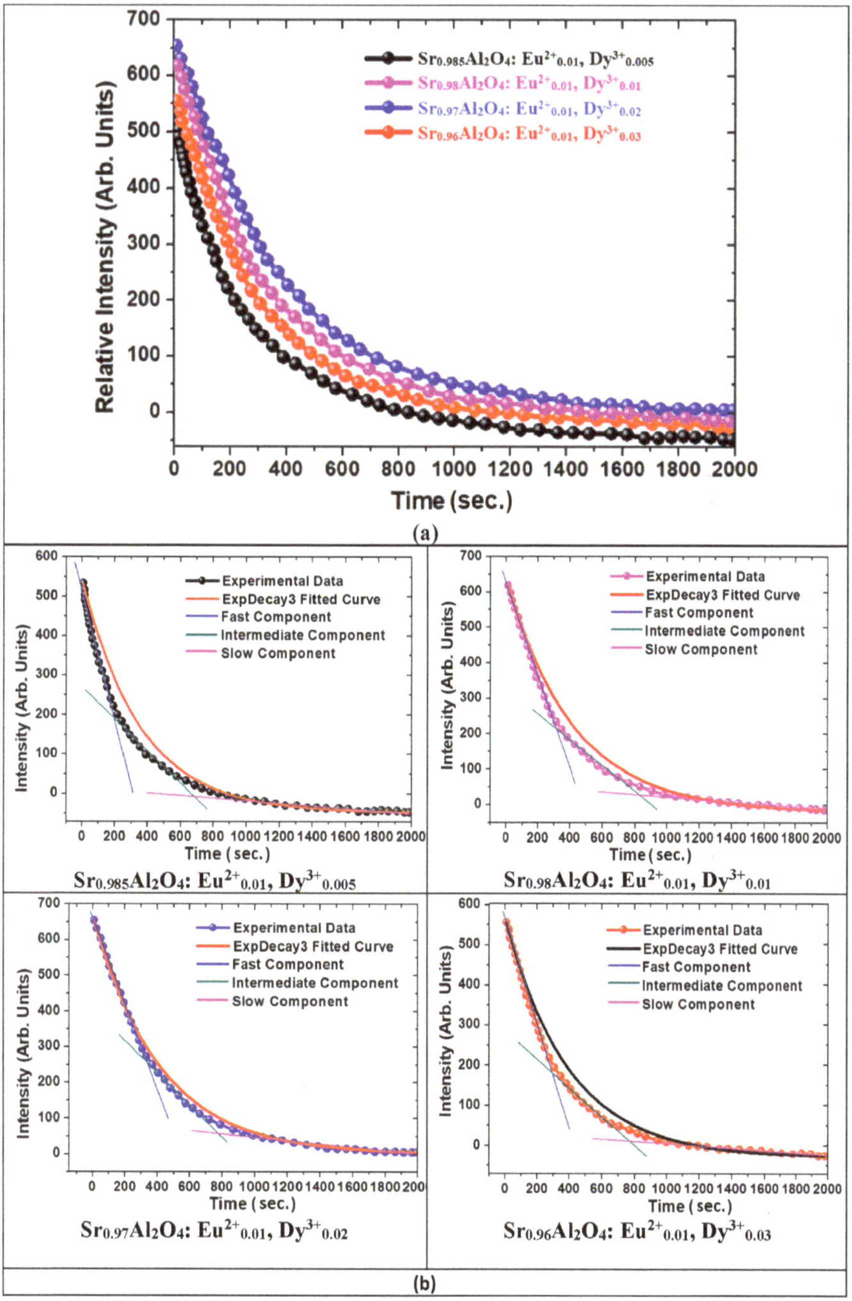

Figure 15.5 (A) Afterglow decay curves of different $Sr_xAl_2O_4$:$Eu^{2+}{}_{0.01}$, $Dy^{3+}{}_y$ ($x = 0.985$ and $y = 0.005$, $x = 0.98$ and $y = 0.01$, $x = 0.97$ and $y = 0.02$, $x = 0.96$ and $y = 0.03$) nanophosphor; (B) ExpDecay3 fitted curves for different phosphors.

Table 15.2 Decay times fitted for exponential characteristics of different $SrAl_2O_4:Eu^{2+}$, Dy^{3+} nanophosphors.

Nanocrystalline	τ_1 (s)	τ_2 (s)	τ_3 (s)
$Sr_{0.985}Al_2O_4:Eu_{0.01}$, $Dy_{0.005}$	71.23	172.93	812.59
$Sr_{0.98}Al_2O_4:Eu_{0.01}$, $Dy_{0.01}$	98.47	211.23	1019.22
$Sr_{0.97}Al_2O_4:Eu_{0.01}$, $Dy_{0.02}$	189.33	617.88	2425.19
$Sr_{0.96}Al_2O_4:Eu_{0.01}$, $Dy_{0.03}$	122.22	515.33	1807.33

4. Conclusions

The $SrAl_2O_4:Eu_{2+}$, Dy_{3+} nanocrystalline powders are explored as new green phosphor with great emission color purity because of the noncentrosymmetric site for the europium ion. The XRD configurations of the integrated phosphors have a place with alpha phase monoclinic design with average crystal size falling in the nano range. The FESEM micrographs expose that $SrAl_2O_4:Eu^{2+}$ phosphors with various Dy^{3+} concentrations show comparative unpredictable morphology. The energized change focused at 528 nm is discovered to be sensitive in nature bringing about green emission. Nonetheless, the phosphorescence characteristics present the highest decay time that supports persistent luminescence characteristics of as-synthesized nanophosphors.

References

[1] E. Wiedemann, Über Fluorescenz und Phosphorescenz, Annalen Der Physik 34 (1888) 446.

[2] C.L. Zhao, et al., Synthesis of $Sr_4Al_{14}O_{25}:Eu^{2+}$, Dy^{3+} phosphor nanometer powders by combustion processes and its optical properties, Mater. Sci. Eng. B 133 (2006) 200.

[3] Z. Qiu, et al., Combustion synthesis of long-persistent luminescent MAl_2O_4: Eu^{2+}, R^{3+} (M = Sr, Ba, Ca, R = Dy, Nd and La) nanoparticles and luminescence mechanism research, Acta Mater. 55 (2007) 2615.

[4] L. Song, S. Zhang, Hydrothermal synthesis and highly visible light-induced photocatalytic activity of zinc-doped cadmium selenide photocatalysts, Chem. Eng. J. 166 (2011) 779.

[5] M. Ayvacikli, et al., Synthesis and optical properties of Er^{3+} and Eu^{3+} doped $SrAl_2O_4$ phosphor ceramic, J. Lumin. 131 (2011) 2432.

[6] C. Yuesheng, et al., Eu^{2+} and Dy^{3+} co-doped $Sr_3Al_2O_6$ red long-afterglow phosphors with new flower-like morphology, Phys. B 403 (2008) 4120.

[7] D.S. Kshatri, A. Khare, Characterization and optical properties of Dy^{3+} doped nanocrystalline $SrAl_2O_4$: Eu^{2+} phosphor, J. Alloys Compd. 588 (2014) 488.

[8] Y. Lin, et al., Preparation of the ultrafine $SrAl_2O_4$:Eu,Dy needle-like phosphor and its optical properties, Mater. Chem. Phys. 65 (2000) 103.

[9] K.V.D. Eeckhout, et al., Persistent luminescence in Eu^{2+}-doped compounds: a review, Materials 3 (2010) 2536.

[10] J. Holsa, et al., Persistent luminescence of Eu^{2+} doped alkaline earth aluminates, MAl_2O_4: Eu^{2+}, J. Alloys Compd. 323 (2001) 326.
[11] C. Zhu, et al., Photoluminescence and gamma-ray irradiation of $SrAl_2O_4$:$Eu^{2+}$and Y_2O_3: Eu^{3+} phosphors, J. Phys. Chem. Solid. 68 (2007) 1721.
[12] T. Peng, et al., Combustion synthesis and photoluminescence of $SrAl_2O_4$:Eu,Dy phosphor nanoparticles, Mater. Lett. 58 (2004) 352.
[13] X.D. Lu, et al., Roles of doping ions in persistent luminescence of $SrAl_2O_4$:Eu^{2+}, RE^{3+} phosphors, J. Mater. Sci. 42 (2007) 6240.
[14] J. Zhang, M. Ge, Effecting factors of the emission spectral characteristics of rare-earth strontium aluminate for anti-counterfeiting application, J. Lumin. 131 (2011) 1765.
[15] X. Xu, L. Caol, D. Jiang, Q. Li, Preparation and luminescent properties of $SrAl_2O_4$:Eu^{2+} ,Dy^{3+}, Key Eng. Mater. 381 (2008) 368.
[16] X.Y. Chen, et al., Porous MAl_2O_4:Eu^{2+} (Eu^{3+}), Dy^{3+} (M=Sr, Ca, Ba) phosphors prepared by Pechini-type sol−gel method, Opt. Mater. 34 (2011) 48.
[17] T. Katsumata, et al., Effects of composition on the long phosphorescent $SrAl_2O_4$: Eu^{2+}, Dy^{3+} phosphor crystals, J. Electrochem. Soc. 144 (1997) L243.

Rare-earth-activated phosphor for laser lighting

16

Vikas Dubey [1], *Neha Dubey* [2], *Jagjeet Kaur* [2] *and Janita Saji* [3]
[1]Department of Physics, Bhilai Institute of Technology Raipur, Naya Raipur, Chhattisgarh, India; [2]Department of Physics, Govt. V.Y.T. PG. Autonomous College Durg, Durg, Chhattisgarh, India; [3]Department of Sciences and Humanities, Faculty of Engineering, Christ (Deemed to be University), Bangalore, Karnataka, India

1. Introduction

Recently, the development of the red color emitting phosphors is in the research hotspot as the tricolor (RGB) luminescent materials for white light emission in solid-state lighting industry [1–6]. A lot of research work has been carried out on Eu^{3+} [7–9] or Sm^{3+} [10–12] ion-doped phosphors varying host matrix for this purpose. But weak absorption near the blue region of these materials due to parity forbiddances and small optical oscillator strength makes those phosphors away from the industrial application. So, the suitable host matrix is in search for efficient and pure red emission.

Hence, in the present work, $Y_2Ba_3B_4O_{12}$ (YBB) borate phosphor doped with Eu^{3+} is investigated as a new red phosphor. However, to our knowledge, a detailed spectroscopic investigation of $Y_2Ba_3B_4O_{12}:Ln^{3+}$ (Ln = Eu) under ultraviolet (UV) or blue excitation has not been reported so far [13–15]. Keeping this in view, this research aims at the estimation of different spectral parameters. The color purity is also calculated for different excitation from the UV region. A suitable host matrix of Eu^{3+} as red-emitting phosphor for industrial application is proposed.

1.1 Experimental

All samples were prepared by the solid-state reaction. Y_2O_3 (99.9%), Eu_2O_3 (99.99%), $BaCO_3$ (A.R.), and H_3BO_3 (A.R.) were used as starting materials. In the reaction process, some consumptions of B_2O_3 obtained by the decomposition of H_3BO_3 at 150–220°C were observed over the melting temperature of 560°C. Accordingly, to obtain $YBa_3B_3O_9:Eu^{3+}$, 7% of excess H_3BO_3 was used. The stoichiometric amounts of starting materials with 7% excess of H_3BO_3 were weighed. After intimately mixing starting materials with ethanol, the mixture was heated at 700°C, then calcined at 1150°C in air and cooled to room temperature for obtaining the samples. The method is known as the modified solid-state method because the previous method reported by Zhang et al. (2009) was prepared the sample in solid-state method [16]. When we compare with solid-state method found that it was less time taken from solid-state

Rare-Earth-Activated Phosphors. https://doi.org/10.1016/B978-0-323-89856-0.00004-3

method. Also, the comparison with other methods, such as hydrothermal method [17], Czochralski method for crystal growth [18], and other solid-state methods reported by many authors [19,20], is not cheaper as compared to our modified solid-state reaction method. Stoichiometric calculations for sample preparation are as follows:

$$Y_2O_3 + 3\ BaCO_3 + 4\ H_3BO_3 = Y_2Ba_3B_4O_{12} + 3\ CO_2 + 6\ H_2O$$

The as-prepared samples were checked by powder X-ray diffraction (XRD). The photoluminescence (PL) emission and excitation spectra were recorded at room temperature by use of a Shimadzu RF-5301 PC spectrofluorophotometer. The excitation source was a xenon lamp.

1.2 Results and discussion

Fig. 16.1 shows the XRD pattern of barium yttrium poly-borate phosphor doped with europium ion, and it is well matched with JCPDS card 48-0309 and has a orthorhombic structure.

1.3 Photoluminescence study

Fig. 16.2 is the emission spectra of YBBO:Eu^{3+} monitored at 254 nm. The following emission was found at 594 nm (orange emission), 614 nm (red emission), and 621 (red emission). The excitation spectra of YBBO:Eu^{3+} phosphor was recorded by monitoring the $^5D_0 \rightarrow {}^7F_2$ transition of Eu^{3+} at 612 nm. The 614 nm (electric dipole transition) emission dominates 594 nm (magnetic dipole transition) emission.

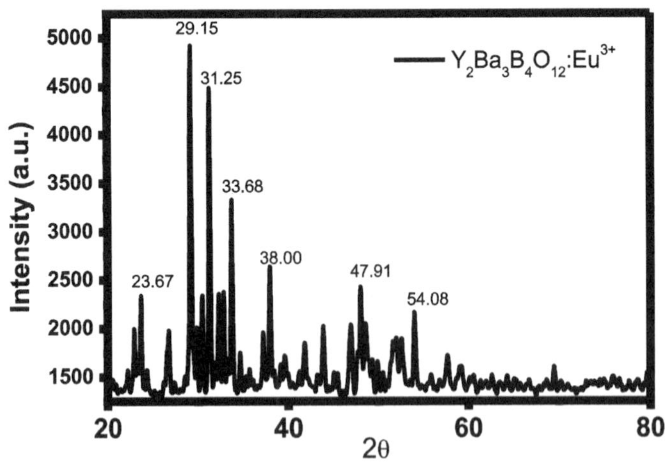

Figure 16.1 XRD pattern of barium yttrium poly-borate phosphor.

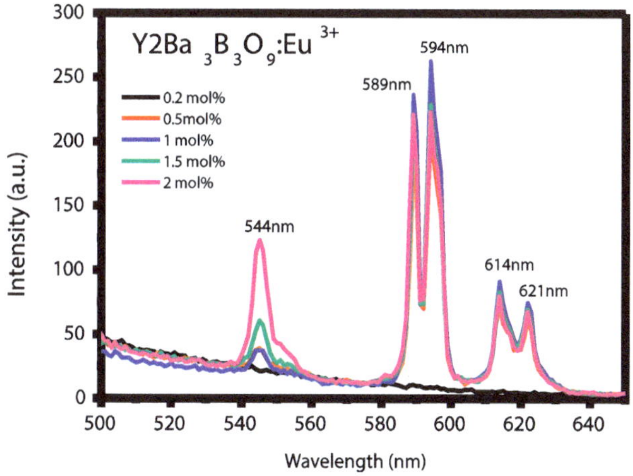

Figure 16.2 PL emission spectra of $Y_2Ba_3B_4O_{12}$:Eu^{3+}-doped phosphor.

1.4 Commission internationale de l'eclairage (CIE) coordinate

Herein, Fig. 16.3 shows the commission internationale de l'eclairage (CIE) 1931 coordinate for Eu^{3+}-doped YBBO phosphor 1 mol% of europium ion and its dominant emission at 614 nm region depicted; it shows the red emission and the coordinates at $x = 0.66$ and $y = 0.34$, which are almost similar to our previous report [21].

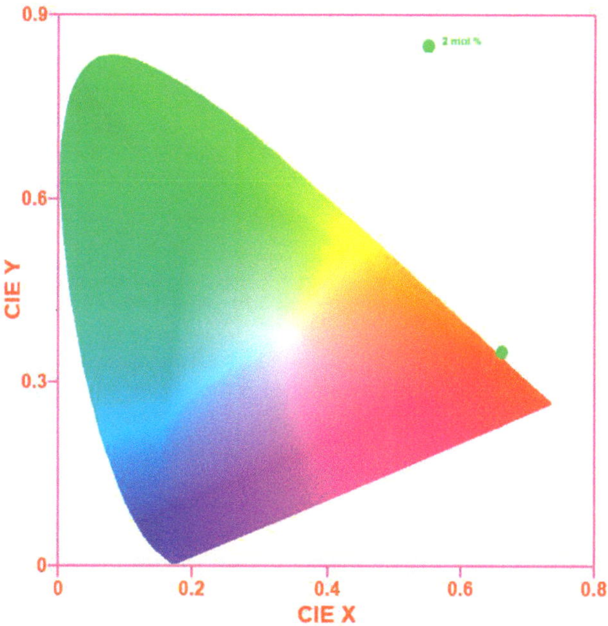

Figure 16.3 CIE coordinate for Eu^{3+}-doped YBBO phosphor.

2. Conclusion

It is concluded that from the previous study related to emission spectra of europium-doped poly-borate phosphor that gives intense emission at 614 nm range (dominant transition of electric dipole $^5D_0 \rightarrow {}^7F_2$) is applicable for laser application. Herein, the 594 nm magnetic dipole transition $^5D_0 \rightarrow {}^7F_1$ is stronger as compared to the electric dipole transition. The phosphor was synthesized by a modified solid-state reaction method, and the crystal structure was determined by X-ray diffraction analysis. It has an orthorhombic structure, and no other impurity phase was found in the sample.

References

[1] Z. Xiaoxia, W. Xiaojun, C. Baojiu, M. Qingyu, D. Weihua, R. Guozhong, Y. Yanmin, Novel Eu^{3+}-doped red-emitting phosphor $Gd_2Mo_3O_9$ for white-light-emitting-diodes (WLEDs), J. Alloys Compd. 433 (2007) 352.

[2] W. Xiao-Xiao, W. Jing, S. Jian-Xin, S. Qiang, G. Meng-Lian, Intense red-emitting phosphors for LED solid-state lighting, Mater. Res. Bull. 42 (2007) 1669.

[3] A.V. Zaushitsyn, V.V. Mikhailin, A.Y. Romanenko, E.G. Khaikina, O.M. Basovich, V.A. Morozov, B.I. Lazoryak, Luminescence study of $LiY_{1-x}Eu_x(MoO_4)_2$, Inorg. Mater. 41 (2005) 766.

[4] S. Neeraj, N. Kijima, A.K. Cheetham, Novel red phosphors for solid-state lighting: the system NaM $(WO_4)_{2-x}(MoO_4)_x$: Eu^{3+} (M Gd, Y, Bi), Chem. Phys. Lett. 387 (2004) 2.

[5] W. Xiao-Xiao, X. Yu-Lun, S. Jian-Xin, S. Qiang, G. Meng-Lian, The potential red emitting $Gd_{2-y}Eu_y(WO_4)_{3-x}(MoO_4)_x$ phosphors for UV InGaN-based light-emitting diode, Mater. Sci. Eng., B 140 (2007) 69.

[6] E. Tomaszewicz, M. Guzik, J. Cybińska, J. Legendziewicz, Spectroscopic Investigation of the Europium(3+) Ion in a New $ZnY_4W_3O_{16}$ Matrix, Helv. Chim. Acta 92 (2009) 2274.

[7] K.S. Sohn, D.H. Park, S.H. Cho, J.S. Kwak, J.S. Kim, Computational evolutionary optimization of red phosphor for use in tricolor white LEDs, Chem. Mater. 18 (2006) 1768–1772.

[8] K.N. Shinde, S.J. Dhoble, A novel reddish-orange phosphor, NaLi2PO4: Eu3+, Luminescence 28 (2013) 93–96.

[9] Z. Zhou, N.F. Wang, N. Zhou, Z.X. He, S. Liu, Y.N. Liu, Z.W. Tian, Z.Y. Mao, H.T. Hintzen, High colour purity single-phased full colour emitting white LED phosphor Sr2V2O7: Eu3+, J. Phys. D 46 (2013) 035104–035110.

[10] F.W. Mo, L.Y. Zhou, Q. Pang, F.Z. Gong, Z.J. Liang, Potential red-emitting NaGd (MO4) 2: R (M = W, Mo, R = Eu3+, Sm3+, Bi3+) phosphors for white light emitting diodes applications, Ceram. Int. 38 (2012) 6289–6294.

[11] Q. Xiao, Q.T. Zhou, M. Li, Synthesis and photoluminescence properties of Sm3+-doped CaWO4 nanoparticles, J. Lumin. 130 (2010) 1092–1094.

[12] S.M. Zhang, B. Zhu, S.F. Zhou, J.R. Qiu, Orange-red upconversion luminescence of Sm3+-doped ZnO-B2O3-SiO2 glass by infrared femtosecond laser irradiation, J. Soc. Inf. Disp. 17 (2009) 507–510.

[13] L. Tian, B.-Y. Yu, C.-H. Pyun, H. Lee Park, S. Mho, New red phosphors $BaZr(BO_3)_2$ and $SrAl_2B_2O_7$ doped with Eu^{3+} for PDP applications, Solid State Commun. 129 (2004) 43–46.

[14] X.Z. Li, C. Wang, X.L. Chen, H. Li, L.S. Jia, L. Wu, Y.X. Du, Y.P. Xu, Syntheses, thermal stability, and structure determination of the novel isostructural $RBa_3B_9O_{18}$ (R = Y, Pr, Nd, Sm, Eu, Gd, Tb, Dy, Ho, Er, Tm, Yb), Inorg. Chem. 43 (2004) 8555.
[15] H.P. You, X.Y. Wu, X.Q. Zeng, G.Y. Hong, C.-H. Kim, C.-H. Pyun, C.-H. Park, Infrared spectra and VUV excitation properties of $BaLnB_9O_{16}$: Re (Ln= La, Gd; Re= Eu, Tb), Mater. Sci. Eng. B 86 (2001) 11.
[16] Y.H. Wang, X. Guo, T. Endo, Y. Murakami, M. Ushirozawa, Identification of charge transfer (CT) transition in (Gd, **Y**) BO_3: Eu phosphor under 100−300 nm, J. Solid State Chem. 177 (2004) 2242.
[17] F. Shen, D.W. He, H.L. Liu, J.H. Xu, VUV excitation properties of LnZr $(BO_3)_2$: Re (Ln= Ba, Sr; Re= Eu, Tb), J. Lumin. 122−123 (2007) 973.
[18] X. ZHANG, H. CHEN, J. KIM, Photoluminescence properties of $(Gd_{1-x}Eu_x)$ $Ba_3B_9O_{18}$ red emitting phosphors, J. Rare Earths 27 (2009) 50.
[19] N. Liu, D. Zhao, L. Yu, K. Zheng, W. Qin, Controlled synthesis and photoluminescence of europium doped barium borate nanorods, nanowires, and flower-like assemblies, Coll. Surf. A Physio Eng. Aspects 363 (2010) 124−129.
[20] K. Sreebunpeng, W. Chewpraditkul, M. Nikl, Luminescence and scintillation properties of advanced $Lu_3Al_5O_{12}$: Pr^{3+} single crystal scintillators, Radiat. Meas. 60 (2014) 42−45.
[21] V. Dubey, R. Tiwari, R.K. Tamrakar, J. Kaur, S. Dutta, S. Das, S. Som, Estimation of spectroscopic parameters and colour purity of the red-light-emitting YBa3B9O18 phosphor: Judd−Ofelt approach, J. Lumin. 180 (2016) 169−176.

Index

Note: Page numbers followed by "*f*" indicate figures and "*t*" indicate tables.

FSC
www.fsc.org
MIX
Papier | Fördert
gute Waldnutzung
FSC® C083411

Kolibri 360 GmbH
Ferdinand-Jühlke-Straße 7
99095 Erfurt, Deutschland
produktsicherheit@kolibri360.de